Nanoengineered Materials for Solid Oxide Cells

Online at: https://doi.org/10.1088/978-0-7503-4064-9

Nanoengineered Materials for Solid Oxide Cells

Edited by
Katherine Develos-Bagarinao
Global Zero Emission Research Center, National Institute of Advanced Industrial Science and Technology, AIST Tsukuba, Japan

IOP Publishing, Bristol, UK

ISBN 978-0-7503-4064-9 (ebook)
ISBN 978-0-7503-4062-5 (print)
ISBN 978-0-7503-4065-6 (myPrint)
ISBN 978-0-7503-4063-2 (mobi)

DOI 10.1088/978-0-7503-4064-9

Version: 20230401

IOP ebooks

British Library Cataloguing-in-Publication Data: A catalogue record for this book is available from the British Library.

Published by IOP Publishing, wholly owned by The Institute of Physics, London

IOP Publishing, No.2 The Distillery, Glassfields, Avon Street, Bristol, BS2 0GR, UK

US Office: IOP Publishing, Inc., 190 North Independence Mall West, Suite 601, Philadelphia, PA 19106, USA

This book is dedicated to my teachers, advisers, colleagues, and collaborators, who have been instrumental in my pursuit of knowledge and stimulated my curiosity for the unknown; and to my family—Baggy and Aya, my sisters—Karen and Karla, and my parents—Jun and Genny, who had passionately taught me the value of perseverance and education.

Contents

Preface

Solid oxide cells (SOCs) are one of the key technologies receiving global scientific and technological interest due to their enormous potential to alleviate the environmental problems associated with the widespread use of fossil fuels. SOCs can provide highly efficient, low-carbon power generation and constitute the core technology used to produce hydrogen, syngas, and synthetic fuels in combination with renewable electricity.

In recent years, the nanoengineering of materials used for SOCs has emerged as a versatile tool for the production of high-performance components such as electrodes, interlayers, and electrolytes with far superior properties than those of conventional materials. Over the years, several methods have been explored to tune various material structures at the nanoscale via 1D, 2D, and 3D nanostructures fabricated using advanced thin-film technology and other related nanoprocessing techniques. These developments are further complemented by significant advancements in nanocharacterization tools that can be used to elucidate the complex mechanisms governing the behavior of nanostructured materials. This strategic approach has enabled the discovery of novel structures and provided an understanding of fundamental mechanisms at the nanoscale, which are crucial in the development of next-generation devices. The use of thin-film technology and related nanotechnology techniques in the materials development of energy conversion devices such as SOCs has significantly proliferated over the last decade, but until now, no book published on this topic has consolidated the advances achieved to date into one volume. While there is a consensus among research communities that the nanoengineering of materials could enable far superior properties than those of conventional materials, there is also a general lack of awareness regarding current research activities that specifically address the issues of integrating such nanomaterials into practical devices. This book aims to bridge this gap and contribute to a better understanding of the breakthroughs achieved in this area.

With this goal in mind, I have invited active researchers and experts in the field of SOC materials development to contribute chapters covering specific topics related to the development and application of nanomaterials. Ten chapters have been selected for this volume; each chapter provides a comprehensive survey of the state-of-the-art experimental approaches and offers perspectives on the current progress achieved as well as future directions in the area. This volume covers the following topics in great detail: an overview of the various approaches used to fabricate nanostructured air electrodes and their implementation in full cells (chapter 1); the influence of surface chemistry and microstructure on the stability and performance of air electrodes (chapter 2); multilayering strategies and characterization tools used to tune the heterointerfaces of air electrodes (chapter 3); interface engineering to improve the activity and stability of SOC air and fuel electrodes (chapter 4); recent progress on atomic layer deposition, which is used to modify various cell components (chapter 5); the development of novel electrode materials and nano-oxide composite electrolytes which are different from conventional materials (chapter 6); a redox exsolution approach for tailoring nanostructured fuel electrodes (chapter 7); an infiltration technique for engineering nanostructured fuel electrodes (chapter 8); microstructural modifications of the electrolyte/electrode interfaces via numerical and experimental approaches (chapter 9); and last but not least, an overview of synchrotron x-ray radiation techniques as advanced characterization tools for probing nanoscale structures (chapter 10).

I hope that this book will be of great interest to groups working on the development of next-generation energy conversion devices, especially those working on the materials development of solid oxide cells but also those developing oxygen separation membranes and memristive devices. Furthermore, I hope that it will be a useful reference not only for academics and researchers actively working in the field, but also for people in the industry who are currently developing systems and devices for commercial applications.

On a final note, I would like to acknowledge the valuable support and hard work of all the contributors to this volume, without which this project would have been impossible. My sincere thanks go to the Institute of Physics Publishing (specifically Caroline Mitchell and Robert Trevelyan) for providing this wonderful opportunity to work on this project.

Katherine Develos-Bagarinao
Tsukuba, Japan
October 2022

Acknowledgements

I would like to thank my colleagues at the National Institute of Advanced Industrial Science and Technology (AIST) for initiating me in this exciting field and providing the opportunity for me to explore new ideas and approaches. I am also grateful to Professor Harumi Yokokawa (The University of Tokyo) for our insightful discussions and encouraging me to pursue this book project. Lastly, I would like to acknowledge the enthusiastic collaboration of all contributors, without whom this book would not have been possible.

List of contributors

Federico Baiutti
Catalonia Institute for Energy Research-IREC, Spain
Department of Materials Chemistry, National Institute of Chemistry, Slovenia

Mónica Burriel
Laboratoire des Matériaux et du Génie Physique (LMGP) UMR 5628 CNRS, France

Fjorelo Buzi
Catalonia Institute for Energy Research-IREC, Spain

Ozden Celikbilek
Université Grenoble Alpes, CNRS, France

Gang Chen
Northeastern University, China

Ming Chen
DTU Energy, Denmark

Yan Chen
South China University of Technology, China

Juan de Dios Sirvent
Catalonia Institute for Energy Research-IREC, Spain

Katherine Develos-Bagarinao
National Institute of Advanced Industrial Science and Technology, Japan

Dario Ferreira Sanchez
Paul Scherrer Institute, Switzerland

Shuai He
University of St Andrews, United Kingdom

Kevin Huang
University of South Carolina, USA

Maxime Hubert
CEA-Liten, France

Tomohiro Ishiyama
National Institute of Advanced Industrial Science and Technology, Japan

Haruo Kishimoto
National Institute of Advanced Industrial Science and Technology, Japan

Jérôme Laurencin
CEA-Liten, France

Aline Léon
EIFER, Germany

Federico Monaco
European Synchrotron Radiation Facility (ESRF), France

Bhaskar Reddy Sudireddy
DTU Energy, Denmark

Takaaki Shimura
The University of Tokyo, Japan

Albert Tarancón
Catalonia Institute for Energy Research-IREC, Spain
and
ICREA, Spain

Miao Yu
DTU Energy, Denmark

Xiangling Yue
University of St Andrews, United Kingdom

Mengzhen Zhou
South China University of Technology, China

Editor biography

Katherine Develos-Bagarinao

Katherine Develos-Bagarinao is a Senior Researcher in the Global Zero Emission Research Center at the National Institute of Advanced Industrial Science and Technology, Tsukuba, Japan. Her current research focuses on the development of nanostructured and thin-film materials for SOC applications, as well as advanced characterization techniques used to understand the degradation mechanisms governing the behaviors of surfaces and heterointerfaces. She obtained her PhD from Yamagata University (1999), and an MS and a BS in Physics from the University of the Philippines (1996 and 1994, respectively). Between 2008 and 2009 she was a visiting scientist in the Department of Materials Science and Metallurgy of the University of Cambridge. She currently serves as an editorial board member for the journal *Scientific Reports*, which is published by Nature Portfolio.

Author biographies

Federico Baiutti

Federico Baiutti holds a MSc in Materials Engineering from the University of Trieste and a PhD in Chemistry from the University of Stuttgart. He has carried out PhD (2011–15) and postdoctoral (2015–16) research at the Max-Planck Institute for Solid State Research (Stuttgart, Germany) in oxide thin films, high-temperature superconductivity, and interface effects. He was a visiting researcher at Cambridge University, UK (2018) and Brookhaven National Labs, USA (2011). His current affiliations are as a research scientist at the Catalonia Institute for Energy Research (IREC), Barcelona, Spain (since 2019) and an associate researcher at the National Institute of Chemistry, Ljubljana, Slovenia (since 2020). His main research interests are related to the fabrication of thin films by physical methods and to fundamental studies of ceramic chemistry and related functionalities for low- and high-temperature energy applications. Federico's current activities are focused on the development of novel electrode materials based on nanostructuring and nanoengineering for high-temperature fuel cells, electrolyzers, electrocatalysis, and microdevices for energy storage.

Mónica Burriel

Mónica Burriel is a CNRS Research Scientist and leads the Oxides for Nanoionic Devices group at the Materials and Physical Engineering Laboratory (LMGP, CNRS, Grenoble-INP) in Grenoble, France. After completing her PhD at the ICMAB Institute, Barcelona, Spain, she worked in the Karlsruhe Institute of Technology, Germany, Imperial College London, UK and IREC, Barcelona, Spain. She recently obtained her Habilitation degree from the Grenoble Alpes University (2021). She has an extensive expertise in the effects of structure, microstructure, and composition on the functional properties of oxide thin films. Her current research into thin films for electrochemical and memristive micro-devices focuses on the deposition (by atomic layer deposition and metal–organic chemical vapour deposition), structural/microstructural control, and optimization of the electronic and ionic transport of functional oxides.

Fjorelo Buzi

Fjorelo Buzi holds a MSc degree in Chemical Sciences with a focus on physical and inorganic chemistry for materials and energy from the University of Genova, Italy. He is a predoctoral researcher at IREC, Spain. His research focuses on the development of nano-structured thin films for solid oxide fuel cells for long-term electrical energy storage. In particular, his work includes the fabrication and study of novel nanocomposite layers as air electrodes.

Ozden Celikbilek

Ozden Celikbilek obtained her PhD from the Grenoble Alpes University in 2016 on the development of air electrodes for solid oxide fuel cells. She subsequently held a postdoctoral research fellow position at Imperial College London from 2018 to 2021. She investigated metal/metal-oxide heterointerfaces for high-performance solid-state energy devices using *in situ* and *ex situ* surface analysis tools. She is currently working as a CNRS postdoctoral researcher at the Materials and Physical Engineering Laboratory (LMGP, CNRS, Grenoble-INP) focusing on the development of thin films for reversible micro-SOC applications.*

Gang Chen

Gang Chen is an Associate Professor in the School of Metallurgy at Northeastern University, Shenyang, China. His current research focuses on the development of new lithium compound electrode ceramic fuel cells that can operate at low temperatures as well as high-performance oxide anode materials for solid oxide cell applications. He obtained his PhD from Hirosaki University (2011), his MS from Dalian University of Technology (2008), and his BS from Chang'an University (2004). Between November 2011 and December 2013, he worked as a research fellow in the Fuel Cell Materials Group of the National Institute of Advanced Industrial Science and Technology. He currently serves as an editorial board member for the journal *Carbon Resources Conversion* of KeAi.

Ming Chen

Ming Chen holds a full professor position at the Department of Energy Conversion and Storage, Technical University of Denmark, Denmark. He completed his PhD at the Swiss Federal Institute of Technology (ETH Zurich) on the topic of solid oxide fuel cells. His research interests lie in the characterization and modelling of solid oxide cells, high-temperature corrosion, thermodynamics, diffusion, and phase field modelling.

Yan Chen

Yan Chen is a professor at the School of Environment and Energy at the South China University of Technology. She received her BS and ME degrees from Peking University and her PhD from the Massachusetts Institute of Technology. Prior to her current position, she worked as a postdoctoral fellow at MIT. Her research focuses on the rational design of new materials to enable high-performance, economical energy and environmental devices based on understanding and controlling surface and interface properties under extreme conditions (electric polarization,

* Currently at Université Grenoble Alpes, CEA-Liten, France.

reactive environments, radiation, etc). She is also working on the application of ion-beam implantation in low-dimensional materials for energy and electronics.

Juan de Dios Sirvent

Juan de Dios Sirvent holds a BSc in Nanoscience and Nanotechnology from the Autonomous University of Barcelona (2016) and an MSc in Applied Materials Chemistry from the University of Barcelona (2017). He later joined the Nanoionics and Fuel Cells team at IREC, where he is currently pursuing his PhD in Materials Science. His work focuses on the study of nanostructured thin-film perovskite-based materials for high-temperature energy production and storage in SOCs. Some of his research interests also include the use of high-throughput methodologies for the study of complete material libraries and the integration of ceramic oxides into silicon microdevice systems.

Shuai He

Shuai He graduated as a Bachelor in 2014 from the School of Materials Science and Engineering, South China University of Technology. In 2015, he studied as a PhD candidate under the supervision of Professor San Ping Jiang from the Department of Chemical Engineering of Curtin University, Australia and then engaged in postdoctoral research in the research group of Dr Xiangling Yue in the University of St Andrews, UK. He has been devoted to the research and development of functional energy materials and solid oxide fuel cells/electrolyzers as well as the interface characterization of solid materials using electron microscopy. He has published more than 30 research papers as the first or co-author in academic journals such as *Appl. Catal. B*, *ACS Appl. Mater. Interfaces*, *J. Mater. Chem. A*, *Adv. Sci.*, *ACS Catal.*, *J. Power Sources*, which have been cited more than 1000 times, and his H-factor is 19.

Kevin Huang

Kevin Huang currently holds a SmartState-endowed chair and is a director of the Solid Oxide Fuel Cell center at the University of South Carolina. His research interests include reversible fuel cells, rechargeable batteries, gas separation membranes, and multiscale computational modeling. He has been published in 220+ peer-reviewed journal papers, two books, and three book chapters and has been granted 14 US patents. He is a recipient of the 2018 Breakthrough Leadership in Research Award; the 2017 Educational Foundation Award for Research in Science, Mathematics, and Engineering; the 2015 College of Engineering and Computing Research Achievement Award; and the 2014 University of South Carolina Breakthrough Stars Award.

Maxime Hubert

Maxime Hubert is a scientist at the French Alternative Energies and Atomic Energy Commission (CEA). He graduated from the Grenoble Institute of Technology (INPG) in Materials Science and obtained his PhD at the Grenoble Alpes University (France) in 2017 for his work on the degradation of SOCs. He has since conducted postdoctoral research using x-ray micro- and nanotomography at the European Synchrotron Radiation Facility. His research interests are currently focused on the improvement of SOC performance and durability using advanced characterization techniques and multiscale modelling.

Tomohiro Ishiyama

Tomohiro Ishiyama is a senior researcher in the Research Institute for Energy Conservation at the National Institute of Advanced Industrial Science and Technology, Tsukuba, Japan. His current research focuses on the elucidation of the degradation mechanism of solid oxide fuel cell materials. He is also working on the development of a new type of fuel cell using phosphate glass proton conductors. He obtained his PhD from Osaka University (2014) by writing a thesis on a new fabrication technique for proton-conducting phosphate glasses.

Haruo Kishimoto

Haruo Kishimoto is the team leader of the Fundamentals of Ionic Devices Research Team at the Global Zero Emission Research Center (GZR) of the National Institute of Advanced Industrial Science and Technology, Tsukuba, Japan. His research field is high-temperature solid-state ionics, in particular, high-temperature electrochemical devices such as solid oxide fuel cells (SOFCs) and solid oxide electrolyzer cells (SOECs). He received his PhD in Engineering from Osaka University in 2002.

Jérôme Laurencin

Jérôme Laurencin is a senior scientist at the French Alternative Energies and Atomic Energy Commission (CEA), France, where he leads a research group on the modeling and characterizations of SOCs. After completing a Master's degree in materials science and engineering, he obtained his PhD from the Grenoble Alpes University by writing a dissertation on the performance and durability of SOFCs. He received his habilitation in 2013 on the modeling of high-temperature electrochemical devices. He has been working on the field of SOCs for more than 15 years at the CEA. His research activities are related to modelling coupled with advanced material and mechanical characterizations. With his research group, he has adapted methods based on synchrotron x-ray radiation for the microstructural and physicochemical characterizations of SOCs. He has developed a multiscale and multiphysics modeling framework that accounts for electrochemical and mechanical cell behavior. His current research interests aim to understand the complex relationships between the electrode microstructure and the fundamental properties of materials in order to optimize the cell durability and robustness in both the electrolysis mode and the fuel cell mode. He has participated in several national and European projects as work package leader. He is the author or co-author of more than 90 articles in peer-reviewed scientific journals and two book chapters, and he holds five patents.

Aline Léon

Aline Léon studied Physics and Material Science in her PhD at the Laval University of Québec (Canada) and holds a habilitation from the Josef Fourier University of Grenoble, France. Her fields of research include advanced characterization techniques and energy materials, in particular, those related to hydrogen technology in terms of purity, storage, and production using high-temperature electrolysis. She is currently an R&D project manager at the European Institute for Energy Research (EIFER). Within the Low Carbon Hydrogen Systems research and innovation team, she is responsible for German/European project management, funding acquisition, the R&D activities of the high-temperature laboratory, supervision, and education about hydrogen, including lectures and seminars.

Federico Monaco

Federico Monaco is a postdoctoral researcher at the European Synchrotron Radiation Facility working on phase contrast and chemical nanotomography.* He studied energy and nuclear engineering at the Polytechnic School of Turin, where he also worked as a research assistant on the production of synthetic fuels. He obtained his PhD in electrochemistry on the degradation of SOCs at the French Alternative and Atomic Energies Commission. His research interests are currently focused on the physicochemical characterization of energy materials, in particular, SOCs and Li-ion batteries.

* Currently at the Fundamental Research Division of CEA (STEP laboratory)

Dario Ferreira Sanchez

Dario Ferreira Sanchez has been working at the Swiss Light Source at the Paul Scherrer Institut since 2015 as a beamline scientist at the microXAS beamline. He has also worked on synchrotron hyperspectral chemical imaging in space and time investigations on a great variety of complex, hierarchical, and heterogeneous materials. Hyperspectral chemical imaging allows a broad range of scientific communities to gain unrivaled insights into the chemical complexity of hierarchical, heterogeneous materials, including corresponding chemical reaction pathways and kinetics (reactivity).

Takaaki Shimura

Takaaki Shimura received his DEng degree in 2015 from the Department of Mechanical Engineering, School of Engineering, The University of Tokyo and worked as a project researcher at the Institute of Industrial Science of the same university from 2015 to 2018. He was an assistant professor at the Department of Mechanical Systems Engineering, Tokyo University of Agriculture and Technology from 2018 to 2021 and is currently a project assistant professor at the Institute of Engineering Innovation, School of Engineering, The University of Tokyo. His main research interests include the mechanism of degradation of SOC electrodes, the characterization of 3D microstructure, and the electrochemical modeling and simulation of electrodes.

Bhaskar Reddy Sudireddy

Bhaskar Reddy Sudireddy (M) is an associate professor at the Department of Energy Conversion and Storage, Technical University of Denmark, Denmark. He was awarded his master's degree and PhD in the field of Materials Science and Engineering by the Indian Institute of Technology Madras, Chennai, India. His research mainly focuses on the synthesis and development of inorganic materials and the development and optimization of dense as well as multilayer/hierarchical porous component manufacturing processes for various applications—ranging from separation membranes to solid oxide cells and piezoelectrics.

Albert Tarancón

Albert Tarancón holds an MSc and a PhD in Physics from the University of Barcelona (2001, 2007) and MEng in Materials Science from the Polytechnic University of Catalonia (2007). He worked as research associate at the IMB-CSIC (ES) and as a visiting researcher at the University of Oslo (Norway), Imperial College, London (UK), and Caltech (USA). In 2010, Albert gained a Ramon y Cajal Fellowship and joined the Catalonia Institute for Energy Research (IREC) as the Head of the Nanoionics and Fuel Cells Group. He is currently the ICREA Professor at IREC and leads a group of 20 + people devoted to nanomaterials for alternative energy technologies and their applicability to powering portable devices and hydrogen/synthetic fuel production. He has been the principal investigator of ten EU projects, including one ERC-CoG and two coordinated H2020 projects, attracting a total of 20+ M€. He is currently an editor of *J. Phys. Energy* (IOP) and *J. Eur. Ceram. Soc.* (Elsevier).

Miao Yu

Miao Yu is a PhD student at the Department of Energy Conversion and Storage, Technical University of Denmark, under the supervision of Professor Ming Chen. She holds an MSc in Energy Conversion and Storage from the Technical University of Denmark and a BSc in Building Environment and Energy Application Engineering from Chang'an University, Xi'an, China. Her PhD project focuses on improving the performance and durability of SOCs and interconnects at intermediate temperatures.

Xiangling Yue

Xiangling Yue is a Rutherford Fellow at the School of Chemistry, University of St Andrews. She was awarded a master's degree in chemical engineering by the Dalian Institute of Chemical Physics, Chinese Academy of Sciences in 2009 and in 2014, she obtained a PhD in Chemistry at the University of St Andrews. Her PhD thesis focused on development of alternative materials for high-temperature electrolysis. After her PhD, she continued to work with professor John Irvine as a postdoctoral researcher for a few years before being awarded the EPSRC UKRI Innovation Fellowship to start her independent research in 2018. She has developed expertise in SOC manufacture, ceramic processing, electrochemical characterisations, and the microstructural engineering of the solid-state materials used in fuel cells and electrolysers for low-carbon energy storage and conversion, in particular for applications involving CO_2 utilisation. She is also working on biomass combustion and understanding the biomass ash deposition process in order to alleviate ash accumulation in biomass boilers and plants, which is a project undertaken in close collaboration with industry.

Mengzhen Zhou

Mengzhen Zhou is currently pursuing her PhD degree at the School of Environment and Energy at the South China University of Technology. Her research focuses on the cathodes of solid oxide fuel/electrolysis cells.

Glossary

p_{O_2}	Oxygen partial pressure
$V_O^{\bullet\bullet}$	Oxygen vacancies
3D	Three-dimensional
ALD	Atomic layer deposition
A-PBC	A-site deficient $PrBa_{0.94}Co_2O_{5+\delta}$
APHAXPES	Ambient-pressure hard x-ray photoelectron spectroscopy
APT	Atom probe tomography
APUs	Auxiliary power units
AP-XPS	Ambient-pressure x-ray photoelectron spectroscopy
ASR	Area-specific resistance (Ω cm^2)
BCO	$BaCoO_{3-x}$
BL	Barrier Layer
BMCNO	$PrBaMn_{1.7}Co_{0.1}Ni_{0.2}O_{5+\delta}$
BSCF	$Ba_{1-x}Sr_xCo_{1-y}Fe_yO_{3-\delta}$
BSCFZY	$Ba_{0.5}Sr_{0.5}Co_{0.6}Fe_{0.2}Zr_{0.1}Y_{0.1}O_{3-\delta}$
BSFM	$Ba_{0.3}Sr_{0.7}Fe_{0.9}Mn_{0.1}O_{3-\delta}$
CB	Conduction band
CHP	Combined heat and power
COBRA	Coherent Bragg rod analysis
CR	Crystal reconstruction
CTE	Coefficient of thermal expansion
CVD	Chemical vapor deposition
c_v	Oxygen vacancy concentration
D	Oxygen diffusivity
D^*	Oxygen self-diffusion coefficient (cm^2 s^{-1})
DCMS	Direct current magnetron sputtering
DFT	Density functional theory
DOS	Electron density of states
DPB	Double phase boundary
DRM	Dry reforming of methane
DRT	Distribution of relaxation times
D_v	Oxygen vacancy diffusivity
EELS	Electron energy loss spectroscopy
EIS	Electrochemical impedance spectroscopy
E_k	Apparent activation energy k
ESM	Electrochemical strain microscopy
$E_{Sr,vac}$	Vacancy formation energy of Sr
EXAFS	Extended x-ray absorption fine structure
FE	Fuel electrodes
FEM	Finite element method
FIB-SEM	Focused ion beam-scanning electron microscopy
FSM	$SrFe_{0.75}Mo0.25O_{3-\delta}$
FT-IR	Fourier transform infrared spectroscopy
GDC	Gadolinia-doped ceria (alternatively, CGO), $Gd_xCe_{1-x}O_{2-\delta}$
GLAD	Glancing angle deposition
GSTF	Gradient structured thin films
HAADF STEM	High-angle annular dark field scanning transmission electron microscopy

HOR	Hydrogen oxidation reaction
HRXRD	High-resolution x-ray diffraction
IE	Isotope exchange
IE-APT	Isotope exchange atomic probe tomography
IEDP	Isotope exchange depth profiling
IT-SOC	Intermediate temperature solid oxide cell
IV-IP	Current/voltage-current/power
k^*	Oxygen surface exchange coefficient (cm s^{-1})
k_{chem}	Chemical oxygen surface exchange coefficient (cm s^{-1})
KFM	Kelvin force microscopy
k^{q}	Electrical oxygen surface exchange coefficient (cm s^{-1})
L2NO4	$La_2NiO_{4\pm\delta}$
LAO	$LaAlO_3$
LC95	$La_{0.95}CoO_{3-\delta}$
LCCFC	Lithium compounds ceramic fuel cell
LCeNT	$La_{0.8}Ce_{0.1}Ni_{0.4}Ti_{0.6}O_3$
LCNT	$La_{0.43}Ca_{0.37}Ni_{0.06}Ti_{0.94}O_{3-\gamma}$
LCO	$LiCoO_2$
LCTCF	$La_{0.43}Ca_{0.37}Ti_{0.8}Co_{0.1}Fe_{0.1}O_{3-\delta}$
LEIS	Low energy ion scattering spectroscopy
LMO	$LiMnO_2$
LNCM-523	$LiNi_{0.5}Co_{0.2}Mn_{0.3}O_2$
LNCM-622	$LiNi_{0.6}Co_{0.2}Mn_{0.2}O_2$
LNCM-811	$LiNi_{0.83}Co_{0.11}Mn_{0.06}O_2$
LNF	$LaNi_{0.6}Fe_{0.4}O_{3-\delta}$
LNO	$La_2NiO_{4+\delta}$
LNZ	Lithium nickel zinc oxide
LPNO	$La_{2-x}Pr_xNiO_{4+\delta}$
LSC	$La_{1-x}Sr_xCoO_3$
LSC_{113}	$La_{1-x}Sr_xCoO_3$
LSC_{214}	$LaSrCoO_{4-\delta}$
LSC_{214}	$La_{2-x}Sr_xCoO_4$
LSC64, LSC_{113}	$La_{0.6}Sr_{0.4}CoO_{3-\delta}$
LSC82	$La_{0.8}Sr_{0.2}CoO_{3-\delta}$
LSCF	$La_{1-x}Sr_xCo_{1-y}Fe_yO_3$
LSCM	$(La_{0.75}Sr_{0.25})Cr_{0.5}Mn_{0.5}O_3$
LSCNi	$(La_{0.7}Sr_{0.3})(Cr_{0.85}Ni_{0.15})O_{3-x}$
LSCNi-Fe	$(La_{0.7}Sr_{0.3})(Cr_{0.85}Ni_{0.15-x}Fe_x)O_{3-\delta}$
LSCoT	$La_{0.3}Sr_{0.7}Co_{0.07}Ti_{0.93}O_{3-\delta}$
LSCr	$La_{1-x}Sr_xCrO_3$
LSCrFeCo	$La_{0.3}Sr_{0.7}Cr_{0.3}Fe_{0.6}Co_{0.1}O_{3-\delta}$
LSCrT	$(La_{0.75}Sr_{0.25})Cr_{0.5}Ti_{0.5}O_3$
LSCuT	$La_{0.43}Sr_{0.37}Cu_{0.12}Ti_{0.88}O_{3-\delta}$
LSF	$La_{1-x}Sr_xFeO_3$
LSF_{113}	$La_{0.8}Sr_{0.2}FeO_{3-\delta}$
LSF_{214}	$(La,Sr)_2FeO_{4-\delta}$
LSFCNb	$La_{0.5}Sr_{0.5}Fe_{0.8}Cu_{0.15}Nb_{0.05}O_{3-\delta}$
LSFN	$La_{0.6}Sr_{0.4}Fe_{0.8}Ni_{0.2}O_{3-\delta}$
LSM	$La_{1-x}Sr_xMnO_3$
LSMF	$La_{1.2}Sr_{0.8}Mn_{0.4}Fe_{0.6}O_{4-\delta}$

LSNT	$La_{0.4}Sr_{0.4}Ni_{0.06}Ti_{0.94}O_{3-\delta}$, $La_{0.52}Sr_{0.28}Ni_{0.06}Ti_{0.94}O_3$
LST	$La_{0.4}Sr_{0.4}TiO_{3-\delta}$
LSTN	$La_{0.4}Sr_{0.4}Ti_{0.94}Ni_{0.06}O_{3-\delta}$
MDC	$Mo_{0.1}Ce_{0.9}O_{2+\delta}$
MIEC	Mixed ionic electronic conductor
MLLS	Multiple linear least squares
MPD	Maximum power density
NAP-XPS	Near-ambient-pressure x-ray photoelectron spectroscopy
NBMO	$NdBaMn_2O_{5+\delta}$
NCAL	$Ni_{0.8}Co_{0.15}Al_{0.05}LiO_{2-\delta}$
NEXAFS	Near-edge x-ray absorption fine structure
Ni/GDC	Nickel/gadolinia-doped ceria
Ni/SDC	Nickel/samaria-doped ceria
Ni/YSZ	Nickel/yttria-stabilized zirconia
NLM	$La_{0.9}Mn_{0.8}Ni_{0.2}O_3$
NPs	Nanoparticles
NSC_{113}	$Nd_{0.5}Sr_{0.5}CoO_{3-\delta}$
NSC_{214}	$Nd_{0.8}Sr_{1.2}CoO_{4\pm\delta}$
$NSC_{214/113}$	$Nd_{0.8}Sr_{1.2}CoO_{4\pm\delta}/Nd_{0.5}Sr_{0.5}CoO_{3-\delta}$
ECR	Electrochemical conductivity relaxation
OCV	Open circuit voltage
OE	Oxygen electrodes
OER	Oxygen evolution reactions
ORR	Oxygen reduction reactions
O-SOCs	Oxygen ion-conducting solid oxide cells
OTR	Optical transmission relaxation
PBCC	$PrBa_{0.8}Ca_{0.2}Co_2O_{5+\delta}$
PBCFN	$PrBaCo_{1.6}Fe_{0.2}Nb_{0.2}O_{5+\delta}$
PBCM	$Pb_{0.5}Ba_{0.5}Co_{0.1}Mn_{0.9}O_x$
PBMCo	$PrBaMn_{1.7}Co_{0.3}O_{5+\delta}$
PBMF	$PrBaMn_{1.7}Fe_{0.3}O_{5+\delta}$
PBMO	$PrBaMn_2O_{5+\delta}$
PEMS	Plasma-enhanced magnetron sputtering
pFIB	Plasma focused ion beam
PLD	Pulsed laser deposition
PNM	$PrNi_{0.5}Mn_{0.5}O_3$
PSCFN	$Pr_{0.4}Sr_{0.6}Co_{0.2}Fe_{0.7}Nb_{0.1}O_{3-\delta}$
PSFM	$(Pr_{0.4}Sr_{0.6})_3(Fe_{0.85}Mo_{0.15})_2O_7$
P-SOCs	Proton-conducting solid oxide cells
PVD	Physical vapor deposition
QCM	Quartz crystal microbalance
RFMS	Radio frequency magnetron sputtering
R_p	Polarization resistance (Ω cm^2)
RP	Ruddlesden–Popper
R-PCECs	Reversible protonic ceramic electrochemical cells
R_s	Surface polarization resistance (Ω cm^2)
RVE	representative volume element
R_Ω	Ohmic resistance (Ω cm^2)
SAED	Selected area electron diffraction
SCF	$Sr_{0.8}Ce_{0.2}FeO_3$

SCN10	$SrCo_{0.9}Nb_{0.1}O_{3-\delta}$
ScSZ	Scandia-stabilized zirconia
SDC	Samaria-doped ceria, $Sm_xCe_{1-x}O_{3-\delta}$
SEM	Scanning electron microscopy
SEM-WDS	Scanning electron microscopy-wavelength dispersive x-ray spectroscopy
SERS	Surface enhanced Raman spectroscopy
SIMS	Secondary ion mass spectrometry
SMT	Semiconductor-to-metal transition
SOCs	Solid oxide cells
SOECs	Solid oxide electrolysis cells
SOFCs	Solid oxide fuel cells
SPM	Scanning probe microscopy
SSC	$Sm_{0.5}Sr_{0.5}CoO_{3-\delta}$
SSOFC	Symmetrical solid oxide fuel cell
SSPM	Scanning surface potential microscopy
STEM	Scanning transmission electron microscopy
STEM-EDS or EDX	Scanning transmission electron microscopy-energy dispersive x-ray spectroscopy
STF	$SrTi_{1-x}Fe_xO_3$
STFN	$Sr_{0.95}(Ti_{0.3}Fe_{0.7-x}Ni_x)O_{3-\delta}$
STM/STS	Scanning tunneling microscopy and spectroscopy
STO	$SrTiO_3$
SYTN	$Sr_{0.92}Y_{0.08}Ti_{1-x}Ni_xO_{3-\delta}$
SZ	$SrZrO_3$
SZY	Yttrium-doped $SrZrO_3$
t	Tolerance factor
TEM	Transmission electron microscopy
TGA-DSC	Thermogravimetric analyzer-differential scanning calorimetry
ToF-SIMS	Time-of-flight secondary ion mass spectrometry
TPB	Triple-phase boundary
TPD	Temperature programmed oxidation
TPR	Temperature programmed reduction
TRF-XAS	Total-reflection fluorescence x-ray absorption spectroscopy
TXM	Transmission x-ray microscopy
UV-Vis	Ultraviolet–Visible absorption spectra
VAN	Vertically aligned nanocomposite
VB	Valence band
XANES	X-ray absorption near-edge structure
XAS	X-ray absorption spectroscopy
XCT	X-ray computed tomography
XPS	X-ray photoelectron spectroscopy
XRD	X-ray diffraction
XRF	X-ray fluorescence
XRR	X-ray reflectivity
YSZ	Yttria-stabilized zirconia, $(YO_{1.5})_{0.16}(ZrO_2)_{0.84}$
ZDC	$Zr_{0.35}Ce_{0.65}O_{2-\delta}$

Chapter 1

Advances in nanoengineered air electrodes: towards high-performance solid oxide cells

Juan de Dios Sirvent, Fjorelo Buzi, Federico Baiutti and Albert Tarancón

Solid oxide fuel and electrolysis cells are the most efficient technologies for clean electricity production from hydrogen and hydrogen generation, respectively. Nonetheless, the presence of thermally activated processes—particularly critical for the air electrode—represents one of the bottlenecks of the technology for operation at more convenient low-to-intermediate temperatures. Among the possible solutions to this limitation, the use of electrodes based on engineered nanostructures has been demonstrated to improve the performance at much lower operational temperatures. Despite this, the origin of this enhancement and the long-term stability of such nanostructured electrodes is still under debate. In this chapter, we present an overview of the different approaches currently followed to fabricate nanostructured electrodes (at the volume and surface level) as well as the reported performance of these nanostructures under operational conditions. In the last section, we compile a selection of successful cases in which nanoengineered air electrodes have been implemented in full cells and discuss their performance; such electrodes are considered to be pioneering works and the seed for the next generation of solid oxide devices based on nanomaterials.

1.1 Introduction

The efficiency of solid oxide cells (SOCs) and in particular solid oxide fuel cells (SOFCs) is determined by the different internal losses that are present in the device [1, 2]: (i) ohmic resistance resulting from ionic conduction, (ii) mass transport at the electrode (bulk or surface ambipolar diffusion), (iii) electrode reactions. As the operational temperature decreases, electrode overpotentials play a more significant role in the performance of the cell due to their thermally activated nature [3]. Therefore, decreasing the activation energy and polarization resistance of the air electrode is of the foremost importance for lowering the operational temperature of the whole SOC.

doi:10.1088/978-0-7503-4064-9ch1

Traditionally, the air electrodes employed in SOCs present a porous microstructure that facilitates the transport of the oxygen gas species towards the electrolyte–electrode interface. The main active region governing the electrochemical reactions at the air electrode is referred to as the triple phase boundary (TPB) [4] in which the electrode, electrolyte, and the gas atmosphere are in direct contact. The TPB reaction occurs via the dissociation of molecular oxygen at the electrode–electrolyte interface, the insertion of the oxide ions into the electrolyte layer, and the transport of the electronic charge through the air electrode. The exact mechanism of the electrochemical reaction depends on the material employed and its microstructure, thus more than one process might play a role (e.g. the additional transport of adsorbed oxygen through the electrode surface into the TPB) [5]. The reaction area can be extended beyond the TPB by employing mixed ionic–electronic conducting (MIEC) electrode materials. These materials allow the geometry to be changed from a TPB to a double phase boundary (DPB), in which the electrochemical reaction occurs across the whole surface of the air electrode. This mechanism extends the reaction's active region, as the oxygen transport can occur across the electrode's bulk or surface. A scheme of the general transport mechanisms of both TPB and DPB-based air electrodes is presented in figure 1.1.

The parameters that describe the kinetics of the electrochemical performance of MIEC air electrode materials are the oxygen surface exchange coefficient (k) and the oxygen diffusion coefficient (D). An approximately linear correlation between these two parameters has been experimentally demonstrated for perovskite-based MIECs employed as air electrodes [6–9], highlighting the strong influence of the ionic transport properties (in particular, oxygen vacancy availability and mobility) on the surface electrochemical activity. However, those additional factors, including the

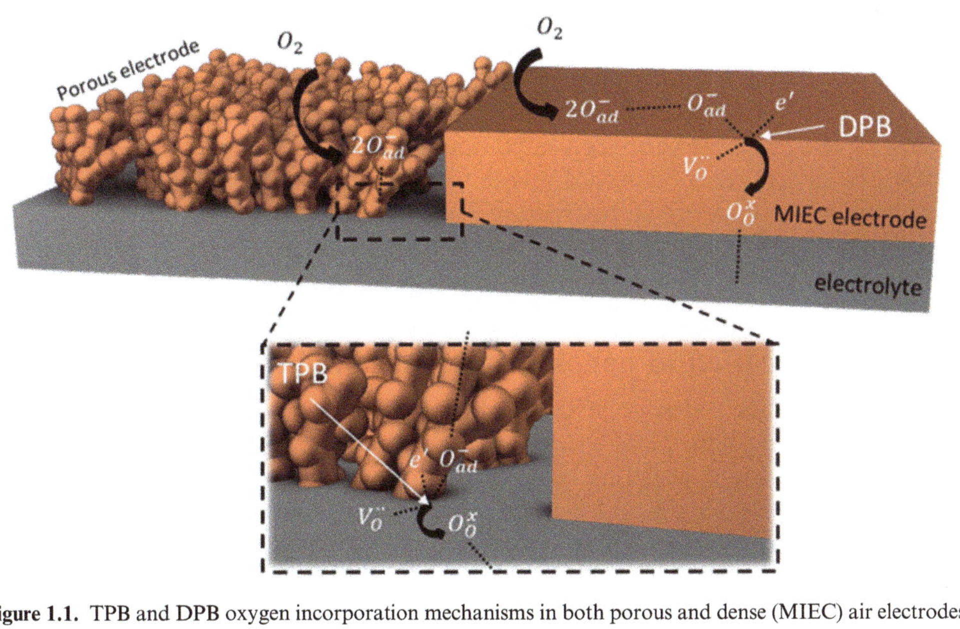

Figure 1.1. TPB and DPB oxygen incorporation mechanisms in both porous and dense (MIEC) air electrodes.

surface chemical composition or the presence of secondary phase aggregates, may lead to strong variability in the surface exchange coefficient, as will be discussed later in the chapter.

Historically, oxide-based air electrodes in SOCs consisted of a porous ceramic layer of an electronic conductor, typically based on manganite perovskites such as $La_{1-x}Sr_xMnO_3$ (LSM) [10], which limited the oxygen reduction reaction (ORR) at the TPB. More recently, composites were introduced to improve the performance, allowing the fabrication of two-phase systems that combine materials with good oxygen affinity (e.g. $La_{1-x}Sr_xMnO3$ (LSM)) and fast ionic conductivity (such as yttria-stabilized zirconia (YSZ) or gadolinia-doped ceria (GDC)) [11–13]. Finally, the use of MIECs to expand the electrochemically active area to the whole electrode surface has become the commonest solution. Among these, lanthanum cobalt- and iron-based oxides (e.g. $La_{1-x}Sr_xCo_{1-y}Fe_yO_3$ (LSCF)) were rapidly found to be the most promising air electrode materials, representing the current technological state of the art [14]. Despite this, perovskite-based MIECs still present limitations caused by secondary reactions with the electrolyte (e.g. the generation of strontium zirconates) [15] and structural instabilities (i.e. surface Sr segregation) [16], which might result in the formation of insulating phases at the electrode–electrolyte interface or the electrode–gas interface. This incompatibility was partially solved by the use of diffusion barrier layers (BLs) based on doped ceria [17–19].

In recent years, advances in vacuum-assisted physical and chemical deposition methods for the deposition of thin-film oxides have opened up the possibility of fabricating functional layers for electrode applications [20, 21], e.g. in the development of novel micro solid oxide systems [22–24]. As is further described in the following sections, the implementation of thin-film-based electrodes allows the fabrication of nanostructured SOC materials and devices with enhanced electrochemical properties that are not achievable at the bulk level. Examples of the advantages of these nanostructured electrodes include enhanced oxygen conduction at the phase contact [25, 26], faster electrochemical activity for the surface reaction [27], and improved stability [28, 29]. In this chapter, we present an overview on the interesting features that nanoengineered materials offer for improving the performance of the oxygen reaction in SOCs, with a specific focus on thin-film materials. An account of other types of heterostructure is also given. Ultimately, research into the use of nanostructured solid oxide air electrodes may lead in the upcoming years to the next generation of nanostructured enhanced SOCs.

1.2 The fabrication of nanostructured electrodes

Nanostructured electrodes for SOCs can be fabricated using a variety of techniques. Among them, thin-film methods are of major relevance, since they allow precise control of composition, structure, and morphology down to the nanometer scale [30] and can be used for the deposition of advanced systems and heterostructures. In the following section, we will provide a description of the methods used for electrode nanostructuring, focusing in particular on advanced thin-film deposition techniques.

1.2.1 Thin-film methods

The deposition method plays a key role in the properties of the grown material in terms of its morphological, chemical, and mechanical features. Moreover, multiple deposition processes can be integrated in the fabrication of more and more versatile nanostructures, depending on the main application they are built for. First, a brief review of the commonest techniques used for the growth of thin films at the nanoscale (<1 μm) for SOC devices will be discussed. As a starting point, every deposition method follows three main steps that include: the synthesis of the deposited material (e.g. the target, the gas precursor, etc.); the transfer of the chemical species from the source to the substrate via a gaseous, liquid, or semi-molten phase; and the adhesion of the source materials to the substrate and the subsequent growth of the film. For the purpose of this paragraph, deposition methods that involve a gaseous precursor will be explained in more detail [31]. Among such techniques, it is possible to distinguish between physical vapor deposition (PVD) and chemical vapor deposition (CVD) techniques: for the former, the vapor phase consists of atoms and molecules that ultimately condense on a substrate, while for the latter, the vapor undergoes a chemical reaction that typically occurs on the substrate surface during the deposition.

Some of the most common methods for fabricating nanoengineered air electrodes, such as pulsed laser deposition (PLD), radio-frequency magnetron sputtering (RFMS) and CVD/atomic layer deposition (ALD) are schematically represented in figure 1.2. With respect to physical vapor deposition methods, PLD shows interesting features that attract attention for the fabrication of nanoscale electrodes in SOCs [32]. The basic step of this deposition route is the ablation of a solid target material by exposure to laser light, resulting in a highly energetic vapor mass (plume) facing the substrate, which is placed at a relatively short distance from the target. The laser energy density threshold used to generate the plume varies, depending on the target material. The stoichiometric transfer of material from the solid target to the substrate arises from collisions between chemical species in the plume and high-kinetic-energy (atoms, molecules, ions, electrons) on the substrate plane that act as a driving force for the following film growth. The quality of the film obtained strongly depends on the parameters controlled during the process, namely, the substrate temperature, background gas, pressure, laser energy, pulse frequency, and target–substrate distance [33–36]. Good stoichiometric transfer of material from the target to the substrate is the most relevant characteristic of PLD deposition, together with precise control over the growth rate, good process flexibility, and relatively fast deposition [33, 34]. The deposition system—see figure 1.2(a)—consists of a laser, a vacuum chamber, the target and substrate holders (which may be rotated by motors), a substrate heater that can reach temperatures of up to 1000 °C (depending on the apparatus), and different gas gauges for performing film growth under different atmospheres and at different pressures. Moreover, a system of optical lenses is usually interposed between the laser source and the high-vacuum chamber in which the deposition takes place. This optical system allows the laser beam to be properly focused on the target surface with a specific spot area, which fixes the energy fluence.

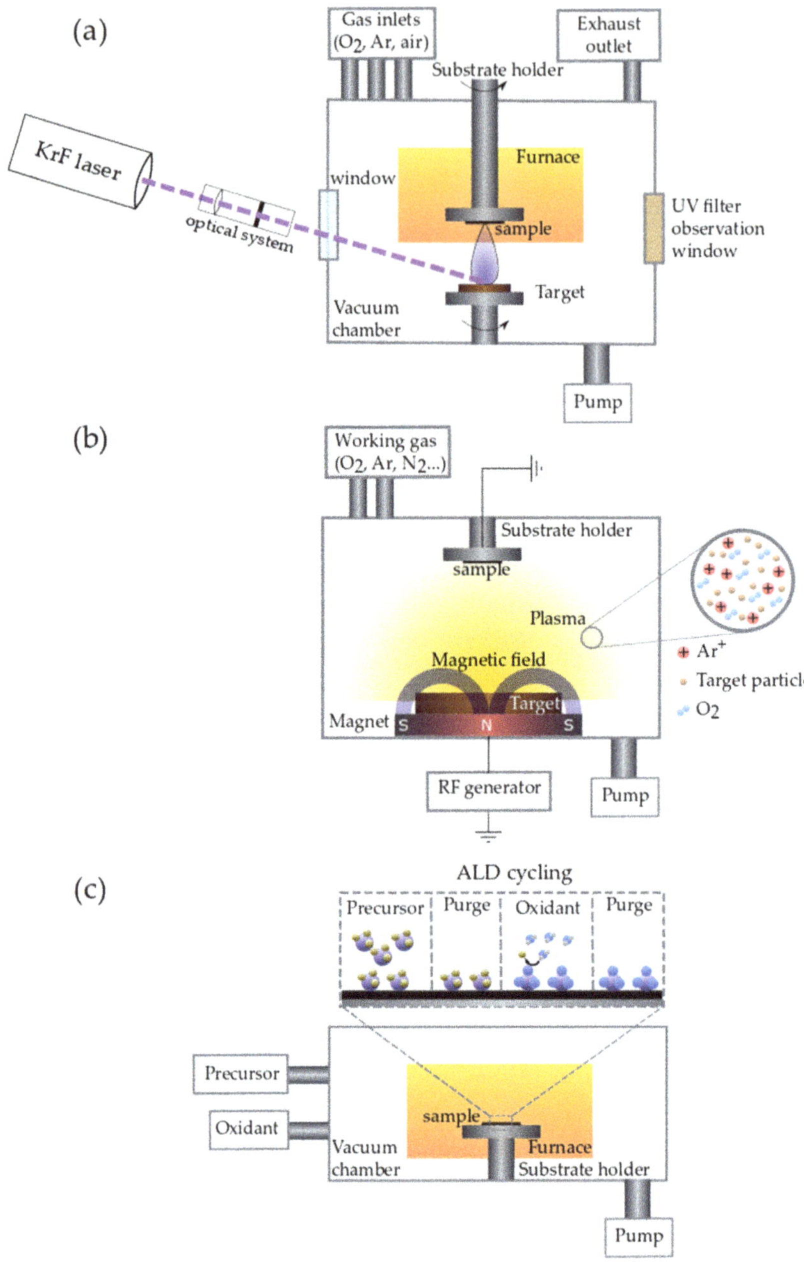

Figure 1.2. Schematic representation of the nanoengineering methods discussed: (a) the PLD, (b) RFMS, and (c) ALD deposition methods.

An interesting systematic study of the PLD growth of LSCF perovskite thin films [37] showed how different deposition conditions can have a strong impact on the final morphology and quality of the deposited layer. The substrate temperature and gas pressure (in this case, oxygen pressure) strongly influence the final layer in terms of the crystallinity and level of porosity of the films, which is very important, as demonstrated in multiple experimental works involving perovskite-based air electrodes, interlayers, and nanocomposites [38–40]. The target-to-substrate distance also has an impact on the internal stresses that occur during the film growth as well as in the adhesion of the film onto the substrate [41, 42]. The deposition rate, determined mainly by the pulse frequency and energy fluence, can change the time range in which the deposited particles rearrange in the film. In the experimental study performed on LSCF by Plonczak *et al* [37], laser energy was said to have an influence on the roughness of the layer by changing the number of particles ablated from the target [37]. Yet, PLD systems manifest some drawbacks that need to be overcome, such as the narrow area of homogeneity guaranteed by the plume and particle ejection that can be detrimental for the final microstructure and the electrical properties of the fabricated electrodes. For these reasons, more sophisticated PLD systems have been designed to overcome these weaknesses, and a careful choice of deposition parameters can ensure the production of higher-quality films and better operational device performance. New potential approaches include cross-beam PLD techniques that use the dynamic interaction between two plasma plumes to suppress particulate formation, thereby improving the final film quality [43], and matrix-shaped PLD, which exploits an additional system consisting of a micro-lens array and a Fourier lens to split the incoming beam into a multicomponent beam directed onto the target in order to achieve better control over the homogeneity of the film during growth over large areas [44].

Another widely used group of physical vapor deposition techniques is based on sputtering. This technique is characterized by the removal of material from a solid target through bombardment with energetic ions in a metastable gaseous phase (plasma) originated from an inert gas. In the field of SOC devices, *Magnetron Sputtering* is a specific sputtering system that allows the fabrication of a relatively wide range of nanostructures [45]. The equipment is composed of an electric field generator, a target plate (cathode), a rotating substrate holder, a vacuum chamber, and a non-reactive/reactive gas to form the plasma, depending on the film required. When a magnetic field is applied to the target plate, atoms in the chamber atmosphere undergo ionization, and the energetic particles collide with the target surface and transfer chemical species onto the substrate under rotation. As in the case of PLD, some of the main deposition parameters are: the substrate temperature, the gas composition and pressure, and the target–substrate distance. More specific to this technique is the importance of controlling the nature of the power supplied as well as the substrate's rotational speed, which lead to good homogeneity and distribution of the plasma. Regarding the power supply system, direct-current magnetron sputtering (DCMS) and RFMS—see figure 1.2(b)—are the most common variants for electronic conductive and non-conductive targets, respectively. In addition, based on the chemical nature of the gas in the chamber, reactive magnetron sputtering can represent an appealing option for depositing oxide thin films in SOCs. Other variants can be found, such as the glancing angle deposition (GLAD) technique and plasma-enhanced magnetron sputtering (PEMS), which increment the versatility of this method [46, 47]. It is common to employ sputtering methods for the fabrication of functional layers (e.g. a Ni/YSZ anode layer in an anode-supported SOFC [48]) or barrier layers (e.g. gradient YSZ/GDC barrier layers [49]) during the implementation of SOC devices. An entirely sputtered SOFC stack was even developed using this deposition process by tailoring parameters to develop the electrodes, electrolyte, and interlayers with the desired properties via a sequential mechanism [50].

CVD techniques can be classified into a large number of existing variants, but all of them share major principles: first of all, the material deposition occurs through a chemical reaction between volatile precursor compounds and the substrate surface. Hence, the overall mechanism includes the diffusion of reactants in a gas phase towards a deposition chamber, chemical or physical absorption of the reactants on the substrate surface, a chemical reaction with the formation of a non-volatile solid thin film that nucleates and grows on top of the substrate, and finally the desorption of volatile by-products and their removal by a carrier gas. A major classification is based on the type of reaction involved: thermal decomposition, reduction, exchange reactions, disproportionation reactions, and coupled reactions [51]. The main unique features of the CVD approach are its high flexibility and its ability to produce simple or relatively complex films at a lower temperature than those of the PVD methods discussed above. In particular, these techniques allow for the fabrication of more homogeneous and conformal layers than those produced by physical methods, even on top of complex geometries.

ALD is an intriguing subclass within the CVD family. The ALD process follows a CVD route—see figure 1.2(c)—but has the advantage of enabling a self-limiting chain of reactions in order to grow ultrathin films through the building of monoatomic layers at a relatively low temperature [52]. For a more detailed discussion, the reader is referred to the chapter dedicated to ALD methods (see chapter 5) in this book. The self-limiting growth mechanism allows monoatomic layers to be deposited during each deposition cycle. This makes it possible to achieve outstanding step coverage in three dimensions independently of the roughness and geometry of the sample to be treated, with great control over the final thickness. Moreover, secondary phase formation is blocked, since the self-limiting mechanism hinders uncontrolled chemical reactions. Ideally, this mechanism leads to the formation of defect-free dense layers. Additionally, the strong chemical binding involved in the process ensures good adhesion and robustness. Another key benefit is given by the relatively low deposition temperatures (from room temperature up to 300 °C) which, particularly in the case of air electrodes, can help to avoid unwanted degradation mechanisms that occur at higher temperatures during preparation. A major drawback is caused by the organometallic compounds used as material precursors, which limit the choice of film composition (typically, binary oxides) [53]. The use of ALD in SOC fabrication has increased in recent years; it is used either for depositing thinner ionic conductive layers, e.g. YSZ or $Ce_{1-x}Gd_xO_2$ (CGO) [54] or for enhancing the electrochemical activity of air or fuel electrodes by decorating their surfaces with ultrathin layers of simple oxides (e.g. ZrO_2) or more complex ones (e.g. perovskites ABO_3). Such layers hinder deleterious phenomena (e.g. dopant segregation) that occur in the device when it is operated at high temperatures [55].

1.2.2 Other common methods

1.2.2.1 The infiltration of templates

Infiltration is based on the introduction of a solution of nitrate-based precursors onto a porous scaffold for further pyrolysis [56, 57]. This method is typically employed for the surface modification of a mechanically stable backbone by means of surface decoration with an active catalyst [58]. For a more detailed explanation and discussion of the infiltration process, the reader is referred to chapter 8 in this book. Since the infiltrated solution can be fired at a relatively low temperature, the resulting composite is finely distributed and also morphologically tunable [59, 60]. The method can be efficiently employed in order to fabricate nanometer-thin layers on top of a backbone while maintaining a very high cost efficiency. This has been demonstrated by Niu *et al*, for example, who highlighted a decrease of one order of magnitude in the area specific resistance of LSCF cathodes (polarization resistance 0.048 Ω cm^2 at 650 °C) after catalyst infiltration [61]. A number of methods can be utilized for the fabrication of the scaffold for infiltration, with the goal of maximizing the porosity and the density of active sites. Very recently, Cao *et al* have reported a porous doped-zirconia vertical microchannel scaffold fabricated by freeze-casting and functionalized by a $La_{1-x}Sr_xCoO_3$ (LSC) nanolayer, which was capable of delivering record performance under electrolysis conditions (cell current density 5.96 A cm^{-2} at 800 °C [62]). Templating, in which a sacrificial replica is used, represents an interesting approach for scaffold fabrication. Here, soft- and hard-template methods permit the fabrication of mesoporous scaffolds with a maximized surface area (>100 m^2 g^{-1}) and a uniform and open pore network [63]. Although thermal stability is often limited, silica-based hard templating has been shown to be effective in forming thermally stable scaffolds, which can be utilized at high temperatures in SOC cells due to self-limited grain growth kinetics [64–66].

1.2.2.2 The exsolution of metal nanoparticles

The reduction of metallic cations in oxides together with the further migration of generated metallic nanostructures to the surface is called exsolution. This exsolution process results in finely dispersed nanoparticles (NPs) of the active metals firmly anchored to the oxide lattice. In particular, exsolution has been recently used for electrode applications in which NPs are the catalysts for the redox reactions [67–72]. For a more detailed explanation and discussion of the exsolution process, the reader is referred to chapter 7 of this book.

In general, nanoparticles are formed by applying reducing conditions (e.g. H_2 atmospheres and high temperatures) to noble-metal-doped transition-metal oxides. This *thermal reduction* and its alternative, *electrochemical reduction*, are known to be the main synthetic approaches so far. They both require relatively high temperatures, but while the former is limited by the slow diffusion rate of ions into the electrode structure at the operating temperature, which implies a slower reduction rate and is hence more time-consuming and expensive, the latter is voltage-driven and allows higher reaction rates with the formation of remarkably richer and finer NPs [71]. Nevertheless, low-temperature approaches and new driving forces for the reduction process are to be investigated further. Moreover, the metal nanostructures that can be chosen for exsolution are limited by their ability to be readily reduced, but compared to other methods of nanoparticle formation (e.g. infiltration), the exsolution mechanism guarantees better control over the particle size, distribution, and morphological stability [70]. The improvement of different perovskites through NPs exsolved from the *A*- and the *B*-site has been reported. Among these, noble metals such as Ag [68], Pt, and Ru [69] improved air electrode performances. Kim *et al* reported an improvement in a Nb-, Ta-doped $BaCoO_{3-x}$ (BCO) electrode due to the use of exsolved Ag nanoparticles which catalyzed the surface exchange rate and promoted the charge transfer reaction at the interface with Ag oxide particles [68]. Marasi *et al* demonstrated the superior performances of common $La_{1-x}Sr_xFeO_3$ (LSF) electrodes with Pt and Ru *B*-site doping [69]. The effect of Pt doping is reported to influence the concentration of oxygen vacancies, yielding a higher ORR rate. Ru doping, on the other hand, allows higher stability at high temperatures in reducing conditions, creating the possibility of fabricating all-perovskite devices with LSF-Pt as the cathode and LSF-Ru as the anode, in which the formation of exsolved Ru NPs would boost the hydrogen oxidation reaction (HOR) rate. Most interestingly, exsolved air electrodes have also been reported to exhibit promising resistance to degradation, showing remarkably stable values of the area specific resistance (ASR) over time [73]. A recent publication by Yildiz's group has also described the changes in the upper surface of exsolved oxides which lead to cation deficiencies [74]. These changes could have implications for the oxygen diffusion and surface reaction chemistry of the electrode oxide and hence should be explored when dealing with exsolved structures.

1.3 Air electrode performance enhancement using nanostructures

Using the nanostructure engineering techniques described in the previous section, local modifications of relevant materials can be applied at the nanoscale, leading to changes in the electrochemical behavior of SOC electrodes. Different approaches can be followed in order to induce electrochemical changes in the electrode by nanostructuring processes, as summarized in figure 1.3 [67, 75–77]. Generally, such approaches can involve bulk/volume or surface modifications. In the following sections, we provide an overview of the main mechanisms and concrete examples of materials engineering with applicability to SOC electrodes.

1.3.1 Nanostructure engineering at the volume level

1.3.1.1 Homo-interface engineering

The chemistry of single-phase materials can be fine-tuned by micro- and nanostructure modification, e.g. by changing the density and composition of grain boundaries, by locally inducing strain fields, and by modifying the chemistry of the surface exposed to the atmosphere.

In general, polycrystalline materials are characterized by the presence of extended defects (twins, dislocations, grain boundaries, etc.), which can be described on the basis of two different regions, namely, a defect core and a space-charge region.

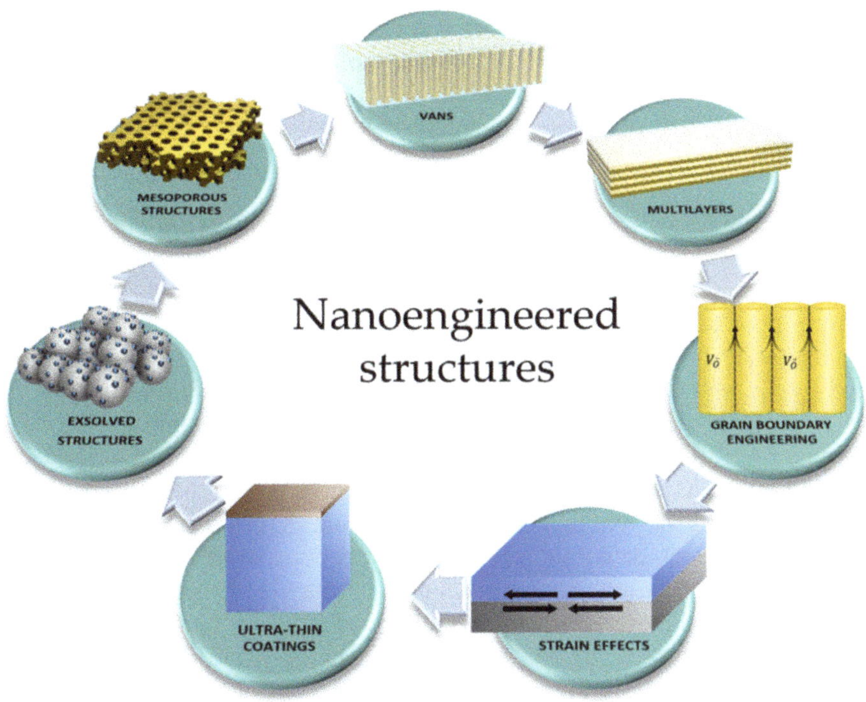

Figure 1.3. The most representative nanoengineered structures currently being studied for air electrode applications in solid oxide cell systems.

As the microstructural adjacency of two grains implies the loss of the bulk periodic structure (core), a local change in the defect chemistry is induced. This is accompanied by the formation of an internal electric field [78], which is often caused by the segregation of charged defects. In order to satisfy the condition of electroneutrality in the lattice, the mobile charged species in the region adjacent to the grain boundary (or dislocation) core rearrange with respect to the concentration level in the bulk, giving rise to space-charge layers. For further details of the defect chemistry of grain boundaries, the reader is referred to excellent textbooks more focused on the topic [79], as well as to the exemplary works e.g. on the $SrTiO_3$ (STO) system [80, 81].

As the oxygen diffusivity is dependent on the oxygen vacancy diffusivity (D_v) and the concentration (c_v):

$$D \propto D_v \cdot c_v, \tag{1.1}$$

one can expect 2D defects to influence vacancy diffusion by locally modifying either the vacancy mobility (thus, D_v) or the vacancy concentration. Hence, several reports in the literature address ionic transport differences—either enhanced or worsened—along interfaces and space-charge-related zones with respect to the bulk [25, 82–86]. The presence of 2D defects has been proven to be a source of faster ionic conduction in, for example, LSM [25, 87], but it hinders diffusion in, for example, STO [88].

The way in which these phenomena can be generalized to ceramic materials and the mechanisms responsible for fast ionic conduction in grain boundaries remain major topics of debate in the community. One of the possible models proposes that the conduction of oxygen vacancies through grain boundaries occurs via the grain boundary core. The work published by Polfus *et al* [89] reports density functional theory (DFT) calculations for the diffusion of oxygen vacancies through grain boundaries in the LSM system according to a Mott–Schottky model. According to these calculations, an increase in the oxygen vacancy concentration in the core is obtained as a result of lowering the vacancy segregation energy. Conversely, the concentration of oxygen vacancies calculated for the space-charge layer does not differ substantially from the value in the bulk, since the core potential is mostly screened by largely available electronic charges. The calculated values for oxygen diffusivity result in an enhancement of up to five orders of magnitude for the grain boundary core diffusivity with respect to the bulk. One major limitation of space-charge modeling in MIEC systems is that typical models (Gouy–Chapman or Mott–Schottky, depending on whether the majority ionic defect is considered mobile or not) do not take defect–defect interactions into account. While these hypotheses are valid for undoped or slightly doped systems, they may lead to inconsistencies in the case of heavily doped materials, such as the typical MIEC materials used for oxygen electrodes. More complex approaches, based e.g. on Poisson–Cahn theory, have been shown to more satisfactorily describe the ionic conductivity, at least for highly-doped ionic conductors such as ceria or zirconia systems [90–93]. On these grounds, the possibility of fast ionic conduction in space-charge layers around LSM dislocations has recently been proposed by Börgers *et al* [87], among others. Their results discard the dislocation core as the main path for oxygen vacancy diffusion. It should be pointed out that this model requires the defect core to be negatively charged—as opposed to the widely accepted picture of a positive core—so that oxygen vacancies accumulate in the adjacent region.

Another local phenomenon that has been proven to have an impact on the electrochemical performance of air electrodes is the presence of elastic strain. Interface strain results from the accommodation of phases with different lattice parameters or crystallographic orientations. Strain has been reported to have a potentially positive effect on the reaction and transport properties of oxygen in ceramic oxides by playing a role in reducing the activation energy of the oxygen reduction reactions, as well as in modifying the formation or migration enthalpy of the mobile defects [94, 95]. Strain is a dimensionless unit which can have a *tensile* nature (a positive value) or a *compressive* nature (a negative value). In general, several computational studies have demonstrated that lattice strain influences many properties of the electrode material, such as surface and bulk oxygen transport, as shown in the work developed by Han and Yildiz [96]. They reported the effect of tensile strain on oxygen adsorption and migration in $LaCoO_3$. Their findings show an anisotropic enhancement of oxygen vacancy migration in the $[1\bar{1}0]$ direction, whereas the mobility of adsorbed atomic oxygen is hindered as a result of an increase in the adsorption energy. According to their analysis, the presence of tensile stress induces a weakening of the Co–O bond, which then leads to the stronger adsorption of oxygen. Similarly, Cai *et al* reported a change in the electronic structure of LSC under strain [97]. Their results show an increase in the density of states around the Fermi level, which is particularly enhanced in the case of tensile strain, pointing to facile electron transfer as the cause of enhanced performance during oxygen exchange reactions. The work published by Yu *et al* shows how dopant segregation in LSCF can also be altered by means of strain engineering, leading to improved electrode stabilities—i.e. reduced Sr segregation—under compressive strain [98]. Another work by Jalili *et al* shows similar strain-derived effects in LSM [99]. They report Sr enrichment, the facile formation of oxygen vacancies, and the transition of the electronic surface structure into metallic-like behavior for tensile strained films. In addition, lattice strain can also lead to a reduction in the migration kinetics of dopant cations, thus inhibiting potential degradation mechanisms by segregating dopants [100, 101].

1.3.1.2 Heterointerface engineering

Tuning the electrochemical properties of air electrodes through nanostructuring may also be achieved by combining different functional materials that, when implemented together, overcome the performance limitations of single-phase materials. This improvement could be the result of combining the individual properties of each phase, but it could also be a result of the presence of new local effects emerging at the interfaces. Hence, the use of heterostructured nanomaterials extends the application of local defect effects to more complex interfaces and electrode systems. These functional materials might exhibit a general intermixed nanocomposite structure, in which the two phases do not exhibit a preferential geometry or orientation in the device. In contrast, they could also be fabricated using a planar and ordered structure in the form of multilayers or superlattices. In addition, the second constituent might be used as a protective coating against degradation processes, or could even take the form of a traditional electrode tuned by the addition of secondary electrochemically active species in the surface. As a result of this, the application of different active materials in a single air electrode architecture seems to be a promising approach for the development of superior air electrodes. Furthermore, the appearance of synergistic transport and catalytic processes at the interface level is an exciting open field of research. In line with this, a particularly interesting case of nanocomposite might be found in vertically aligned nanostructures (VANs). This type of heterointerface is formed from a self-assembled nanocomposite array of columnar grains that exhibits alternation of the two constituent phases, i.e. the matrix and the nanopillars. Such a structure is particularly advantageous in the case of electrodes due to its off-plane geometry, which maximizes the TPB density and offers a pathway for vertical mass transport with no tortuosity.

1.3.1.3 Examples of nanoengineered structures at the bulk level

In the introductory section, an outline of the mechanisms used for interface engineering via bulk effects in air electrodes was given. In the following section, we provide specific examples of electrode systems whose interfaces have been engineered via bulk effects.

1.3.1.3.1 Grain-boundary-engineered structures

It is well known that in ionic conductors such as acceptor-doped ceria and zirconia, the ionic transport is hindered by the presence of grain boundaries [102–104]. This has also been proved for perovskite materials such as acceptor-doped STO [80, 105]. Additionally, a blocking effect has recently been observed for oxygen conduction through dislocations in STO [85, 88] and ceria-based systems [86]. Some recent works have complementarily shown the relevance of ionic conduction along grain boundaries in some systems such as LSM, which leads to an enhancement of up to several orders of magnitude in the electrochemical properties of LSM when it is used as an electrode. Several authors have reported an improvement in the performance of polycrystalline LSM thin-film performance in comparison to conventional bulk values, as first reported by independent works by Saranya *et al* [106] and Navickas *et al* [25]. On the basis of isotopic exchange depth profiling (IEDP) and finite element method (FEM) simulations, oxygen mass transport coefficients were extracted for both bulk and grain boundary contributions. These results were later confirmed by impedance spectroscopy measurements [107, 108], showing that the conduction of the oxygen species through the grain boundary is predominant with respect to the bulk grain contribution. Moreover, in both studies a dependence of the ORR contribution on the grain boundary density was reported, i.e. the higher the density the more—electrochemically active—grain boundary sites are present in the surface. Another work by Chiabrera *et al* relates the introduction of local changes in composition at the grain boundary level to a direct modification of the electrochemical properties of LSM thin films through GB engineering [109]. The resulting local nonstoichiometry (in particular *B*-site deficiency) is related to an enhancement of the electrochemical performance of the films. Additionally, a compositional work published by Saranya *et al* demonstrates, by means of isotope exchange depth profiling–secondary ion mass spectrometry (IEDP-SIMS), the presence of faster oxygen diffusion through grain boundaries in $La_{1-x}Sr_xMn_{1-y}Co_yO_3$ (LSMC) for a wide range of Mn and Co compositions [84]. Interestingly, this work raises the possibility of enhanced oxygen diffusion through grain boundaries in other systems—with higher ionic diffusivity—besides LSM.

A publication recently released by Thoréton *et al* [110] explores the apparent grain boundary contribution in diffusion profiles of LSCF mixed conductors. This work gives an excellent overview of transport through grain boundaries in most typical oxides (e.g. manganites, chromites, and Sr-free perovskites) and reports a record value—by several orders of magnitude—for the grain boundary product ($D_{gb}\delta$) of LSCF, as measured by IEDP. Remarkably, the reported analysis was performed for the lower temperature regime (300 °C–500 °C), because at higher temperatures, the oxygen exchange profile becomes indistinguishable from the bulk grain regions (Harrison type A diffusion [111]). This would explain why ionic conduction through grain boundaries in MIEC materials has been a rarely observed phenomenon, almost exclusively seen in poor ionic conductor materials such as manganites and chromites [25, 107, 109, 112]. This study paves the way for investigations of fast ionic conduction through grain boundaries in other systems, suggesting the establishment of a general phenomenon with great potential in thin-film applications and low-temperature devices.

A particularly critical aspect of dealing with oxygen transport through grain boundaries is the correct assessment of local transport kinetics. Several experimental techniques deal with ion transport in ionic and mixed conductors. In general, they are based on indirect observation of the change in the material's physicochemical properties. Examples of these are electrical conductivity relaxation experiments [113–115], electrochemical impedance spectroscopy [116–118], optical relaxation [119], and color changes [120]. However, these global techniques do not yield information about the local grain boundary kinetics. The latter can be extracted by oxygen IEDP-SIMS [121, 122]. Nevertheless, SIMS is representative of an averaged contribution of grain boundaries, so direct confirmation of the phenomenon remains a challenge. In some recent studies, fast conduction through single grain boundaries in nanocrystalline $La_{1-x}Sr_xCrO_3$ (LSCr) films was directly measured, for the first time, by atom probe tomography (APT) [112]. Combined with oxygen isotopic exchange (isotopic exchange–atom probe tomography (IE-APT)), this tool allows the direct observation of the ionic behavior in the grain boundary environment and a local analysis of the conduction pathway of the oxygen ions through the nanostructure. This study presents a fine reconstruction of the oxygen distribution along the grain boundaries and cores, as can be observed in figure 1.4(b). The highly precise profile obtained by this technique, coupled with FEM simulations, makes it possible to perform an accurate calculation of the oxygen transport parameters. The kinetics through the grain boundaries was proven to be around three to four orders of magnitude greater for both the oxygen diffusivity and surface exchange coefficient [112], compared to the bulk values. This is summarized in figure 1.4(d), which compiles a set of relevant data on the oxygen surface exchange coefficient of different nanoengineered electrodes and state-of-the-art materials. In this way, the local kinetics of LSCr was found to be even faster than that of prototypical LSM (100- and 10-fold increases in D and k at 640 °C, respectively). These values demonstrate how nanostructuring via grain boundary engineering can result in the conversion of an electronic conductor into a promising mixed ionic–electronic conductor that can be applied in solid oxide systems. The validity of the IE-APT approach was later extended to LSM [123]. In this study, the authors also proposed a comparison between the core diffusion and the space-charge diffusion model, concluding that both are in principle able to adequately explain the experimental data set, provided that a space-charge zone of less than 1 nm is considered.

Chen *et al* [119] studied the evolution of the surface oxygen exchange coefficient upon the crystallization of different MIEC thin films. By performing *in situ* optical transmission relaxation experiments, they were able to obtain the surface exchange coefficient upon crystallization for a set of initially amorphous MIEC oxides. In this work, a novel *in situ* characterization method for studying the electrochemical properties of air electrodes was developed. Moreover, the authors also observed an enhancement of one to two orders of magnitude in the surface exchange coefficient compared to those of highly crystalline films. Huber *et al* [124] studied the trade-off between improving the oxygen transport properties and reducing the lateral electronic conduction in a model LSM electrode. According to their analysis, for

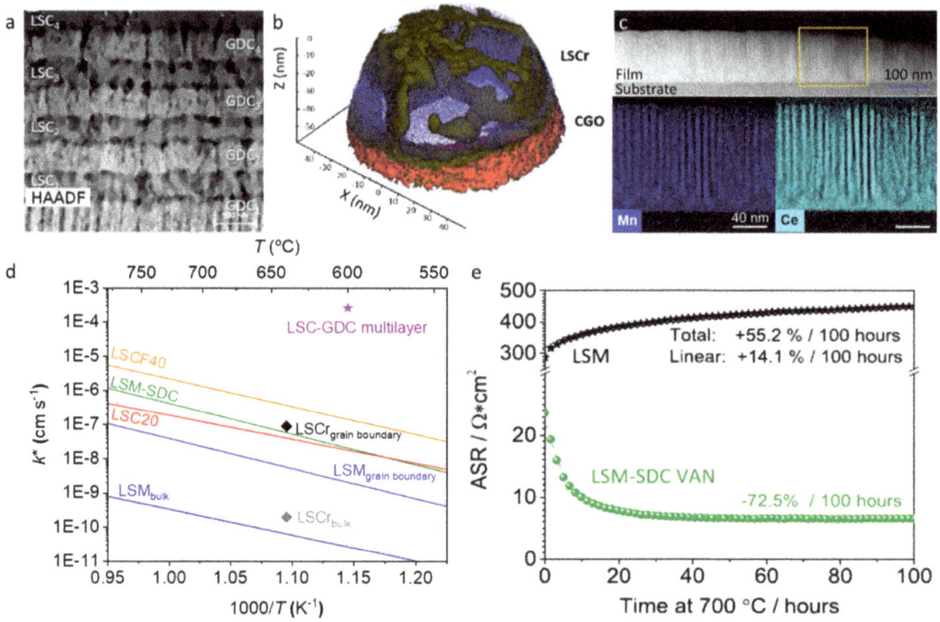

Figure 1.4. Air electrode engineered structures produced via bulk mechanisms: (a) scanning transmission electron microscope high-angle annular dark field (STEM-HAADF) image of a LSC–GDC quadrilayer. Reprinted from [130], Copyright (2020), with permission from Elsevier. (b) APT visualization of fast oxygen conduction through LSCr grain boundaries [112], John Wiley & Sons. © 2021 The Authors. Advanced Materials published by Wiley-VCH GmbH. (c) Low-magnification transmission electron microscopy (TEM) micrograph and energy-dispersive x-ray (EDX) signals (Mn, Ce) of an LSM–samarium-doped ceria (SDC) VAN structure [28] Copyright © 2021, The Author(s). With permission of Springer. (d) Arrhenius plot of the oxygen exchange coefficients of reported nanoengineered structures: LSCr [112], LSC–GDC [130], and LSM–SDC [28] compared to published values for state-of-the-art materials: LSM [25], LSC20 bulk [8], and LSCF40 [139]. (e) ASR evolution over time of LSM and an LSM–SDC VAN structure during a degradation treatment at 700 °C [28] Copyright © 2021, The Author(s). With permission of Springer.

a hypothetical 100 nm-thick LSM film with a grain size of 30 nm, the current should be collected in spaces of no more than 100 μm at 700 °C. This could hinder the performance of lateral devices or planar systems and should be considered for realistic applications.

1.3.1.3.2 Strain-engineered systems

Strain engineering plays a key role in tuning the electrochemical properties of existing well-studied air electrodes based on oxide thin films in SOCs. A recent work showed that the kinetics of the ORR could be improved through epitaxial strain induced by a lattice mismatch between the grown film and the substrate [96–99, 125, 126], while higher stability of the oxide-based cathode materials and subsequent enhanced durability of the cell under operational conditions could mainly be achieved by tailoring the dopant cation size to hinder segregation processes occurring at higher temperatures [126, 127].

Furthermore, many experimental efforts have focused on finding evidence for these enhanced properties in model perovskite systems ABO_3 [97–99, 125, 126]. Thin films with limited thicknesses from ≈20 nm to 50 nm were used in order to maintain a good quality of epitaxy while avoiding strain relaxation at higher temperatures. Complementary, innovative *ex situ* and *in situ* characterization methods for depth profiling and surface sensitive analysis were adapted to such low film thicknesses and to different operating conditions with the aim of fully comprehending the phenomena connecting lattice strain with the enhancement of the ORR kinetics. In particular, studies of LSC showed that film subjected to tensile strain grown on (100) STO achieved faster oxygen surface exchange (≈four times) and oxygen diffusion (≈ten times) compared to films subjected to compressive strain grown on (100) $LaAlO_3$ (LAO) in a temperature range of 280 °C–475 °C and for different film compositions (LSC82, LSC64) [125]. This was explained through a change in the film thermodynamics that results in lower energy barriers related to oxygen mobility plus an enhancement in oxygen vacancy formation. As mentioned, the electronic structure is also fundamental, and strained films, particularly films of typical perovskite materials ($La_{1-x}Sr_xMO_3$, M = Mn, Fe, Co) subjected to tensile strain showed an enhancement in the electronic density of states (DOS) near the Fermi level above room temperature. In this case, a transition from the semiconducting state (with an energy bandgap of 0.8–1.5 eV) to the metallic state (without a bandgap) was reported, indicating a highly active surface for electron transfer in oxygen reduction [97, 99]. It is also worth mentioning that a correlation has been drawn between strain formation in the material and surface chemistry. In particular, segregation processes at operational temperatures were investigated in detail, and strains were reported to modify the diffusivity of Sr cations in Sr-doped perovskites by changing the Gibbs free energy of migration of Sr cations from the bulk to the surface [97–99]. The role of such surface segregation in common air electrodes (typically based on doped perovskites) is one of the main causes of performance degradation at operational temperatures. Perovskite-based air electrodes are widely adopted for SOCs because of their capability to change composition by incorporating a range of cation dopants in the *A*- or *B*-sites, which plays an important role in modulating the oxygen vacancies in the material. Alkali, alkali-earth, and transition metals are mainly used as dopants in these sites. In order to hinder segregation, recent studies proved that a smaller size mismatch between the host and the dopant reduces this degradation phenomenon. An interesting study of the lanthanum manganite-based system with different *A*-site cation dopants (Ca, Sr, Ba) demonstrated that a bigger cation size mismatch between La^{3+} (host) and Sr^{2+}, Ba^{2+} (dopants) cations contributes significantly to a constant rearrangement at high temperature that behaves as a driving force for segregation and secondary phase formation at the electrode surface. On the other hand, the smaller cation radius of Ca is closer to that of the host cation, and it can be accommodated in the lattice more effectively, limiting its diffusion toward the surface [127]. Hence, the choice of dopant elements in the cathode and a preliminary study of the strains in the cathode lattice itself (caused by its constituent elements) is fundamental to ensure not only good oxygen reduction activity but also high durability.

1.3.1.3.3 Multilayers and superlattices

Another class of nanostructure corresponds to the use of layered architectures in the form of multilayers and superlattices. Further details of property tuning using multilayers are discussed in chapter 3 in this book. Chen *et al* fabricated superlattices of Ruddlesden–Popper (RP) LSC_{214} and single-perovskite LSC_{113} as promising electrodes for the ORR [128]. In their work, they proposed that the expected high oxygen kinetics of LSC_{214} (favored by the presence of oxygen interstitials in the lattice) could be coupled with the high performance of LSC_{113} in terms of electronic conduction. Their results reveal an electronic activation for LSC_{214}—stemming from the interface coupling with LSC_{113}—which is expected to enhance the charge transfer to oxygen. Moreover, an investigation of Sr segregation in an LSC_{214}–LSC_{113} heterointerface carried out by Feng *et al* [100] suggests promising performance against degradation for these nanostructured electrodes.

For their part, Develos-Bagarinao *et al* have shown that interesting electrochemical properties arise from the use of LSC and GDC multilayers [129]. These findings suggest an increase in the oxygen vacancy concentration in the LSC/GDC interface, resulting in values of the oxygen surface exchange coefficient of up to 10^{-5} cm s^{-1} in the 500 °C–700 °C range. In a more recent work [130], they observed a direct correlation between the increase in the surface exchange coefficient and the number of LSC–GDC layers stacked. The results reported for four LSC–GDC bilayers, presented in figures 1.4(a) and (d), showed a remarkable value of $k \approx 10^{-4}$ cm s^{-1} at 600 °C and improved performance upon degradation compared to those of single LSC–GDC bilayers. Even so, there is an initial loss in performance during the first 100 h of testing, which the authors ascribe to the thermal evolution of the microstructure; the ASR of the multilayer seems to reach a plateau beyond this point, stabilizing the electrochemical activity of the electrode. They propose that the multilayer structure might restrict the segregation of Sr species to the surface, which is known to be one of the degradation mechanisms in air electrodes for SOCs. Further analyses beyond the reported work would be probably needed to confirm such an interesting possibility. For a more detailed discussion of multilayers, the reader is further referred to chapter 3 in this book.

1.3.1.3.4 Vertically aligned nanostructures and nanocomposites

The use of VANs and other nanocomposites has recently attracted the attention of the research community [67, 75, 77, 131, 132]. The combination of a conventional electrode (i.e. a good electronic or mixed conductor with high catalytic activity for the ORR) with a good ionic conductor, e.g. doped ceria species, is commonly accepted as a promising strategy for the fabrication of next-generation electrodes [133]. Among other studies, early works by Murray and Barnett [11] showed the promising properties of this kind of heterostructure. They explored the properties obtained by mixing LSM and GDC in a composite electrode, showing a decrease of up to one order of magnitude for an LSM-GDC mixture at 50 wt%. In a recent publication by Tarancón's group, the use of LSM and SDC in a VAN configuration is explored using a functional air electrode for SOFCs [28]. The obtained VAN, shown in figure 1.4(c), surpasses the electrochemical performance of single-phase

LSM and even stabilizes at a lower ASR upon 100 h of thermal degradation—see figure 1.4(e). In this particular case, the local microstructure and its oxygen mass transport properties were analyzed by APT in a pioneering way, which revealed fast oxygen conduction and a high cationic intermixing of the VAN's constituent components. Interestingly, this cationic intermixing formed medium-entropy oxides (substitution of La for Ce and Sm in the perovskite phase) that the authors were able to correlate with the observed unprecedented thermal stability of the nanocomposite. This was confirmed by DFT calculations, which showed that the introduction of Ce and Sm reduced the elastic instability of the Sr atoms in the lattice, which is the source of its tendency to segregate in regular doped perovskites such as nominal LSM. Moreover, the chemical analysis performed by low-energy ion spectroscopy (LEIS) revealed the absence of additional Sr segregated species in the surface of the electrode, hence demonstrating that the suppression of Sr segregation was the reason for the outstanding thermal stability. A VAN structure based on RP-LSC_{214} and single-perovskite LSC_{113} has also been reported by Ma *et al* [134], leading to a decrease of one order of magnitude of the ASR with respect to the parent materials and also revealing apparent stabilization upon Sr segregation. Nonetheless, the electrochemical activity of the heterostructures was reported to be thermally unstable at temperatures higher than 400 °C. More recent studies of mixed-phase LSC_{214}–LSC_{113} nanocomposites have been reported by Maier's group [135, 136]. In these, Stämmler *et al* investigated the formation and performance of these nanocomposites under different growth conditions and at different crystallite sizes, reaching an enhancement of one order of magnitude with respect to the VAN structure discussed. A separate work by Develos-Bagarinao *et al* [137] explored the use of a nanocomposite based on LSCF and CGO as a functional layer for use in air electrode architectures. They proposed a dense nanocomposite of vertically aligned LSCF and CGO columns in combination with a top layer of porous LSC, reaching values as low as ~36 mΩ cm^2 at 600 °C, which makes it a highly interesting electrode for use in intermediate-temperature SOCs (IT-SOCs). Another study involving LSCF–CGO nanocomposites [138] reported an overall enhancement in the oxygen transport parameters compared to those of the single phases. In particular, the authors of this study observed a slightly lower value for the LSCF contribution, but a thirtyfold increase for the CGO, which they attributed to a catalytic spill-over process for the incorporation of oxygen, probably in combination with a surface cleaning effect which avoided the formation of blocking impurity layers in the CGO.

1.3.2 Nanostructure engineering at the surface level

Surface modification of air electrode materials is emerging as a powerful solution for enhancing their electrochemical performance by improving their surface exchange coefficient and stability.

1.3.2.1 Decoration

Surface decoration by means of the deposition of a highly active catalyst layer can be successfully achieved by thin-film coating. ALD in particular allows for the conformal deposition of nanometer-thick films (typically from 1 nm to 20 nm) that do not significantly affect the bulk electrode properties while enhancing its electrochemical properties by tuning the surface.

In the field of air electrode materials, ALD has been revealed to be an effective process either for improving ORR kinetics by adding an extra functional active layer on the surface [140, 141] that can act as a catalyst for oxygen incorporation or for enhancing the stability against ageing at high temperature, in particular by preventing detrimental surface phenomena [29, 142, 143]. Regarding the first case, complex 'super-cycle' depositions are required in order to grow active films, which are usually quaternary oxides (e.g. doped perovskites); hence, a multistep process involving multiple precursors is required. Inevitably, longer deposition times are needed, but different compositions can be obtained by modulating the number of deposition cycles used for each precursor [144]. The employment of ALD for such goals is still under development, yet recent experimental works have shown how effectively the electrochemical activity of LSCF electrodes was improved by the deposition of a nanoscale layer of LSC with an ALD-treated surface, compared to an electrode with a bare cathode surface. The role of the ALD layer was to ensure higher oxygen surface exchange rate via an ultrathin layer, while the underlying 'backbone' improved bulk diffusion [140]. Other ways of increasing catalytic activity via ALD also include the incorporation of transition metals into the cathode surface lattice to increase the concentration of oxygen vacancies and enhance the charge transfer mechanism in the electrode. In particular, Co and Ni cations from ALD monolayers were able to penetrate a lanthanum-ferrite-based perovskite lattice, enhancing its activity [141].

Notably, the widest use of ALD in air electrode surface engineering so far involves the deposition of binary oxides as protective layers against degradation mechanisms—see figures 1.5(a) and (b). Among these oxides, ZrO_2, HfO_2, TiO_2, Al_2O_3, and CeO_2 were the most explored materials. ALD binary oxide layers improved electrode stability by suppressing Sr segregation (a common cation dopant in perovskite-based electrodes and the main cause of their resistance increase, as previously discussed). In the case of ZrO_2 coatings, ultrathin layers helped to segregate Sr on the surface via $SrZrO_3$ formation, thus avoiding the blockage of active sites for oxygen adsorption. In particular, it was reported that the formation of isolated clusters of $SrZrO_3$ on the surface avoids the coarsening of insulating secondary phases that form during the segregation process [29, 142]. One main reason for the inhibition of the Sr segregation effect is the local reduction of the oxygen vacancy concentration in the outer surface of the electrode, on the interface electrode/ALD layer. An accumulation of oxygen vacancies in this specific region acts as a driving force (together with lattice deformation) for the segregation of Sr^{2+} on the surface [143].

It is crucial to mention the important role of ALD in maintaining the beneficial effect on the overall long-term performance of the electrode. Hence, ALD layers of binary oxides result in a favorable effect only when deposited at lower thicknesses (1–5 nm), otherwise counterproductive effects caused by the formation of large areas of insulating phases (e.g. the uncontrolled presence of $SrZrO_3$) or an excessive reduction in the oxygen vacancy concentration at the surface can negatively impact the electrode's electrochemical activity. More experimental studies are needed to

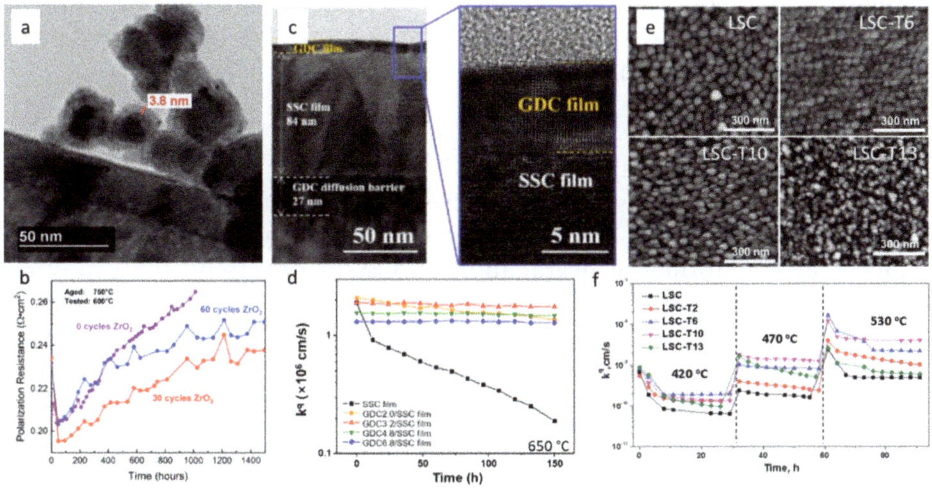

Figure 1.5. Air electrode structures engineered via surface mechanisms: (a) scanning transmission electron microscope (STEM) image of an SSC–GDC porous electrode coated with a zirconia layer deposited by ALD. Reproduced from [143] with permission of The Royal Society of Chemistry. (b) Electrode ASR evolution over time for different zirconia layer thicknesses. Reproduced from [143] with permission of The Royal Society of Chemistry. (c) TEM image of an SSC electrode coated with GDC by PLD. Adapted with permission from [145]. Copyright (2020) American Chemical Society. (d) k evolution over time for electrodes coated with different GDC thicknesses at 650 °C. Adapted with permission from [145]. Copyright (2020) American Chemical Society. (e) Atomic force microscopy (AFM) micrographs taken from above showing different Ti-substituted LSC electrodes (0%, 6%, 10%, and 13% for LSC, LSC-T6, LSCT-10, and LSC-T13, respectively). Reproduced from [152] with permission of The Royal Society of Chemistry. (f) k evolution over time at increasing temperature steps for the different Ti-based LSC electrodes. Reproduced from [152] with permission of The Royal Society of Chemistry.

implement this novel approach to air electrode nanostructuring, in parallel with the development of new materials that can be deposited using ALD. It is worth mentioning that the fabrication of such functional protective layers can be achieved by other thin-film techniques—although these may result in lower conformality compared to that obtained using ALD. In one study, Choi *et al* proposed the use of a heterostructured $Sr_{0.5}Sm_{0.5}CoO_3$–$Ce_{1-x}Gd_xO_2$ (SSC–GDC) electrode fabricated by PLD to prevent the formation of strontium segregated species in the surface [145]. The presence of the GDC overlayer (with a thickness ranging from 2 nm to 7 nm, as seen in figure 1.5(c)) kept the electrochemical performance practically unaltered during a degradation test lasting 150 h (figure 1.5(d)). The reasons for this behavior were associated with electrostatic attraction based on coulombic forces between charged defects, which is one of the factors that gives rise to segregation phenomena. Charged defects redistribute along the GDC/SSC interface, reaching an equilibrium. In this way, the positively charged defects (oxygen vacancies) responsible for the attraction of negatively charged defects (doped Sr) are decreased at the GDC/SSC interlayer, drastically reducing the detrimental effect that leads to the formation of Sr compounds on the electrode surface.

Ultimately, a deep understanding of the intimate atomic-level bonding that occurs at the interface between the air electrode and ultrathin layers is fundamental for the application of these nanostructures in realistic SOC devices.

1.3.2.2 Modification of surface chemical composition

Dedicated investigations have highlighted a correlation between the chemical composition of the outermost layers and the oxygen reaction rate. In this regard, recent works have provided specific evidence of a correlation between surface Sr enrichment and performance degradation in LSC. Conversely, the exposure of *B*-sites plays an important role in achieving a faster oxygen exchange rate. For example, a combined experimental and theoretical work by Mastrikov *et al* that focused on LSM predicted an improvement by five orders of magnitude in the oxygen reduction kinetics of MnO_2-terminated surfaces compared to (La,Sr)O termination [146]. Fleig's group observed a substantial change in the oxygen exchange reaction activity of LSC, depending on the concentration of Sr in the termination layer [27]. Their studies revealed lower activity on pristine LSC—characterized by a higher concentration of Sr at the surface—compared to that of H_2O-treated LSC. In addition, they observed the reconstruction of a Sr-rich surface layer when the electrodes were annealed for more than 1 h at a temperature greater than 550 °C. This was supported in a recent work by Cao *et al* addressing *ab initio* DFT calculations for the ORR in a model LSC electrode [147]. Their work predicted that the use of CoO_2 termination results in vacancy-assisted oxygen dissociation, leading to faster incorporation by a factor of 10^2–10^3 than in the case of a Sr-terminated surface.

Nonetheless, even though the addition of, for example, extra cobalt oxide species in the surface might seem to be an appealing strategy for improving the performance of LSC-based electrodes, it has been reported that this can result in a detrimental effect [148], making it difficult to properly find a trade-off between the exposed Co and the enhancement in the electrochemical properties. A strong limitation on surface composition engineering is particularly imposed by the thermodynamic stability of highly active BO_2 termination, as the AO surface is typically the stable one in MIEC materials [7]. This situation is intensified by the tendency of Sr to migrate to the surface, driven by electrostatic and elastic forces [149], forming an insulating layer. The removal of excess Sr and *B*-site exposure can typically be obtained in MIEC materials by chemical etching in mild acids, as demonstrated by Jung *et al* for model $SrTiFe_xO_3$ and by Kubicek *et al* for LSC [150]. One recent study by Acosta *et al* developed a high-surface-area LSCF honeycomb structure, obtained by the selective acid etching of a VAN starting material [151]. Acid etching also exposes highly active *B*-site termination in LSCF, leading to a strongly increased surface exchange coefficient (one order of magnitude at less than 400 °C). The termination was shown to be stable at temperatures up to 450 °C.

Tsvetkov *et al* studied the possibility of hindering Sr migration by modifying the bulk chemical composition: Ti-modified LSC electrodes [152] (figure 1.5(e)) showed an eightfold increase in the surface exchange coefficient with remarkable retention over time at 530 °C (figure 1.5(f)). Some studies have reported a certain degree of reversibility in the dopant segregation phenomena, recovering the electrochemical activity of the electrode either after applying a cathodic bias [153] or after a thermal treatment at more than 800 °C [154]. In both cases, the performance of the electrode could be differentiated as either a passivated state (the presence of SrO species segregated on the surface) or an activated state (recrystallization of Sr on the lattice).

1.4 Implementation of thin-film air electrode layers in full devices

The functional layers and nanoengineered structures explained in previous paragraphs have been implemented in full SOFC devices. The main purpose of developing more sophisticated electrodes or interlayers through nanostructuring strategies is to enhance their performance while overcoming issues related to chemical and thermal stability, in order to achieve a commercial breakthrough that can make SOFC devices competitive with other power sources and energy storage systems [155].

Nanoengineered air electrodes have successfully been integrated into anode-supported cells, which showed greater performance in terms of current density. Porous LSC has been deposited onto a self-assembled highly ordered LSCF–GDC VAN layer, as reported in figure 1.6(a). The combination of the two layers resulted in higher electrochemical performance of the cell, which exhibited a current density at 0.7 V in the range of 3 A cm^{-2} at 700 °C (see figure 1.6(b)) and a higher interfacial stability over 250 h of testing [137]. An enhancement in SOC performance has been

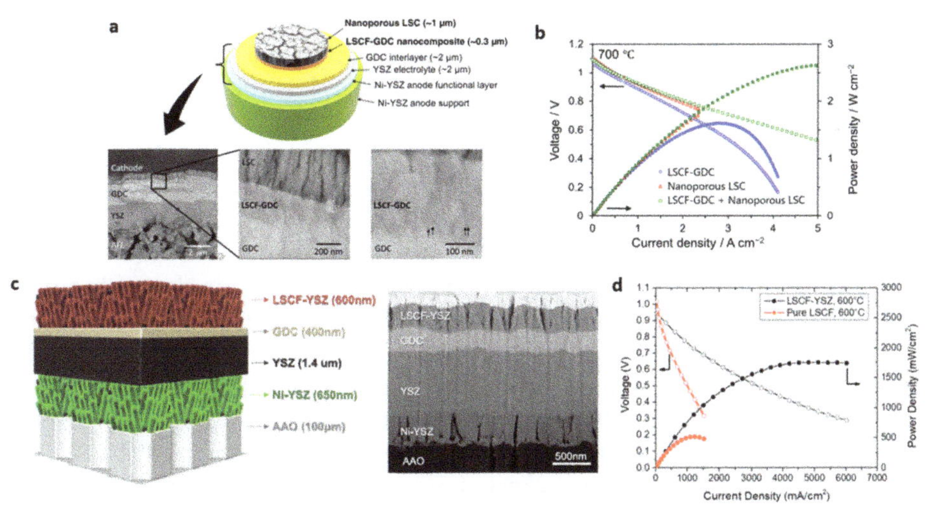

Figure 1.6. Exemplary cases of nanoengineered structures implemented in a full cell: (a) schematic and SEM micrographs of an LSCF–GDC VAN in an anode-supported cell [137] Copyright © 2021, The Author(s). With permission of Springer. (b) *I–V* and *I–P* curves of the LSCF–GDC VAN-based SOFCs [137] Copyright © 2021, The Author(s). With permission of Springer. (c) Schematic of an LSCF–YSZ nanofiber-based cell with a cross-sectional SEM micrograph of the cell. Adapted with permission from [50]. Copyright (2020) American Chemical Society. (d) *I–V* and *I–P* curves of the LSCF–YSZ-based SOFC. Adapted with permission from [50]. Copyright (2020) American Chemical Society.

reported for a cell that used an air electrode with a unique nanofibrous microstructure composed of LSCF and YSZ produced via magnetron sputtering (figure 1.6(c)), which helped oxygen transport through the entire electrode and allowed outstanding peak power densities of ≈1.7 W cm^{-2} and ≈2.5 W cm^{-2} at 600 °C (see figure 1.6(d)) and 650 °C, respectively [50]. In a recent work by C. Su *et al*, a mesoporous $Ba_{0.5}Sr_{0.5}Co_{0.8}Fe_{0.2}O_3$ (BSCF) electrode prepared via a top-down synthetic process was demonstrated to help lower the sintering temperature for the material by increasing the porosity level. In this way, the electrolyte and the air electrode were co-fired at a lower critical temperature that did not allow secondary phase formation at the interface. The enhancement in the oxygen diffusion through the pores boosted the power density peak, which ranged from 460 mW cm^{-2} at 800 °C for BSCF-0 (pore volume: 0.001 42 cm^3 g^{-1}) to 1800 mW cm^{-2} at the same temperature for BSCF-5 (pore volume: 0.0201 cm^3 g^{-1}) [156]. A study conducted by Noh *et al* demonstrated the importance of structural design when fabricating air electrodes by comparing the performances of two devices fabricated using a single-phase LSC PLD-deposited film electrode and a gradient-structured thin film. In the latter, two LSC–GDC layers deposited at different oxygen pressures plus an extra LSC layer deposited at room temperature were employed as air electrodes. Both devices showed an outstanding peak in power densities exceeding 500 mW cm^{-2} at 500 °C, and the gradient-structured device was particularly efficient below 600 °C. Hence, this work highlights the importance of the structural design of the gradient-structured thin-film (GSTF) electrode in overcoming the decrease in power density at lower operating temperatures, while retaining enhanced ORR activity compared to the single-phase LSC [157].

As mentioned in the previous section, ALD thin films can be employed in order to enhance the chemical, thermal, and mechanical stability of common air electrodes. Evidence of their beneficial effects under certain conditions (e.g. choice of the material for the ALD coating, deposition parameters, and, most importantly, the thickness produced) has been confirmed in full devices. An ALD-CeO_2 coating was used on top of a Pt air electrode layer in an anodized aluminum-oxide-based fuel cell, showing a peak in the power density of 800 mW cm^{-2} at 500 °C [158]. ALD processes have also been used to tune air electrode surfaces through the deposition of more complex and electrochemically active oxides. Anode-supported devices have been tested using an LSCF air electrode and ALD-deposited LSC functional nanolayers with thicknesses ranging from 1 to 12 nm. The resulting outputs of the devices show an increase in power density from 224 mW cm^{-2} at 600 °C for the bare LSCF electrode to a maximum of 394 mW cm^{-2}, which was achieved by a forty-supercycle LSC nanofilm (corresponding to ≈8 nm) [140].

Complementing the coated surfaces produced by thin-film deposition, nanoparticle exsolution techniques have been also widely investigated to improve the ORR activity of common perovskite-based air electrodes in SOC devices. A layered device comprising a simple perovskite nanorod-decorated *A*-site-deficient double perovskite $PrBa_{0.94}Co_2O_{5+\delta}$ (A-PBC) has been fabricated and tested in a Ni-YSZ anode-supported SOC device, showing enhanced catalytic activity with a maximum power density of 1.1 W cm^{-2} at 700 °C, which appears to be almost double the power density of 0.62 W cm^{-2} achieved by the respective double perovskite A-PBC [159]. Exsolution techniques have also been adopted in full cells for common perovskite materials such as LSF with exsolved Fe and SrO nanorods and nanoparticles [160].

A critical aspect of integrating air electrodes into full SOC devices is the interlayer region between the air electrode and the electrolyte, where the effects of cation interdiffusion may lead to the formation of insulating secondary phases [161]. For this reason, electrolyte-dense barrier layers (typically ceria-based) are employed. The use of thin-film technology is particularly advantageous here, as it allows a fully dense layer to be obtained without the need for high-temperature sintering processes (up to 1500 °C for classical deposition methods such as screen printing or tape casting [15, 162]). Thin-film barrier layers were first introduced by Mai *et al* [163], who utilized reactive sputtering for the fabrication of a 0.3 μm CGO layer, resulting in an SOFC current density of 1.7 A cm^{-2} at 0.7 V and 700 ° C. Over the years, magnetron sputtering was subsequently optimized for barrier layer deposition in terms of thickness, deposition conditions, and applied bias [49, 164, 165]. An interesting solution has recently been proposed by Coppola *et al*, in which room-temperature sputtering followed by mild annealing (800 °C) was employed for the fabrication of a high-performance CGO barrier layer (current density 0.86 A cm^{-2} and $V = 800$ mV at 700 °C [166]). PLD layers have been also shown to lead to a remarkable decrease in bulk cation interdiffusion [17, 18] and to boost the device performance [19]. Tarancón's group reported the use of a dense CGO barrier layer deposited by PLD (1 μm thick) sintered at different temperatures between 1100 °C and 1400 °C to evaluate the optimal conditions for preventing secondary phase formation between the LSCF electrode and the YSZ electrolyte. Their work revealed that for PLD-deposited BLs heat treated at 1150 ° C, the output power density of an SOFC device is increased by about 70% compared to those of the as-deposited BL and screen-printed BLs [21]. Bernadet *et al* produced dense CGO BLs with thicknesses ranging from 200 to 800 nm. A complete anode-supported device was then tested for 14 000 h, which exhibited an extremely low and stable degradation rate of 0.54% V kh^{-1} at 750 °C over 5800 h. A decrease to 0.47% V kh^{-1} at 720 °C over 6100 h was also shown and good results obtained for the thinnest BL (200 nm) [167]. More recently, following the approach of Yoon *et al* [20], Machado *et al* proposed the combination of an SDC BL with a dense VAN layer produced by PLD (LSM–SDC) in an anode-supported commercial SOFC, with a total thickness as low as 400 nm and without post-annealing treatments [168]. A high initial power density of 1.00 W cm^{-2} was obtained at 750 °C. The device was tested for more than 1500 h and showed a moderate degradation rate of 2.8% V kh^{-1}.

All these encouraging results obtained for nanostructured air electrodes integrated into full SOC cells (including large-area commercial cells in some cases) anticipate an increasing impact on the performance of this field.

1.5 Conclusions

SOFC electrode nanostructuring has become a major field of research and technological development. Fundamental materials research into the ORR has established nanostructuring as an effective way of enhancing oxygen reduction kinetics. In this regard, *true* size effects have been reported beyond the *trivial* size effects directly related to higher surface-to-volume ratio characteristic of nanomaterials. These true size effects comprise, among others, charge redistribution in space-charge zones, the reduction of activation barriers for defect formation, migration and reaction due to strain, and increased surface chemical stability. Such effects may be achieved by materials engineering using grain boundaries, dislocations, heterostructures, surface decoration, or high entropy induced at the nanoscale. On top of this, the introduction of advanced materials such as MIEC conductors represents a new degree of freedom that reinforces the role of bulk oxygen transport pathways within the volume of the electrode and therefore increases interest in achieving fast diffusivity, which can be enhanced by nanostructuring.

In this chapter, we have presented an overview of the mechanisms and strategies used for SOFC air electrode engineering, as well as a brief account of fabrication techniques. In this regard, it has become evident that electrode nanostructuring goes hand-in-hand with the development of advanced thin-film technology. While classical approaches such as hard templating and infiltration are still relevant, complex thin-film architectures fabricated using advanced techniques have the capability to fully exploit the nanoionic effects underlying enhanced ORRs. Nowadays, such thin films can be fabricated with high control over composition; they range from dense interlayers to porous electrodes and ultrathin coatings. However, the real milestone for the ultimate implementation of thin films in commercial devices will only come with a significant cost reduction of the fabrication processes and an improvement in their scalability.

Despite the current limited applicability of the strategies presented in this chapter at a commercial level, we have presented extremely successful case studies of thin-film implementation in real cells. This great improvement in the performance of existing commercial cells due to the use of thin films (mostly by means of the deposition of simple oxide films as diffusion barrier layers or catalysts) represents a big step forward and encourages further work in the field. Indeed, since the implementation of more complex electrode nanostructures is still in its infancy, disruptive advances are expected in the near future, which will eventually lead to high-performance all-thin-film devices.

Acknowledgements

The authors acknowledge the support from the European Union's Horizon 2020 research and innovation program under grant agreement No. 824072 (HARVESTORE) and No. 101017709 (EPISTORE), and from the Generalitat de Catalunya (2021-SGR-00750, NANOEN).

References

[1] Chan S H, Khor K A and Xia Z T 2001 A complete polarization model of a solid oxide fuel cell and its sensitivity to the change of cell component thickness *J. Power Sources* **93** 130–40

[2] Kim J, Virkar A V, Fung K, Mehta K and Singhal S C 1999 Polarization effects in intermediate temperature, anode-supported solid oxide fuel cells *J. Electrochem. Soc.* **146** 69–78

[3] Gao Z, Mogni L V, Miller E C, Railsback J G and Barnett S A 2016 A perspective on low-temperature solid oxide fuel cells *Energy Environ. Sci.* **9** 1602–44

[4] Adler S B 2004 Factors governing oxygen reduction in solid oxide fuel cell cathodes *Chem. Rev.* **104** 4791–843

[5] Jung W and Tuller H L 2011 A new model describing solid oxide fuel cell cathode kinetics: model thin film $SrTi_{1-x}Fe_xO_{3-\delta}$ mixed conducting oxides-a case study *Adv. Energy Mater.* **1** 1184–91

[6] Kilner J 1996 Surface exchange of oxygen in mixed conducting perovskite oxides *Solid State Ion.* **86–88** 703–9

[7] Wang L, Merkle R, Mastrikov Y A, Kotomin E A and Maier J 2012 Oxygen exchange kinetics on solid oxide fuel cell cathode materials—general trends and their mechanistic interpretation *J. Mater. Res.* **27** 2000–8

[8] De Souza R A and Kilner J A 1999 Oxygen transport in $La_{1-x}Sr_xMn_{1-y}Co_yO_{3\pm\delta}$ perovskites *Solid State Ion.* **126** 153–61

[9] De Souza R A 2006 A universal empirical expression for the isotope surface exchange coefficients (k^*) of acceptor-doped perovskite and fluorite oxides *Phys. Chem. Chem. Phys.* **8** 890

[10] Jiang S P 2003 Issues on development of $(La,Sr)MnO_3$ cathode for solid oxide fuel cells *J. Power Sources* **124** 390–402

[11] Murray E P and Barnett S A 2001 $(La,Sr)MnO_3$–$(Ce,Gd)O_{2-x}$ composite cathodes for solid oxide fuel cells *Solid State Ion.* **143** 265–73

[12] Hart N T and Brandon N P 2001 Functionally graded cathodes for solid oxide fuel cells *J. Mater. Sci.* **36** 1077–85

[13] Jørgensen M J and Mogensen M Impedance of solid oxide fuel cell LSM/YSZ composite cathodes *J. Electrochem. Soc.* **148** A433–42

[14] Takeda Y, Kanno R, Noda M and Yamamoto O 1986 Perovskite electrodes for high temperature solid electrolyte fuel cells *Bull. Inst. Chem. Res.* **64** 157–69 http://hdl.handle.net/2433/77164

[15] Lu Z, Darvish S, Hardy J, Templeton J, Stevenson J and Zhong Y 2017 $SrZrO_3$ Formation at the interlayer/electrolyte interface during $(La_{1-x}Sr_x)_{1-\delta}Co_{1-y}Fe_yO_3$ cathode sintering *J. Electrochem. Soc.* **164** F3097–103

[16] Willy Poulsen F and van der Puil N 1992 Phase relations and conductivity of Sr- and La-zirconates *Solid State Ion.* **53–56** 777–83

[17] Knibbe R, Hjelm J, Menon M, Pryds N, Søgaard M, Wang H J and Neufeld K 2010 Cathode–electrolyte interfaces with CGO barrier layers in SOFC: cathode–electrolyte interfaces in IT-SOFCs *J. Am. Ceram. Soc.* **93** 2877–83

[18] De Vero J C, Bagarinao K D, Kishimoto H, Ishiyama T, Yamaji K, Horita T and Yokokawa H 2017 Effect of Gd-doped ceria interlayer microstructure on the interdiffusion behavior between $La_{0.6}Sr_{0.4}Co_{0.2}Fe_{0.8}O_{3-\delta}$ cathodes and yttria-stabilized zirconia electrolytes *ECS Trans.* **78** 971–81

[19] De Vero J C, Bagarinao K D, Ishiyama T, Kishimoto H, Yamaji K, Horita T and Yokokawa H 2017 Effect of $SrZrO_3$ formation at LSCF-cathode/GDC-interlayer interfaces on the electrochemical properties of solid oxide fuel cells *ECS Trans.* **75** 75–81
[20] Yoon J, Cho S, Kim J-H, Lee J, Bi Z, Serquis A, Zhang X, Manthiram A and Wang H 2009 Vertically aligned nanocomposite thin films as a cathode/electrolyte interface layer for thin-film solid oxide fuel cells *Adv. Funct. Mater.* **19** 3868–73
[21] Morales M, Pesce A, Slodczyk A, Torrell M, Piccardo P, Montinaro D, Tarancón A and Morata A 2018 Enhanced performance of gadolinia-doped ceria diffusion barrier layers fabricated by pulsed laser deposition for large-area solid oxide fuel cells *ACS Appl. Energy Mater.* **1** 1955–64
[22] Garbayo I, Pla D, Morata A, Fonseca L, Sabaté N and Tarancón A 2014 Full ceramic micro solid oxide fuel cells: towards more reliable mems power generators operating at high temperatures *Energy Environ. Sci.* **7** 3617–29
[23] Huang H, Nakamura M, Su P, Fasching R, Saito Y and Prinz F B 2007 High-performance ultrathin solid oxide fuel cells for low-temperature operation *J. Electrochem. Soc.* **154** B20
[24] Evans A, Bieberle-Hütter A, Rupp J L M and Gauckler L J 2009 Review on micro-fabricated micro-solid oxide fuel cell membranes *J. Power Sources* **194** 119–29
[25] Navickas E *et al* 2015 Fast oxygen exchange and diffusion kinetics of grain boundaries in Sr-doped $LaMnO_3$ thin films *Phys. Chem. Chem. Phys.* **17** 7659–69
[26] Chiabrera F, Garbayo I, Pla D, Burriel M, Wilhelm F, Rogalev A, Núñez M, Morata A and Tarancón A 2019 Unraveling bulk and grain boundary electrical properties in $La_{0.8}Sr_{0.2}Mn_{1-y}O_{3\pm\delta}$ thin films *APL Mater.* **7** 013205
[27] Rupp G M, Téllez H, Druce J, Limbeck A, Ishihara T, Kilner J and Fleig J 2015 Surface chemistry of $La_{0.6}Sr_{0.4}CoO_{3-\delta}$ thin films and its impact on the oxygen surface exchange resistance *J. Mater. Chem.* A **3** 22759–69
[28] Baiutti F *et al* 2021 A high-entropy manganite in an ordered nanocomposite for long-term application in solid oxide cells *Nat. Commun.* **12** 2660
[29] Zhang Y, Wen Y, Huang K and Nicholas J D 2020 Atomic layer deposited zirconia overcoats as on-board strontium getters for improved solid oxide fuel cell nanocomposite cathode durability *ACS Appl. Energy Mater.* **3** 4057–67
[30] Garbayo I, Baiutti F, Morata A and Tarancón A 2019 Engineering mass transport properties in oxide ionic and mixed ionic-electronic thin film ceramic conductors for energy applications *J. Eur. Ceram. Soc.* **39** 101–14
[31] Pederson L R, Singh P and Zhou X-D 2006 Application of vacuum deposition methods to solid oxide fuel cells *Vacuum* **80** 1066–83
[32] Noh H-S, Son J-W, Lee H, Song H-S, Lee H-W and Lee J-H 2009 Low temperature performance improvement of SOFC with thin film electrolyte and electrodes fabricated by pulsed laser deposition *J. Electrochem. Soc.* **156** B1484
[33] Lin Y and Chen X (ed) 2016 *Advanced Nano Deposition Methods* (Weinheim: Wiley-VCH Verlag GmbH & Co. KGaA)
[34] Eason R (ed) 2006 *Pulsed Laser Deposition of Thin Films: Applications-Led Growth of Functional Materials* (Hoboken, NJ: Wiley-Interscience)
[35] Ogugua S N, Ntwaeaborwa O M and Swart H C 2020 Latest development on pulsed laser deposited thin films for advanced luminescence applications *Coatings* **10** 1078

[36] Xu M, Yu J, Song Y, Ran R, Wang W and Shao Z 2020 Advances in ceramic thin films fabricated by pulsed laser deposition for intermediate-temperature solid oxide fuel cells *Energy Fuels* **34** 10568–82
[37] Plonczak P, Bieberle-Hütter A, Søgaard M, Ryll T, Martynczuk J, Hendriksen P V and Gauckler L J 2011 Tailoring of $La_xSr_{1-x}Co_yFe_{1-y}O_{3-\delta}$ nanostructure by pulsed laser deposition *Adv. Funct. Mater.* **21** 2764–75
[38] Siebenhofer M, Huber T M, Friedbacher G, Artner W, Fleig J and Kubicek M 2020 Oxygen exchange kinetics and nonstoichiometry of pristine $La_{0.6}Sr_{0.4}CoO_{3-\delta}$ thin films unaltered by degradation *J. Mater. Chem.* A **8** 7968–79
[39] De Vero J C, Develos-Bagarinao K, Kishimoto H, Ishiyama T, Yamaji K, Horita T and Yokokawa H 2018 Enhanced stability of solid oxide fuel cells by employing a modified cathode–interlayer interface with a dense $La_{0.6}Sr_{0.4}Co_{0.2}Fe_{0.8}O_{3-\delta}$ thin film *J. Power Sources* **377** 128–35
[40] Garbayo I, Esposito V, Sanna S, Morata A, Pla D, Fonseca L, Sabaté N and Tarancón A 2014 Porous $La_{0.6}Sr_{0.4}CoO_{3-\delta}$ thin film cathodes for large area micro solid oxide fuel cell power generators *J. Power Sources* **248** 1042–9
[41] Evans A, Martynczuk J, Stender D, Schneider C W, Lippert T and Prestat M 2015 Low-Temperature micro-solid oxide fuel cells with partially amorphous $La_{0.6}Sr_{0.4}CoO_{3-\delta}$ cathodes *Adv. Energy Mater.* **5** 1400747
[42] Zeljković S, Ivas T, Infortuna A and Gauckler L J 2014 Optimization of parameters and microstructural properties of $Ba_{0.5}Sr_{0.5}Co_{0.8}Fe_{0.2}O_{3-\delta}$ thin films grown by pulsed laser deposition (PLD) *J. New Mater. Electrochem. Syst.* **17** 257–63
[43] Tselev A, Gorbunov A and Pompe W 2001 Cross-beam pulsed laser deposition: general characteristic *Rev. Sci. Instrum.* **72** 2665–72
[44] Akkan C K 2014 Matrix shaped pulsed laser deposition: new approach to large area and homogeneous deposition *Appl. Surf. Sci.* **302** 149–52
[45] Liang J, Liu Q, Li T, Luo Y, Lu S, Shi X, Zhang F, Asiri A M and Sun X 2021 Magnetron sputtering enabled sustainable synthesis of nanomaterials for energy electrocatalysis *Green Chem.* **23** 2834–67
[46] Nuchuay P *et al* 2020 Effect of oxygen plasma treatment on ito nanorods films prepared by glancing angle deposition *AIP Conference Proceedings* 2279 (Woodbury, NY: AIP Publishing), 120010
[47] Kim H, Kim H, Lambert T H and Lin S 2020 Reductive electrophotocatalysis: merging electricity and light to achieve extreme reduction potentials *J. Am. Chem. Soc.* **142** 2087–92
[48] Solovyev A A, Lebedynskiy A M, Shipilova A V, Ionov I V, Smolyanskiy E A, Lauk A L and Remnev G E 2019 Effect of magnetron sputtered anode functional layer on the anode-supported solid oxide fuel cell performance *Int. J. Hydrog. Energy* **44** 30636–43
[49] Coddet P, Amany M-L, Vulliet J, Caillard A and Thomann A-L 2019 YSZ/GDC bilayer and gradient barrier layers deposited by reactive magnetron sputtering for solid oxide cells *Surf. Coat. Technol.* **357** 103–13
[50] Lee Y H, Ren H, Wu E A, Fullerton E E, Meng Y S and Minh N Q 2020 All-sputtered, superior power density thin-film solid oxide fuel cells with a novel nanofibrous ceramic cathode *Nano Lett.* **20** 2943–9
[51] Carlsson J-O and Martin P M 2010 Chemical vapor deposition *Handbook of Deposition Technologies for Films and Coatings* (Norwich, NY: William Andrew Publishing) ch 7 pp 314–63

[52] Johnson R W, Hultqvist A and Bent S F 2014 A brief review of atomic layer deposition: from fundamentals to applications *Mater. Today* **17** 236–46
[53] Shim J H *et al* 2019 Atomic layer deposition for surface engineering of solid oxide fuel cell electrodes *Int. J. Precis. Eng. Manuf.-Green Technol.* **6** 629–46
[54] Ha S, Su P-C and Cha S-W 2013 Combinatorial deposition of a dense nano-thin film YSZ electrolyte for low temperature solid oxide fuel cells *J. Mater. Chem.* A **1** 9645
[55] Rahmanipour M, Cheng Y, Onn T M, Donazzi A, Vohs J M and Gorte R J 2017 Modification of LSF-YSZ composite cathodes by atomic layer deposition *J. Electrochem. Soc.* **164** F879–84
[56] Hernández E, Baiutti F, Morata A, Torrell M and Tarancón A 2018 Infiltrated mesoporous oxygen electrodes for high temperature co-electrolysis of H_2O and CO_2 in solid oxide electrolysis cells *J. Mater. Chem.* A **6** 9699–707
[57] Wu T, Zhang W, Li Y, Zheng Y, Yu B, Chen J and Sun X 2018 Micro-/nanohoneycomb solid oxide electrolysis cell anodes with ultralarge current tolerance *Adv. Energy Mater.* **8** 1802203
[58] Ding D, Li X, Lai S Y, Gerdes K and Liu M 2014 Enhancing SOFC cathode performance by surface modification through infiltration *Energy Environ. Sci.* **7** 552
[59] Anelli S, Moreno-Sanabria L, Baiutti F, Torrell M and Tarancón A 2021 Solid oxide cell electrode nanocomposites fabricated by inkjet printing infiltration of ceria scaffolds *Nanomaterials* **11** 3435
[60] Wang W, Gross M D, Vohs J M and Gorte R J 2007 The stability of LSF-YSZ electrodes prepared by infiltration *J. Electrochem. Soc.* **154** B439
[61] Niu Y *et al* 2021 Enhancing oxygen reduction activity and Cr tolerance of solid oxide fuel cell cathodes by a multiphase catalyst coating *Adv. Funct. Mater.* **31** 2100034
[62] Cao J, Li Y, Zheng Y, Wang S, Zhang W, Qin X, Geng G and Yu B 2022 A novel solid oxide electrolysis cell with micro-/nano channel anode for electrolysis at ultra-high current density over 5 A cm^{-2} *Adv. Energy Mater.* **12** 2200899
[63] Baiutti F, Blanco-Portals J, Anelli S, Torruella P, López-Haro M, Calvino J, Estradé S, Torrell M, Peiró F and Tarancón A 2021 Tailoring the transport properties of mesoporous doped cerium oxide for energy applications *J. Phys. Chem.* C **125** 16451–63
[64] Almar L, Andreu T, Morata A, Torrell M, Yedra L, Estradé S, Peiró F and Tarancón A 2014 High-surface-area ordered mesoporous oxides for continuous operation in high temperature energy applications *J. Mater. Chem.* A **2** 3134
[65] Anelli S, Baiutti F, Hornés A, Bernadet L, Torrell M and Tarancón A 2019 Improved mesostructured oxygen electrodes for highly performing solid oxide cells for co-electrolysis of steam and carbon dioxide *J. Mater. Chem.* A **7** 27458–68
[66] Anelli S, Hernández E, Bernadet L, Sun X, Hagen A, Baiutti F, Torrell M and Tarancón A 2020 Co-electrolysis of steam and carbon dioxide in large area solid oxide cells based on infiltrated mesoporous oxygen electrodes *J. Power Sources* **478** 228774
[67] Acosta M, Baiutti F, Tarancón A and MacManus-Driscoll J L 2019 Nanostructured materials and interfaces for advanced ionic electronic conducting oxides *Adv. Mater. Interfaces* **6** 1900462
[68] Kim J H, Kim J K, Seo H G, Lim D, Jeong S J, Seo J, Kim J and Jung W 2020 Ex-solved Ag nanocatalysts on a Sr-free parent scaffold authorize a highly efficient route of oxygen reduction *Adv. Funct. Mater.* **30** 2001326

[69] Marasi M, Panunzi A P, Duranti L, Lisi N and Di Bartolomeo E 2022 Enhancing oxygen reduction activity and structural stability of $La_{0.6}Sr_{0.4}FeO_{3-\delta}$ by 1 mol% Pt and Ru *B*-site doping for application in all-perovskite IT-SOFCs *ACS Appl. Energy Mater.* **5** 2918–28
[70] Chen Y *et al* 2017 A robust and active hybrid catalyst for facile oxygen reduction in solid oxide fuel cells *Energy Environ. Sci.* **10** 964–71
[71] Hua B, Li M, Sun Y-F, Li J-H and Luo J-L 2017 Enhancing perovskite electrocatalysis of solid oxide cells through controlled exsolution of nanoparticles *ChemSusChem* **10** 3333–41
[72] Zhang X, Zhang W, Zhang L, Meng J, Meng F, Liu X and Meng J 2017 Enhanced electrochemical property of $La_{0.6}Sr_{0.4}Co_{0.8}Fe_{0.2}O_3$ as cathode for solid oxide fuel cell by efficient *in situ* polarization-exsolution treatment *Electrochim. Acta* **258** 1096–105
[73] Liang M, Zhu Y, Song Y, Guan D, Luo Z, Yang G, Jiang S P, Zhou W, Ran R and Shao Z 2022 A new durable surface nanoparticles-modified perovskite cathode for protonic ceramic fuel cells from selective cation exsolution under oxidizing atmosphere *Adv. Mater.* **34** 2106379
[74] Wang J *et al* 2022 Exsolution-driven surface transformation in the host oxide *Nano Lett.* **22** 5401–8
[75] Zhao C, Li Y, Zhang W, Zheng Y, Lou X, Yu B, Chen J, Chen Y, Liu M and Wang J 2020 Heterointerface engineering for enhancing the electrochemical performance of solid oxide cells *Energy Environ. Sci.* **13** 53–85
[76] Connor P A *et al* 2018 Tailoring SOFC electrode microstructures for improved performance *Adv. Energy Mater.* **8** 1800120
[77] Chen A, Su Q, Han H, Enriquez E and Jia Q 2019 Metal oxide nanocomposites: a perspective from strain, defect, and interface *Adv. Mater.* **31** 1803241
[78] Maier J 1986 On the conductivity of polycrystalline materials *Ber. Bunsenges. Phys. Chem.* **90** 26–33
[79] Maier J 2004 *Physical Chemistry of Ionic Materials: Ions and Electrons in Solids* (Chichester: Wiley)
[80] De Souza R A 2009 The formation of equilibrium space-charge zones at grain boundaries in the perovskite oxide $SrTiO_3$ *Phys. Chem. Chem. Phys.* **11** 9939
[81] De Souza R A 2015 Oxygen diffusion in $SrTiO_3$ and related perovskite oxides *Adv. Funct. Mater.* **25** 6326–42
[82] Gregori G, Merkle R and Maier J 2017 Ion conduction and redistribution at grain boundaries in oxide systems *Prog. Mater Sci.* **89** 252–305
[83] Huber T M, Opitz A K and Fleig J 2015 Oxygen reduction via grain boundary transport in thin film platinum electrodes on yttria stabilized zirconia *Solid State Ion.* **273** 8–12
[84] Saranya A M, Morata A, Pla D, Burriel M, Chiabrera F, Garbayo I, Hornés A, Kilner J A and Tarancón A 2018 Unveiling the outstanding oxygen mass transport properties of mn-rich perovskites in grain boundary-dominated $La_{0.8}Sr_{0.2}(Mn_{1-x}Co_x)_{0.85}O_{3\pm\delta}$ nanostructures *Chem. Mater.* **30** 5621–9
[85] Marrocchelli D, Sun L and Yildiz B 2015 Dislocations in $SrTiO_3$: easy to reduce but not so fast for oxygen transport *J. Am. Chem. Soc.* **137** 4735–48
[86] Sun L, Marrocchelli D and Yildiz B 2015 Edge dislocation slows down oxide ion diffusion in doped CeO_2 by segregation of charged defects *Nat. Commun.* **6** 6294
[87] Börgers J M, Kler J, Ran K, Larenz E, Weirich T E, Dittmann R and De Souza R A 2021 Faster diffusion of oxygen along dislocations in $(La,Sr)MnO_{3+\delta}$ is a space-charge phenomenon *Adv. Funct. Mater.* 2105647

[88] Metlenko V, Ramadan A H H, Gunkel F, Du H, Schraknepper H, Hoffmann-Eifert S, Dittmann R, Waser R and De Souza R A 2014 Do dislocations act as atomic autobahns for oxygen in the perovskite oxide $SrTiO_3$? *Nanoscale* **6** 12864–76
[89] Polfus J M, Yildiz B and Tuller H L 2018 Origin of fast oxide ion diffusion along grain boundaries in Sr-doped $LaMnO_3$ *Phys. Chem. Chem. Phys.* **20** 19142–50
[90] Mebane D S 2015 A variational approach to surface cation segregation in mixed conducting perovskites *Comput. Mater. Sci.* **103** 231–6
[91] Mebane D S and De Souza R A 2015 A generalised space-charge theory for extended defects in oxygen-ion conducting electrolytes: from dilute to concentrated solid solutions *Energy Environ. Sci.* **8** 2935–40
[92] Tong X, Mebane D S and De Souza R A 2020 Analyzing the grain-boundary resistance of oxide-ion conducting electrolytes: Poisson–Cahn vs Poisson–Boltzmann theories *J. Am. Ceram. Soc.* **103** 5–22
[93] Zurhelle A F, Tong X, Klein A, Mebane D S and De Souza R A 2017 A space-charge treatment of the increased concentration of reactive species at the surface of a ceria solid solution *Angew. Chem.* **129** 14708–12
[94] Yildiz B 2014 'Stretching' the energy landscape of oxides—effects on electrocatalysis and diffusion *MRS Bull.* **39** 147–56
[95] Wen K, Lv W and He W 2015 Interfacial lattice-strain effects on improving the overall performance of micro-solid oxide fuel cells *J. Mater. Chem.* A **3** 20031–50
[96] Han J W and Yildiz B 2011 Enhanced one dimensional mobility of oxygen on strained $LaCoO_3$ (001) surface *J. Mater. Chem.* **21** 18983
[97] Cai Z, Kuru Y, Han J W, Chen Y and Yildiz B 2011 Surface electronic structure transitions at high temperature on perovskite oxides: the case of strained $La_{0.8}Sr_{0.2}CoO_3$ thin films *J. Am. Chem. Soc.* **133** 17696–704
[98] Yu Y, Ludwig K F, Woicik J C, Gopalan S, Pal U B, Kaspar T C and Basu S N 2016 Effect of Sr content and strain on Sr surface segregation of $La_{1-x}Sr_xCo_{0.2}Fe_{0.8}O_{3-\delta}$ as cathode material for solid oxide fuel cells *ACS Appl. Mater. Interfaces* **8** 26704–11
[99] Jalili H, Han J W, Kuru Y, Cai Z and Yildiz B 2011 New insights into the strain coupling to surface chemistry, electronic structure, and reactivity of $La_{0.7}Sr_{0.3}MnO_3$ *J. Phys. Chem. Lett.* **2** 801–7
[100] Feng Z *et al* 2014 Anomalous interface and surface strontium segregation in $(La_{1-y}Sr_y)_2CoO_{4\pm\delta}/La_{1-x}Sr_xCoO_{3-\delta}$ heterostructured thin films *J. Phys. Chem. Lett.* **5** 1027–34
[101] la O' G J, Savinell R F and Shao-Horn Y 2009 Activity enhancement of dense strontium-doped lanthanum manganite thin films under cathodic polarization: a combined AES and XPS study *J. Electrochem. Soc.* **156** B771–81
[102] Guo X and Waser R 2006 Electrical properties of the grain boundaries of oxygen ion conductors: acceptor-doped zirconia and ceria *Prog. Mater Sci.* **51** 151–210
[103] Guo X and Maier J 2001 Grain boundary blocking effect in zirconia: a Schottky barrier analysis *J. Electrochem. Soc.* **148** E121
[104] De Souza R A, Pietrowski M J, Anselmi-Tamburini U, Kim S, Munir Z A and Martin M 2008 Oxygen diffusion in nanocrystalline yttria-stabilized zirconia: the effect of grain boundaries *Phys. Chem. Chem. Phys.* **10** 2067
[105] Zhang Z, Sigle W, De Souza R A, Kurtz W, Maier J and Rühle M 2005 Comparative studies of microstructure and impedance of small-angle symmetrical and asymmetrical grain boundaries in $SrTiO_3$ *Acta Mater.* **53** 5007–15

[106] Saranya A M, Pla D, Morata A, Cavallaro A, Canales-Vázquez J, Kilner J A, Burriel M and Tarancón A 2015 Engineering mixed ionic electronic conduction in $La_{0.8}Sr_{0.2}MnO_{3+\delta}$ nanostructures through fast grain boundary oxygen diffusivity *Adv. Energy Mater.* **5** 1500377

[107] Chiabrera F, Morata A, Pacios M and Tarancón A 2017 Insights into the enhancement of oxygen mass transport properties of strontium-doped lanthanum manganite interface-dominated thin films *Solid State Ion.* **299** 70–7

[108] Usiskin R E, Maruyama S, Kucharczyk C J, Takeuchi I and Haile S M 2015 Probing the reaction pathway in $(La_{0.8}Sr_{0.2})_{0.95}MnO_{3+\delta}$ using libraries of thin film microelectrodes *J. Mater. Chem.* A **3** 19330–45

[109] Chiabrera F *et al* 2019 Engineering transport in manganites by tuning local nonstoichiometry in grain boundaries *Adv. Mater.* **31** 1–8

[110] Thoreton V, Niania M, Druce J, Tellez H and Kilner J 2022 Oxygen diffusion in ceramic mixed conducting $La_{0.6}Sr_{0.4}Co_{0.2}Fe_{0.8}O_{3-\delta}$: the role of grain and twin boundaries *J. Electrochem. Soc.* **169** 044513

[111] Harrison L G 1961 Influence of dislocations on diffusion kinetics in solids with particular reference to the alkali halides *Trans. Faraday Soc.* **57** 1191

[112] Baiutti F *et al* 2021 Direct measurement of oxygen mass transport at the nanoscale *Adv. Mater.* **33** 2105622

[113] Yasuda I and Hishinuma M 1995 Chemical diffusion in polycrystalline calcium-doped lanthanum chromites *J. Solid State Chem.* **115** 152–7

[114] van der Haar L M, den Otter M W, Morskate M, Bouwmeester H J M and Verweij H 2002 Chemical diffusion and oxygen surface transfer of $La_{1-x}Sr_xCoO_{3-\delta}$ studied with electrical conductivity relaxation *J. Electrochem. Soc.* **149** J41

[115] Kim G, Wang S, Jacobson A and Chen C 2006 Measurement of oxygen transport kinetics in epitaxial $La_2NiO_{4+\delta}$ thin films by electrical conductivity relaxation *Solid State Ion.* **177** 1461–7

[116] Jamnik J 2003 Impedance spectroscopy of mixed conductors with semi-blocking boundaries *Solid State Ion.* **157** 19–28

[117] Jamnik J and Maier J 2001 Generalised equivalent circuits for mass and charge transport: chemical capacitance and its implications *Phys. Chem. Chem. Phys.* **3** 1668–78

[118] Fleig J, Kim H R, Jamnik J and Maier J 2008 Oxygen reduction kinetics of lanthanum manganite (LSM) model cathodes: partial pressure dependence and rate-limiting steps *Fuel Cells* **8** 330–7

[119] Chen T, Harrington G F, Masood J, Sasaki K and Perry N H 2019 Emergence of rapid oxygen surface exchange kinetics during *in situ* crystallization of mixed conducting thin film oxides *ACS Appl. Mater. Interfaces* **11** 9102–16

[120] Viernstein A, Kubicek M, Morgenbesser M, Walch G, Brunauer G C and Fleig J 2019 High-temperature photochromism of Fe-doped $SrTiO_3$ caused by UV-induced bulk stoichiometry changes *Adv. Funct. Mater.* **29** 1900196

[121] Fischer E and Hertz J L 2012 Measurability of the diffusion and surface exchange coefficients using isotope exchange with thin film and traditional samples *Solid State Ion.* **218** 18–24

[122] De Souza R A, Zehnpfenning J, Martin M and Maier J 2005 Determining oxygen isotope profiles in oxides with time-of-flight SIMS *Solid State Ion.* **176** 1465–71

[123] Chiabrera F, Baiutti F, Diercks D, Cavallaro A, Aguadero A, Morata A and Tarancón A 2022 Visualizing local fast ionic conduction pathways in nanocrystalline lanthanum manganite by isotope exchange-atom probe tomography *J. Mater. Chem.* A **10** 2228–34
[124] Huber T M, Navickas E, Sasaki K, Yildiz B, Hutter H, Tuller H and Fleig J 2018 Interplay of grain size dependent electronic and ionic conductivity in electrochemical polarization studies on Sr-doped $LaMnO_3$ (LSM) thin film cathodes *J. Electrochem. Soc.* **165** F702–9
[125] Kubicek M, Cai Z, Ma W, Yildiz B, Hutter H and Fleig J 2013 Tensile lattice strain accelerates oxygen surface exchange and diffusion in $La_{1-x}Sr_xCoO_{3-\delta}$ thin films *ACS Nano* **7** 3276–86
[126] Li Z, Li M and Zhu Z 2022 Perovskite cathode materials for low-temperature solid oxide fuel cells: fundamentals to optimization *Electrochem. Energy Rev.* **5** 263–311
[127] Lee W, Han J W, Chen Y, Cai Z and Yildiz B 2013 Cation size mismatch and charge interactions drive dopant segregation at the surfaces of manganite perovskites *J. Am. Chem. Soc.* **135** 7909–25
[128] Chen Y, Cai Z, Kuru Y, Ma W, Tuller H L and Yildiz B 2013 Electronic activation of cathode superlattices at elevated temperatures—source of markedly accelerated oxygen reduction kinetics *Adv. Energy Mater.* **3** 1221–9
[129] Develos-Bagarinao K, De Vero J, Kishimoto H, Ishiyama T, Yamaji K, Horita T and Yokokawa H 2018 Multilayered LSC and GDC: an approach for designing cathode materials with superior oxygen exchange properties for solid oxide fuel cells *Nano Energy* **52** 369–80
[130] Develos-Bagarinao K, Budiman R A, Ishiyama T, Yamaji K and Kishimoto H 2020 Leveraging catalytic effects of heterointerfaces through multilayering for superior cathode performance *J. Power Sources* **480** 229094
[131] Lee S and MacManus-Driscoll J L 2017 Research update: fast and tunable nanoionics in vertically aligned nanostructured films *APL Mater.* **5** 042304
[132] Huang J, MacManus-Driscoll J L and Wang H 2017 New epitaxy paradigm in epitaxial self-assembled oxide vertically aligned nanocomposite thin films *J. Mater. Res.* **32** 4054–66
[133] Cho S, Kim Y N, Lee J, Manthiram A and Wang H 2012 Microstructure and electrochemical properties of $PrBaCo_2O_{5+\delta}/Ce_{0.9}Gd_{0.1}O_{1.95}$ vertically aligned nanocomposite thin film as interlayer for thin film solid oxide fuel cells *Electrochim. Acta* **62** 147–52
[134] Ma W, Kim J J, Tsvetkov N, Daio T, Kuru Y, Cai Z, Chen Y, Sasaki K, Tuller H L and Yildiz B 2015 Vertically aligned nanocomposite $La_{0.8}Sr_{0.2}CoO_3/(La_{0.5}Sr_{0.5})_2CoO_4$ cathodes —electronic structure, surface chemistry and oxygen reduction kinetics *J. Mater. Chem.* A **3** 207–19
[135] Stämmler S, Merkle R and Maier J 2017 Oxygen exchange activity of La–Sr–Co–O PLD films and quantification of $(La,Sr)CoO_{3-\delta}/(La,Sr)_2CoO_{4+\delta}/O_2$ triple phase boundary activity *J. Electrochem. Soc.* **164** F454–63
[136] Stämmler S, Merkle R, Stuhlhofer B, Logvenov G, Hahn K, van Aken P A and Maier J 2017 Phase constitution, Sr distribution and morphology of self-assembled La–Sr–Co–O composite films prepared by PLD *Solid State Ion.* **303** 172–80
[137] Develos-Bagarinao K, Ishiyama T, Kishimoto H, Shimada H and Yamaji K 2021 Nanoengineering of cathode layers for solid oxide fuel cells to achieve superior power densities *Nat. Commun.* **12** 3979
[138] Druce J, Téllez H, Ishihara T and Kilner J A 2015 Oxygen exchange and transport in dual phase ceramic composite electrodes *Faraday Discuss.* **182** 271–88

[139] Benson S J 1999 Oxygen transport and degradation processes in mixed conducting perovskites *PhD Thesis* (London: University of London)
[140] Choi H J *et al* 2018 Surface tuning of solid oxide fuel cell cathode by atomic layer deposition *Adv. Energy Mater.* **8** 1802506
[141] Paige J M, Cheng Y, Pepin P A, Curran C D, Sun D, Chen M U, McIntosh S, Vohs J M and Gorte R J 2019 Surface modification of SOFC cathodes by Co, Ni, and Pd oxides *Solid State Ion.* **341** 115051
[142] Wen Y, Yang T, Lee D, Lee H N, Crumlin E J and Huang K 2018 Temporal and thermal evolutions of surface Sr-segregation in pristine and atomic layer deposition modified $La_{0.6}Sr_{0.4}CoO_{3-\delta}$ epitaxial films *J. Mater. Chem.* A **6** 24378–88
[143] Schmauss T A, Railsback J G, Lu M Y, Zhao K Y and Barnett S A 2019 ZrO_2 atomic layer deposition into $Sr_{0.5}Sm_{0.5}CoO_{3-\delta}$–$Ce_{0.9}Gd_{0.1}O_{2-\delta}$ solid oxide fuel cell cathodes: mechanisms of stability enhancement *J. Mater. Chem.* A **7** 27585–93
[144] Chao C-C, Kim Y B and Prinz F B 2009 Surface modification of yttria-stabilized zirconia electrolyte by atomic layer deposition *Nano Lett.* **9** 3626–8
[145] Choi M, Ibrahim I A M, Kim K, Koo J Y, Kim S J, Son J-W, Han J W and Lee W 2020 Engineering of charged defects at perovskite oxide surfaces for exceptionally stable solid oxide fuel cell electrodes *ACS Appl. Mater. Interfaces* **12** 21494–504
[146] Kuklja M M, Kotomin E A, Merkle R, Mastrikov Y A and Maier J 2013 Combined theoretical and experimental analysis of processes determining cathode performance in solid oxide fuel cells *Phys. Chem. Chem. Phys.* **15** 5443
[147] Cao Y, Gadre M J, Ngo A T, Adler S B and Morgan D D 2019 Factors controlling surface oxygen exchange in oxides *Nat. Commun.* **10** 1–15
[148] Choi H J, Bae K, Jang D Y, Kim J W and Shim J H 2015 Performance degradation of lanthanum strontium cobaltite after surface modification *J. Electrochem. Soc.* **162** F622–6
[149] Kim D, Bliem R, Hess F, Gallet J-J and Yildiz B 2020 Electrochemical polarization dependence of the elastic and electrostatic driving forces to aliovalent dopant segregation on $LaMnO_3$ *J. Am. Chem. Soc.* **142** 3548–63
[150] Kubicek M, Limbeck A, Frömling T, Hutter H and Fleig J 2011 Relationship between cation segregation and the electrochemical oxygen reduction kinetics of $La_{0.6}Sr_{0.4}CoO_{3-\delta}$ thin film electrodes *J. Electrochem. Soc.* **158** B727–34
[151] Acosta M *et al* 2022 Surface chemistry and porosity engineering through etching reveal ultrafast oxygen reduction kinetics below 400 °C in *B*-site exposed $(La,Sr)(Co,Fe)O_3$ thin-films *J. Power Sources* **523** 230983
[152] Tsvetkov N, Lu Q and Yildiz B 2015 Improved electrochemical stability at the surface of $La_{0.8}Sr_{0.2}CoO_3$ achieved by surface chemical modification *Faraday Discuss.* **182** 257–69
[153] Huber A-K, Falk M, Rohnke M, Luerssen B, Amati M, Gregoratti L, Hesse D and Janek J 2012 In situ study of activation and de-activation of LSM fuel cell cathodes—electrochemistry and surface analysis of thin-film electrodes *J. Catal.* **294** 79–88
[154] Tripković Đ, Küngas R, Mogensen M B and Hendriksen P V 2019 Surface recrystallization—an underestimated phenomenon affecting oxygen exchange activity *J. Mater. Chem.* A **7** 11782–91
[155] Zarabi Golkhatmi S, Asghar M I and Lund P D 2022 A review on solid oxide fuel cell durability: latest progress, mechanisms, and study tools *Renew. Sustain. Energy Rev.* **161** 112339

[156] Su C, Xu X, Chen Y, Liu Y, Tadé M O and Shao Z 2015 A top-down strategy for the synthesis of mesoporous $Ba_{0.5}Sr_{0.5}Co_{0.8}Fe_{0.2}O_{3-\delta}$ as a cathode precursor for buffer layer-free deposition on stabilized zirconia electrolyte with a superior electrochemical performance *J. Power Sources* **274** 1024–33

[157] Noh H-S, Yoon K J, Kim B-K, Je H-J, Lee H-W, Lee J-H and Son J-W 2014 The potential and challenges of thin-film electrolyte and nanostructured electrode for yttria-stabilized zirconia-base anode-supported solid oxide fuel cells *J. Power Sources* **247** 105–11

[158] Shin J W, Oh S, Lee S, Yu J G, Park J, Go D, Yang B C, Kim H J and An J 2019 Ultrathin atomic layer-deposited CeO_2 overlayer for high-performance fuel cell electrodes *ACS Appl. Mater. Interfaces* **11** 46651–7

[159] Lu F, Xia T, Li Q, Wang J, Huo L and Zhao H 2019 Heterostructured simple perovskite nanorod-decorated double perovskite cathode for solid oxide fuel cells: highly catalytic activity, stability and CO_2-durability for oxygen reduction reaction *Appl. Catal. B: Environ.* **249** 19–31

[160] Thalinger R, Gocyla M, Heggen M, Klötzer B and Penner S 2015 Exsolution of Fe and SrO nanorods and nanoparticles from lanthanum strontium ferrite $La_{0.6}Sr_{0.4}FeO_{3-\delta}$ materials by hydrogen reduction *J. Phys. Chem.* C **119** 22050–6

[161] Zhang J, Ricote S, Hendriksen P V and Chen Y 2022 Advanced materials for thin-film solid oxide fuel cells: recent progress and challenges in boosting the device performance at low temperatures *Adv. Funct. Mater.* **32** 2111205

[162] Szász J, Wankmüller F, Wilde V, Störmer H, Gerthsen D, Menzler N H and Ivers-Tiffée E 2018 Nature and functionality of $La_{0.58}Sr_{0.4}Co_{0.2}Fe_{0.8}O_{3-\delta}/Gd_{0.2}Ce_{0.8}O_{2-\delta}/Y_{0.16}Zr_{0.84}O_{2-\delta}$ interfaces in SOFCs *J. Electrochem. Soc.* **165** F898–906

[163] Mai A, Haanappel V, Tietz F and Stover D 2006 Ferrite-based perovskites as cathode materials for anode-supported solid oxide fuel cells. Part II. Influence of the CGO interlayer *Solid State Ion.* **177** 2103–07

[164] Uhlenbruck S, Moskalewicz T, Jordan N, Penkalla H-J and Buchkremer H P 2009 Element interdiffusion at electrolyte–cathode interfaces in ceramic high-temperature fuel cells *Solid State Ion.* **180** 418–23

[165] Sønderby S, Klemensø T, Christensen B H, Almtoft K P, Lu J, Nielsen L P and Eklund P 2014 Magnetron sputtered gadolinia-doped ceria diffusion barriers for metal-supported solid oxide fuel cells *J. Power Sources* **267** 452–58

[166] Coppola N, Polverino P, Carapella G, Sacco C, Galdi A, Montinaro D, Maritato L and Pianese C 2020 Optimization of the electrical performances in solid oxide fuel cells with room temperature sputter deposited $Gd_{0.1}Ce_{0.9}O_{1.95}$ buffer layers by controlling their granularity via the in-air annealing step *Int. J. Hydrog. Energy* **45** 12997–3008

[167] Bernadet L, Baiutti F, Martinelli F, Torrell M, Morata A, Montinaro D and Tarancón A 2021 Thin film barrier layers with increased performance and reduced long-term degradation in SOFCs *ECS Trans.* **103** 1177–85

[168] Machado M, Baiutti F, Bernadet L, Morata A, Nuñez M, Ouweltjes J P, Fonseca F C, Torrell M and Tarancón A 2022 Functional thin films as cathode/electrolyte interlayers: a strategy to enhance the performance and durability of solid oxide fuel cells *J. Mater. Chem.* A **10** 17317–25

IOP Publishing

Nanoengineered Materials for Solid Oxide Cells

Katherine Develos-Bagarinao

Chapter 2

Recent advances in the understanding of the influence of surface chemistry and microstructure on the functional properties of solid oxide cell air electrodes

Ozden Celikbilek and Mónica Burriel

The quest for intermediate-temperature solid oxide cell (SOC) air electrodes has led to novel compositions and microstructures, new surface activation strategies, as well as smart nanostructures composed either of materials with different functionalities or single compositions with distinct grain cores surrounded by high ionically conducting grain boundaries. In this chapter, we report the most novel concepts and the latest experimental advances in the development of air electrodes with better performance and stability. We also focus on the current, often contradictory, understanding of the intrinsic and extrinsic factors affecting the oxygen reduction reaction (ORR), the origin of the detrimental surface segregation, and the various concepts proposed to limit common electrode degradation processes. Finally, we also give some hints about the expected prospects in the field. The advent of new multimodal *in situ* and *operando* techniques, combined with combinatorial and machine-learning approaches, allows us to envisage a significant improvement in air electrodes' performance in the near future.

2.1 Introduction

2.1.1 Solid oxide cells

Fuel cells are electrochemical devices that are able to directly convert the chemical energy of fuels into electrical energy (and heat for some applications). A variety of fuel cells exist, usually classified as a function of their operational temperature range and the electrolyte material used. Among them, SOCs operate at high temperatures, typically at 600 °C–1000 °C, and work efficiently in both the solid oxide fuel cell (SOFC) mode and the solid oxide electrolysis mode (SOEC). In the fuel cell mode,

doi:10.1088/978-0-7503-4064-9ch2

they are not only able to operate with H_2 but also with hydrocarbon-based fuels, as the processes of reforming and the electrochemical oxidation of hydrocarbons are activated at high temperatures in the presence of catalysts. This flexibility creates the opportunity to use most hydrocarbon fuels in either their gaseous or liquid state, provided that they are properly cleaned and reformed into simple fuels such as H_2 and CO.

Despite the numerous advantages of SOCs, the requirement for high operational temperatures presents a disadvantage, as it leads to long start-up times, high operating costs, and degradation of the cell components. New electrode and electrolyte materials, material combinations, and other optimization strategies are being developed to reduce the operating temperature to the intermediate range (500 °C–700 °C). This chapter focuses on air electrode materials and aims to give an overview of the most recent advances related to the influence of surface chemistry and microstructure on their functional properties. This understanding will ultimately lead to the selection of the most suitable materials compositions, nano- and microstructures, as well as other external factors for improved SOC performance.

2.1.2 Solid oxide cell air electrodes

The materials used as SOC air electrodes must fulfill a number of stringent requirements: they must possess high electronic conductivity, good electrochemical activity for the ORR (equation (2.1)), high thermal stability, and compatibility with the most common electrolytes (e.g. yttria-stabilized zirconia (YSZ)) under the operating conditions.

$$\frac{1}{2}O_2(g) + 2e^{-} \leftrightarrow O^{2-} \tag{2.1}$$

For traditional SOFCs working at high temperatures (800 °C–1000 °C), the most widely used state-of-the-art air electrode material is $La_{1-x}Sr_xMnO_{3\pm\delta}$ (LSM). LSM has a perovskite-type structure with the general formula $ABO_{3\pm\delta}$ (figure 2.1), in which A and B are cations with a total charge of +6. The perovskite structure, in

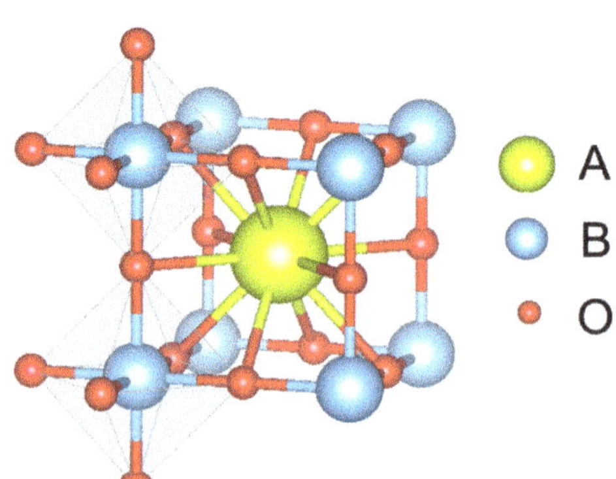

Figure 2.1. Representation of an ABO_3-type perovskite with the $Pm\bar{3}m$ space group showing the twelvefold oxygen coordination for the A-site cation and the sixfold oxygen coordination for the B-site cation.

which A is the larger cation with twelvefold oxygen coordination and B is the smaller cation with sixfold oxygen coordination, can be tailored by partial or total substitution of one or both cations, resulting in a variation in the materials properties. This versatility results in a wide range of different materials functionalities and applications. At lower temperatures, the polarization losses increase markedly; therefore, at lower temperatures, LSM no longer provides the required performance when used as an air electrode. Consequently, a large number of different perovskite compositions and new oxide families have been proposed, from which the most promising results have been obtained for single perovskites and for oxides from the perovskite-related Ruddlesden–Popper (RP) family which have the general formula $A_{n+1}B_nO_{3n+1}$ [1]. The search for appropriate materials to operate at lower temperatures introduces an additional requirement, i.e. good mixed ionic and electronic conductivity, which results in an increase in the area over which the ORR takes place. The characteristic active width for the ORR directly depends on the oxygen surface exchange and the oxygen diffusion coefficients of the given material.

Doped lanthanum cobaltites that have perovskite structures have been widely studied due to their high mixed ionic and electronic conductivity and exceptional oxygen reduction activity, $La_{1-x}Sr_xCo_{1-y}Fe_yO_{3-\delta}$ and $La_{1-x}Sr_xCoO_{3-\delta}$ being the most extensively investigated and promising compositions [2]. The substitution of Sr^{2+} for La^{3+} in the perovskite structure leads to an increase in the valence of cobalt and the formation of oxygen vacancies, by which both the catalytic activity and the oxygen ion conductivity are increased, resulting in improved cathode performance [3]. However, there is a considerable mismatch between the thermal expansion coefficient (TEC) of LSC and that of the common electrolytes, such as YSZ and $Ce_{1-x}Gd_xO_2$ (CGO). However, by partially substituting Fe for Co, this mismatch is reduced, and stability is improved [4]. In particular, $La_{0.6}Sr_{0.4}Co_{0.2}Fe_{0.8}O_{3-\delta}$ (LSCF) has been presented as a good compromise between high electronic and ionic conductivity, good catalytic properties, and stability. Despite being the state-of-the-art cathode materials for the intermediate temperature range, Sr-substituted perovskite cathodes face performance degradation over time associated with a number of different possible mechanisms, such as cathode–electrolyte interaction, Cr poisoning (from the stainless-steel interconnect materials), sulfur poisoning, coarsening of the cathode microstructure, loss of conductivity, formation of an insulating silicate layer in a humid atmosphere, in addition to Sr surface segregation.

With the aim of trying to avoid the abovementioned problems, in particular those related to Sr enrichment on the surface, a large number of different families have been proposed as alternative materials including layered ferrites, layered cuprates, pyrochlore-structured ruthenates, bismuth oxides based on the $Bi_4V_2O_{11}$ Aurivillius family (BIMEVOX-based composites), rutile-structured compounds, spinel-structured compounds, as well as compounds with layered structure, i.e. the A-site ordered double perovskites (mainly pure or partially substituted $\mathit{Re}BaCo_{2-x}B'_xO_{5+\delta}$ (RBCO) cobaltites, Re being a rare-earth element) and the Ruddlesden–Popper phases (in particular, the $\mathit{Ln}_2NiO_{4+\delta}$ nickelates with Ln = La, Nd, or Pr). The double perovskite structure can be obtained by a 50% substitution of the A-site (of the

$ABO_{3-\delta}$ structure) by a lower-valence A' cation, resulting in the alternation of A and A' cations along the c-axis and giving rise to high oxygen diffusivity along the ab planes. On the other hand, the $n = 1$ $Ln_2NiO_{4+\delta}$ Ruddlesden–Popper phase consists of an alternating perovskite–rock salt structure along the c-axis, which can accommodate over-stoichiometric oxygen, resulting in high oxygen diffusivity with no need for alkaline-earth (e.g. Sr) substitution [5].

2.1.3 Parameters and techniques used to evaluate the performance of air electrodes

The intrinsic performance of an air electrode is evaluated using AC impedance spectroscopy at a given temperature and in a specific gas atmosphere. The real part of the impedance spectrum measures the resistance of the system, and it is typically normalized to the electrode area for reporting purposes. The so-called area specific resistance (ASR) value should be as low as possible for better performance. However, this value alone does not allow for a direct comparison between different compositions, as it is not only governed by the oxygen surface exchange and diffusion coefficients of the material but also by the microstructural properties of the electrode. Thus, a number of different techniques have been designed and are commonly used to evaluate and compare the phenomenological oxygen transport coefficients of the different materials. As explained in detail in [6, 7] the three most common techniques are the following: (1) the chemical experiment in which the gas oxygen partial pressure is abruptly changed; the re-equilibration within the sample is measured by electrical conductivity relaxation (ECR) or by time-resolved x-ray diffraction [8], from which the chemical diffusion coefficient D_{chem} and surface exchange coefficient k_{chem} can be obtained; (2) the tracer experiment, in which the oxygen tracer distribution of the gas phase is changed (isotope exchange), which is typically combined with secondary ion mass spectrometry (SIMS) analysis of the sample [9], from which the oxygen tracer coefficients D^* and k^* are obtained; and (3) the electrical experiment, in which an outer electrical potential gradient is applied as a driving force which is commonly measured by electrochemical impedance spectroscopy (EIS) and from which D^{q} and k^{q} are obtained. It should be noted that the diffusion coefficients obtained from the three techniques are inherently different and, in general, cannot be directly compared [6].

In addition, to evaluate and understand the differences in catalytic activity and long-term stability, a variety of surface-sensitive elemental analysis methods are coupled with structural, microstructural, and electrochemical characterization techniques. Each of these elemental characterization techniques works on a different principle, and they provide complementary information from different depth scales and element sensitivities. SIMS, low-energy ion scattering spectroscopy (LEIS), and x-ray photoelectron spectroscopy (XPS) are the surface analysis techniques most commonly used to probe the surface chemistry of materials. Whereas LEIS has the unique advantage of being able to measure the elemental composition of the outermost surface atomic layer, it has a relatively poor lateral resolution (a few μm) and a detection limit which varies from a few percent of a monolayer for the light elements to 0.1%–1% for the heavier elements [10]. SIMS, on the other hand, gives chemical information about a few atomic layers; it has a unique detection limit at the ppm level and a lateral resolution from the micron to submicron scale. In addition to providing quantitative chemical composition, XPS gives information about the oxidation states of the constituent elements. It has a surface sensitivity of approximately 1–10 nm and a detection limit of about 0.1–1 at. %. All three techniques can be combined with a sputter gun to perform dynamic profiling. It is worth mentioning that heavier elements such as La have been reported to display preferential sputtering due to strong matrix effects [11, 12]. Therefore, the techniques that show strong matrix effects (i.e. LEIS and SIMS) are suggested for qualitative information. Other complementary techniques such as high-resolution transmission electron microscopy (HR-TEM) can help with the interpretation of the results.

Finally, the microstructure and evolution of air electrodes are typically observed in top and cross-sectional view by scanning electron microscopy (SEM) and transmission electron microscopy (TEM). These techniques are often coupled with energy-dispersive x-ray spectroscopy (EDS) to obtain chemical information in different regions or grains. Next-generation characterization techniques aim to improve the detection limits and lateral and depth resolution as well as to operate at different atmospheres and temperatures. Some of the more relevant techniques used for the characterization of the air electrode include near-ambient pressure x-ray photoelectron spectroscopy (NAP-XPS) and atmospheric pressure x-ray photoelectron spectroscopy (AP-XPS) [13–19], synchrotron x-ray studies (i.e. synchrotron radiation photoelectron spectroscopy (SRPES)), total x-ray reflection fluorescence (TXRF), high-resolution micro-beam diffraction [20, 21], and atom probe tomography [22]. *In situ* and *operando* characterization techniques, which monitor changes at high temperatures, under relevant atmospheres and/or under applied electric fields, are becoming essential for the advancement of knowledge and our understanding of complex electrochemical processes.

2.2 Understanding the effects of surface chemistry on the oxygen reduction reaction

2.2.1 The origin of surface cation segregation and strategies to limit it

Perovskite oxides with the general formula ABO_3 offer two possible (001) surface termination sites; AO and BO_2 (figure 2.2). The B-site cations are usually selected from among the transition-metal cations. The A-site cations are selected from the larger-radius alkali, alkaline-earth, and rare-earth elements from the lanthanoid series. Lattice distortions are created by the use of a wide selection of elements to occupy both sites. These distortions are compensated either by creating anion vacancies or by tilting the octahedra. However, the surface structure of a crystal is different than its bulk due to the cleaving of the bulk atoms on the surface. Thus, surface atoms have a lower coordination number than the bulk atoms. This difference on the surface disrupts the structural symmetry and increases the surface energy compared to that of the bulk, so much so that the surface cannot exist and has to rearrange itself to reduce the high surface energy [23, 24]. The surface atomic

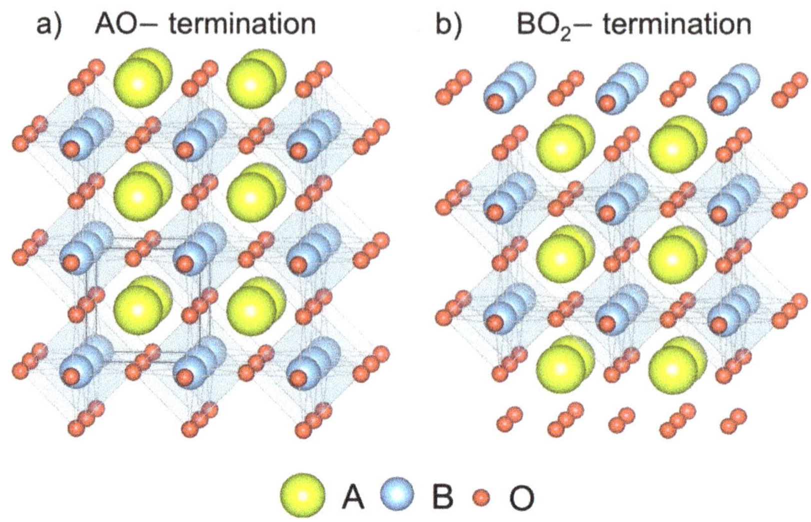

Figure 2.2. Schematics of surfaces with (a) *A*O- and (b) BO_2-termination layers in (001) ABO_3 perovskite.

arrangement in perovskite oxides usually leads to surfaces enriched with one of the *A*-site bulk atoms in oxidizing conditions and with one of the *B*-site metal nanoparticles in reducing conditions. The extent of the surface rearrangement is strongly related to the material stoichiometry, the selection of *A*-site or *B*-site elements, and the film strain as well as other environmental factors such as temperature, oxygen partial pressure, applied bias, etc.

To date, many studies have highlighted the elastic and electrostatic interactions as the driving force for surface cation segregation [25, 26]. The main drive behind the creation of the elastic energy is attributed to the size mismatch between the host and the substitution cations. In the case of the electrostatic interactions, the surface polarity must be considered. For example, in an ABO_3 perovskite in which both *A* and *B* elements have charges of 3+, the terminating layer on the surface leaves either a positively or negatively charged surface (the *A*O plane has a positive charge ($A^{3+}O^{2-}$) and the BO_2 plane has a negative charge ($B^{3+}O_2{}^{2-}$)). In Sr-doped perovskites such as (LaSr)BO_3 (where *B* is Co, Fe, Mn, etc), it has been suggested that neutralization of the surface charge happens either by changing the *B*-site element charge state or by replacing the surface termination through depleting the La at the surface and increasing the Sr concentration [27]. Notably, the surface charge of the SrO-terminated surface (+0.52e) in $La_{1-x}Sr_xCo_{0.2}Fe_{0.8}O_{3-\delta}$ has been found to be smaller than that of a LaO-terminated surface (+1.35e) [28].

It has been suggested that elastic or electrostatic effects alone cannot quantitatively explain the amount of observed surface segregation [29, 30]. Using a coupled experimental and computational approach, Lee *et al* showed the combined role of both the elastic and electrostatic energy on the segregation mechanism of *A*-site doped $LnMnO_3$ thin films (dopant cations: Ca, Sr, Ba and host cations *Ln*: La and Sm) [30]. Due to increased surface elastic energy, a larger size dopant induced a larger surface segregation in the following order: Ba > Sr > Ca (the ionic radius of La^{3+} is 1.36 Å, that of Ca^{2+} is 1.34 Å, that of Sr^{2+} is 1.44 Å, and that of Ba^{2+} is 1.61 Å). Furthermore, a change in the oxygen partial pressure during annealing stimulated the redistribution of point charge defects and localized electrons associated with the electrostatic interactions and altered the chemical expansion associated with the elastic energy. The chemically expanded lattice due to the low oxygen partial pressure and the reduction of the surface positive charge were found to reduce the segregation level. Following this work, Kwon *et al* reported that the size of *B*-site cations also has an influence on the *A*-site cation segregation and suggested the use of *B*-site elements with smaller sizes such as Co or Ni to limit the *A*-site cation segregation [31].

Using the same *A*-site doped $LaMnO_3$ model thin-film system (dopant cations: Ca, Sr, Ba), Kim *et al* showed that a transition can be made between electrostatically and elastically dominant segregation regimes via electrochemical polarization [32]. Using a specific electrochemical cell configuration that allows effective oxygen pressures ranging from 10^{-16} to 10^{15} atm to be obtained, the authors observed a progressive increase in surface segregation under cathodic polarization, which was attributed to the electrostatic attraction of positively charged dopants towards the oxygen-vacancy-enriched surface under reducing conditions. On the other hand, the increase in surface segregation under anodic polarization has been attributed to elastic strain.

Ostrovskiy *et al* also studied the effect of oxygen partial pressure on the Sr surface segregation of dense LSCF pellets and observed a volcano-like trend in the surface coverage of LSCF as a function of pO_2 [33]. The sample was somewhat stable under pure O_2 and N_2 gas environments; however, it showed the highest SrO coverage in air. The authors also studied the surface coverage under varying partial pressures of CO_2 atmosphere. $SrCO_3$ was found to be the main segregating phase instead of SrO, but it did not agglomerate to the same extent as SrO. The same study also explored the influence of *A*-site deficiency (5%) on Sr segregation. Under 1% and 50% O_2 at 700 °C, the sample surface remained stable; however, under 50% CO_2, the segregation was observed once again. A few other studies also confirmed that less surface coverage can be obtained with *A*-site deficiency [28, 34].

Reducing the surface oxygen vacancies through additives has been proposed as another route for surface stabilization. Tsvetkov *et al* showed that the addition of more reducible cations than Co (such as Hf^{4+}, Ti^{4+}, and Zr^{4+}) on the surface of $La_{0.8}Sr_{0.2}Co_{3-\delta}$ (LSC82) helped suppress the Sr segregation toward the surface [13]. This was linked with the reduction of positively charged oxygen vacancies and therefore a reduction in the electrostatic attraction of negatively charged Sr toward the surface (Sr^{2+} ion in La^{3+} lattice site in Kröger–Vink notation: Sr_{La}'). Koo *et al* suggested that isovalent *B*-site doping with 2 wt.% Hf or Zr in $SrTi_{0.5}Fe_{0.5}O_{3-\delta}$ effectively suppressed Sr segregation by increasing the Sr–O bond length [35]. In the same study, tensile strain was also found to inhibit Sr segregation and to increase the surface exchange rate. A similar conclusion was drawn for $La_{1-x}Sr_xCoO_{3-\delta}$ ($x = 0.2$ and 0.4) epitaxial thin films [36]. A ~4*x* faster surface exchange coefficient and a ~10*x* faster diffusion coefficient were measured for films under tensile strain.

The faster surface exchange coefficient was speculated to be related to the reduced Sr segregation at the surface. On the contrary, Lee *et al* did not find a link between faster surface exchange and reduced Sr segregation in strained $La_{0.6}Sr_{0.4}Co_{3-\delta}$ (LSC64) thin films [37]. They rather found that the enhanced ORR kinetics was linked to the electronic structure, where the presence of tensile strain effectively increased the oxygen vacancy concentration.

Yu *et al* showed that the amount of Sr in strained $La_{1-x}Sr_xCo_{0.2}Fe_{0.8}O_{3-\delta}$ thin films affected the extent of Sr segregation [38]. After extensive structural and surface characterizations, the authors found that Sr segregation decreased as the Sr content was reduced from $x = 0.4$–0.3 but increased with decreasing Sr content from $x = 0.3$–0.2. Regarding the strain effects, in contrast to the previous studies, compressive strain was found to lower the Sr segregation whereas tensile strain enhanced it. Therefore, given the variety of results documented in the literature, it is not straightforward to draw a general conclusion about the effects of strain on segregation; each material composition and environment has to be independently evaluated.

2.2.2 Influence of the surface termination on the electrode surface exchange

The surface of a perovskite can be terminated in various low-Miller-index planes that may contain various concentrations of *A*- and *B*-site elements. A number of state-of-the-art surface-sensitive characterization techniques such as LEIS [39, 40], NAP-XPS and AP-XPS [13–19], synchrotron x-ray studies [20, 21] and high-resolution scanning transmission electron microscopy (STEM) [21, 41] have confirmed the surface passivation due to cation segregation. Additionally, theoretical studies and experimental observations have provided evidence that *A*O-terminated surfaces are less reactive toward the ORR [21, 42–47]. On the other hand, a few studies have revealed contradicting results, showing that the presence of a few atomic layers of *A*O termination (e.g. SrO, La_2O_3) improved the ORR activity [48, 49]. In fact, density functional theory studies of SrO-terminated $SrTiO_3$ (STO) and Fe-substituted STO have proved that SrO-terminated surfaces can still actively participate in molecular oxygen dissociation through surface oxygen vacancy sites [42]. These sites provide access for molecular oxygen to interact with the electronically conducting transition-metal–oxygen network, where the electron transfer occurs for oxygen dissociation.

As can be understood, control of the surface termination layer has become a topic of great interest. STO in particular has been among the most studied oxides in various fields, as it is considered a model perovskite structure, and it allows for each of these surface terminations, i.e. SrO or TiO_2, to be consistently obtained. For instance, selective surface etching with hydrofluoric acid (HF) leads to a TiO_2 termination, whereas growing a monolayer of Sr by pulsed laser deposition (PLD) leads to SrO termination [50, 51]. Recently, a new strategy for controlling the surface termination was reported for perovskite manganite powders obtained using the hydrothermal synthesis method [21]. It was concluded that non-stoichiometry and specific hydrothermal conditions are both key factors responsible for tuning the surface termination of perovskite oxides. However, it is worth mentioning that further studies are necessary to verify the surface stability of the different terminations under the operating conditions of solid oxide air electrodes.

Sr-free electrode materials have thus gained further attention as promising air electrode materials for SOCs. Materials such as $LaNiO_3$, $La_2NiO_{4+\delta}$, $Pr_2NiO_{4+\delta}$, and their derivatives have been widely studied, as they possess good catalytic activity and high electrical conductivity [52–55]. In terms of their outermost surface termination, $La_2NiO_{4+\delta}$ (L2NO4), has been found to be LaO terminated, and $Pr_{2-x}La_xNiO_{4+\delta}$ ($x = 0, 1, 2$) have been found to be both La and Pr terminated, with the same cation ratios at the surface and in the bulk [40, 56]. However, several studies have shown different sources for degradation under varying polarization and atmospheric conditions, including the formation of new phases due to structural instability and poisoning by chromium in humidified atmospheres [57, 58].

The high density of interstitials in these Ruddlesden–Popper oxides suppresses the oxygen vacancies. Thus, the oxygen dissociation mechanism is different than the oxygen-vacancy-mediated one in Sr-containing perovskites. Akbay *et al* [59] and Staykov *et al*'s [42] theoretical calculations revealed the fundamental differences in the ORR mechanism between the SrO- and LaO-terminated surfaces of L2NO4 and STO. The authors highlighted the differences in the electronic structures of Sr and La, whereby Sr leads to inactive properties for molecular oxygen dissociation, whereas La can directly participate in oxygen dissociation through its occupied *d* orbitals.

2.2.3 Intrinsic factors affecting the surface exchange

Understanding and accelerating oxygen exchange and transport in complex oxides have been challenging, since there are multiple aspects of chemical and structural disorder to consider. These disorders have revealed a realm full of potential for superior materials performance. Nevertheless, they sometimes also revealed contradictory results for similar materials and compositions and unforeseeable outcomes. In the following, we present some of the studies that investigated the role of film strain, film thickness, grain size, and grain orientation in terms of the surface exchange properties.

Yan *et al* studied the surface exchange rate, k_{chem}, of tensilely and compressively strained and relaxed $La_{0.7}Sr_{0.3}MnO_3$ (LSM73) thin films with thicknesses ranging from 50 to 600 nm [60]. An activation energy of 1.5 eV was found for 50 and 100 nm thick strained films, and 0.75 eV was observed for 300 and 600 nm thick relaxed films between 600 °C and 800 °C (figure 2.3(a) and (b)). The 50 nm thick tensilely strained films grown on $SrTiO_3$ (100) substrates showed k_{chem} values an order of magnitude higher than those of compressively strained films grown on $NdGaO_3$ (NGO) (110) substrates. However, in the case of the 600 nm thick relaxed films, the k_{chem} value of the film grown on STO was only twice that of the film grown on an NGO substrate.

The effect of the presence of grain boundaries on the ion transport properties has been studied by several groups. Oxygen exchange and diffusion kinetics were

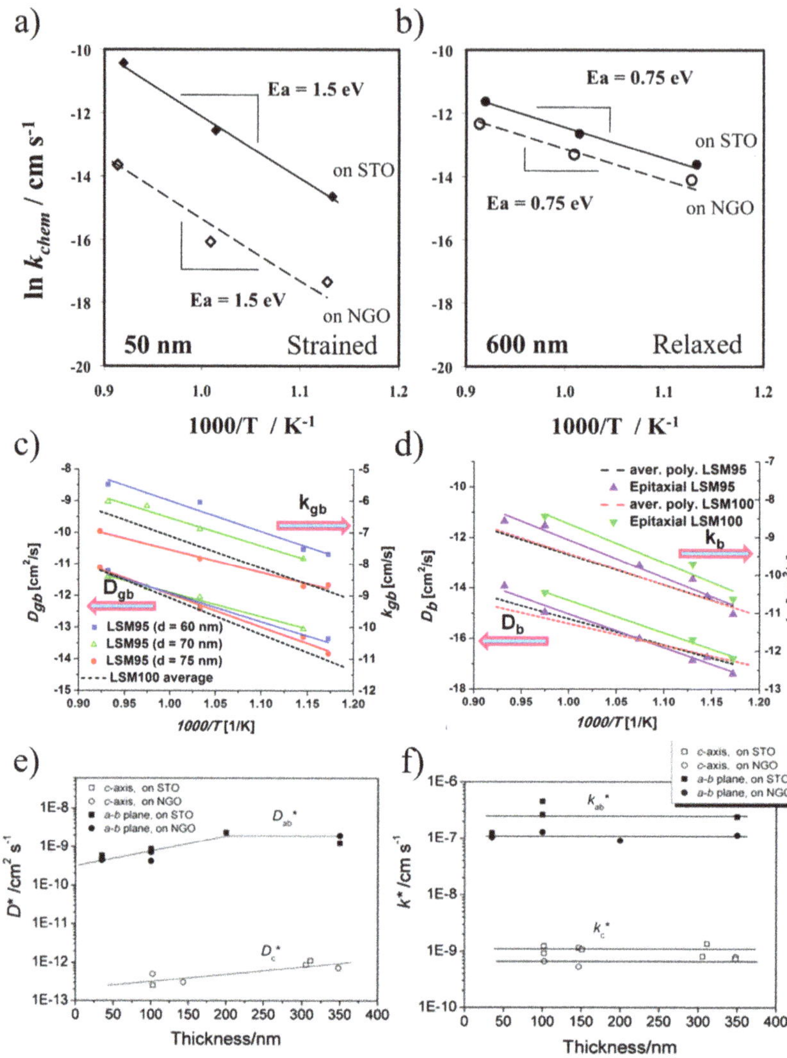

Figure 2.3. (a and b) Arrhenius plots of the k_{chem} values of LSM73 films 50 nm and 600 nm thick deposited on STO (100) and NGO (110). *Ea* refers to activation energies. Reprinted with permission from [60]. Copyright 2012, American Chemical Society. (c and d) Arrhenius plots of grain boundary (D_{gb}, k_{gb}) and bulk (D_{b}, k_{b}) oxygen transport values. Reproduced from reference [61] with permission of The Royal Society of Chemistry. (e and f) D^* and k^* values measured along the *ab*- and *c*-planes as a function of film thickness for L2NO4 films deposited on STO (100) and NGO (110) exchanged at ~430 °C. Reprinted with permission from [5] with permission of The Royal Society of Chemistry.

measured on LSM thin films grown on a YSZ substrate with controlled grain sizes (from 30 to 75 nm) and epitaxial films grown on an STO substrate [61]. For the polycrystalline films, the ^{18}O isotope exchange profiles indicated that both grain (G) and grain boundary (GB) oxygen transport were present as two parallel paths. Whereas the oxygen mass transport of the epitaxial films provided values similar to the G values of the polycrystalline films, the oxygen exchange and diffusion at the GBs were two to three orders of magnitude faster compared to those in the grains (figure 2.3(c) and (d)). Similar findings were obtained in an independent study carried out for polycrystalline LSM films, which was additionally supported by EIS measurements, and in which GB engineering was proposed as a new strategy for converting a purely electronic conductor into a mixed ionic–electronic conducting material [62]. In a subsequent study by Fleig's group, the authors separately investigated the influence of grain size on the ionic and electronic conductivities of LSM films [63]. The grain sizes were varied between 30 and 60 nm and the film thicknesses were varied between 120 and 640 nm. It was found that the electronic in-plane conductivity was reduced by the GBs, whereas the ionic conductivity and oxygen surface exchange were enhanced by the GBs, in agreement with the previous study.

In addition, various studies have examined the effects of grain orientation on the oxygen ion exchange and diffusion kinetics using different materials, both on thin films and pellets, but so far, no universal conclusion has been reached. For instance, an increase of up to about two orders of magnitude in both D^* and k^* has been observed in epitaxial L2NO4 thin films along the *ab*-plane compared to those of *c*-axis-oriented grains (figure 2.3(e) and (f)) [5]. On the contrary, later studies of epitaxial $(La_{1-x}Sr_x)_2CoO_4$ thin films found similar k^* values regardless of the film orientation [64]. This was linked to the presence of Sr-rich precipitates on the surface and surface rearrangement processes occurring during post-annealing. The combination of these factors was assumed to override any effect of grain orientation. A similar observation was made for LSCF thin films deposited on GDC-buffered single-crystal YSZ substrates [65]. Films grown on (111) oriented substrates initially showed higher k^* values compared to those grown on (100) oriented substrates. However, the former exhibited a higher surface segregation after annealing at 800 °C for one month, overriding their high surface activity.

The influence of grain orientation was also reported to be contradictory in the bulk pellets. For instance, LSM did not show any influence of grain orientation on k^* using focused-ion beam (FIB)/SIMS on single grains [66]. However, other groups used electron backscattering diffraction (EBSD) to determine whether Sr-rich secondary-phase precipitates appeared in higher abundance in certain crystal orientations in LSC64 pellets [67]. Grains oriented in the {001} direction showed over three times more Sr precipitates than those oriented in the {101} and {111} directions. Another study used a high-temperature environmental SEM to monitor the effect of atmosphere (i.e. pure oxygen, pure water, and ambient lab air) on the grain surfaces of LSCF pellets [68]. As in the case of LSC, the crystal orientation was found to affect the nature and the growth rate of Sr-rich particles, however a precise crystal orientation could not be proposed due to the pseudosymmetry of LSCF.

2.2.4 Extrinsic factors affecting the surface exchange

Large discrepancies in the extracted k values can be observed in the literature, even for materials considered to have the same stoichiometry. Other than the intrinsic materials properties, environmental factors such as thermal history and film processing, the particular measurement technique, the operational conditions (e.g. temperature, atmosphere, and applied bias), and chemical contamination during manufacture and/or storage can potentially lead to different oxygen surface exchange coefficients. In the following, we discuss the most relevant cases.

2.2.4.1 Thermal history and processing

To date, only a handful of works have mentioned the influence of thermal history and film processing on the surface chemistry. Thermal history refers to any heat treatment, from film deposition to post-thermal annealing. A study published by Kubicek *et al* was among the first to show that the surface cation distribution, in particular, the Sr content, changes depending on the thermal history and processing [69]. In the study, columnar LSC64 thin films were grown via PLD on single-crystal YSZ (100) at 450 °C and 650 °C and under 0.4 mbar oxygen partial pressure. The cation distribution from the surface to the bulk of the film was measured by SIMS and inductively coupled plasma optical emission spectrometry (ICP-OES), showing a higher Sr content near the surface of films grown at high deposition temperatures and after thermal annealing. In a subsequent study, the authors examined the surfaces of LSC64 films similarly deposited at 450 °C and 650 °C using XPS [29]. They identified the chemical species found on the surfaces of the deposited films and after annealing, thus revealing how these surfaces evolved under different thermal treatments. Both studies reported that the higher catalytic activity and stability of the low-temperature grown film were associated with a lower degree of Sr segregation and phase separation at the surface.

Contrary to these studies, Cavallaro *et al* did not find a considerable change in the surface chemistry of LSC82 thin films grown by PLD on single-crystal Si substrates under an oxygen partial pressure of 0.04 mbar and at varying deposition temperatures between 200 °C and 750 °C (figure 2.4(a)) [70]. The low-temperature-grown films were amorphous and did not show any grain formation, whereas the high-temperature grown ones showed a columnar microstructure. The authors pointed out that this discrepancy may be due to differences in the Sr content, deposition pressure, and substrate when compared to previous studies. Nevertheless, the oxygen transport properties measured by IEDP-SIMS showed substantially higher mass transport coefficients for the films grown at low temperatures than for those grown at high temperatures (figure 2.4(b)). It was speculated that the non-crystalline films could provide a different oxygen diffusion pathway than those of fully crystalline ones. This hypothesis was supported by the fact that the activation energy for the oxygen self-diffusion enthalpy was lower for the low-temperature-grown films.

In a subsequent study, Celikbilek *et al* investigated the effects of thermal history on LSC64 thin films using three complementary surface analysis techniques; LEIS, SIMS, and XPS [71]. The study looked at three LSC64 films grown by PLD at 0.04 mbar of oxygen partial pressure on a single-crystal Si substrate at 100 °C, at 100 °C followed by post-annealing at 600 °C, and finally at 750 °C. The film grown at 100 °C was amorphous and did not show any grain formation. The post-annealed film showed a randomly oriented granular microstructure, whereas the film grown at

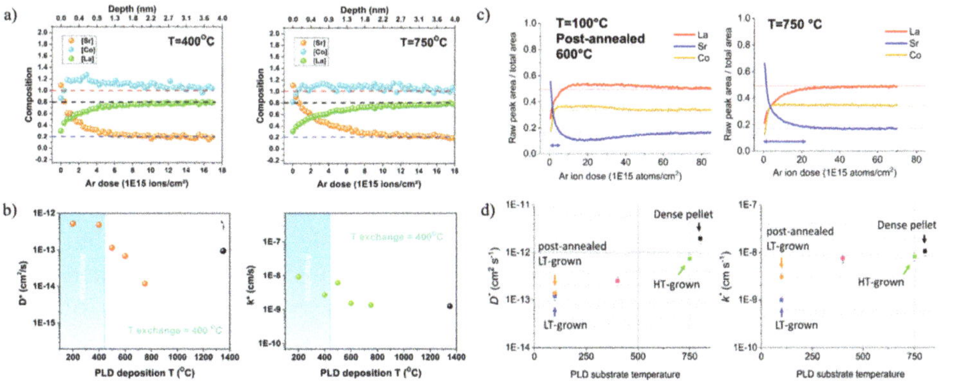

Figure 2.4. (a) LEIS depth profiles showing the outermost surface composition of ~300 nm thick LSC82 films grown by PLD at 400 °C and 750 °C. Reproduced from [70] with permission of The Royal Society of Chemistry. (b) D^* and k^* values measured for ~300 nm thick LSC82 films that were isotopically exchanged in an $^{18}O_2$ enriched atmosphere under the same conditions (T = 400 °C, annealing time = 15 min). Reproduced from [70] with permission of The Royal Society of Chemistry. (c) LEIS depth profiles of LSC64 films grown at 100 °C and then post-annealed at 600 °C and 750 °C. Reprinted with permission from [71]. Copyright (2020) American Chemical Society. Further permissions related to the material excerpted should be directed to the ACS. (d) D^* and k^* values measured for ~200 nm thick LSC64 film that have been isotopically exchanged in an $^{18}O_2$ enriched atmosphere under the same conditions (T = 400 °C, annealing time = 15 min). LT denotes low temperature (100 °C). Reprinted with permission from [71]. Copyright (2020) American Chemical Society. Further permissions related to the material excerpted should be directed to the ACS.

750 °C showed a columnar microstructure with textured growth. The films showed distinct differences in surface chemistry, as shown in the LEIS spectra in figure 2.4(c). In addition, contrary to the study by Cavallaro *et al*, here films grown at low substrate temperatures showed inferior oxygen transport kinetics compared to those grown at high temperatures (figure 2.4(d)). The authors speculated that the contradictory findings might be related to several differences between the two studies, such as (i) microstructural defects in the forms of cracks or porosities, (ii) a different cation segregation mechanism, (iii) differences in Sr stoichiometry, i.e. LSC82 vs LSC64. Nevertheless, after the film grown at the low temperature was post-annealed at 600 °C, the initially measured poor ASR performance improved, and became comparable to that of the film grown at the high temperature. Moreover, the films showed half the degradation rate at 500 °C for 100 h and outperformed the film grown at high temperature after about 70 h.

Similarly, a more recent work by Develos-Bagarinao *et al* focused on the effects of deposition temperature and the long-term annealing of LSC64 films on CGO polycrystalline substrates [72]. The films were grown by PLD under 0.4 mbar of oxygen partial pressure at room temperature, 300 °C, and 500 °C, and then all films were post-annealed at 700 °C for 300 h. The cation distribution was analyzed using XPS, SIMS, and STEM-EDX. In line with the previous works, this study also found marked differences in cation distribution in the as-deposited films. However, the Sr cation segregation showed an opposite trend compared to previous reports, showing higher amounts of Sr on the surface at lower deposition temperatures. After long-term annealing, however, all surfaces evolved to final states that had similar compositions. In contrast to the results found in other studies, the catalytic activity and stability of these films were instead affected by the chemical restructuring at the electrode–electrolyte interface rather than the surface restructuring. Sr diffusion across the electrode–electrolyte interface was promoted at high deposition temperatures and was further accelerated by long-term annealing at 700 °C. The common result of all these studies of the LSC-based perovskite family is that films grown at lower temperatures tend to degrade at a reduced rate compared to films grown at higher substrate temperatures. As point analysis could not be performed by any of the abovementioned techniques due to instrument detection limitations, obtaining the answer to why these films degrade more slowly is not straightforward. However, the results support the hypothesis that the strontium segregation is limited by poor crystalline grains and grain boundaries.

Following the demonstration of the novel *in situ* impedance spectroscopy technique used during PLD (IPLD), it has become possible to measure the impedance of pristine sample surfaces [45]. In this way, the exposure of the film surfaces to various air contaminants and the surface restructuring that occurs upon cooling the PLD chamber and the further reheating process in the impedance rigs are avoided. Using the IPLD setup, Siebenhofer *et al* compared the surface exchange resistance of LSC64 films grown at 450 °C, 500 °C, 550 °C, and 600 °C [73]. The films were grown on single-crystal YSZ (001) substrates under 0.04 mbar of oxygen partial pressure. Figure 2.5(a) shows the surface exchange resistances of the films measured at temperatures up to the deposition temperature. For instance, the film grown at 500 °C was measured at 450 °C and 500 °C, whereas that grown at 600 °C was measured at 450 °C, 500 °C, 550 °C, and 600 °C. All data points were found to

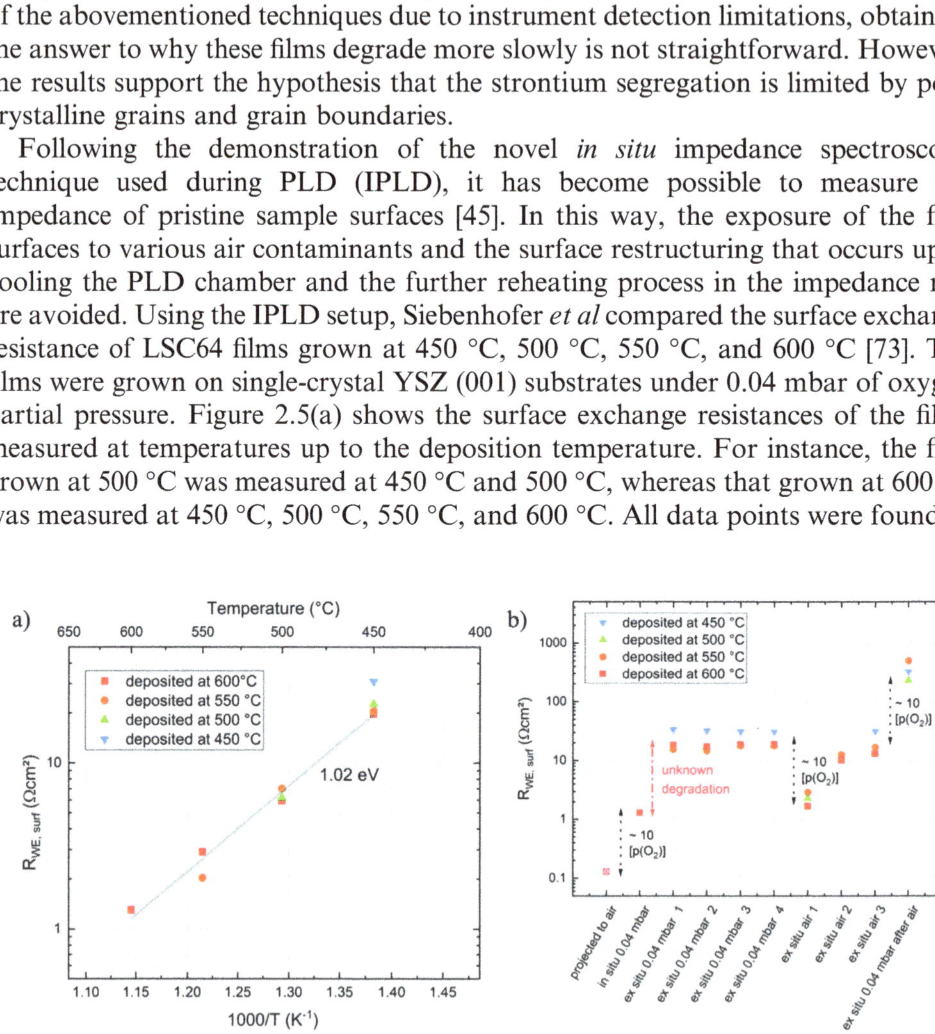

Figure 2.5. (a) A comparison of the surface exchange resistances ($R_{\mathrm{WE,surf}}$) of LSC64 films deposited at 450 °C, 500 °C, 550 °C, and 600 °C as measured by AC impedance inside the PLD chamber. The measurements were performed at temperatures up to the deposition temperature of the film. (b) $R_{\mathrm{WE,surf}}$ of the films grown at different deposition temperatures measured at 600 °C. The experimental modifications are described below the graph. Reprinted from [73] with permission of the Royal Society of Chemistry.

lie close to the activation energy of 1.02 eV, which was calculated for the sample grown at 600 °C. This result meant that the thermal history of the sample had little effect on the intrinsic catalytic activity of LSC64. On the other hand, once the films were exposed to different atmospheres, the degradation of the surface exchange resistance became evident (figure 2.5(b)). A factor of ten increase was observed after the samples were transferred to a laboratory impedance rig while keeping the same experimental parameters (i.e. temperature and oxygen partial pressure). This 'unknown' degradation was attributed to the environmental contributions such as poisoning of the surface with air contaminants. After being exposed to air, a decrease by a factor of ten was observed. In addition, the samples quickly degraded in a short period of time due to further chemical structuring and Sr segregation on the surface.

Apart from the LSC-based perovskite family, Perry's group has investigated the effects of thermal history on the surface exchange kinetics of $SrTi_{0.65}Co_{0.35}O_{3-\delta}$, $SrTi_{0.65}Fe_{0.35}O_{3-\delta}$, and Ruddlesden–Popper $Sr_2Ti_{0.65}Fe_{0.35}O_{4\pm\delta}$ thin films [74–76]. Continuous oxygen surface exchange measurements using optical transmission relaxation (OTR) method enabled higher k_{chem} values to be observed for films grown below and up to the crystallization temperature (~550 °C) compared with the highly crystalline films grown at high temperatures (750 °C).

In a more recent study, the same group created a new experimental setup to eliminate the inconsistencies in k_{chem} values measured by different techniques. The validity and reliability of the k_{chem} measurements were ensured by simultaneously measuring the same region of the same film using different techniques. In this way, any inconsistencies due to intersample variations or thermal history differences were avoided. First, $SrTi_{0.65}Co_{0.35}O_{3-\delta}$ thin films were analyzed to compare the values obtained by OTR and ECR (figure 2.6(a) and (b)). Both techniques yielded similar values within the error range. Subsequently, the k_{chem} values of another sample were simultaneously measured by AC impedance spectroscopy (AC-IS) and OTR on an as-grown amorphous film during crystallization (figure 2.6(c)). The values obtained by AC impedance spectroscopy were initially an order of magnitude higher than those measured by the OTR; however, a rapid decrease was observed during the second measurement cycle. Annealing at 590 °C for ~22 h after the first measurement cycle also resulted in a sharp decline (figure 2.6(d)). The higher k_{chem} values initially observed using AC-IS were thus thought to be coincidental, and the decrease was attributed to the degradation of the current-collecting layer deposited on top of the thin film.

In summary, several examples with strikingly different findings for the surface exchange kinetics have been presented, even for films with similar stoichiometry and prepared using the same technique. It is evident that the differences in the processing parameters of these samples, including the substrates (monocrystalline or polycrystalline) and oxygen partial pressures that can affect the crystallinity and the microstructure of the films should be highlighted and further investigated. The cooling of the films inside the PLD chamber, reheating in an impedance rig for electrochemical tests, and interactions with the air and other undesirable contaminants that may pass from the rigs to the materials are all important parameters that needs further attention in order to understand and limit the possible degradation sources.

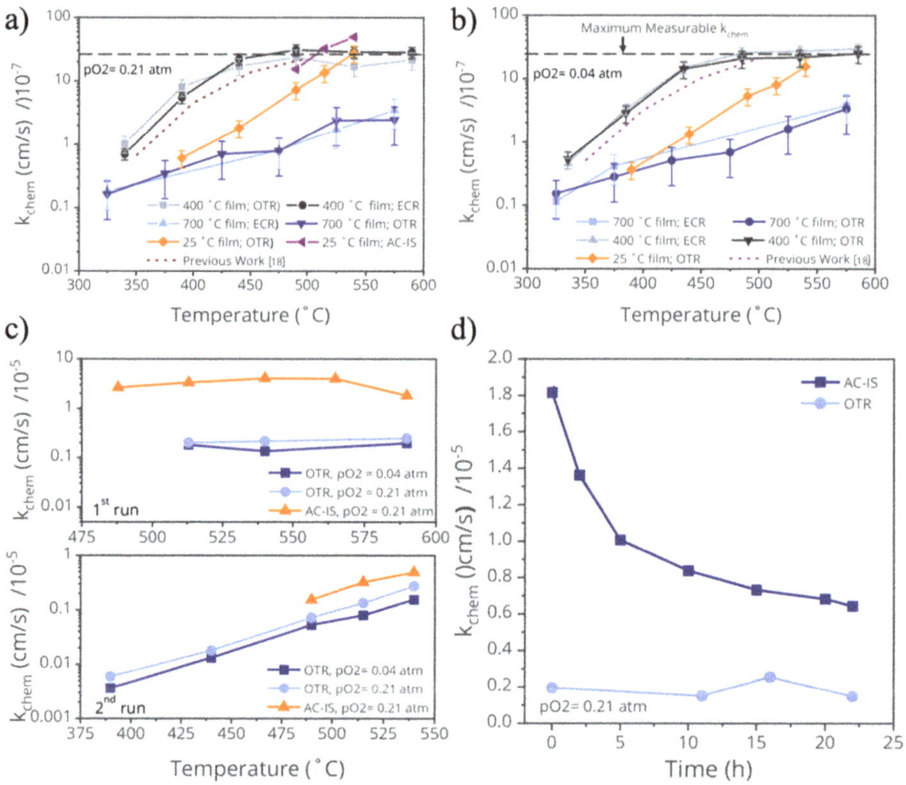

Figure 2.6. Values of k_{chem} as measured by OTR, ECR, and AC-IS techniques at (a) pO_2 = 0.21 atm and (b) 0.04 atm. (c) Values of k_{chem} measured by cross-cell AC-IS and OTR techniques at pO_2 = 0.21 and 0.04 atm during the first and second measurement cycles. (d) Comparison between the k_{chem} values measured by AC-IS and OTR as a function of time during intermediate annealing at 590 °C. Reprinted with permission from [76]. Copyright (2020) American Chemical Society.

2.2.4.2 Humidified atmosphere

Several studies have shown contrasting findings for the impacts of dry and humid atmospheres on the oxygen exchange kinetics of different materials. Sha *et al* showed the significant positive impact of oxygen partial pressure and water vapor on the oxygen transport properties of $(La_{0.8}Sr_{0.2})_{0.95}Cr_{0.5}Fe_{0.5}O_{3-\delta}$ between 600 °C and 900 °C [77]. Samples isotopically exchanged in a water vapor atmosphere (pO_2 < 1 mbar, $pH_2{}^{18}O$ = 30 mbar) showed an increase of up to two orders of magnitude in their surface exchange kinetics compared with those of samples exchanged in dry ($p^{18}O_2$ = 200 mbar) and wet (pO_2 = 200 mbar, $pH_2{}^{18}O$ = 30 mbar) oxygen atmospheres. The activation energy for the surface exchange under a water vapor atmosphere was also found to be remarkably lower (~0.8 eV) compared to the dry oxygen atmosphere value (~1.5 eV). Nenning *et al* reached a similar conclusion for LSF and $SrTi_{0.3}Fe_{0.7}O_{3-\delta}$ thin-film electrodes, showing that the exchange rate was up to two orders of magnitude higher at 418 °C in humid oxidizing ($^{16}O_2 + H_2{}^{18}O$) and humid reducing ($H_2 + H_2{}^{18}O$) atmospheres compared with dry conditions [78].

However, simultaneous AC impedance spectroscopy measurements also revealed that $k^{*}/k^{q} > 100$ under such circumstances. The authors highlighted that trace amounts of $H_2{}^{18}O$ might have been present in the ${}^{18}O_2$ bottles and that even a residual amount may have enhanced k^{*} tremendously. In this case, higher k^{*} does not necessarily mean increased SOFC cathode performance.

In contrast, a high-temperature environmental scanning electron microscopy study able to measure samples from room temperature to 1000 °C in atmospheres of pure oxygen, pure water, and ambient lab air showed a significantly increased strontium segregation rate on LSCF surfaces in the presence of pure water, as compared to the rates in pure O_2 and ambient air [68]. In line with this finding, Bucher *et al* observed the negative impact of a humid atmosphere on the oxygen exchange kinetics of LSCF and LSC64 at 600 °C for 1000 h [79, 80]. In a dry O_2-Ar atmosphere, they observed a twofold decrease in k_{chem}, which they ascribed to Sr enrichment at the surface. However, during a subsequent treatment in a humidified O_2–Ar atmosphere, they saw a further tenfold decrease in k_{chem} [79]. The faster degradation of the film in the presence of humidity was attributed to a reaction between Si (present in glass seals and thermal insulation materials) and Sr. In a different study, Joo *et al* found an accelerated oxygen incorporation kinetics in short-term tests, but a faster degradation rate in long-term tests under humid atmospheres at 600 °C [81].

The surface degradation mechanism caused by impurities in the gas phase (H_2O, SO_2, and CO_2) was clarified by Zhang *et al*'s study of $La_{0.5}Sr_{0.5}FeO_{3-\delta}$ thin-film electrodes [82]. Time-resolved *in situ* AP-XPS was coupled with EIS to monitor the ASR degradation at 650 °C as a function of water vapor pressure. In agreement with previous studies, the authors observed that water vapor caused hydroxylation of the surface and segregation and precipitation of lattice Sr on the surface, leading to increased ASR. However, they attributed the main cause of ASR degradation to the hydroxylation of the surface, rather than the additional Sr segregation under humid environments.

Thoréton *et al* obtained results that conflicted with those of the IEDP experiments of Sha *et al* and Nenning *et al* under dry and humid conditions [83]. The apparent k^{*} of LSCF was found to be higher in dry oxygen at 350 °C. After further estimations, they found that the surface exchange coefficient attributed to molecular oxygen was about one order of magnitude lower in humid conditions than in dry conditions.

These findings for different materials are puzzling given their opposite behaviors. Thus, Huang *et al* specifically chose two materials which show fundamentally different reaction mechanisms, namely LSCF and 5% *A*-site deficient LSM, to study oxygen dissociation kinetics using gas-phase isotopic oxygen exchange [84]. For LSM between 400 °C and 550 °C, the exchange was promoted in the presence of water and CO_2, whereas for LSCF, the exchange rate was found to decrease by an order of magnitude. The authors proposed that water and CO_2 could potentially be blocking the available exchange sites in LSCF, thus decreasing the oxygen dissociation ability. Attention was drawn to the different catalytic properties of Co, Fe, and Mn, the different oxygen vacancy concentrations of both materials, and the combination of all these factors to explain the differences in the exchange processes. Clearly, there is a lack of consensus regarding the impact of humidity on the exchange kinetics of perovskite-related materials. Its effect must be independently considered for each material at different timescales and for varying temperatures and gas compositions.

2.2.4.3 Applied bias

IEDP/SIMS experiments are carried out as a function of temperature and pO_2 and are based on self-change without an electrochemical driving force [85]. However, early studies showed substantial oxygen exchange at gold electrodes coated on BIMEVOX materials under 80 mA of applied current compared with the oxygen exchange of samples measured under no bias [86]. In another study, Huber *et al* studied the influence of anodic and cathodic bias on $La_{0.75}Sr_{0.25}Cr_{0.5}Mn_{0.5}O_{3-\delta}$ thin films grown on a single-crystal YSZ (111) substrate by PLD [87]. Strontium depletion was observed under cathodic polarization, whereas the segregation of strontium occurred in subsequent anodic polarization.

Cohn *et al* developed an *in operando* ^{18}O isotope exchange technique based on isothermal isotope exchange (IEE) and modified it to measure surface exchange processes under cathodic polarization from 0 mV to −700 mV [88]. IEE takes advantage of powders with particle sizes that are much smaller than the critical thickness L_c ($L_c = D/k$), making it possible to extract the oxygen surface exchange coefficients in a surface exchange-controlled regime. It uses Crank's solution to the diffusion equation, just as the IEDP/SIMS and ECR techniques do, but for a diffusion process in powder particles assumed to be spheres. The authors were able to monitor the exchange profiles of various powder samples, namely LSCF, 5% *A*-site deficient LSM, 5% *A*-site deficient LSC82, and 5% *A*-site deficient LSF. The exchange profile of LSM was found to be highly dependent on the applied bias from 450 °C to 700 °C, whereas the applied bias was found to have little effect on LSCF at 600 °C. LSC82 and LSF, both mixed ionic–electronic conductors, had very similar exchange profiles between 450 °C and 700 °C, although under zero bias LSC82 showed slightly higher values. However, it was concluded that under cathodic polarization at 600 °C, LSF becomes advantageous over LSC82.

In order to monitor the change of surface chemistry under bias, initial studies explored synchrotron-based *in situ* XPS [87, 89, 90]. It was demonstrated that the surface composition of perovskites can be reversibly controlled by applying a small electrical bias [89]. However, it is important to note that the crystal lattice of perovskites collapses at high temperatures and under ultrahigh vacuum reducing conditions. Recent advances in NAP-XPS, combined with EIS, have revealed how real cells respond in an environment close to actual operational conditions [17, 91]. Comparing the Sr core-level spectra of LSC64, LSF, and $SrTi_{0.7}Fe_{0.3}O_{3-\delta}$ thin-film electrodes under oxidizing and reducing atmospheres, Sr-enriched species at the surface were revealed in accordance with findings reported in the literature. Electrochemical polarization did not significantly affect the relative amount of these species. In terms of valence changes, only Fe and Co appeared to change valence states. In an oxidizing atmosphere, Fe was found to have an oxidation state of +3 irrespective of the applied bias. In a reducing atmosphere, it was found to have oxidation states of +3, +2, and 0; the relative amounts largely depend on the applied bias.

In summary, it is apparent from the abovementioned studies that these materials respond differently in terms of surface exchange activity and surface segregation behavior under applied bias, i.e. anodic and cathodic polarization. With the advancement of surface characterization techniques that have *in situ/operando* capabilities, it has become possible to measure samples in conditions close to real operating conditions.

2.3 Tuning the surface chemistry of engineered nanostructures for better performance and stability

In the previous sections, our current understanding of the effects of surface chemistry on catalytic activity was summarized. In this section, we discuss the various surface tuning methods that have been examined to date for better cathode performance and stability. These include deliberate surface decoration, adjusting the temperature of operation for optimal surface stability, cleaning contaminated surfaces after long-term operation, and using advanced microstructures.

2.3.1 Surface decoration

The foremost step in beginning a strategy for increased oxygen reduction activity has been to deliberately decorate surfaces using PLD. Mutoro *et al* investigated the surfaces of LSC82 thin films grown by PLD on a single-crystal YSZ (001) substrate at 650 °C and under 0.013 mbar. The films were then decorated with a few nanometers of La, Sr, and Co hydroxides/oxides/carbonates at 80 °C [92]. The electrical surface exchange coefficient (k^q) significantly increased with Sr decoration and reduced with Co decoration. On the other hand, La decoration did not lead to a significant change. Surface topography investigations by AFM indicated different amounts of surface coverage for these oxides. In particular, the coverage of the surface particles with Co decoration was large. Surfaces with La decoration showed a few large particles, corresponding to merely 3%–4% of the investigated area, while more uniform and homogeneous coverage was obtained using Sr decoration. The authors speculated that the high activity with Sr-decorated surfaces might be related to the formation of a highly active interface between LSC82 and the Sr-rich surface particles.

A similar approach to surface decoration by PLD was later taken by Rupp *et al* [45]. The authors were the first to demonstrate the effect of the surface termination on the oxygen reduction kinetics by measuring the AC impedance while the films were still in the PLD chamber [45]. For this, LSC64 thin films were grown on single-crystal YSZ (001) and then their surfaces were modified by depositing a few nanometers thick SrO, Co_3O_4 and La_2O_3 films at 450 °C under an oxygen partial pressure of 0.5 mbar (figure 2.7(a)). The surface resistance of an as-prepared sample decreased by a factor of three after Co decoration, which then levelled off with further depositions. However, an immediate adverse effect was observed even with small amounts of Sr. Continuous depositions led to an increase by orders of

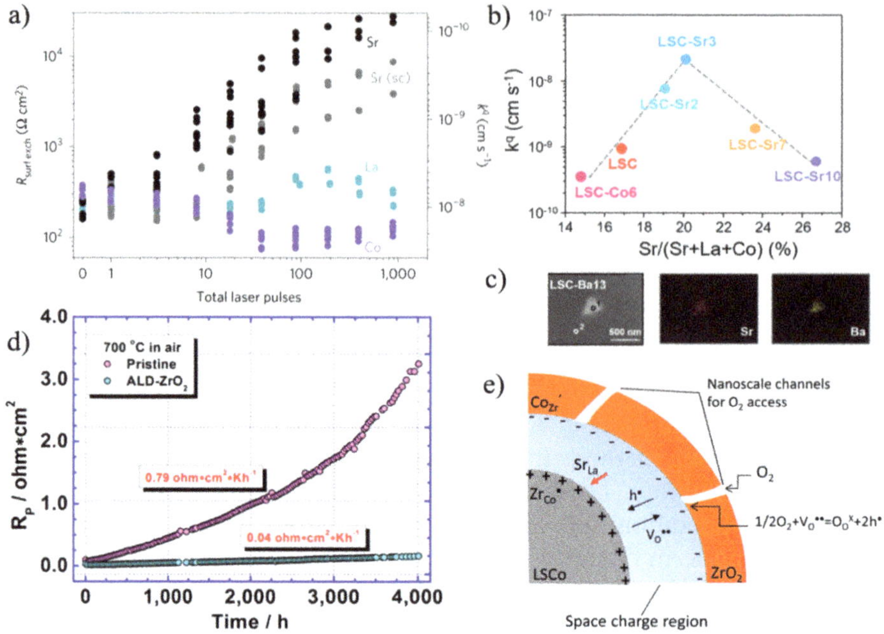

Figure 2.7. (a) The surface exchange resistance ($R_{\text{surf exch}}$) and k^q dependence resulting from the decoration of SrO, Co_3O_4, and La_2O_3 on LSC64 [45] Copyright © 2017, Nature Publishing Group. With permission of Springer. (b) The relationship between k^q and the surface Sr content. [93] John Wiley & Sons. [© 2018 WILEY-VCH Verlag GmbH & Co. KGaA, Weinheim]. (c) SEM image of LSC-Ba13 and the elemental distributions of Sr and Ba measured by Auger electron spectroscopy [93] John Wiley & Sons, copyright 2018 WILEY-VCH Verlag GmbH & Co. KGaA, Weinheim. (d) A comparison of the long-term polarization resistance (R_p) evolutions of pristine and ZrO_2-coated LSC64 films. Reprinted with permission from [94]. Copyright (2013) American Chemical Society. (e) Schematic of the defect model at the film surface, showing the space charge layer created at the LSC/ZrO_2 interface. Reprinted with permission from [94]. Copyright (2013) American Chemical Society.

magnitude in the surface resistance compared to the initial value. La, on the other hand, did not show a considerable effect. It is worth mentioning that Co and La depositions were thought to form islands instead of layer-by-layer growth.

The apparent difference between the results of Mutoro *et al* [92] and those of Rupp *et al* [45] may be related to several factors: (i) differences in the concentration of Sr in the studied films (LSC82 vs LSC64); (ii) differences in the thermal history and processing, i.e. deposition temperature and oxygen partial pressure; and (iii) whether the measurements were performed *in situ* or *ex situ*. Therefore, the discrepancy between similar studies from different laboratories may be related to the differences in the structural and microstructural evolutions of decorated surfaces exposed to different thermal and atmospheric backgrounds.

The infiltration (or impregnation) method has been demonstrated by many groups as a way of decorating surfaces for different applications. In this technique, a wet solution of metal nitrates, chlorides, or sulfides is infiltrated onto the film and subsequently dried and/or annealed to form metal-oxide catalysts at the nanometer length scale. Li *et al* explored the influence of SrO and CoO_x decoration by wet impregnation using nitrate solutions of different concentrations on the surfaces of LSC64 thin films [93]. The authors reported that the resistance was reduced 30 times by a 3% excess Sr surface ratio (measured by XPS) compared with the resistance of the pristine sample. However, further impregnation with Sr solution reset the improvement and led to a decrease in the k^q values (figure 2.7(b)). Decoration with 6% CoO_x similarly impeded the ORR activity at the surface and resulted in a degradation of the performance. Using SEM and Auger electron spectroscopy, the authors then monitored the elemental cation distribution of annealed LSC64 film surfaces (600 °C in air for 12 h) before and after Sr impregnation. The surface maps showed SrO islands on the annealed samples. After impregnation, the Sr/(Sr+La) ratio was found to be similar both on the islands and the island-free regions, suggesting that further Sr segregation was effectively suppressed. In the same study, BaO impregnation was also tested, since Ba ions possess the same electric charge as Sr ions. Elemental maps after Ba impregnation revealed that all the Ba migrated to the SrO islands formed after long annealing (figure 2.7(c)). The authors suggested that extrinsic Sr (or Ba) in the liquid phase can easily migrate and localize on the surface defective sites such as defects, grain boundaries, or dislocations, thereby suppressing the intrinsic Sr segregation. The enhancement of k^q by 3% excess Sr was thought to originate from the *in situ* formation of perovskite-like structures on the surface. Adding more Sr possibly led to an excessive accumulation of extrinsic SrO, preventing the surface structure from transforming into more conductive perovskite-like structures.

In addition to the evaluation performed using $La_{1-x}Sr_xCoO_{3-\delta}$, the effect of surface decoration on the surface exchange coefficient has been evaluated on Fe-doped STO single crystals using IEDP experiments [95]. Alkaline-earth (Ca, Sr, Ba) metal-oxide compounds a few tens of nanometers thick were deposited on Fe-doped STO by physical vapor deposition. The films homogeneously covered the surface but showed small-scale porosity. The measured k^* values exhibited an increase of up to three orders of magnitude with BaO and SrO coatings compared to the value measured for the bare surface of Fe-doped STO.

More recently, atomic layer deposition (ALD) has started to replace or complement the established deposition techniques due to its ability to control the surface composition with atomic-level precision and to coat the surfaces in a conformal manner [49, 94, 96, 97]. For a more detailed discussion of ALD techniques, the reader is referred to chapter 5 in this book. For instance, Paige *et al* reported a change in ASR values with the addition of transition-metal (Pd, Ni, and Co) oxide catalysts on $La_{0.8}Sr_{0.2}FeO_3$ (LSF82), LSC82, and LSCF films [96]. The authors found that a small negative effect resulted from the addition of a PdO layer for all cathodes, whereas the addition of Co_3O_4 had no effect on the Co-containing cathodes: LSC and LSCF. Both Co_3O_4 and NiO addition slightly improved the ASR of LSF82. The authors speculated that the metals were most likely not incorporated into the perovskite lattice and therefore did not contribute to increasing the concentration of surface oxygen vacancies.

The beneficial role of ALD-coated films for long-term stability is conceivably related to the limited particle growth during high-temperature operation. However, it has been suggested that the ALD layer plays an important role not only in the microstructural integrity but also at the atomic level involving cation exchange. Gong *et al* compared the long-term stability of LSC64 films impregnated on a porous $La_{0.8}Sr_{0.2}Ga_{0.83}Mg_{0.17}O_{3-\delta}$ scaffold with and without a conformal ~5 nm thick ZrO_2 ALD surface coating [94]. Impedance measurements at 700 °C over 4000 h showed a factor of 18 reduction in the polarization resistance and a 45% reduction in the ohmic resistance compared with those of the pristine sample (figure 2.7(d)). To explain the positive outcome, the authors suggested a defect chemistry model based on the exchange of Co and Zr at the LSC/ZrO_2 interface. According to their model, as depicted in figure 2.7(e), a negatively charged surface with Co'_{Zr} is created at the ZrO_2 side, while a positively charged surface layer (Zr'_{Co}) is created at the LSC side. This space charge region would repel the negatively charged Sr'_{La} from the interface. Moreover, the electrons supplied by the external circuit would reduce the positive charge accumulated at the cathode surface, which would effectively lower the segregation rate of strontium at the surface.

All the abovementioned studies reached different conclusions regarding the influence of surface decoration on the activity and stability of air electrodes. To this end, Tsvetkov *et al* took a systematic approach and coated LSC82 film surfaces with binary oxides that had either lower (V^{5+}) or higher (Nb^{5+}, Ti^{4+}, Zr^{4+}, Hf^{4+}, Al^{3+}) oxygen vacancy formation enthalpies than that of pristine LSC82 [13]. It was found that less reducible cations (Nb^{5+}, Ti^{4+}, Zr^{4+}, Hf^{4+}, Al^{3+}) decreased the surface oxygen vacancy concentration, leading to an increased Co oxidation state at the LSC surface. Meanwhile, the addition of V and excess Co led to higher surface degradation. As an example, a sample with Ti coating was investigated by AP-XAS. Ti coordination was found to be different than that of TiO_2 but similar to that of a perovskite lattice, suggesting that Ti occupied the Co sites in LSC. According to these findings, the oxygen surface vacancy concentration ($V_O^{\bullet\bullet}$) was decreased, resulting in more oxidized Co (figure 2.8(a)), and the electrostatic attraction of Sr'_{La} towards the surface was suppressed. The empirical findings of Tsvetkov *et al* supported the hypothesis of Gong *et al* that a higher oxygen vacancy concentration at the surface does not always facilitate the ORR. This finding is demonstrated more clearly in the plot shown in figure 2.8(b), which shows a volcano-like behavior of k^q that depends on the oxygen vacancy formation enthalpy of binary-oxide-decorated surfaces compared with k^q behavior of pristine LSC82.

Prior to this systematic study by Tsvetkov *et al*, some authors showed that environmental factors (such as sulfur or chromium poisoning) adversely affect the surface oxygen exchange rate by changing the surface chemistry [99–101]. For instance, Ni *et al* showed a chromium poisoning effect on 5% *A*-site deficient LSCF surfaces by intentionally introducing Cr by impregnation [102]. A monoclinic $SrCrO_4$ phase was formed at the LSCF grain boundaries, resulting in a reduction of up to two orders of magnitude in the conductivity, oxygen surface exchange rate, and diffusivity of LSCF.

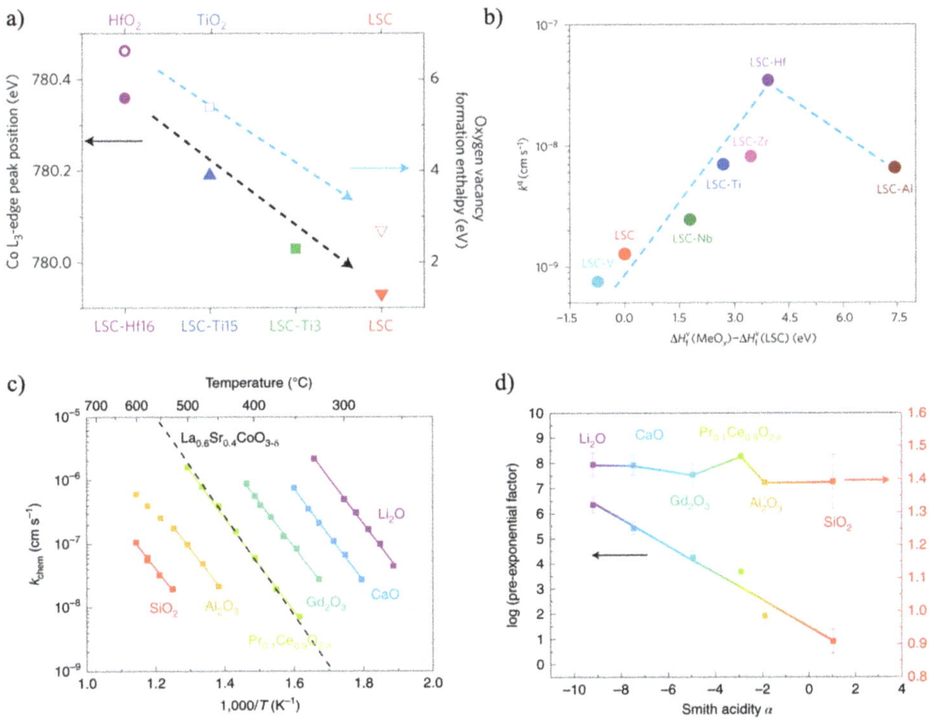

Figure 2.8. (a) The oxidation state of Co based on Co $L2,3$-edge XAS on LSC dense thin film at 300 °C and 0.76 torr [13] Copyright © 2016, Nature Publishing Group. With permission of Springer. (b) The dependence of oxygen surface exchange kinetics on the reducibility of the LSC surface. [13] Copyright © 2016, Nature Publishing Group. With permission of Springer. (c) The influence of surface decoration of $Pr_{0.1}Ce_{0.9}O_2$ on k_{chem} using oxides in different acidity scales. [98] Copyright © 2020, The Author(s), under exclusive licence to Springer Nature Limited. With permission of Springer. (d) Pre-exponential factors and activation energies of k_{chem} as a function of the acidity factor. [98] Copyright © 2020, The Author(s), under exclusive licence to Springer Nature Limited. With permission of Springer.

Nicollet *et al* methodically studied the infiltration of many binary oxides with different acidities on the surface of $Pr_{0.1}Ce_{0.9}O_{2-\delta}$ [98]. As their acidity scale definition, the authors used the tendency of binary oxides to accept an oxide ion [103]. They selected five oxides ranging from highly basic to highly acidic: Li_2O, CaO, Gd_2O_3, Al_2O_3, and SiO_2 and showed that it was possible to vary k_{chem} by six orders of magnitude without significantly affecting the activation energy (figure 2.8(c) and (d)). The authors showed that the acidity of the surface affected the electronic structure of $Pr_{0.1}Ce_{0.9}O_{2-\delta}$, resulting in a higher surface electron concentration with the addition of basic oxides, and a lower surface electron concentration with the addition of acidic oxides. However, they argued that the apparent change in the surface electron concentration was not sufficient to explain the changes in k_{chem}. They hypothesized that the electron density may have an indirect effect on k_{chem}. Other possibilities were also mentioned, such as changes in Pr–O and Ce–O bond strength, as well as changes in the surface electron concentration that extend over the entire surface rather than around the discrete additive particles.

Surface decorations of thin and thick films with many different oxide materials at the nanometer length scale exhibit a complex behavior. It is clear that the choice of deposition technique, i.e. solution-based or physical/chemical vapor-based and the deposition temperature make a difference in the surface catalytic activity of these oxides. These are possibly related to varying degrees of surface and structural defects, the amount of coverage, and the chemical interactions including cation swapping in the vicinity of the film/decoration oxide.

2.3.2 Operational temperature

Published reports do not necessarily find a decrease in the stability of the electrodes with an increase in the operating temperature. For instance, Opitz *et al* simultaneously looked at the electrode resistance (using AC impedance spectroscopy) and the surface composition of LSC64 films (using NAP-XPS) [14]. Compared to room-temperature measurements under ultrahigh vacuum conditions, NAP-XPS measurements at 400 °C in 0.5 mbar of oxygen demonstrated the initial surface reconstruction process. It was found that surface hydroxyls or adventitious carbon decomposed, and the formation of very small amounts of SrO was observed. Upon further heating to 700 °C, the amount of SrO species increased. Concurrent EIS measurements and NAP-XPS showed an interesting trend in the degradation rate with increasing temperature (figure 2.9(a) and (b)). At 400 °C, the initial ASR was 308 Ω cm^2, which then increased to 490 Ω cm^2 within 3.5 h (a degradation of 17%/h). Upon increasing the temperature to 479 °C, the ASR recorded was 127 Ω cm^2, but this rapidly increased to 491 Ω cm^2 within 45 min (a degradation of 382%/h). However, at 600 °C and above, the degradation rate dropped to double digits as similarly observed at 400 °C, although the ASR continued to increase. The highest degradation rate was observed at the same time as the steepest increase in the amount of SrO. At higher temperatures, such as 600 °C, the authors stated that the surface had already reached a sort of a 'saturation state.' Measuring for longer times at 600 °C or increasing the temperature to 700 °C mildly changed the surface, which explained the slower degradation rate.

Tripković *et al* took a different approach to demonstrate the 'reactivation' of surfaces by thermal treatments at high temperatures [104]. They studied bulk pellets of *A*-site deficient $(La_{0.6}Sr_{0.4})_{0.99}CoO_{3-\delta}$ and $(La_{0.6}Sr_{0.4})_{0.98}FeO_{3-\delta}$, and measured k_{chem} by ECR in oxidizing atmospheres (0.2–0.1 bar pO_2) from 650 °C to 900 °C. The activation energy and D_{chem} values for both samples were in general unaffected by heating and cooling and were in good agreement with the published values. However, a large scatter was observed for k_{chem} values, especially for LSF. The effect of thermal treatment on the oxygen transport properties of LSF was investigated at 650 °C. A sample was heat treated at 700 °C, 800 °C, 900 °C, and 1000 °C in sequence for 10 h and the sample temperature was brought down to 650 °C after each treatment for the ECR measurements. D_{chem} remained unchanged, but higher k_{chem} values were observed after annealing at temperatures above 700 °C.

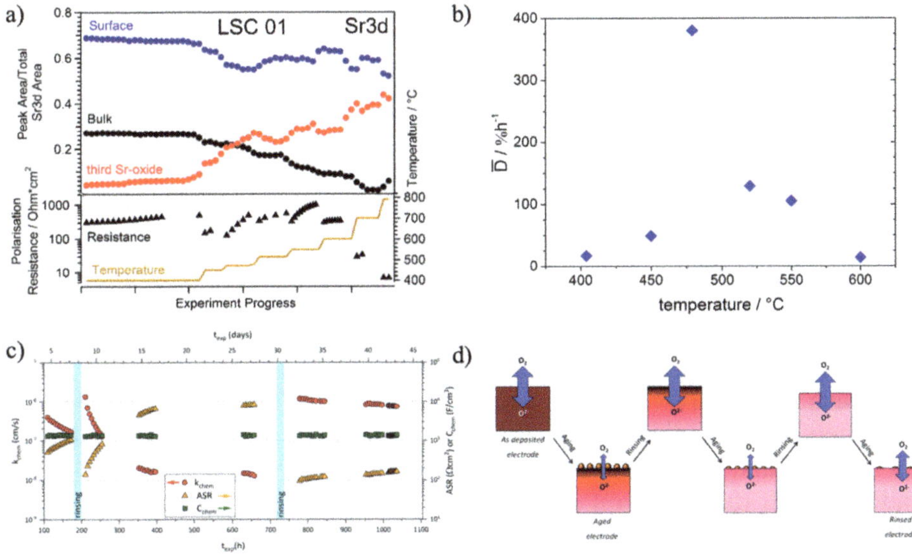

Figure 2.9. (a) The evolution of the three Sr 3d species in XPS Sr core-level spectra and the electrode polarization resistance as a function of temperature in 0.5 mbar O_2 [14] Copyright © 2018, The Author(s). With permission of Springer. (b) The average degradation rate of the surface resistance as a function of temperature [14] Copyright © 2018, The Author(s). With permission of Springer. (c) The stability of LSF thin films at 491 °C with intermittent rinsing in water. Reproduced from [106] with permission of The Royal Society of Chemistry. (d) A sketch of the proposed passivation and activation mechanisms. Reproduced from [106] with permission of The Royal Society of Chemistry.

Prolonged measurements at 650 °C without any further high-temperature treatments resulted in the degradation of k_{chem}. The authors suggested that samples can be shuffled between high- and low-temperature annealing to 're-activate' the sample surfaces that have been 'passivated' by aging during low-temperature operations. In a subsequent study, the authors followed the surface chemistry changes at specific times during the sequence [105]. The secondary-phase precipitates (i.e., SrO, $SrCO_3$, and $SrSO_4$) that form after prolonged treatment in the low-temperature regime (630 °C–675 °C) dissolve back into the lattice after a high-temperature treatment at 1000 °C. This resulted in an increase of two orders of magnitude in k_{chem}.

2.3.3 Surface cleaning/resetting

Etching or rinsing thin-film surfaces with water or dilute acids are promising strategies for 'resetting' the contaminated surfaces to their original states. For instance, chemical etching with a 10% diluted buffered HF (NH_4F:HF = 7:1) solution was shown to clean a 5–10 nm thick Sr-enriched layer on $SrTi_{1-x}Fe_xO_{3-\delta}$ electrodes [107]. ASR values measured after etching exhibited significantly reduced values compared to those before etching, and no change in the activation energy was observed. However, following post-annealing at 650 °C for 5 h, an increase in ASR that approached the values measured before etching was observed.

Wang *et al* took a different approach to illustrate the positive effects of etching on perovskite surfaces [108]. Co_3O_4 and $La_{0.3}Sr_{0.7}CoO_3$ (LSC37) composite powders were prepared by mechanically grinding the mixture and then the sample was etched in an ethylene glycol solution at 80 °C for 1.5 h. After the chemical etching, the sample was annealed at 650 °C for another 1.5 h. Investigations using multiple techniques e.g. SEM, TEM, XAS, XPS, and extended x-ray absorption fine structure (EXAFS) indicated a different chemical interaction between the Co_3O_4 and the LSC37 in the etched composite compared to the composite without etching in terms of microstructure, electronic structure, and surface chemical bonding, meaning that the composite was restructured after chemical etching. In terms of electrochemical activity, the chemically restructured composite was shown to possess higher oxygen evolution and oxygen reduction capabilities compared to the those of physically mixed composite.

Following a previous work reported in reference [104], Tripković *et al* investigated the variation in k_{chem} values after rinsing dense thin films of LSF in deionized water [106]. The ASR and k_{chem} were measured by AC impedance spectroscopy at 491 °C for extended periods (figure 2.9(c)) and their surfaces were analyzed by XPS in the as-deposited, aged, and rinsed states. The k_{chem} decreased exponentially within seven days of aging. After the first rinsing, a seven times higher k_{chem} was achieved initially, but the films deteriorated quickly. The sample did not seem to deteriorate after further aging for ten days, when it was assumed to have reached its passivated state. After the second rinsing, identical k_{chem} values to those after the first rinsing were obtained. More interestingly, the film seemed to deteriorate much more slowly than before. The Sr core-level XPS spectra were recorded to check the Sr local binding environment near the surface for the three samples, i.e. as-deposited, aged after seven days of electrochemical testing, and rinsed twice after 46 days of electrochemical testing. A small increase in the amount of surface-bound Sr species was detected in the aged sample compared to the as-deposited one (50%–54%), but this value was reduced to 22% in the rinsed sample. As illustrated in figure 2.9(d), it was suggested that the surface of the sample builds up detrimental secondary phases, which can be removed from the surface by rinsing it with water. Prolonged testing results in more Sr segregation to the surface. After the second rinsing, the already highly Sr-depleted film shows a slower degradation rate, since the driving force for Sr segregation becomes smaller.

In contrast, Cavallaro *et al* studied the rinsing of $SrCo_{0.95}Sb_{0.05}O_{3-\delta}$ thin films and powders with water and acetone. LEIS measurements indicated that the powder surface was enriched with Sr after being dipped in water and dried at 100 °C. In addition, XPS and TEM analyses revealed the presence of a decomposed surface layer composed of the $Sr_6Co_5O_{15}$ phase after exposure to water [109]. IEDP-SIMS experiments exhibited a substantially different ^{18}O isotope-diffusion concentration profile for thin films cleaned with water and acetone, confirming the surface decomposition with water.

2.3.4 Advanced microstructures

The development of advanced microstructures involves tailoring microstructures at the nanoscale as well as grains, grain boundaries, and dislocations [110].

Develos-Bagarinao *et al* studied a multilayer microstructure composed of subsequent layers of LSC64 and CGO (figure 2.10(a)) [111]. This design, comprising varying thicknesses of multilayered CGO/LSC films yielded k^* values approximately 1.5 to 2.5 times higher than that of single-phase LSC. The authors observed a mutual diffusion of La into the CGO and Ce into the LSC phase. Using STEM coupled with electron energy loss spectroscopy (EELS), Ce was found to exist preferentially in the 3+ oxidation state at the multilayer interfaces. The enhancement of the oxygen exchange kinetics was attributed to cation intermixing and to the creation of more oxygen vacancies in the vicinity of the LSC/CGO interface, providing more active sites for oxygen incorporation. For a more detailed discussion of the multilayered approach, the reader is referred to chapter 3 in this book.

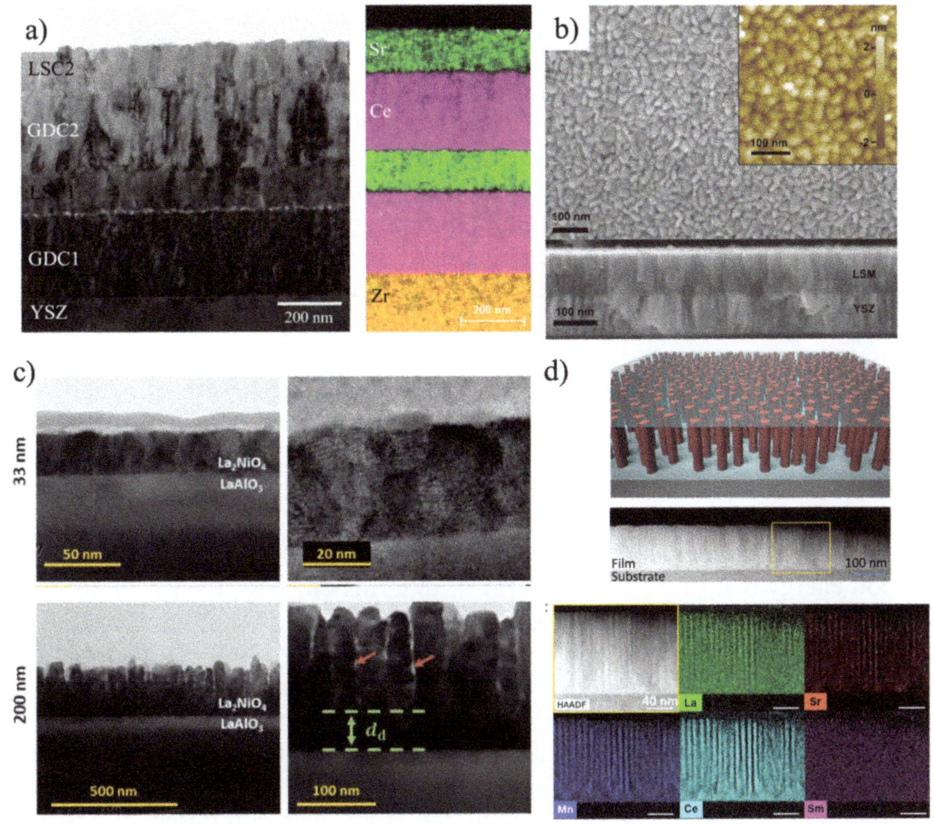

Figure 2.10. (a) A cross-sectional TEM image of the multilayer CGO/LSC design. Reprinted from [111], Copyright (2018), with permission from Elsevier. (b) The top view and cross-sectional view of the LSM columnar-type nanostructures [62] John Wiley & Sons, copyright 2015 WILEY-VCH Verlag GmbH & Co. KGaA, Weinheim]. (c) The cross-sectional TEM image of the 33 nm and 200 nm thick L2NO4 films. Reproduced from [113] with permission of The Royal Society of Chemistry. (d) A sketch of the LSM-SDC VAN structure, a cross-sectional TEM image of the VAN film, and color maps obtained using TEM-EDS [22] Copyright © 2021, The Author(s). With permission of Springer.

As previously mentioned, another study postulated that using a film design composed of vertically aligned nanocolumnar grains with a high concentration of strain-induced defects, an electronic conductor could be converted to a mixed ionic and electronic conductor (figure 2.10(b)) [62]. The study used LSM as model electrodes and aimed to differentiate between the G and GB contributions using EIS and simulations based on a finite-element model. Both the D^*_{GB} and k^*_{GB} values were found to be orders of magnitude higher than the bulk grain contribution. The increased k^*_{GB} were thought to be related to the coupling of the oxygen adsorption and dissociation steps at the GBs. In addition, in a later study carried out by Tarancón's group, the 'grain boundary engineering' concept was extended to other perovskite compositions, such as the $La_{0.8}Sr_{0.2}(Mn_{1-x}Co_x)_{0.85}O_{3\pm\delta}$ family, for which a substantial enhancement of the grain boundary oxygen mass transport properties was measured over the entire range of compositions [112].

Advanced microstructures showing porosity at the nanometer length scale also allowed for strong surface exchange properties [72, 113]. For instance, L2NO4 thin films grown by pulsed injection metal organic chemical vapor deposition (PI-MOCVD) showed faster surface exchange kinetics from dense thin films (33 nm) to thicker nanocolumnar (200 and 540 nm) microstructures (figure 2.10(c)). This was attributed to a very large increase in the specific surface area, the presence of other grain orientations, together with the influence of additional features such as kinks and edges present at the sides of the columns [113].

Recently, self-assembled vertically aligned nanocomposite (VAN) thin films grown by PLD have attracted a lot of attention, as they have yielded significant performance enhancements when used as air electrodes. For instance, epitaxial VAN films based on 5% *A*-site deficient LSCF and MgO were grown by PLD [114, 115]. Subsequently, the MgO was selectively etched out in an acetic acid solution to obtain high-surface-area electrodes. The etching procedure transformed the surface from one consisting predominantly of *A*-site terminations to one with mixed *A*- and *B*-site terminations, which led to an order of magnitude increase in the oxygen exchange kinetics between 300 °C and 450 °C and a reduction in the activation energy from 1 eV to 0.4 eV [115]. In terms of ASR, a value of 275 Ω cm^2 was recorded for a 70 nm thick etched LSCF film, while 85 Ω cm^2 was recorded for a 210 nm thick etched film at 450 °C [114]. Similarly, using LSCF/CGO VAN films, Develos-Bagarinao *et al* reported very high current densities in a full fuel cell, reaching up to ~2.2 and ~4.7 A cm^{-2} at 650 °C and 700 °C, respectively at 0.7 V [116]. Baiutti *et al* revealed that VAN structures (figure 2.10(d)) not only show high performance but also better stability compared to those of their single-phase counterparts [22]. The VAN structure in the study was composed of LSM and $Ce_{0.8}Sm_{0.2}O_2$ (SDC). LSM-SDC VAN films initially showed a tenfold lower ASR that that of single-phase LSM. In stability tests at 700 °C, the VAN film showed a 70% decrease in the ASR in the first 30 h of testing, with good stability during the next 70 h. On the other hand, single-phase LSM showed a 50% increase in the ASR. Using atom probe tomography (APT), the authors showed evidence of cationic intermixing between the LSM and SDC columns, in which Ce and Sm replaced La in LSM and La migrated into the SDC. From the DFT calculations, it was determined that in particular, Ce migration into LSM could lead to Sr–O over-bonding relative to the La–O polyhedral, thereby decreasing the elastic driving force for Sr surface segregation. The local high entropy achieved through cation intermixing is believed to influence long-term stability. According to ^{18}O oxygen tracer mapping using APT, the maximum ^{18}O concentration was localized in the SDC columns, indicating a preferential oxygen diffusion pathway. Thus, better performance and long-term durability were linked with the cation intermixing and work sharing among columnar grains.

2.4 Conclusions and prospects

In this chapter, we focused on materials for solid oxide cell air electrodes, mainly perovskite oxides, paying special attention to the influence of their microstructure and surface chemistry on their functional properties, and discussed the current understanding and recent developments in this area. Both internal and external factors play a key role in the surface exchange, oxygen diffusion, and polarization resistance which are measured for these materials at high temperatures and under operational conditions. Most SOC air electrodes contain Sr as a substitute cation in the *A*-site, which is introduced into the structure to increase the ionic conductivity of the perovskite. However, Sr tends to segregate to the surface, blocking its catalytic surface activity and leading to performance degradation over time. Through a number of representative examples, we have highlighted that even materials of the same stoichiometry can yield very different catalytic activity due to differences in synthesis conditions, e.g. deposition technique, deposition temperature, atmosphere, etc. In addition, the differences observed from one laboratory to another might be due to additional processes or contaminants to which the material is exposed. These could include post-thermal treatments, exposure to pollutants such as CO and SO_2 in the atmosphere, issues with the purity of the precursors and solvents, cross-contamination from previous depositions and measurements, or storage conditions.

Although the extent of this segregation largely depends on the composition of the material, the film microstructure is just as important. Furthermore, film crystallinity, grain orientation, and grain size, which are intricately linked to the thermal history and processing of the film, can also lead to important changes in its functional properties. Although *ex situ* structural, chemical, and microscopy measurements have formed the basis of our current understanding, recent findings obtained using *in situ* and/or coupled multimodal techniques have shown that the surface activity is actually affected by various combinations of the abovementioned points.

Today, the design of materials at the atomic and nano scales, using novel strategies such as surface decoration, advanced microstructures, and control of the atomic surface termination have gained further attention as ways of improving films' performance and durability. In addition, it has become even more important to study these materials under the relevant operational conditions, under cathodic and anodic biases, in humid and dry atmospheres, and at the most suitable temperature range and working environment to avoid degradation. In the near future, high-throughput deposition, characterization, and computation will be combined to screen a large number of candidate materials under various operational conditions. The use of artificial-intelligence-assisted simulation is expected to help reveal important composition–structure–property relationships and will guide further research in the field. Additionally, the development of new experimental setups and coupled multimodal approaches will allow researchers to investigate these materials under operational conditions without further exposing them to atmospheric or temperature variations, providing valuable information in the search for high-performance SOC air electrodes.

Acknowledgments

This project has received funding from the European Union's Horizon 2020 research and innovation program under grant agreements No. 824072 (HARVESTORE), No. 681146, and No. 101017709 (EPISTORE).

References

[1] Irvine J *et al* 2021 Roadmap on inorganic perovskites for energy applications *J. Phys. Energy* **3** 031502

[2] Tai L W, Nasrallah M M, Anderson H U, Sparlin D M and Sehlin S R 1995 Structure and electrical-properties of $La_{1-x}Sr_xCo_{1-y}Fe_yO_3$. Part 1. The system $La_{0.8}Sr_{0.2}Co_{1-y}Fe_yO_3$ *Solid State Ion.* **76** 259–71

[3] Zhu Z, Shi Y, Aruta C and Yang N 2018 Improving electronic conductivity and oxygen reduction activity in Sr-doped lanthanum cobaltite thin films: cobalt valence state and electronic band structure effects *ACS Appl. Energy Mater.* **1** 5308–17

[4] Tai L W, Nasrallah M M, Anderson H U, Sparlin D M and Sehlin S R 1995 Structure and electrical properties of $La_{1-x}S_xCo_{1-y}Fe_yO_3$. Part 2. The system $La_{1-x}Sr_xCo_{0.2}Fe_{0.8}0_3$ *Solid State Ion.* **76** 273–83

[5] Burriel M, Garcia G, Santiso J, Kilner J A, Chater R J and Skinner S J 2008 Anisotropic oxygen diffusion properties in epitaxial thin films of $La_2NiO_{4+\delta}$ *J. Mater. Chem.* **18** 416–22

[6] Maier J 1998 On the correlation of macroscopic and microscopic rate constants in solid state chemistry *Solid State Ion.* **112** 197–228

[7] Santiso J and Burriel M 2011 Deposition and characterisation of epitaxial oxide thin films for SOFCs *J. Solid State Electrochem.* **15** 985–1006

[8] Moreno R, García P, Zapata J, Roqueta J, Chaigneau J and Santiso J 2013 Chemical strain kinetics induced by oxygen surface exchange in epitaxial films explored by time-resolved X-ray diffraction *Chem. Mater.* **25** 3640–7

[9] Chater R, Carter S, Kilner J and Steele B 1992 Development of a novel SIMS technique for oxygen self-diffusion and surface exchange coefficient measurements in oxides of high diffusivity *Solid State Ion.* **53–56** 859–67

[10] Cushman C V, Brüner P, Zakel J, Major G H, Lunt B M, Smith N J, Grehl T and Linford M R 2016 Low energy ion scattering (LEIS). A practical introduction to its theory, instrumentation, and applications *Anal. Methods* **8** 3419–39

[11] Zameshin A A, Yakshin A E, Sturm J M, Brongerma H H and Bijkerk F 2018 Double matrix effect in low energy ion scattering from La surfaces *Appl. Surf. Sci.* **440** 570–9

[12] Niania M, Sharpe M, Webb R and Kilner J A 2020 The surface of complex oxides: ion beam based analysis of energy materials *Nucl. Instrum. Methods Phys. Res.* B **480** 27–32

[13] Tsvetkov N, Lu Q, Sun L, Crumlin E J and Yildiz B 2016 Improved chemical and electrochemical stability of perovskite oxides with less reducible cations at the surface *Nat. Mater.* **15** 1010–6
[14] Opitz A K *et al* 2018 The chemical evolution of the $La_{0.6}Sr_{0.4}CoO_{3-\delta}$ surface under SOFC operating conditions and its implications for electrochemical oxygen exchange activity *Top. Catal.* **61** 2129–41
[15] Hwang J, Rao R R, Katayama Y, Lee D, Wang X R, Crumlin E, Venkatesan T, Lee H N and Shao-Horn Y 2018 CO_2 reactivity on cobalt-based perovskites *J. Phys. Chem.* C **122** 20391–401
[16] Stoerzinger K A, Hong W T, Crumlin E J, Bluhm H and Shao-Horn Y 2015 Insights into electrochemical reactions from ambient pressure photoelectron spectroscopy *ACC. Chem. Res.* **48** 2976–83
[17] Nenning A, Opitz A K, Rameshan C, Rameshan R, Blume R, Hävecker M, Knop-Gericke A, Rupprechter G, Klötzer B and Fleig J 2016 Ambient pressure XPS study of mixed conducting perovskite-type SOFC cathode and anode materials under well-defined electrochemical polarization *J. Phys. Chem.* C **120** 1461–71
[18] Wen Y, Yang T, Lee D, Lee H N, Crumlin E J and Huang K 2018 Temporal and thermal evolutions of surface Sr-segregation in pristine and atomic layer deposition modified $La_{0.6}Sr_{0.4}CoO_{3-\delta}$ epitaxial films *J. Mater. Chem.* A **6** 24378–88
[19] Hong W T, Stoerzinger K A, Crumlin E J, Mutoro E, Jeen H, Lee H N and Shao-Horn Y 2016 Near-ambient pressure XPS of higherature surface chemistry in $Sr_2Co_2O_5$ thin films *Top. Catal.* **59** 574–82
[20] Chang K C, Ingram B, Ilavsky J, Lee S, Fuoss P and You H 2017 Synchrotron x-ray studies of model SOFC cathodes, part I: thin film cathodes *Solid State Ion.* **311** 118–26
[21] Wang X *et al* 2021 Manipulating surface termination of perovskite manganate for oxygen activation *Adv. Funct. Mater.* **31** 2006439
[22] Baiutti F *et al* 2021 A high-entropy manganite in an ordered nanocomposite for long-term application in solid oxide cells *Nat. Commun.* **12** 1–11
[23] Hess F, Staykov A, Yildiz B and Kilner J 2020 Solid oxide fuel cell materials and interfaces ed W Andreoni and S Yip *Handbook of Materials Modeling.* (Cham: Springer)
[24] Tasker P W 1979 The stability of ionic crystal surfaces *J. Phys. C: Solid State Phys.* **12** 4977–84
[25] Koo B, Kim K, Kim J K, Kwon H, Han J W and Jung W C 2018 Sr segregation in perovskite oxides: why it happens and how it exists *Joule* **2** 1476–99
[26] Chen K and Jiang S P 2020 Surface segregation in solid oxide cell oxygen electrodes: phenomena, mitigation strategies and electrochemical properties *Electrochem. Energy Rev.* **3** 730–65
[27] Harrison W A 2011 Origin of Sr segregation at $La_{1-x}Sr_xMnO_3$ surfaces *Phys. Rev.* B **83** 1–5
[28] Ding H, Virkar A V, Liu M and Liu F 2013 Suppression of Sr surface segregation in $La_{1-x}Sr_xCo_{1-y}Fe_yO_{3-\delta}$: a first principles study *Phys. Chem. Chem. Phys.* **15** 489–96
[29] Cai Z, Kubicek M, Fleig J and Yildiz B 2012 Chemical heterogeneities on $La_{0.6}$ $Sr_{0.4}CoO_{3-\delta}$ thin films—correlations to cathode surface activity and stability *Chem. Mater.* **24** 1116–27
[30] Lee W, Han J W, Chen Y, Cai Z and Yildiz B 2013 Cation size mismatch and charge interactions drive dopant segregation at the surfaces of manganite perovskites *J. Am. Chem. Soc.* **135** 7909–25

[31] Kwon H, Lee W and Han J W 2016 Suppressing cation segregation on lanthanum-based perovskite oxides to enhance the stability of solid oxide fuel cell cathodes *RSC Adv.* **6** 69782–9

[32] Kim D, Bliem R, Hess F, Gallet J J and Yildiz B 2020 Electrochemical polarization dependence of the elastic and electrostatic driving forces to aliovalent dopant segregation on $LaMnO_3$ *J. Am. Chem. Soc.* **142** 3548–63

[33] Ostrovskiy E, Huang Y L and Wachsman E D 2021 Effects of surface chemical potentials on cation segregation *J. Mater. Chem.* A **9** 1593–602

[34] Lee W and Yildiz B 2013 Factors that influence cation segregation at the surfaces of perovskite oxides *ECS Trans.* **57** 2115–23

[35] Koo B, Kwon H, Kim Y, Seo G and Han W 2018 Enhanced oxygen exchange of perovskite oxide surfaces through strain-driven chemical stabilization *Energy Environ. Sci.* **11** 71–7

[36] Kubicek M, Cai Z, Ma W, Yildiz B, Hutter H and Fleig J 2013 Tensile lattice strain accelerates oxygen surface exchange and diffusion in $La_{1-x}Sr_xCoO_{3-\delta}$ thin films *ACS Nano* **7** 3276–86

[37] Lee D, Jacobs R, Jee Y, Seo A, Sohn C, Ievlev A V, Ovchinnikova O S, Huang K, Morgan D and Lee H N 2017 Stretching epitaxial $La_{0.6}Sr_{0.4}CoO_{3-\delta}$ for fast oxygen reduction *J. Phys. Chem.* C **121** 25651–8

[38] Yu Y, Ludwig K F, Woicik J C, Gopalan S, Pal U B, Kaspar T C and Basu S N 2016 Effect of Sr content and strain on sr surface segregation of $La_{1-x}Sr_xCo_{0.2}Fe_{0.8}O_{3-\delta}$ as cathode material for solid oxide fuel cells *ACS Appl. Mater. Interfaces* **8** 26704–11

[39] Rupp G M, Téllez H, Druce J, Limbeck A, Ishihara T, Kilner J and Fleig J 2015 Surface chemistry of $La_{0.6}Sr_{0.4}CoO_{3-\delta}$ thin films and its impact on the oxygen surface exchange resistance *J. Mater. Chem.* A **3** 22759–69

[40] Druce J, Téllez H, Burriel M, Sharp M D, Fawcett L J, Cook S N, McPhail D S, Ishihara T, Brongersma H H and Kilner J A 2014 Surface termination and subsurface restructuring of perovskite-based solid oxide electrode materials *Energy Environ. Sci.* **7** 3593–9

[41] Yoon H I, Lee D K, Bin Bae H, Jo G Y, Chung H S, Kim J G, Kang S J L and Chung S Y 2017 Probing dopant segregation in distinct cation sites at perovskite oxide polycrystal interfaces *Nat. Commun.* **8** 1–7

[42] Staykov A, Téllez H, Akbay T, Druce J, Ishihara T and Kilner J 2015 Oxygen activation and dissociation on transition metal free perovskite surfaces *Chem. Mater.* **27** 8273–81

[43] Lee C W, Behera R K, Okamoto S, Devanathan R, Wachsman E D, Phillpot S R and Sinnott S B 2011 Stabilization mechanisms of $LaFeO_3$ (010) surfaces determined with first principles calculations *J. Am. Ceram. Soc.* **94** 1931–9

[44] Kotomin E A, Mastrikov Y A, Merkle R and Maier J 2020 First principles calculations of oxygen reduction reaction at fuel cell cathodes *Curr. Opin. Electrochem.* **19** 122–8

[45] Rupp G M, Opitz A K, Nenning A, Limbeck A and Fleig J 2017 Real-time impedance monitoring of oxygen reduction during surface modification of thin film cathodes *Nat. Mater.* **16** 640–5

[46] Li Y, Zhang W, Zheng Y, Chen J, Yu B, Chen Y and Liu M 2017 Controlling cation segregation in perovskite-based electrodes for high electro-catalytic activity and durability *Chem. Soc. Rev.* **46** 6345–78

[47] Cao Y, Gadre M J, Ngo A T, Adler S B and Morgan D D 2019 Factors controlling surface oxygen exchange in oxides *Nat. Commun.* **10** 1–15

[48] Rahmanipour M, Cheng Y, Onn T M, Donazzi A, Vohs J M and Gorte R J 2017 Modification of LSF–YSZ composite cathodes by atomic layer deposition *J. Electrochem. Soc.* **164** F879–84
[49] Cheng Y, Raman A S, Paige J, Zhang L, Sun D, Chen M U, Vojvodic A, Gorte R J and Vohs J M 2019 Enhancing oxygen exchange activity by tailoring perovskite surfaces *J. Phys. Chem. Lett.* **10** 4082–8
[50] Pai Y Y, Tylan-Tyler A, Irvin P and Levy J 2018 Physics of $SrTiO_3$-based heterostructures and nanostructures: a review *Reports Prog. Phys.* **81** 036503
[51] Kawasaki M, Takahashi K, Maeda T, Tsuchiya R, Shinohara M, Ishiyama O, Yonezawa T, Yoshimoto M and Koinuma H 1994 Atomic control of the $SrTiO_3$ crystal surface *Science (80-.).* **266** 1540
[52] Sharma R K, Burriel M, Dessemond L, Bassat J-M and Djurado E 2016 $La_{n+1}Ni_nO_{3n+1}$ ($n = 2$ and 3) phases and composites for solid oxide fuel cell cathodes : facile synthesis and electrochemical properties *J. Power Sources* **325** 337–45
[53] Sharma R K, Burriel M, Dessemond L, Martin V, Bassat J-M and Djurado E 2016 An innovative architectural design to enhance the electrochemical performance of $La_2NiO_{4+\delta}$ cathodes for solid oxide fuel cell applications *J. Power Sources* **316** 17–28
[54] Amow G, Davidson I J and Skinner S J 2006 A comparative study of the Ruddlesden–Popper series, $La_{n+1}Ni_nO_{3n+1}$ ($n = 1$, 2 and 3), for solid-oxide fuel-cell cathode applications *Solid State Ion.* **177** 1205–10
[55] Staykov A, Nguyen T, Akbay T and Ishihara T 2022 Oxygen reduction reaction and electronic properties of LnO-terminated surfaces of Pr_2NiO_4 and La_2NiO_4 *J. Phys. Chem.* C **126** 7390–9
[56] Druce J, Ishihara T and Kilner J 2014 Surface composition of perovskite-type materials studied by Low Energy Ion Scattering (LEIS) *Solid State Ion.* **262** 893–6
[57] Khamidy N I, Laurencin J, Ferreira Sanchez D, Monaco F, Charlot F and Djurado E 2020 Durability of nanostructured $LaPrNiO_{4+\delta}$ electrode for solid oxide cells: electrochemical, microstructural, and structural investigation *J. Power Sources* **450** 227724
[58] Egger A, Schrödl N, Gspan C and Sitte W 2017 $La_2NiO_{4+\delta}$ as electrode material for solid oxide fuel cells and electrolyzer cells *Solid State Ion.* **299** 18–25
[59] Akbay T, Staykov A, Druce J, Téllez H, Ishihara T and Kilner J A 2016 The interaction of molecular oxygen on LaO terminated surfaces of La_2NiO_4 *J. Mater. Chem.* A **4** 13113–24
[60] Yan L and Salvador P A 2012 Substrate and thickness effects on the oxygen surface exchange of $La_{0.7}Sr_{0.3}MnO_3$ thin films *ACS Appl. Mater. Interfaces* **4** 2541–50
[61] Navickas E *et al* 2015 Fast oxygen exchange and diffusion kinetics of grain boundaries in Sr-doped $LaMnO_3$ thin films *Phys. Chem. Chem. Phys.* **17** 7659–69
[62] Saranya A M, Pla D, Morata A, Cavallaro A, Canales-Vázquez J, Kilner J A, Burriel M and Tarancón A 2015 Engineering mixed ionic electronic conduction in $La_{0.8}Sr_{0.2}MnO_{3+\delta}$ nanostructures through fast grain boundary oxygen diffusivity *Adv. Energy Mater.* **5** 1500377
[63] Huber T M, Navickas E, Sasaki K, Yildiz B, Hutter H, Tuller H and Fleig J 2018 Interplay of grain size dependent electronic and ionic conductivity in electrochemical polarization studies on Sr-doped $LaMnO_3$ (LSM) thin film cathodes *J. Electrochem. Soc.* **165** F702–9
[64] Chen Y, Téllez H, Burriel M, Yang F, Tsvetkov N, Cai Z, McComb D W, Kilner J A and Yildiz B 2015 Segregated chemistry and structure on (001) and (100) surfaces of $(La_{1-x}Sr_x)$ $2CoO_4$ override the crystal anisotropy in oxygen exchange kinetics *Chem. Mater.* **27** 5436–50

[65] Develos-Bagarinao K, De Vero J, Kishimoto H, Ishiyama T, Yamaji K, Horita T and Yokokawa H 2019 Oxygen surface exchange properties and surface segregation behavior of nanostructured $La_{0.6}Sr_{0.4}Co_{0.2}Fe_{0.8}O_{3-\delta}$ thin film cathodes *Phys. Chem. Chem. Phys.* **21** 7183–95

[66] Fearn S, Rossiny J C H, Kilner J A and Evans J R G 2012 Measurement of oxygen transport in $La_{0.8}Sr_{0.2}MnO_3$ perovskite grains *Solid State Ion.* **211** 51–7

[67] Pişkin F, Bliem R and Yildiz B 2018 Effect of crystal orientation on the segregation of aliovalent dopants at the surface of $La_{0.6}Sr_{0.4}CoO_3$ *J. Mater. Chem.* A **6** 14136–45

[68] Niania M, Podor R, Ben Britton T, Li C, Cooper S J, Svetkov N, Skinner S and Kilner J 2018 In situ study of strontium segregation in $La_{0.6}Sr_{0.4}Co_{0.2}Fe_{0.8}O_{3-\delta}$ in ambient atmospheres using high-temperature environmental scanning electron microscopy *J. Mater. Chem.* A **6** 14120–35

[69] Kubicek M, Limbeck A, Fromling T, Hutter H and Fleig J 2011 Relationship between cation segregation and the electrochemical oxygen reduction kinetics of $La_{0.6}Sr_{0.4}CoO_{3-\delta}$ thin film electrodes *J. Electrochem. Soc.* **158** B727–34

[70] Cavallaro A, Pramana S S, Ruiz-Trejo E, Sherrell P C, Ware E, Kilner J A and Skinner S J 2018 Amorphous-cathode-route towards low temperature SOFC *Sustain. Energy Fuels* **2** 862–75

[71] Celikbilek O, Cavallaro A, Kerherve G, Fearn S, Chaix-Pluchery O, Aguadero A, Kilner J A and Skinner S J 2020 Surface restructuring of thin-film electrodes based on thermal history and its significance for the catalytic activity and stability at the gas/solid and solid/solid interfaces *ACS Appl. Mater. Interfaces* **12** 34388–401

[72] Develos-Bagarinao K, Celikbilek O, Budiman R A, Kerherve G, Fearn S, Skinner S J and Kishimoto H 2022 On the role of surfaces and interfaces on electrochemical performance and long-term stability of nanostructured LSC thin film electrodes *J. Mater. Chem.* A **10** 2445–59

[73] Siebenhofer M, Huber T M, Friedbacher G, Artner W, Fleig J and Kubicek M 2020 Oxygen exchange kinetics and nonstoichiometry of pristine $La_{0.6}Sr_{0.4}CoO_{3-\delta}$ thin films unaltered by degradation *J. Mater. Chem.* A **8** 7968–79

[74] Chen T, Harrington G F, Sasaki K and Perry N H 2017 Impact of microstructure and crystallinity on surface exchange kinetics of strontium titanium iron oxide perovskite by *in situ* optical transmission relaxation approach *J. Mater. Chem.* A **5** 23006–19

[75] Chen T, Harrington G F, Masood J, Sasaki K and Perry N H 2019 Emergence of rapid oxygen surface exchange kinetics during *in situ* crystallization of mixed conducting thin film oxides *ACS Appl. Mater. Interfaces* **11** 9102–16

[76] Skiba E J, Chen T and Perry N H 2020 Simultaneous electrical, electrochemical, and optical relaxation measurements of oxygen surface exchange coefficients: $Sr(Ti,Fe)O_{3-d}$ film crystallization case study *ACS Appl. Mater. Interfaces* **12** 48614–30

[77] Sha Z, Cali E, Shen Z, Ware E, Kerherve G and Skinner S J 2021 Significantly enhanced oxygen transport properties in mixed conducting perovskite oxides under humid reducing environments *Chem. Mater.* **33** 8469–76

[78] Nenning A, Navickas E, Hutter H and Fleig J 2016 Water-induced decoupling of tracer and electrochemical oxygen exchange kinetics on mixed conducting electrodes *J. Phys. Chem. Lett.* **7** 2826–31

[79] Bucher E, Sitte W, Klauser F and Bertel E 2011 Oxygen exchange kinetics of $La_{0.58}Sr_{0.4}Co_{0.2}Fe_{0.8}O_3$ at 600 °C in dry and humid atmospheres *Solid State Ion.* **191** 61–7

[80] Bucher E, Sitte W, Klauser F and Bertel E 2012 Impact of humid atmospheres on oxygen exchange properties, surface-near elemental composition, and surface morphology of $La_{0.6}Sr_{0.4}Co_{3-\delta}$ *Solid State Ion.* **208** 43–51
[81] Joo J H, Merkle R and Maier J 2011 Effects of water on oxygen surface exchange and degradation of mixed conducting perovskites *J. Power Sources* **196** 7495–9
[82] Zhang D, Machala M L, Chen D, Guan Z, Li H, Nemsak S, Crumlin E J, Bluhm H and Chueh W C 2020 Hydroxylation and cation segregation in $(La_{0.5}Sr_{0.5})FeO_{3-\delta}$ electrodes *Chem. Mater.* **32** 2926–34
[83] Thoréton V, Niania M and Kilner J 2021 Kinetics of competing exchange of oxygen and water at the surface of functional oxides *Phys. Chem. Chem. Phys.* **23** 2805–11
[84] Huang Y L, Pellegrinelli C and Wachsman E D 2017 Oxygen dissociation kinetics of concurrent heterogeneous reactions on metal oxides *ACS Catal.* **7** 5766–72
[85] Kilner J A, Skinner S J and Brongersma H H 2011 The isotope exchange depth profiling (IEDP) technique using SIMS and LEIS *J. Solid State Electrochem.* **15** 861–76
[86] Vannier R N, Chater R J, Skinner S J, Kilner J A and Mairesse G 2003 Characterisation of the oxygen transfer in BIMEVOX membranes under applied current conditions *Solid State Ion.* **160** 327–34
[87] Huber A K, Falk M, Rohnke M, Luerßen B, Gregoratti L, Amati M and Janek J 2012 In situ study of electrochemical activation and surface segregation of the SOFC electrode material $La_{0.75}Sr_{0.25}Cr_{0.5}Mn_{0.5}O_{3\pm\delta}$ *Phys. Chem. Chem. Phys.* **14** 751–8
[88] Cohn G and Wachsman E D 2017 In-operando determination of sofc cathode oxygen surface exchange coefficients for enhanced oxygen reduction reaction kinetics *J. Electrochem. Soc.* **164** F3035–44
[89] Mutoro E *et al* 2012 Reversible compositional control of oxide surfaces by electrochemical potentials *J. Phys. Chem. Lett.* **3** 40–4
[90] Huber A K, Falk M, Rohnke M, Luerssen B, Amati M, Gregoratti L, Hesse D and Janek J 2012 In situ study of activation and de-activation of LSM fuel cell cathodes—electrochemistry and surface analysis of thin-film electrodes *J. Catal.* **294** 79–88
[91] Bozzini B *et al* 2019 In situ near-ambient pressure X-ray photoelectron spectroscopy discloses the surface composition of operating $NdBaCo_2O_{5+\delta}$ solid oxide fuel cell cathodes *J. Power Sources* **436** 226815
[92] Mutoro E, Crumlin E J, Biegalski M D, Christen H M and Shao-Horn Y 2011 Enhanced oxygen reduction activity on surface-decorated perovskite thin films for solid oxide fuel cells *Energy Environ. Sci.* **4** 3689
[93] Li Y, Zhang W, Wu T, Zheng Y, Chen J, Yu B, Zhu J and Liu M 2018 Segregation induced self-assembly of highly active perovskite for rapid oxygen reduction reaction *Adv. Energy Mater.* **8** 1801893
[94] Gong Y, Palacio D, Song X, Patel R L, Liang X, Zhao X, Goodenough J B and Huang K 2013 Stabilizing nanostructured solid oxide fuel cell cathode with atomic layer deposition *Nano Lett.* **13** 4340–5
[95] Argirusis C, Wagner S, Menesklou W, Warnke C, Damjanovic T, Borchardt G and Ivers-Tiffe E 2005 Enhancement of oxygen surface exchange kinetics of $SrTiO_3$ by alkaline earth metal oxides *Chem. Phys.* **7** 3523–5
[96] Paige J M, Cheng Y, Pepin P A, Curran C D, Sun D, Chen M U, McIntosh S, Vohs J M and Gorte R J 2019 Surface modification of SOFC cathodes by Co, Ni, and Pd oxides *Solid State Ion.* **341** 115051

[97] Gong Y, Patel R L, Liang X, Palacio D, Song X, Goodenough J B and Huang K 2013 Atomic layer deposition functionalized composite SOFC cathode LSCF-CGO: enhanced long-term stability *Chem. Mater.* **25** 4224–31

[98] Nicollet C, Toparli C, Harrington G F, Defferriere T, Yildiz B and Tuller H L 2020 Acidity of surface-infiltrated binary oxides as a sensitive descriptor of oxygen exchange kinetics in mixed conducting oxides *Nat. Catal.* **3** 913–20

[99] Sasaki K *et al* 2011 Chemical durability of solid oxide fuel cells: influence of impurities on long-term performance *J. Power Sources* **196** 9130–40

[100] Ni N and Skinner S 2016 Combined Cr and Mo poisoning of (La,Sr)(Co,Fe)$O_{3-\delta}$ solid oxide fuel cell cathodes at the nanoscale *Solid State Ion.* **288** 28–31

[101] Ni N, Wang C C, Jiang S P and Skinner S J 2019 Synergistic effects of temperature and polarization on Cr poisoning of $La_{0.6}Sr_{0.4}Co_{0.2}Fe_{0.8}O_{3-\delta}$ solid oxide fuel cell cathodes *J. Mater. Chem.* A **7** 9253–62

[102] Ni N, Cooper S J, Williams R, Kemen N, McComb D W and Skinner S J 2016 Degradation of $(La_{0.6}Sr_{0.4})_{0.95}(Co_{0.2}Fe_{0.8})O_{3-\delta}$ solid oxide fuel cell cathodes at the nanometer scale and below *ACS Appl. Mater. Interfaces* **8** 17360–70

[103] Smith D W 1987 An acidity scale for binary oxides *J. Chem. Educ.* **64** 480–1

[104] Tripković D, Küngas R, Mogensen M B and Hendriksen P V 2019 Surface recrystallization-an underestimated phenomenon affecting oxygen exchange activity *J. Mater. Chem.* A **7** 11782–91

[105] Tripković Đ, Wang J, Küngas R, Mogensen M B, Yildiz B and Hendriksen P V 2022 Thermally controlled activation and passivation of surface chemistry and oxygen-exchange kinetics on a perovskite oxide *Chem. Mater.* **34** 1722–36

[106] Tripković Đ, Küngas R, Mogensen M B and Hendriksen P V 2020 Passivation and activation of $La_{0.6}Sr_{0.4}FeO_3$ thin film electrodes *Phys. Chem. Chem. Phys.* **22** 15418–26

[107] Jung W and Tuller H L 2012 Investigation of surface Sr segregation in model thin film solid oxide fuel cell perovskite electrodes *Energy Environ. Sci.* **5** 5370–8

[108] Wang X, Pan Z, Huang K, Cong Y, Cao R, Sarangi R, Li L, Li G and Feng S 2019 Atomic-scale insights into surface lattice oxygen activation at the spinel/perovskite interface of $Co_3O_4/La_{0.3}Sr_{0.7}CoO_3$ *Angew. Chemie Int. Ed.* **58** 11720–5

[109] Cavallaro A, Wilson G E, Kerherve G, Cali E, van den Bosch C A M, Boldrin P, Payne D, Skinner S J and Aguadero A 2021 Analysis of H_2O-induced surface degradation in $SrCoO_3$-derivatives and its impact on redox kinetics *J. Mater. Chem.* A **9** 24528–38

[110] Acosta M, Baiutti F, Tarancón A and MacManus-Driscoll J L 2019 Nanostructured materials and interfaces for advanced ionic electronic conducting oxides *Adv. Mater. Interfaces* **1900462** 1–15

[111] Develos-Bagarinao K, De Vero J, Kishimoto H, Ishiyama T, Yamaji K, Horita T and Yokokawa H 2018 Multilayered LSC and GDC: an approach for designing cathode materials with superior oxygen exchange properties for solid oxide fuel cells *Nano Energy* **52** 369–80

[112] Saranya A M, Morata A, Pla D, Burriel M, Chiabrera F, Garbayo I, Hornés A, Kilner J A and Tarancón A 2018 Unveiling the outstanding oxygen mass transport properties of Mn-rich perovskites in grain boundary-dominated $La_{0.8}Sr_{0.2}(Mn_{1-x}Co_x)_{0.85}O_{3\pm\delta}$ nanostructures *Chem. Mater.* **30** 5621–9

[113] Stangl A, Riaz A, Rapenne L, Caicedo J M, De Dios Sirvent J, Baiutti F, Jiménez C, Tarancón A, Mermoux M and Burriel M 2022 Tailored nano-columnar La_2NiO_4 cathodes for improved electrode performance *J. Mater. Chem.* A **10** 2528–40

[114] Wells M P *et al* 2021 Route to high-performance micro-solid oxide fuel cells on metallic substrates *ACS Appl. Mater. Interfaces* **13** 4117–25

[115] Acosta M *et al* 2022 Surface chemistry and porosity engineering through etching reveal ultrafast oxygen reduction kinetics below 400 °C in *B*-site exposed (La,Sr)(Co,Fe)O_3 thin-films *J. Power Sources* **523** 230983

[116] Develos-Bagarinao K, Yamaji K, Kishimoto H and Shimada H 2021 Nanoengineering of cathode layers for solid oxide fuel cells to achieve superior power densities *Nat. Commun.* **12** 3979

Chapter 3

Tuning heterointerfaces for high-performance solid oxide cells

Katherine Develos-Bagarinao, Tomohiro Ishiyama and Haruo Kishimoto

Solid oxide cells (SOCs) are essentially multilayers composed of various oxide materials that form intimate contacts with each other. The ionic transport phenomena occurring at heterointerfaces invariably affect the overall performance of SOC devices, due to the inherent interfacial resistance associated with heterointerfaces and the cation diffusion behavior which primarily affects the long-term degradation and durability of the cell. In this chapter, we will focus on the effects of various types of heterointerface on the performance of SOCs, with an emphasis on innovative strategies adopted to tune the properties of said heterointerfaces, including advanced characterization techniques employed for investigating their properties.

This chapter is divided into three main sections. First, we will look at several multilayer strategies adopted for the preparation of air electrodes, which offer the unique potential to tune heterointerfacial properties in order to boost air electrode performance. Second, we will focus on electrode–interlayer–electrolyte heterointerfaces and, in particular, the tuning of interlayer microstructures to enhance charge transfer properties and the stability of cell performance. Third, we will review advanced methods for probing interfacial properties. Finally, we will summarize the key points of the chapter and present perspectives on the future development of this area.

3.1 Tuning electrode heterointerfaces by multilayering

The performance of air electrodes is primarily determined by their oxygen reduction reaction (ORR) kinetics, which essentially begins with the process of interaction between the solid electrode surface and the oxygen gas. SOCs are typically operated at high temperatures of 700 °C–850 °C in order to overcome the slow kinetics at the air electrode, which leads to high electrode overpotentials at lower temperatures [1, 2].

doi:10.1088/978-0-7503-4064-9ch3

Typical air electrode materials employed for SOC applications are mixed ionic and electronic conductors (MIECs), e.g. Sr-doped $LaCoO_3$ ($La_{0.6}Sr_{0.4}CoO_{3-\delta}$ (LSC)) [3–6] or Sr- and Fe-doped $LaCoO_3$ ($La_{0.6}Sr_{0.4}Co_{0.2}Fe_{0.8}O_{3-\delta}$ or LSCF) [7–12]. The ORR process on MIEC surfaces is considered to proceed via a bulk pathway or a surface pathway, as illustrated in figure 3.1. The ORR process proceeds through the following steps:

1. Gas diffusion and adsorption
2. Dissociation
3. Electron transport
4. The incorporation of oxide ions
5. The diffusion of oxide ions
6. Oxide ion transport across the electrode/electrolyte interface

For the bulk pathway, adsorption, the dissociation of oxygen, and charge transfer occur on the surface, whereas for the surface pathway, dissociated oxygen species diffuse across the surface and reach the triple phase boundary (TPB) comprising the gas, electrode, and electrolyte where charge transfer occurs [13, 14]. This is described by the following defect reaction expressed in Kröger–Vink notation:

$$O_2 + 2V_o^{\bullet\bullet} + 4e^- \leftrightarrows 2O_o^x, \tag{3.1}$$

where $V_o^{\bullet\bullet}$, e^-, and O_o^x are doubly charged oxygen vacancies, electrons, and oxide ions in the lattice, respectively.

Previous studies have shown that the ORR process is the rate-limiting process for temperatures between 500 °C and 750 °C [15, 16]. Therefore, air electrode materials should typically have the required properties, such as high ionic and electronic conductivities and high surface exchange coefficients, in order to deliver optimum performance for SOC operation. In this regard, heterointerfaces have a unique

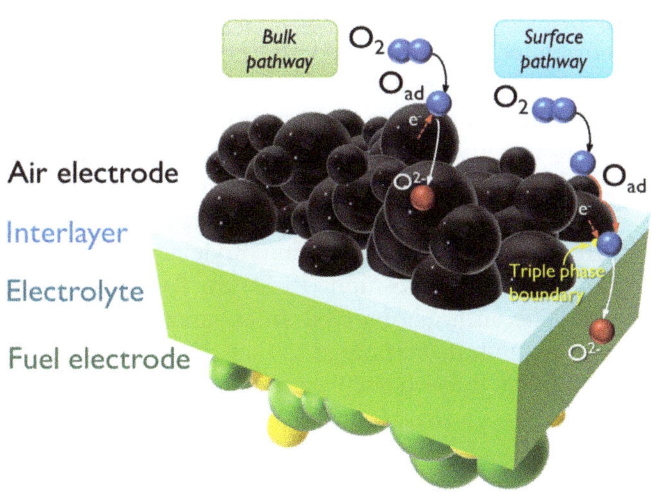

Figure 3.1. Schematic illustration of the oxygen reduction reaction (ORR) process on MIEC-type air electrodes.

potential to tune the performance, as these regions have been demonstrated to be highly active sites for the ORR process. A recent review of heterointerface engineering by Zhao *et al* [17] provides a comprehensive summary of the various types of heterointerface reported in the research literature, although the list is not exhaustive. For a further detailed discussion of interface engineering approaches, the reader is referred to chapter 4 in this book. This section covers the multilayer approach, with a focus on the utilization of this simple yet versatile technique to enhance the performance and stability of SOCs for intermediate-temperature operation.

Conceptually, the multilayer approach takes advantage of the catalytic effects associated with heterointerfaces and the flexibility that allows the overall electrode properties to be tuned via combinations of various materials. Two types of multilayer combinations have been examined in depth to date: multilayered ABO_3/$(AB)_2O_4$ perovskites and perovskite–fluorite heterostructures.

3.1.1 Multilayered perovskite oxides

A well-investigated type of ABO_3/$(AB)_2O_4$ heterointerface is that represented by $La_{0.6}Sr_{0.4}O_3$ (LSC_{113}) and $(La,Sr)_2CoO_4$ (LSC_{214}), which were first reported by Kawada *et al* during their investigation of the surface of polycrystalline LSC_{113} [18]. To investigate the distribution of active sites, the LSC sample was isotopically exchanged (annealed in an enriched gas of ^{18}O oxygen isotope) and subsequently analyzed using secondary ion mass spectrometry (SIMS). However, during the high-temperature sintering of the LSC sample, the temperature exceeded the stability of the LSC_{113} phase, resulting in the precipitation of LSC_{214} as a secondary phase. This serendipitous discovery led to the first observation of the 'anomalous' enhancement of surface exchange as evidenced by the high concentration of ^{18}O tracer at the $LSC_{113/214}$ heterointerface. In a follow-up study in 2008 reported by Sase *et al* [19] using SIMS imaging and depth profiling of the ^{18}O tracer, regions around LSC_{214} precipitates could be directly correlated with enhanced ^{18}O concentration. It was inferred that oxygen exchange occurs predominantly on the periphery of the precipitates and that oxide ions eventually diffuse and become incorporated through the LSC_{113} matrix (figure 3.2). Furthermore, an extremely high surface exchange coefficient (k^*) of 8×10^{-6} cm s^{-1} was obtained, which was about three orders of magnitude higher than that obtained for single-phase LSC_{113}.

Crumlin *et al* performed a similar study using model electrodes comprising LSC_{113} films prepared by pulsed laser deposition (PLD) on (001) yttria-stabilized zirconia (YSZ) substrates [20]. The authors observed an enhancement of the ORR by about three to four orders of magnitude compared to that of bulk LSC_{113}. A comparison of LSC_{113} films ranging from those partially covered by LSC_{214} films to those fully covered by LSC_{214} films (~0.1 to ~15 nm in thickness) is shown in figure 3.3. The active regions for oxygen surface exchange are attributed to the $LSC_{113/214}$ heterointerfaces (figure 3.3(a)). For very low film thicknesses, it is considered that the LSC_{214} forms as nanoislands on the LSC_{113}, thus resulting in a high density of $LSC_{113/214}$ boundaries on the film surface and consequently the highest ORR activity values

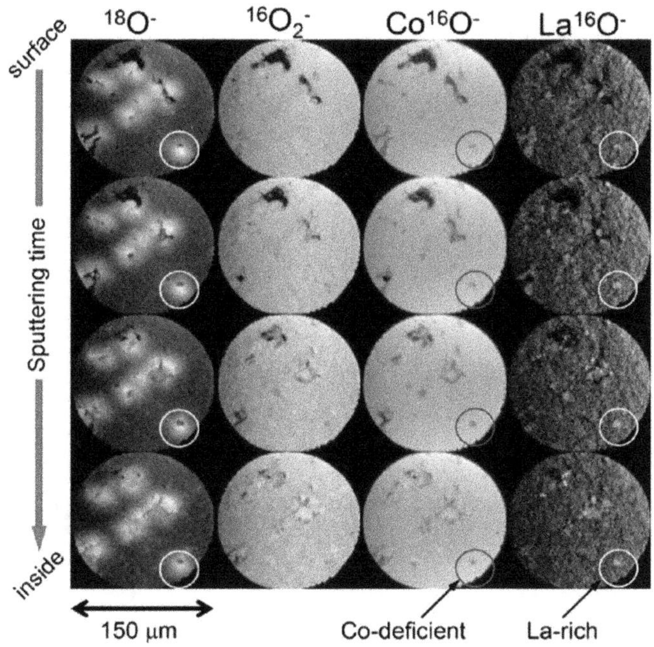

Figure 3.2. Secondary ion imaging of $La_{0.6}Sr_{0.4}CoO_3$ surfaces obtained using the SIMS method. The samples were $^{18}O/^{16}O$ isotopically exchanged at 0.2 bar and 773 K for 180 s. The areas indicated by circles correspond to regions where La-rich and Co-deficient precipitates appeared, assumed to be of $(La,Sr)_2CoO_4$ origin. The ^{18}O concentration on the periphery of the precipitates is observed to be higher. Reprinted from [19], Copyright (2008), with permission from Elsevier.

(figure 3.3(b)). For fully covered films, it is considered that the edges themselves may have an influence on the ORR enhancement, as inferred from the comparison of fully covered $LSC_{113/214}$ with varied electrode diameters (figure 3.3(c)). By decreasing the electrode diameter from ~160 μm to ~35 μm, they found that the entire electrode became highly activated and approached the ASR performance of partially covered samples. Although the exact mechanism was not clarified, the enhancement of the ORR activity at the heterointerfaces was attributed to 'interfacial properties including strain, space charge effects, an increase in oxygen vacancy concentration, or a change in electronic structure.'

With the goal of clarifying the effect of the $LSC_{113/214}$ heterointerfaces on the ORR and oxygen evolution reaction (OER) processes, Zhao *et al* studied model electrodes made from LSC_{113} (~50 nm thick) decorated with ~2 nm thick LSC_{214} particles and evaluated their performance using electrochemical impedance spectroscopy (EIS) [21]. The films were deposited on GDC-buffered (001) YSZ single-crystal substrates using PLD and were evaluated using a self-developed high-temperature microprobe electrochemical platform. As shown in figure 3.4, the impedance spectra of the single-phase LSC_{113} film evaluated at 873 K show a relatively larger arc compared to those of the $LSC_{113/214}$ heterostructure. Extracting the surface exchange coefficient, k_q, from the equivalent circuit fitting of the

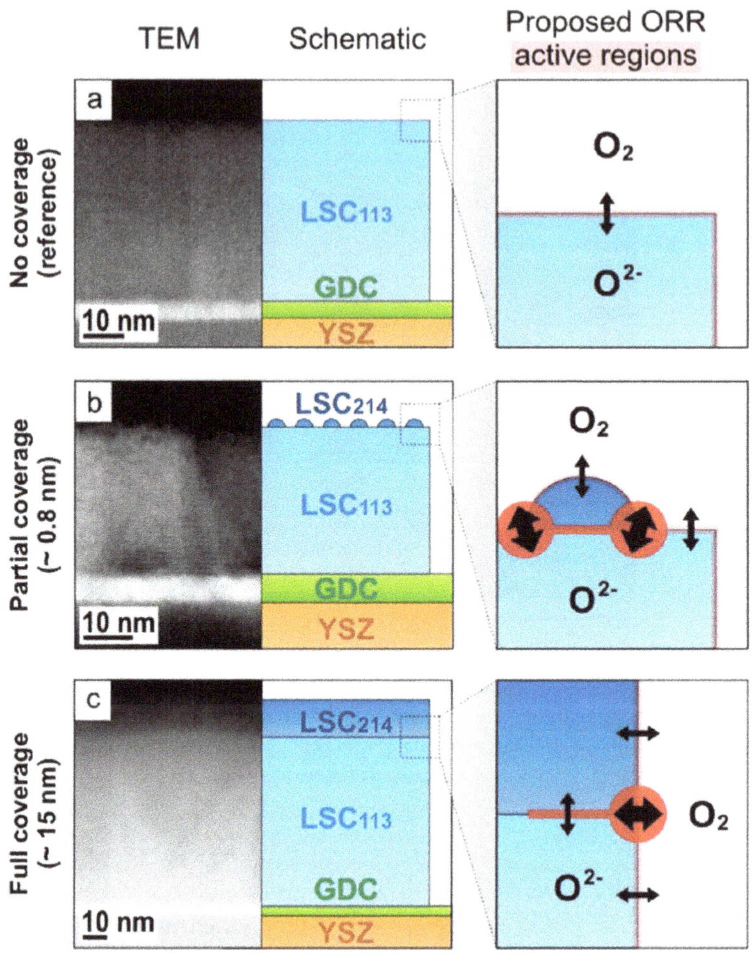

Figure 3.3. Comparison of ORR active regions for the LSC_{113} reference model electrode: (a) partially covered by LSC_{214} (b) fully covered by LSC_{214} (c). The proposed ORR active regions are indicated in red. Reprinted with permission from [20]. Copyright (2010) American Chemical Society.

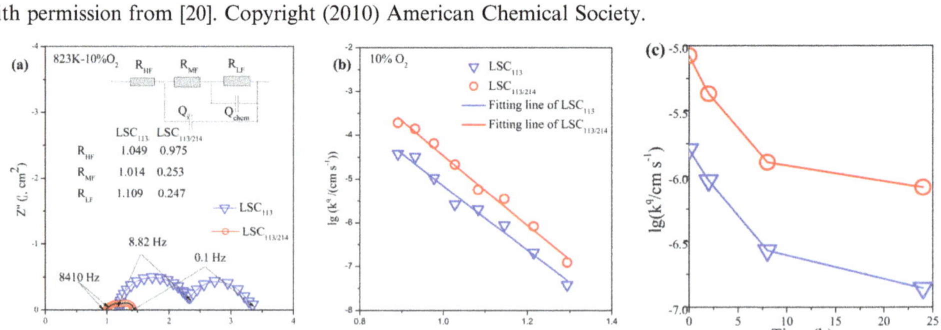

Figure 3.4. (a) Impedance spectra of LSC_{113} and $LSC_{113/214}$ obtained at 923 K and under 10% O_2. Fitting is performed using equivalent circuit analysis, from which the surface exchange coefficient (k^q) is extracted. (b) Temperature dependence of k^q for the two films. (c) Time-dependent degradation of k^q for the two films. Reprinted from [21], Copyright (2018), with permission from Elsevier.

impedance spectra, it was found that the values for $LSC_{113/214}$ were three to ten times higher than those of LSC_{113}. On the other hand, both films exhibited a drastic decrease in the surface exchange coefficient after annealing for 24 h (figure 3.4(c)).

From the evaluation of the films using time-of-flight SIMS (ToF-SIMS) and x-ray photoelectron spectroscopy (XPS) analysis, an apparent Sr enrichment was further observed at the $LSC_{113/214}$ heterointerface, and this phenomenon was correlated with a lower amount of Sr segregation on the surfaces (i.e. less surface-bound Sr), which could explain the improvement in the surface exchange property. In other words, the existence of an active interface for Sr diffusion may serve to 'offset' the usual tendency for Sr surface segregation, which can be detrimental to the ORR/OER processes.

In a related study, Tsuji *et al* studied the surface oxidation state of transition metals in LSC during cathodic polarization using operando total-reflection fluorescence x-ray absorption spectroscopy (TRF-XAS) [22]. A schematic of the

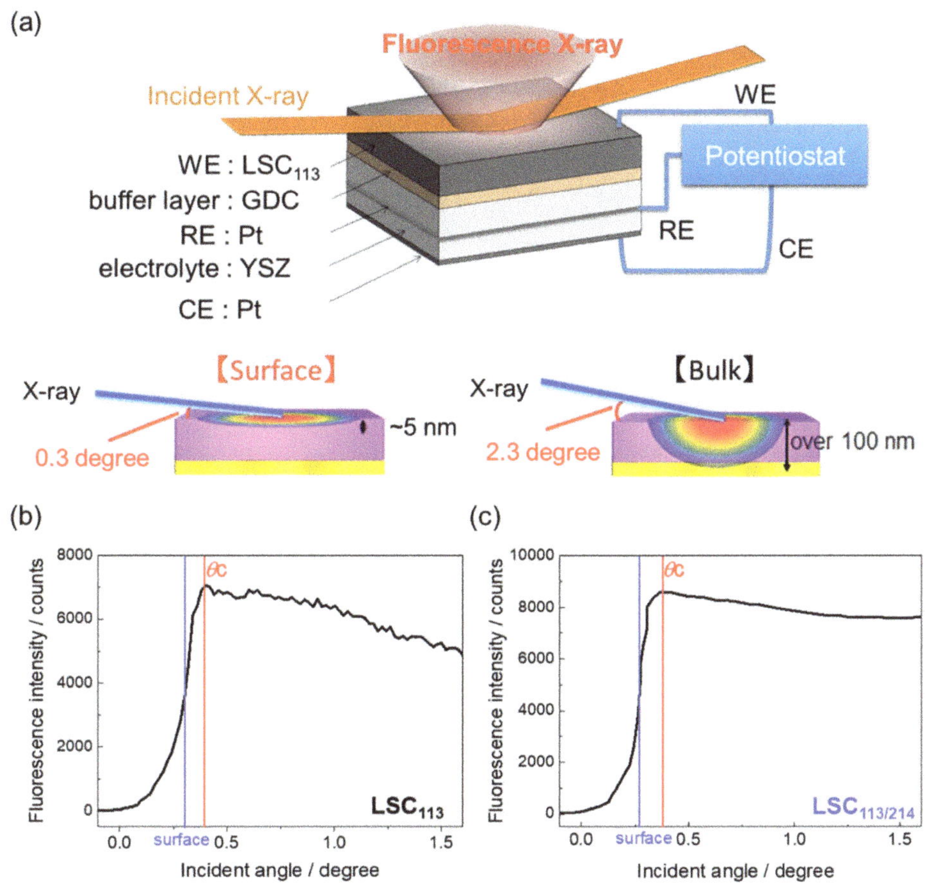

Figure 3.5. (a) Schematic of the operando TRF-XAS setup. By varying the incident angle and measuring the fluorescence intensity, it is possible to probe the nature of the interfaces of the LSC_{113} (a) and $LSC_{113/214}$ (b) samples. Reprinted from [22], Copyright (2020), with permission from Elsevier.

TRF-XAS setup is shown in figure 3.5(a). The samples were LSC thin-film electrodes prepared on polycrystalline YSZ substrates with a GDC interlayer prepared using PLD. LSC_{214} was deposited at various thicknesses (~3 to ~10 nm) on LSC_{113}. XAS and x-ray fluorescence (XRF) measurements were performed at the open-circuit voltage (OCV) and at constant polarizations of −100 mV and −900 mV at 773 K. Figures 3.5(b) and (c) show plots of the fluorescence detector counts as a function of the incident angle for the LSC_{113} without and with LSC_{214}, respectively. By changing the angle of incidence, the penetration depth can be varied; thus, it is possible to differentiate the characteristic spectra measured at the surface (at a depth of ~5 nm), and the bulk of the film (at a depth of >100 nm). Examining the operando Co *K*-edge x-ray absorption near-edge structure (XANES) shown in figure 3.6 for LSC_{113} and $LSC_{113/214}$ at 773 K under a $p(O_2)$ of 10^5 Pa and at various applied potentials, it can be observed that when the cathodic polarization increased, the absorption edge at 7720 eV also shifted towards lower energies. Given that the absorption edge of Co *K*-edge can be correlated with the oxidation state of cobalt as determined by the oxygen nonstoichiometry [23], the shift thus indicates the occurrence of cathodic-polarization-induced oxygen vacancies in LSC.

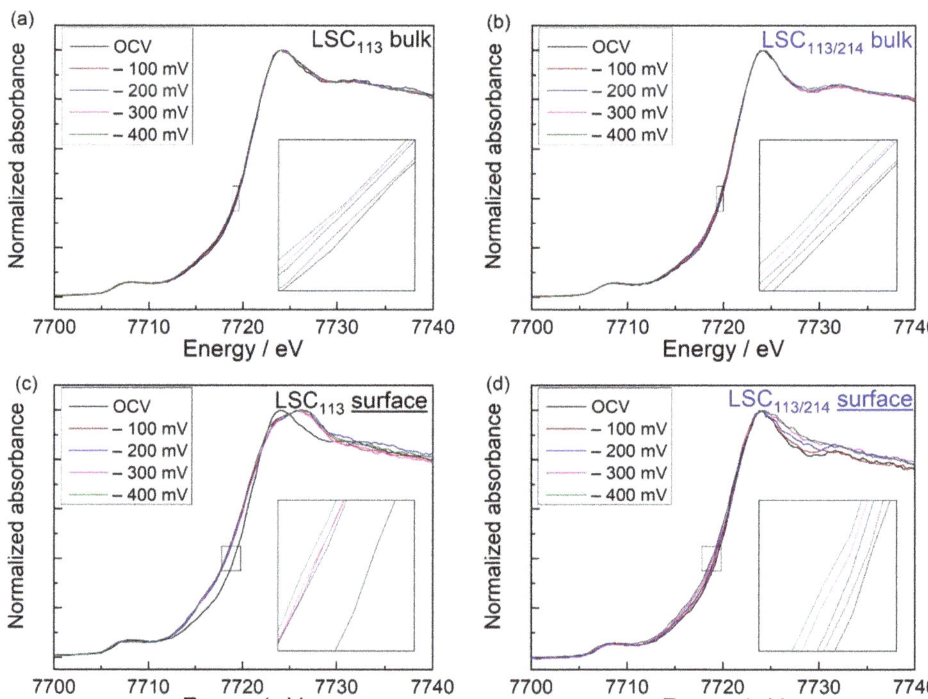

Figure 3.6. Co *K*-edge XANES spectra for LSC_{113} ((a) and (c)) and $LSC_{113/214}$ ((b) and (d)) measured under cathodic polarization at 773 K and a $p(O_2)$ of 10^5 Pa. The spectra were obtained from surface and bulk regions for both samples. A drastic shift toward lower energy values is observed for the surface of LSC_{113}, unlike the gradual transition observed for $LSC_{113/214}$. Reprinted from [22], Copyright (2020), with permission from Elsevier.

A comparison between the bulk spectra of LSC_{113} and $LSC_{113/214}$ shows a similar low-energy shift with cathodic polarization for both samples (figures 3.6(a) and (b)); however, an interesting difference can be observed for the spectra obtained at the surfaces. As shown in figures 3.6(c) and (d), a drastic shift can be seen for the LSC_{113} upon the application of cathodic polarization, whereas that of $LSC_{113/214}$ only shows a gradual transition. Although there was no direct observation of the Sr enrichment on the LSC surfaces, this behavior was attributed to Sr enrichment on the surface of the LSC_{113}, which appears to be significantly suppressed in the presence of the LSC_{214} decoration.

The presence of oxygen vacancies on ABO_3-type surfaces has been considered to be a driving force for Sr segregation behavior for charge compensation [6, 24]; therefore, the induced oxygen vacancies could be considered to be indicative of Sr diffusion to the surface. The improvement in ORR activity with LSC_{214} decoration is consequently attributed to the suppression of this Sr enrichment. A possible reason for the effect of LSC_{214} was attributed by the authors to the low amount of oxygen nonstoichiometry for LSC_{214}, which could hinder the reconfiguration of the LSC_{113} in the vicinity of this phase.

To directly visualize the origin of the enhancement of the performance associated with the heterointerfaces, Zheng *et al* studied a structurally analogous heterostructure of $Nd_{0.8}Sr_{1.2}CoO_{4\pm\delta}/Nd_{0.5}Sr_{0.5}CoO_{3-\delta}$, ($NSC_{214/113}$) [25]. PLD was used to prepare (a) single-phase NSC_{113} layers ~90 nm thick, (b) NSC_{214} layers ~90 nm thick, and (c) $NSC_{214/113}$ multilayers with alternately deposited NSC_{113} and NSC_{214} layers at ~5 nm each, with a total thickness of ~90 nm. These were deposited on (001) YSZ single-crystal substrates. As in the case of the results observed for $LSC_{113/214}$ heterointerfaces, based on electrochemical tests performed over a wide temperature range, an improvement in surface exchange reactions was likewise observed for $NSC_{214/113}$. Specifically, the surface exchange coefficient k_q was about 15–20 times higher for the $NSC_{214/113}$ sample than for either phase alone. In addition, by performing spatially resolved electron energy loss spectroscopy (EELS) line scans across the interface, they studied the characteristic spectra of the O *K*-edge and the Co *L*-edge. The chemical shifts of the Co L_3-edge were then used to qualitatively describe the changes in the Co oxidation state.

As shown in figure 3.7(c), there appears to be a shift toward a lower energy loss value for the Co L_3-edge for the interface (782.25 eV) as compared to those of NSC_{214} (782.75 eV) or NSC_{113} (782.5 eV). This chemical shift is attributed to the change of the Co oxidation state at the interface and is further corroborated by a relatively higher L_3/L_2 intensity ratio (figure 3.7(d)). There are two possible implications: first, the energy barrier for electron transfer is lower, meaning that the energy gap between the valence band and the conduction band is narrower at the heterointerface [26], more electrons are generated for the ORR/OER process, and a decrease in the Co oxidation state can be facilitated through the acceptance of electrons [27–29]; second, the decreased Co oxidation state implies the formation of a higher concentration of oxygen vacancies for charge compensation [24]. For the latter, further evidence is provided by the absence of the O *K*-edge pre-peak shown in figure 3.7(e). The flattening of the pre-peak frequently observed in cobaltites has been associated with

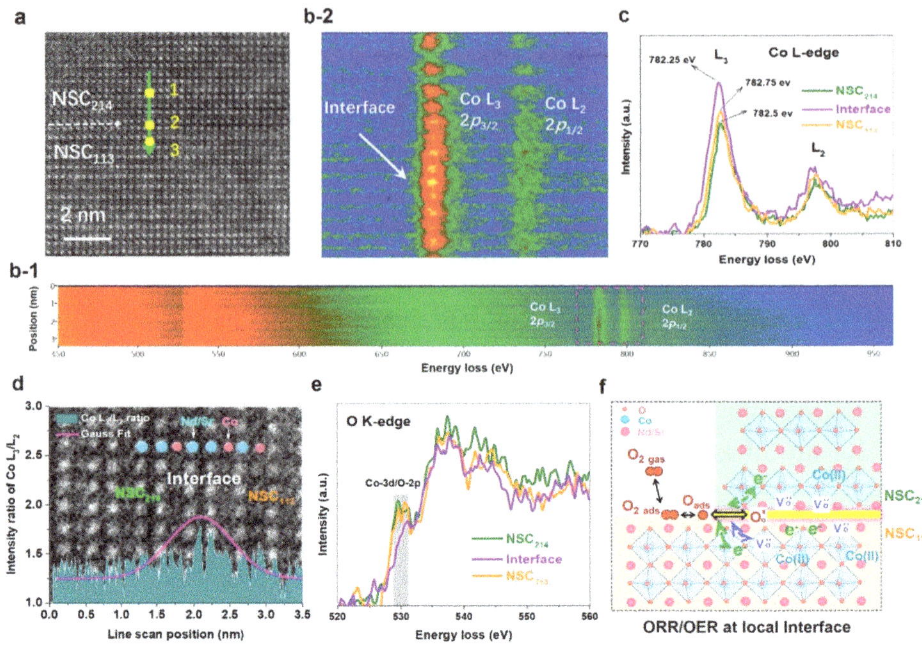

Figure 3.7. The high-angle annular dark field scanning transmission electron microscopy (HAADF-STEM) image in (a) shows the $NSC_{214/113}$ heterointerface, where the region indicated by the green arrow is finely probed using the EELS method (b-1 and b-2). Three specific locations are selected: points 1, 2, and 3 corresponding to NSC_{214}, $NSC_{214/113}$, and NSC_{113} regions, respectively. (c) The representative spectra of the Co L-edge, showing a variation in the intensities of the L_3 and L_2 peaks, further confirmed as a peak in the calculated intensity ratio of L_3/L_2 shown in (d). This is interpreted as a change in the valence of Co at the interface which is correlated by the disappearance of the O k-edge pre-peak in (e). The ORR-OER processes occurring at the $NSC_{214/113}$ heterointerfaces are schematically illustrated in (f). Reprinted from [25], Copyright (2020), with permission from Elsevier.

the hybridization of the Co 3d orbitals with O 2p from the transferred electrons induced by the oxygen vacancies [6, 30–32]. Based on the mechanism presented in figure 3.7(f), due to the presence of oxygen vacancies, the energy barrier for oxygen migration is decreased, whereas oxide ion transport is promoted, which could account for the observed enhancement in the ORR/OER kinetics at the heterointerface.

3.1.2 Multilayered perovskite–fluorite oxides

Another type of multilayer that has been widely adopted comprises a cobaltite perovskite oxide such as LSC, LSCF, $Sm_{0.5}Sr_{0.5}CoO_{3-\delta}$ (SSC), and similar materials combined with a good ionic conducting oxide such as GDC and YSZ; such a multilayer is analogous to the structures of composite electrodes [8, 33–35]. Of particular interest are the mechanisms by which such an interface could promote ORR kinetics. Among these, the study by Choi *et al* investigated the possibility of engineering the distribution of charged defects, particularly the oxygen vacancies on perovskite oxide surfaces, with the goal of suppressing Sr segregation [36]. The samples were model electrodes of SSC films prepared using PLD on GDC-buffered (100) YSZ single-crystal substrates. GDC layers with thickness ranging from ~2 to 6.8 nm were then additionally deposited on the SSC layers. Their hypothesis was that as a result of adding GDC on the SSC surfaces, a redistribution of oxygen vacancies could occur at the heterointerfaces, resulting in reduced oxygen vacancies at the SSC surfaces. In this way, the Sr segregation induced by electrostatic attraction between charged defects could be suppressed, thereby leading to better electrode performance. Using this strategy, the authors reported an approximately elevenfold reduction in the degradation rate of ~8% at 650 °C and an approximately tenfold higher surface exchange k_q for the GDC/SSC film compared to that of a bare SSC film after 150 h.

Figure 3.8 shows the representative O 1*s* photoelectron spectra at three locations of the GDC/SSC film, as indicated in the cross-sectional transmission electron microscopy (TEM) image (figure 3.8(a)). The fitting of the photoelectron spectra was conducted with three binding states: the oxygen hydroxyl group (OH^-, 532.7 eV), oxygen-related defects ($V_O^{\cdot\cdot}$, 531.1 eV), and oxygen in the lattice ($O_O^{\times}$, 529.1 eV). Here, the intensity ratio given by $V_O^{\cdot\cdot}/O_O^{\times}$ is taken as a quantitative parameter for the distribution of positively charged defects or oxygen vacancies in the thickness direction (figure 3.8(c)) [37]. Interestingly, the SSC film exhibits a significant increase in oxygen vacancies toward the surface, unlike the GDC/SSC film which exhibits a flat profile in both the SSC and GDC regions and a gradient across the GDC/SSC interface. The authors interpreted this as strong evidence for the redistribution of charged defects between two films that have different concentrations, resulting in equilibrium. Accompanying this redistribution of positively charged defects (i.e. oxygen vacancies), changes in the valence states of Co and Ce were also evidenced by the EELS mapping. In particular, analysis of the EELS spectra revealed that the concentration of Co^{3+} tends to be higher than that of Co^{2+}at the interface, indicating the oxidation of Co, whereas a higher concentration of Ce^{3+} compared to Ce^{4+} is evidenced at the interface, indicating the reduction of Ce. The oxidation of Co and the reduction of Ce are consistent with the observed oxygen

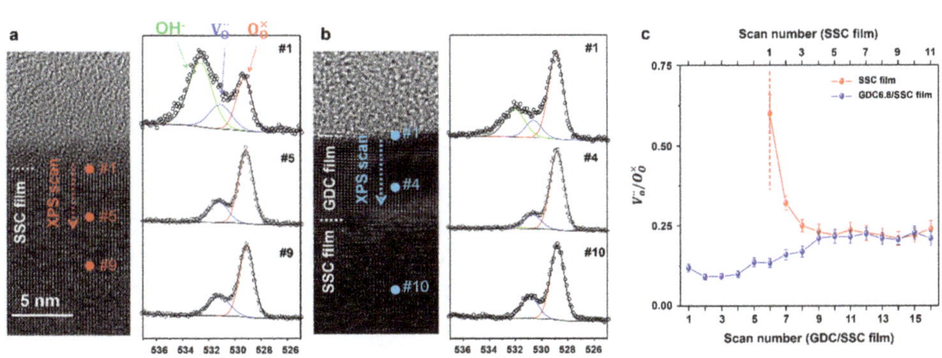

Figure 3.8. Comparison of XPS spectra for bare SSC film (a) and GDC/SSC film (b) obtained in the thickness direction indicated by the arrows. A representative fitting of O 1s spectra is shown for the locations indicated by the numbers on the corresponding cross-sectional TEM images. The calculated oxygen vacancies/oxide in the lattices of the two films are shown as a function of depth in (c). Reprinted with permission from [36] Copyright (2020) American Chemical Society.

vacancy concentration indicated by the XPS results. The difference in the Sr segregation behaviors of the two types of film (*viz.* SSC and GDC/SDC) is thus attributed to such changes in the oxygen vacancy concentration on their surfaces. Furthermore, the results of density functional theory (DFT) analysis strongly supported the experimental results in terms of the charge transfer from the GDC film to the SDC film, a downshift of the O *p*-band center, and an increase in the oxygen vacancy formation energy in the presence of the GDC film.

In another study, Develos-Bagarinao *et al* adopted the multilayer approach by alternately growing LSC and GDC thin-film layers using PLD to form a multitiered nanostructured electrode [38]. Multilayers offer several advantages compared to conventional nanocomposites, which are essentially multilayers distributed within the film structures. First, the cation interdiffusion which normally occurs at elevated temperatures is limited to the vicinity of the planar interface between the two materials; second, whereas grain coarsening inevitably occurs at elevated temperatures, the original interfaces are mostly maintained; third, the multilayer scheme provides a more flexible route for processing by simply adding more layers, and can therefore be adapted for different thin-film techniques; and finally, by obtaining well-defined heterointerfaces, it is possible to obtain accurate information at the nanoscale using spatially resolved characterization techniques such as SIMS and TEM.

Using the PLD technique, Develos-Bagarinao *et al* prepared LSC–GDC bilayers, i.e., two layers of LSC and GDC alternately stacked on (100) YSZ single-crystal substrates as shown in figure 3.9(a). The layers were grown by simply switching the ceramic targets of LSC and GDC and sequentially carrying out the film deposition using identical deposition parameters. The typical nanostructure comprises nano-columnar grains which are oriented normal to the substrate surface. The accompanying scanning transmission electron microscopy-energy dispersive x-ray spectroscopy (STEM-EDX) mapping results of Sr, Zr, and Ce elements show a clear demarcation between the LSC and GDC layers, and no secondary phases are evident (figure 3.9(b)). In addition, the selected area electron diffraction (SAED) shown in figure 3.9(c) shows that the bottom GDC layer (GDC1) is highly oriented due to its epitaxial growth on the YSZ single-crystal substrate, but the top GDC layer (GDC2) has a polycrystalline orientation. The LSC layers, on the other hand, do not show any particular dominant orientation and are comprised of nano-polycrystalline grains. The high-resolution TEM image shown in figure 3.9(d) shows that the topmost LSC layer (LSC2) has a semiamorphous structure.

To determine the surface exchange properties of the multilayer thin-film cathodes, the authors employed the isotope exchange depth profiling technique. Using ^{18}O as the isotopic tracer, SIMS depth profiles were obtained for LSC single layers and LSC–GDC multilayers prepared at different deposition temperatures of 750 °C and 650 °C (denoted as LSC–GDC Multilayers I and II, respectively). From the fitting of the ^{18}O depth profiles using Crank's solution to the diffusion equation [43], values for the apparent surface exchange coefficient, k^*, were obtained and plotted as shown in figure 3.10.

The multilayers exhibit k^* values about two orders of magnitude higher than the values reported for LSC bulk and about one order of magnitude higher than that of

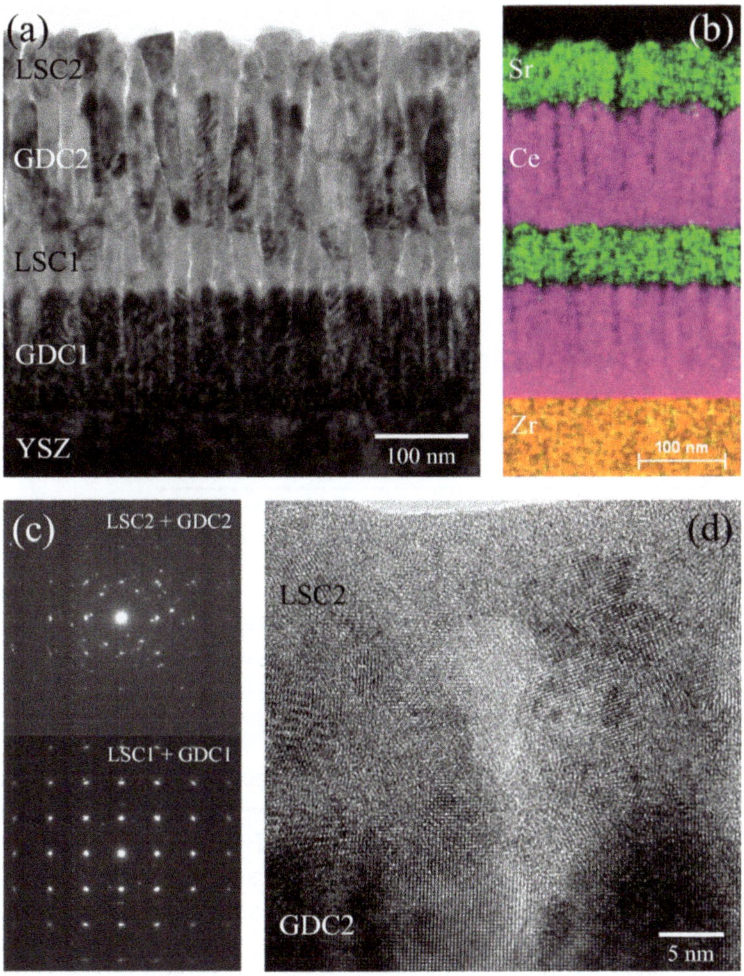

Figure 3.9. Representative cross-sectional TEM image of an LSC/GDC bilayer prepared using PLD on a YSZ single-crystal substrate (a) and its corresponding STEM-EDS elemental mapping showing the Sr (green), Ce (pink), and Zr (orange) distribution in the thin-film structure (b). Selected area electron diffraction (SAED) patterns obtained from the top and bottom LSC/GDC layers (c) indicate differences in the characteristic crystalline structures; the top LSC exhibits a semiamorphous structure (d). Reprinted from [38], Copyright (2018), with permission from Elsevier.

the LSC single layer. This result indicated that the multilayers perform better as air electrodes due to the enhancement of the ORR process, a behavior which was likely facilitated by the presence of additional heterointerfaces.

To further understand the mechanism for the enhanced oxygen surface exchange properties of the multilayers, the electronic properties of the LSC–GDC heterointerfaces were further probed using spatially resolved STEM-EELS analysis. In the typical EELS spectra shown in figure 3.11(a), peaks attributed to the Co L_3/L_2 edges, La M_5/M_4 edges, and Ce M_5/M_4 edges can be observed. The intensities of the M_5 and M_4 peaks

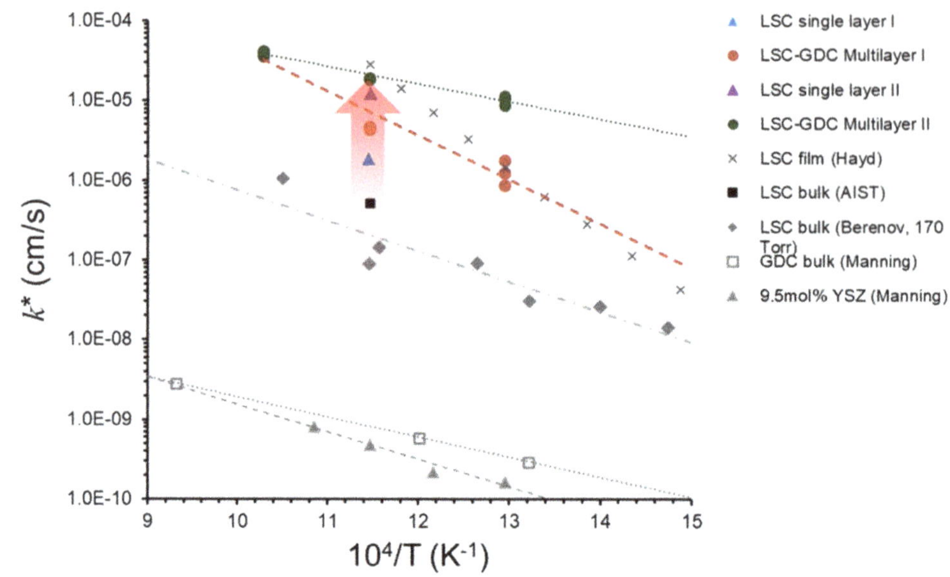

Figure 3.10. Arrhenius-type plot of the apparent surface exchange coefficient k^* of the LSC/GDC multilayers (LSC–GDC Multilayers I and II) in comparison with LSC single layers, published data for LSC film [39], LSC bulk [40], GDC bulk [41], and YSZ single crystal [42]. Reprinted from [38], Copyright (2018), with permission from Elsevier.

provide information about the valence state of Ce; specifically, by deriving the ratio of the M_5/M_4 edges, it is possible to obtain quantitative information about the relative proportion of Ce^{3+} and Ce^{4+} in the sample [44]. EELS spectra were consequently obtained at various locations along the heterointerfaces shown in figures 3.11(b)–(e), along with the computed intensity M_5/M_4 ratios for each heterointerface. It can be seen that the M_5/M_4 ratio exhibits a clear increase towards the LSC/GDC heterointerfaces. This increase indicates that there is a proportionally higher amount of Ce^{3+} present near the LSC/GDC heterointerface compared to that within the GDC layer itself, which mostly contains Ce^{4+}. In the GDC lattice, the reduction of Ce^{4+} to Ce^{3+} is accompanied by the formation of oxygen vacancies due to charge compensation. From this result, it is inferred that there must be a higher concentration of oxygen vacancies at the LSC/GDC heterointerfaces which likely facilitate oxygen incorporation through the multilayers.

Based on the preceding results, a follow-up study was further conducted for LSC–GDC multilayers by systematically increasing the number of LSC and GDC heterointerfaces [45]. To evaluate the electrode performance, symmetrical cells were prepared by depositing LSC/GDC multilayers on both sides of double-polished (100) YSZ single-crystal substrates, and electrochemical impedance measurements were performed at 600 °C in air. The impedance spectra of single-layer LSC, a bilayer cell (two layers of LSC and GDC), a trilayer cell (three layers of each), and a quadrilayer cell (four layers of each) were obtained and analyzed using the distribution of relaxation times (DRT) method [46, 47], as shown in figure 3.12(a). Based on the oxygen partial pressure (pO_2) dependence obtained

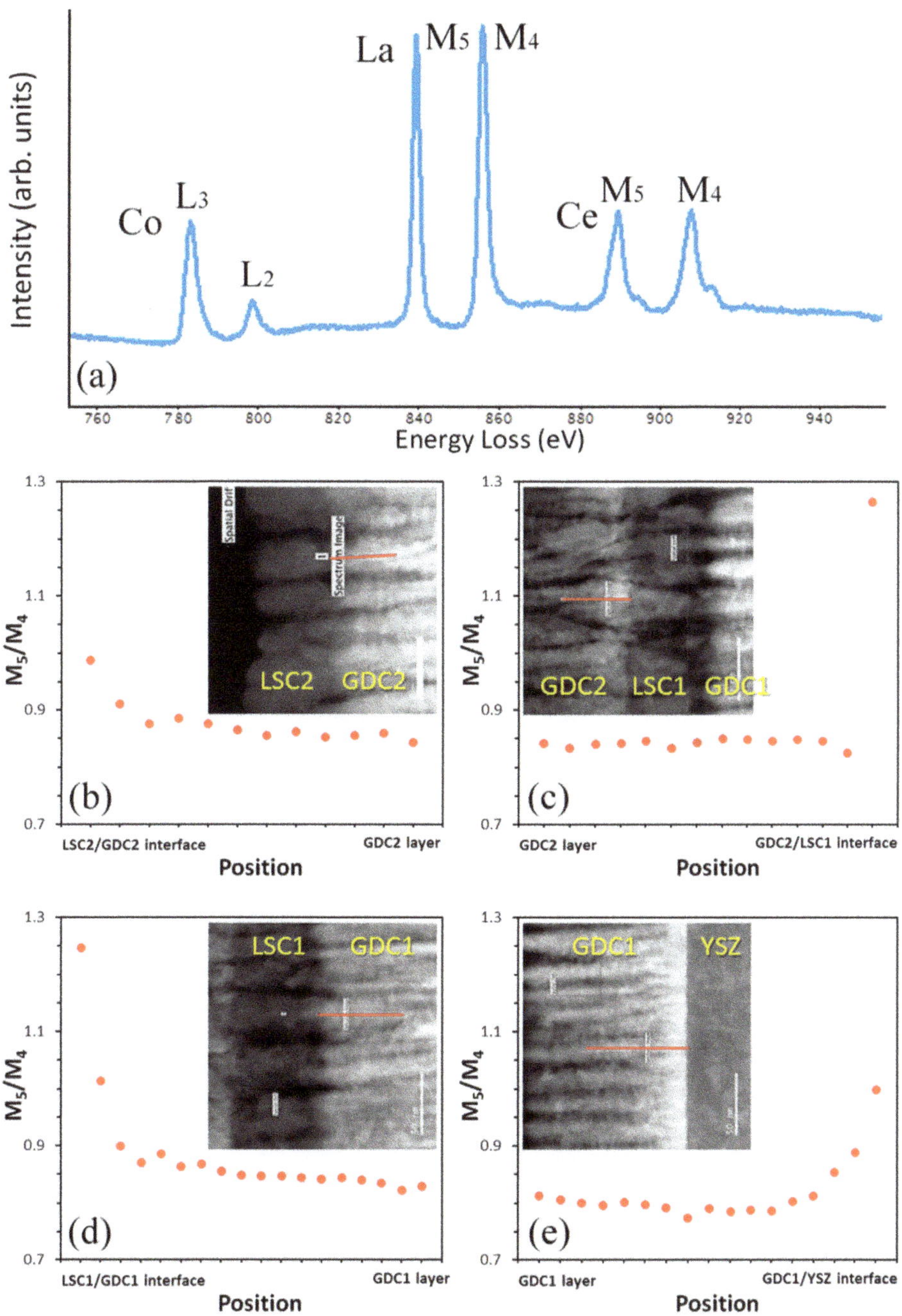

Figure 3.11. Representative EELS spectra for the LSC/GDC bilayer depicting the Co L_3/L_2, La M_5/M_4, and Ce M_5/M_4 edges (a). The Ce M_5/M_4 ratios as a function of position obtained at various locations ((b)–(e)) exhibit higher values in the vicinity of the LSC/GDC interfaces compared to the GDC regions. Reprinted from [38], Copyright (2018), with permission from Elsevier.

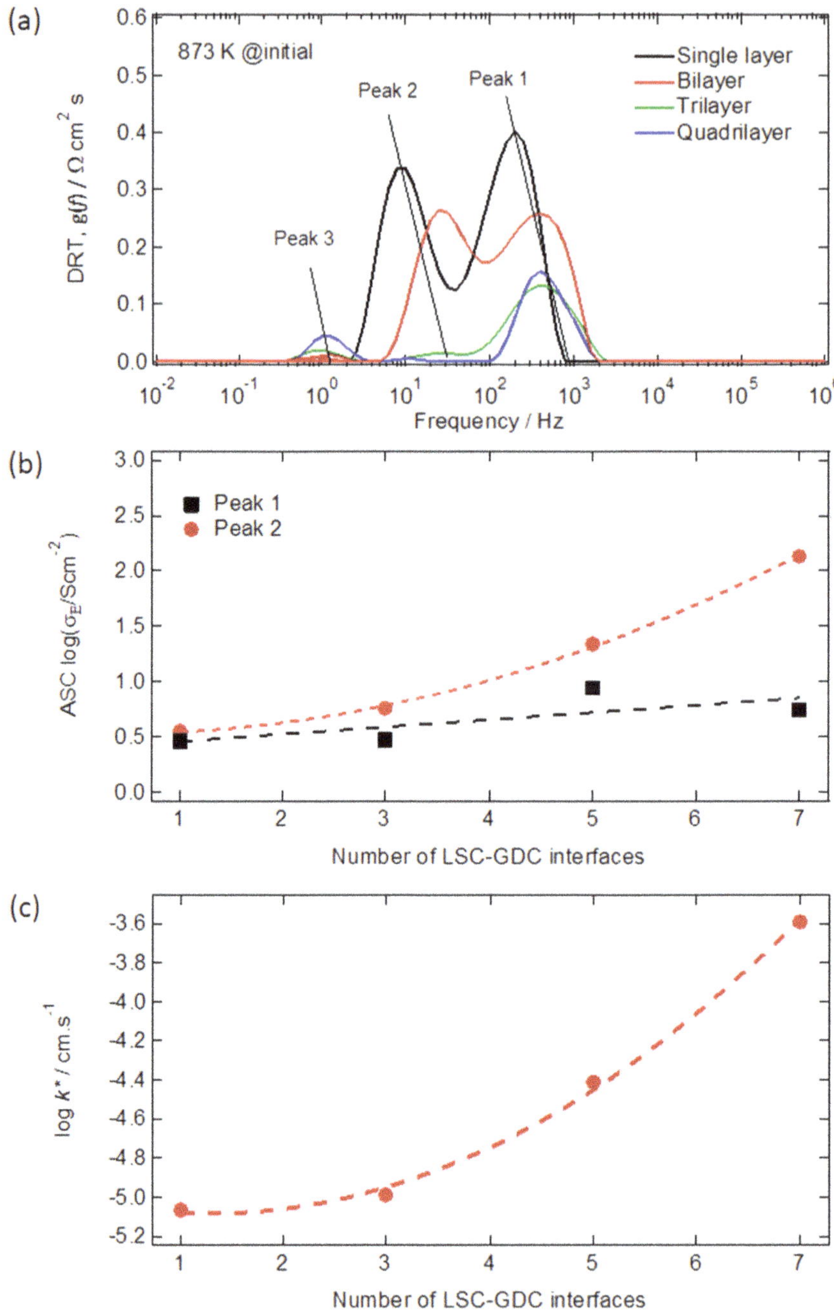

Figure 3.12. Evaluation of the electrochemical performance of LSC–GDC multilayers at 873 K under 0.21 bar of $p(O_2)$. Deconvolution of the impedance spectra using DRT showed three characteristic peaks (Peak 1, Peak 2, and Peak 3) occurring at various frequency ranges (a). The calculated area specific conductivity σ_E values associated with Peak 2 (b) and the surface exchange coefficient k^* (c) showed a monotonic increase with an increasing number of LSC–GDC interfaces. Reprinted from [45], Copyright (2020), with permission from Elsevier.

for the multilayers, the DRT peaks occurring at various frequency range and denoted by Peak 1, Peak 2, and Peak 3 could be attributed to different physical origins, as follows: Peak 1 is related to the electrode/electrolyte reaction, Peak 2 is related to the surface reaction at the multilayer electrode, and Peak 3 is related to the gas conversion. Among these, Peak 2 shows a significant change with an increase in the number of heterointerfaces, as evident from the area specific conductivity (σ_E) and the surface exchange coefficient, k^*, shown in figures 3.12(b) and (c), respectively. Therefore, the existence of additional heterointerfaces is considered to have catalytic effects on the oxygen surface exchange properties of multilayers.

In addition, durability tests were performed at 600 °C for up to 600 h. The polarization resistances of single-layer LSC and quadrilayer LSC–GDC are plotted as a function of time in figure 3.13. A comparison between these two samples indicates that the degradation of oxygen surface exchange properties is substantially suppressed for the LSC–GDC quadrilayer compared to that of the single-layer LSC. This behavior is likely attributed to the suppression of Sr segregation from the LSC perovskite due to the improved stability of its surfaces, as well as the restriction of LSC grain coarsening due to the presence of the GDC layers.

Similar heterostructures comprising alternating layers of LSC and GDC prepared on single-crystal $NdGaO_3$ (NGO) substrates were further studied by Dos Santos-Gómez *et al* [48]. Here, NGO was chosen as the substrate due to its lower electronic conductivity as well as good lattice matching with LSC and GDC. The thicknesses and number of heterointerfaces for several multilayer samples prepared using PLD

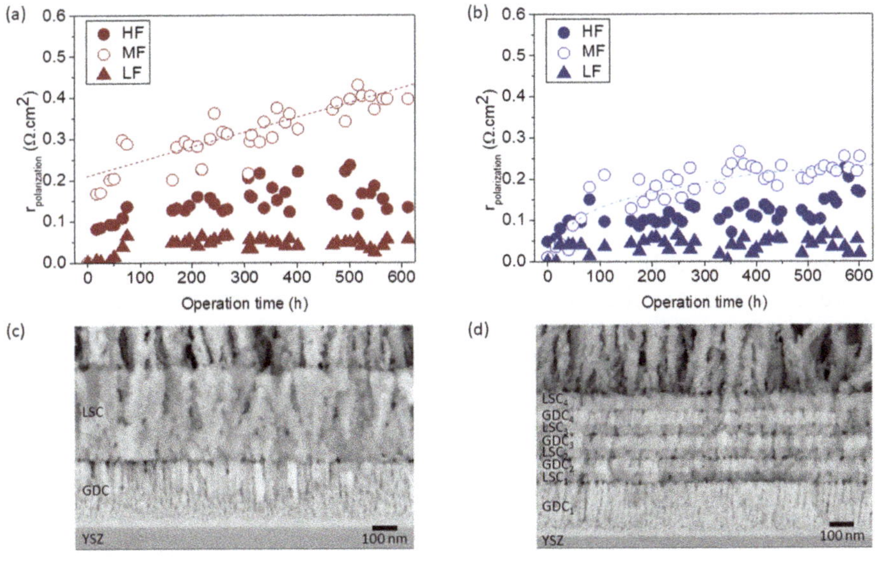

Figure 3.13. Durability evaluation for (a) single-layer LSC and (b) quadrilayer LSC–GDC at 873 K under 0.21 bar of oxygen pressure. The polarization resistance associated with the middle frequency (MF, Peak 2) in the impedance spectra showed a steady increase for the single-layer LSC, in contrast to the plateau behavior exhibited by quadrilayer LSC–GDC at $t > 100$ h. Panels (c) and (d) depict representative cross-sectional SEM images of the two samples after testing. Reprinted from [45], Copyright (2020), with permission from Elsevier.

were systematically varied and examined in order to understand the chemical stability and electrochemical properties of the LSC–GDC heterointerfaces. Two specific kinds of architecture were investigated: LSC–GDC multilayers with LSC termination, and those with GDC termination. In-plane conductivity measurements using the four-probe Van der Pauw method revealed an enhancement in conductivity for low thickness (~50 nm) as compared to higher thickness (~300 nm), even when the number of interfaces was the same. A possible explanation is that the interfacial conductivity becomes dominant when the layers are extremely thin, an effect that diminishes due to the stronger bulk contribution when the thickness of the layer is increased. Furthermore, it was also observed that the conductivity was directly related to the number of layers; e.g. the conductivity ranged from 100 S cm^{-1} for two layers of LSC and GDC (bilayers), compared to 240 S cm^{-1} for a sample that had five layers of each. This was interpreted as an effect of the increase in the number of interfaces, which can be correlated with the concentration and mobility of charge carriers.

By measuring the EIS properties at low temperatures, the different contributions to the electrical properties of the various interfaces were then estimated. Nyquist and Bode plots were obtained at 200 °C for the two types of multilayer, i.e. LSC- and GDC-terminated, each of which had a total thickness of 100 nm, and a solely GDC-terminated sample 50 nm thick. A comparison of samples with and without the GDC termination reveals a significant difference in the electrical behavior; specifically, a large impedance at high frequencies was observed for the latter. This behavior, in turn, is correlated with a possible degradation of the film without the GDC termination. On the other hand, the GDC-terminated samples exhibit two different impedance contributions, which become more prominent at higher temperatures (up to 500 °C). Moreover, the typical relaxation times and conductivity were noted to be quite different from those reported for GDC thin films, which suggests that the observed features can be accounted for by the LSC-related effects in the multilayer. The results further confirm the absence of capacitive effects attributed to any charge-blocking effects at the LSC/GDC interface, which suggests that the topmost GDC layer may be electronic in nature. The overall dominant electronic conductivity exhibited by the multilayers strongly indicates that the GDC components and interfaces are themselves dominated by electronic conduction.

From the XPS analysis of the samples, it was confirmed that LSC-terminated samples exhibit Sr-enriched species including $SrCO_3$, SrO, and $Sr(OH)_2$ on their surfaces, whereas the inner layers remain chemically stable. In this sense, the addition of the GDC layer is helpful in improving the stability of the LSC layer by suppressing Sr segregation, which is consistent with the results reported for LSC–GDC quadrilayers by Develos-Bagarinao *et al* [45]. Therefore, combining LSC and GDC layers into the electrode microstructure offers advantages in terms of improving chemical stability, at least in the case of thin-film electrodes.

The multilayer approach was further adopted by Solovyev *et al* in a study of LSC/GDC/LSC multilayers prepared using reactive magnetron sputtering [49], thereby providing further evidence of its adaptability to other deposition methods. The multilayers were deposited on GDC pellets in a symmetrical cell configuration, as well as in commercial anode-supported cells of NiO–8YSZ on which 8YSZ and GDC bilayer electrolytes were deposited beforehand using reactive magnetron sputtering. The thicknesses of the LSC and GDC layers were varied and included 25, 50, 100, and 200 nm. Among these, the cell utilizing a multilayer with LSC and GDC layers 50 nm thick showed the best performance in terms of power density (2430 mW cm^{-2} at 800 °C) as well as the lowest ASR value (0.091 Ω cm^2). On comparing its performance to that of a reference cell without the multilayers, it was found that the power density improved by about two times, indicating the likely advantage of incorporating multilayers into the electrode structure.

3.2 Tuning electrode–interlayer–electrolyte heterointerfaces

The cell performance can be severely limited by the efficiency and quality of the electrode–electrolyte interface, where the important process of charge transfer occurs. Typical powder processed electrodes, however, have grain sizes in the order of micrometers and also require a sintering step that must be performed at relatively high temperatures to improve adhesion with the underlying electrolyte material. For the latter, whereas sintering at high temperatures could lead to better adherence, the issue of grain coarsening frequently leads to low porosities and a decrease in the TPB density [50]. To increase the TPB density and improve adherence between the electrode and the electrolyte materials, interfacial modification has been adopted by many groups. For practical cells, rare-earth-doped ceria layers such as GDC are frequently employed to function as barrier layers, specifically to mitigate the cation interdiffusion between common types of perovskite electrodes and the state-of-the-art YSZ electrolyte. In this section, we will mainly focus on the development of perovskite-type interlayers which are used to modify the interfaces between air electrodes and electrolytes.

The concept of utilizing a dense interlayer between the electrode and electrolyte interface using the PLD technique was previously introduced by Hildenbrand *et al* [51]. Their samples consisted of symmetrical cells of 3YSZ electrolyte, on which yttrium-doped ceria (YDC) and dense LSCF thin films were prepared using PLD. As a result of introducing the LSCF thin film, the polarization resistance of the cathode was drastically reduced by a factor of three as compared to that without the LSCF dense layer. A similar result was reported for the use of dense $La_2NiO_{4+\delta}$ (LNO), which reduced the ASR by ~35%; this was attributed to an enhancement in the oxygen transfer mechanism. On the other hand, the electronic conductivity of LNO was too low, which was considered to result in the addition of further resistance to the apparent ohmic resistance.

Chrzan *et al* adopted this method for spin-coated interlayers of various perovskite oxide layers including $LaNi_{0.6}Fe_{0.4}O_{3-\delta}$ (LNF), LSCF, and $SrTi_{0.65}Fe_{0.35}O_{3-\delta}$ (STF) [52]. The primary motivation for this work was to establish whether or not the use of a thin perovskite layer could also be improved using a predominantly electronic conducting air electrode. Moreover, the spin-coating method is a low-cost technique and therefore ideally suited to large-scale production. Symmetrical cells based on GDC electrolyte were prepared, in which spin-coated thin perovskite layers of LNF, LSCF, and STF were subsequently contacted by LNF paste as the main electrode layer. The ASR improvement was highest for the cell utilizing an LSCF interlayer (29%), followed by that using LNF (27%), whereas the use of the STF layer resulted in a significant increase in the polarization resistance of the cells. In the case of the STF layer, it was presumed that its poor performance was related to its reaction with the LNF electrode.

This concept was further explored by De Vero *et al*, who utilized dense LSCF layers in combination with dense GDC barrier layers prepared using PLD [53]. Despite using dense, ~2 μm thick GDC barrier layers on YSZ polycrystalline substrates, reference cells utilizing porous LSCF electrodes exhibited extensive formation of the $SrZrO_3$ phase at the LSCF/GDC interfaces after testing at 800 °C for 300 h (figure 3.14). As evidenced by a SEM wavelength dispersive x-ray

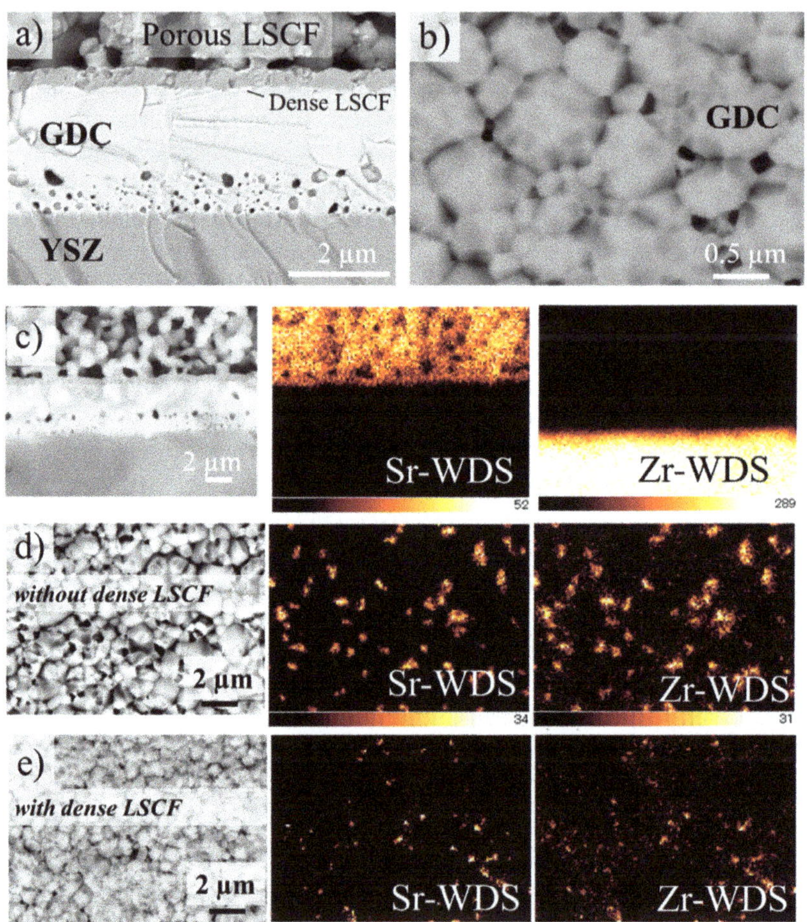

Figure 3.14. Representative cross-sectional SEM micrographs of the cell utilizing a dense LSCF layer inserted between the porous LSCF and the GDC interlayer (a) after testing under OCV conditions for 300 h in air at 800 °C. Subsequent removal of the LSCF layer reveals the mostly intact microstructure of the GDC surface (b). A comparison of the GDC surfaces of cells without dense LSCF (d) and with dense LSCF (e) shows a higher proportion of areas with residual Sr and Zr distribution for the former. Reprinted from [53], Copyright (2017), with permission from Elsevier.

spectroscopy (SEM-WDS) analysis of the cross section and the exposed surfaces of GDC barrier layers (obtained by removing the LSCF layer using diluted nitric acid), the introduction of ~300 nm thick dense LSCF layers prepared using PLD effectively mitigated the formation of the $SrZrO_3$ phase at the LSCF/GDC interfaces. A long-term electrochemical performance evaluation also showed that the polarization resistance did not significantly degrade as a result of the introduction of the dense LSCF layer. According to a SIMS depth profile analysis, the amount of Co which diffused through the GDC layer differed significantly between the cells with and without the LSCF dense layers. Based on this result, one of the possible reasons considered is the amount of cobalt dissolution in the GDC layer, which appeared to be significantly reduced by the use of the LSCF dense layer. The diffusional behavior of transition metals such as Co was reported to affect the Sr diffusion, as the mass transfer of transition-metal ions is accompanied by changes in their valence state, implying the evolution of oxygen [54, 55]. Inserting a dense LSCF layer effectively reduces the available oxygen species, thereby creating a more stable interface with respect to reactions involving oxygen evolution/incorporation accompanying changes in the valence state of transition-metal ions such as Co. By limiting the Sr transport across the GDC layer, the formation of deleterious phases such as $SrZrO_3$ is thereby effectively reduced, resulting in an improvement in long-term stability and an efficient ORR accommodated by the dense LSCF/GDC interface.

Spray pyrolysis was also explored as an effective method for stabilizing the electrode–electrolyte interfaces. In a recent study, Kamecki *et al* demonstrated the use of this method to produce ~500 nm thick layers of SSC, LSC, LSCF, and Pr_6O_{11} (PrO_x) on the electrode–electrolyte interfaces [56]. The annealing temperature of the interlayers was varied between 600 °C and 700 °C. The various spray-pyrolysis layers were then placed in contact with ~30 μm LSCF electrode layers prepared using a screen-printing method for symmetrical cells with a $Ce_{0.8}Gd_{0.2}O_{2-\delta}$ (GDC20) electrolyte and additionally annealed at 600 °C for 2 h.

Regardless of the annealing temperature, the ASR values obtained for the symmetrical cells utilizing interlayers showed an improvement compared to the reference LSCF electrode (no interlayer, sintered at 1050 °C for 2 h). Among the interlayers, PrO_x exhibited the highest ASR values; this was attributed to the relatively dense microstructure of this layer, which might have limited the adsorption or diffusion of gas species. The best performance was obtained for the SSC interlayer among samples annealed at 650 °C; however, LSC outperformed all other samples annealed at the lowest temperature of 600 °C. The behavior of SSC at 600 °C appeared inconsistent with those of the other samples, which showed a systematic decrease with annealing temperature. A plausible explanation for this inconsistency may be related to its apparent amorphicity at this temperature, as indicated by the XRD results, which additionally confirmed traces of $Sr(CO)_3$. Nevertheless, these results indicate that the use of interlayers generally leads to better adhesion of the electrode to the electrolyte, even for low sintering temperatures, which consequently avoids the issues of defects such as cracks or delamination as well as the interdiffusion of elements due to high-temperature sintering.

Based on the results obtained for symmetrical cells, given the lowest ASR value obtained at 600 °C, the utilization of an LSC interlayer was then investigated for anode-supported cells. Anode-supported cells with NiO–GDC20 supports, GDC20 interlayers, and LSC prepared by spray pyrolysis were tested; specifically, the effect of the sintering temperature used for the LSCF electrode was examined for cells sintered at 600 °C and 1050 °C for 2 h. Moreover, modifying the interface with LSC interlayer improved the performance in terms of lower polarization resistance as well as ohmic resistance. The maximum power output obtained was 1215 mW cm^{-2} at 600 °C, which was much higher than that obtained for the reference sample (740 mW cm^{-2}). Finally, durability tests conducted at 0.75 V and a current load of 1 A cm^{-2} indicated stable performance for up to ~60 h. These results therefore indicate that a simple modification of the electrode–electrolyte interfaces via interlayers prepared using spray pyrolysis significantly improves performance and possibly provides better stability than that of conventional electrodes.

3.3 Advanced methods for probing heterointerfaces

Experimental observations of increased ionic conductivities have been reported for multilayer structures such as YSZ [57] and YSZ/$SrTiO_3$ [58]. To examine the interfacial properties of these types of multilayered material in detail, advanced characterization techniques have been developed and explored by various groups to understand the ionic transport behavior within these heterostructures. In this section, we focus primarily on methods which are particularly used to probe the ionic transport behavior in multilayer heterostructures and explore the intriguing phenomena revealed for these systems.

3.3.1 Isotope exchange depth profiling with SIMS analysis

A combination of STEM-EDX/EELS and SIMS was employed by Perkins *et al* to investigate $CeO_2/Ce_{0.85}Sm_{0.15}O_{2-x}$ multilayer thin films grown on MgO (001) substrates by PLD [59]. The thicknesses of CeO_2 and $Ce_{0.85}Sm_{0.15}O_{2-x}$ were about 20 nm and 30 nm, respectively, which were repeated for 50 and 400 cycles of multilayer deposition. STEM-EDX mapping confirmed good quality, compositionally abrupt layers of CeO_2 and $Ce_{0.85}Sm_{0.15}O_{2-x}$ in the structure. STEM-EELS was used to probe the oxygen K edges, Ce $M_{4,5}$, and Sm $M_{4,5}$ edges, and compositional profiles were extracted from the region. The EELS spectra were analyzed using multiple linear least squares (MLLS) fitting to extract the Ce^{3+}/Ce^{4+} ratio. The valence profiles extracted from the spectra were correlated with the STEM-HAADF signal which clearly showed the compositional and structural contrast between adjacent layers. The undoped layers appeared darker in the HAADF image, which primarily contained Ce^{4+} only. The Sm-doped layers, on the other hand, contained both Ce^{3+} and Ce^{4+} at a ratio of approximately 30:70. Furthermore, MLLS fitting of the profiles acquired within each layer confirmed that only Ce^{4+} was present in the undoped layer and that the doped layer contained a mixture of both oxidation states. After annealing the samples at 800 °C in air, both doped and undoped layers exhibited a mixture of Ce^{3+} and Ce^{4+}, though the Ce^{4+} content remained higher within the undoped layer. The authors interpreted this behavior as an indication of some barrier to oxygen diffusion, in which oxygen diffuses from the doped layer into the undoped layer, resulting in the reduction of Ce in the latter.

To examine the oxide ion transport through the multilayer, the samples were isotopically exchanged in ^{18}O gas and were subsequently measured using SIMS line scan analysis. The ^{18}O isotopic fraction diffusion profiles for the doped and undoped CeO_2 along with the calculated fits were then extracted. Due to edge effects in the SIMS analysis, the first few points of the line scan profile were disregarded; thus, it was not possible to estimate the surface exchange coefficient. On the other hand, the diffusion coefficient D^* could still be extracted from the fitting, yielding 2.8×10^{-8} cm^2 s^{-1} and 2.6×10^{-8} cm^2 s^{-1} for the doped CeO_2 and the undoped CeO_2 layer, respectively. Though the proximity of the values obtained for the doped and undoped CeO_2 was unexpected, this behavior was interpreted as a possible effect of the manner in which the tracer oxygen diffused through the multilayer. This was correlated by the electrical conductivities calculated using values obtained by the ion exchange method and the Nernst–Einstein equation, for which similar values of 0.023 S cm^{-1} for CeO_2 and 0.024 S cm^{-1} for doped CeO_2 at 970 K were obtained. This indicated that the conduction method in the samples analyzed at ~973 K was dominated by ionic conductivity. In addition, the overall conductivity was similar to that of bulk samaria-doped ceria. An explanation for the presence of Ce^{3+}/Ce^{4+} states in the multilayers was attributed to a possible effect of the stabilization that occurs at the nanoscale, which differs remarkably from that of bulk oxides. It was concluded that the presence of low-angle grain boundaries at heterointerfaces may prevent the oxidation of the Ce^{3+} in the doped CeO_2 layer in addition to the 'immobility' of oxygen vacancies that hinders full oxidation in the multilayer structure.

One of the main issues for SOC applications is the long-term degradation associated with the formation of secondary phases at heterointerfaces. For commonly utilized electrodes such as LSCF or LSC in conjunction with a state-of-the-art YSZ electrolyte, the formation of a $SrZrO_3$ (SZ) phase due to cation interdiffusion at high operating temperatures and under polarization has been identified as a primary contributory factor to the degradation of cell performance [53, 55, 60]. To elucidate the effect of the SZ phase on the oxide ion transport, Develos-Bagarinao *et al* prepared model heterostructures of GDC/SZ/YSZ and GDC/yttrium-doped $SrZrO_3$ (SZY)/YSZ and investigated the multilayers using the isotope exchange depth profiling technique [61]. Typical cross-sectional TEM images revealed highly oriented polynanocrystalline structures with sharp, well-defined GDC/SZ(Y) interfaces.

The multilayers were then isotopically exchanged with ^{18}O tracer at 600 °C for 5 min and subsequently analyzed using SIMS depth profiling (figure 3.15). The normalized intensities for the various metal cations (primary y-axis) are shown together with the corrected fraction of the ^{18}O concentration (c_{18O}, secondary y-axis) for the two types of multilayer. Focusing on the c_{18O} profiles, it can be seen that for both samples, the profiles are extremely flat across the GDC layers, which suggests very fast oxide ion diffusion through this region. However, an abrupt decrease that

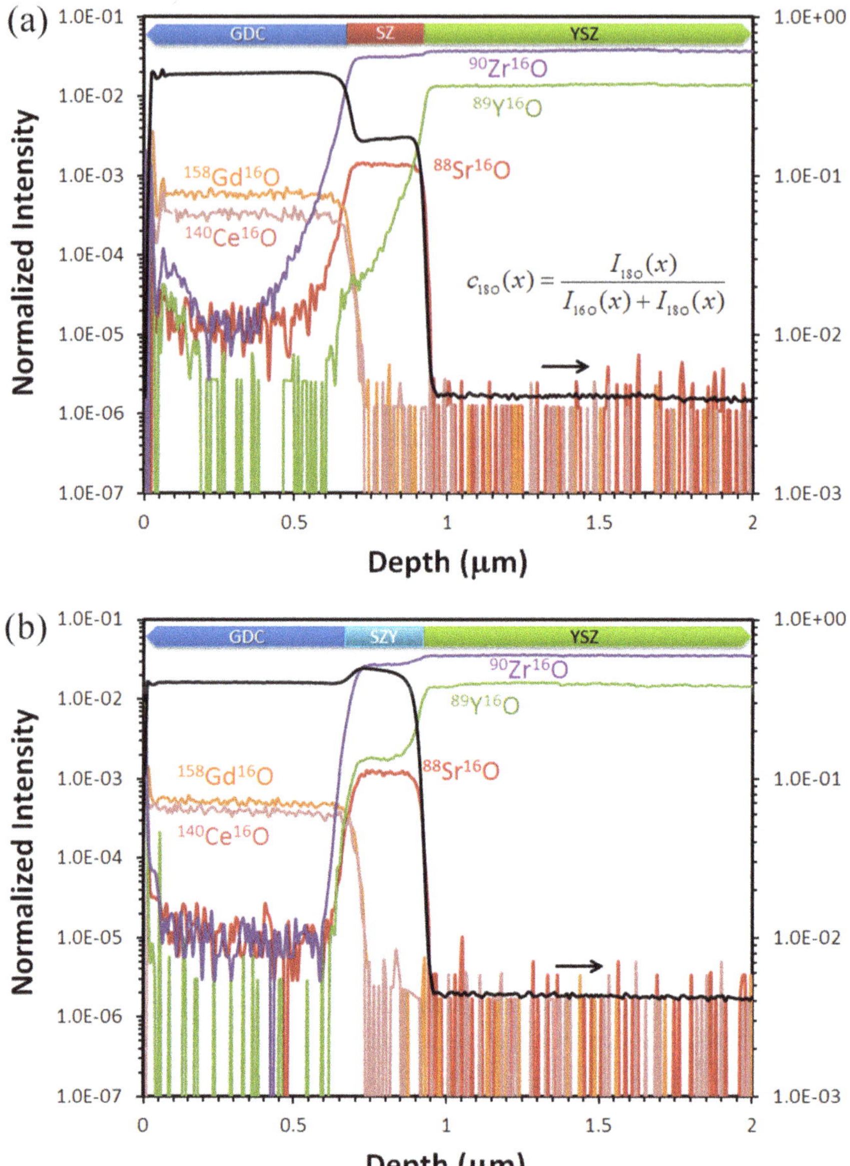

Figure 3.15. SIMS depth profiles showing the normalized intensities of secondary ions (primary *y*-axis) and the corrected fraction of the ^{18}O concentration (c_{18O}, secondary *y*-axis) for GDC/SZ/YSZ and GDC/SZY/YSZ multilayers after isotopic exchange. For both types of multilayer, a drastic drop in the c_{18O} profile can be observed at the SZ(Y)/YSZ interfaces. Reproduced from [61] with permission of the Royal Society of Chemistry.

coincides with the GDC/SZ and GDC/SZY interfaces can be observed in the c_{18O} profiles. This implies the presence of oxide ion blocking effects for both types of interface. In other words, both interfaces behave almost similarly, irrespective of yttrium doping. From the Y profile, some diffusion of Y from YSZ is also evidenced by a broadening of the profile from the YSZ region towards the SZ region, suggesting that the SZ region in the immediate vicinity of the YSZ interface is likely doped by yttrium. In other words, the SZ/YSZ interface may be considered similar to an SZY/YSZ interface.

The preceding SIMS results were further complemented by a detailed STEM-EELS analysis of the O *K*-edge obtained at the SZ(Y)/YSZ heterointerfaces, which revealed changes in the spectra indicative of the probable presence of an interface-induced oxygen disorder in the vicinity of the YSZ region. This disorder is attributed to factors including mismatch strain between SZ(Y) and YSZ layers as well as the diffusion of Y across the interface. Thermodynamic analyses of the SZ–YSZ region also indicated that the interface between these two materials is unstable, leading to the possible formation of Y-rich oxides along the interface, and/or the formation of a Y-rich YSZ region with lower ionic conductivity. These phases are considered to be potential causes of the ionic blocking behavior at such interfaces. The main conclusion is that it is not the SZ(Y) phase *per se* that impedes ionic flow, but its resulting interface with the YSZ. These results demonstrated that the use of analytical methods such as the isotope exchange depth profiling method could yield crucial information for identifying impediments to ionic flow and therefore provide an analysis that complements the other advanced characterization tools used to understand oxide ion transport in these systems.

3.3.2 Isotope labeling and visualization by SIMS

The imaging of oxide ion diffusion provides a clear understanding of the processes that give rise to electrode overpotential, which is one of the major contributions to cell resistance. Using a stable oxygen isotope of ^{18}O as a diffusion tracer, localized regions for active oxygen incorporation can be identified using high-resolution SIMS imaging. Adopting this technique, Horita *et al* studied the TPB region of O_2, the electrode, and the electrolyte using mesh-shaped $La_{0.9}Sr_{0.1}MnO_3$ (LSM) and Au electrodes prepared on YSZ substrates [62]. The LSM mesh was prepared using sputtering and subsequently patterned using photolithography, while the Au mesh was attached to the YSZ electrolyte by firing at a high temperature (1283 K for 2 h). The isotope exchange was performed at 973 K for 600 s at a $p(O_2)$ of 0.1 bar under cathodic polarization (0.5 V). A higher concentration of ^{18}O was observed on the top surfaces of the LSM mesh and the Au mesh, suggesting that these regions are active sites for the ORR. However, removal of the electrodes and subsequent imaging using SIMS further revealed that the LSM/YSZ interfaces are also active for oxygen incorporation, whereas in the case of the Au mesh, it is limited to the TPB regions only. These differences in ORR active sites are attributed not only to the materials characteristics of the electrodes but also to the nature of the electrode/electrolyte interfaces.

This technique was further developed to investigate the ORR processes in LSM/ScSZ composite electrodes by Nagasawa *et al* [63]. This study used a YSZ-electrolyte-supported cell with a Ni/YSZ anode and a scandia-stabilized zirconia (LSM/ScSZ) cathode. The cell was operated under 50% of H_2 and 50% of Ar on the fuel electrode side and under 100% of $^{16}O_2$ on the air electrode side at 1073 K for 10 min at a constant current density of 0.14 A cm^{-2}. The isotope exchange was carried out by switching to $^{18}O_2$ and operating the cell for 3 min. The setup used a customized helium impinging jet to abruptly quench the cell to room temperature. The ^{18}O distribution was then visualized using SIMS imaging (NanoSIMS) with a spatial resolution of 50 nm.

Figure 3.16 shows the SIMS results obtained at various locations and under different operating conditions. Here, Section A was obtained from the reference cathode under OCV conditions, while Sections B and C were obtained under biased

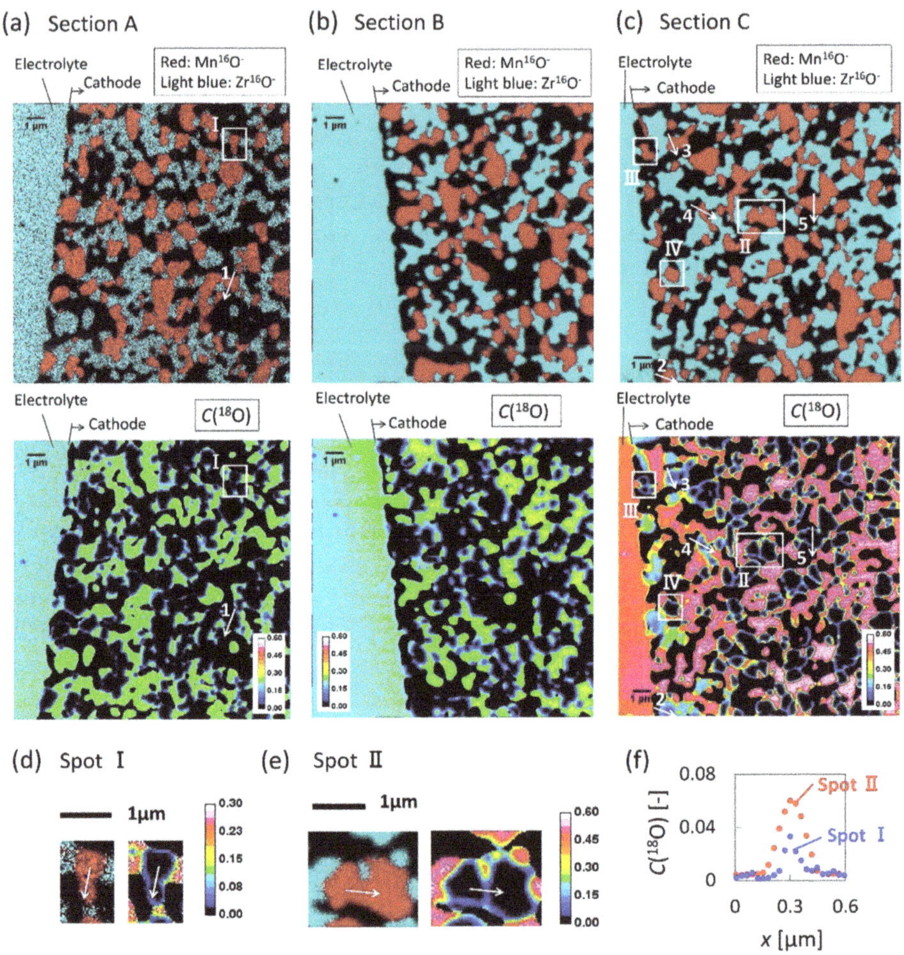

Figure 3.16. Distribution of secondary ions mapped using the SIMS imaging technique. (a) Section A: at the OCV, (b) Section B: biased, away from the current collector, and (c) Section C: biased, directly under the current collector. The corrected fractions of ^{18}O (denoted by $C(^{18}O)$) are depicted in the middle SIMS images. (d)–(e) Magnification of the areas indicated by white rectangles on the SIMS images. (f) $C(^{18}O)$ line scan profiles extracted from the regions marked by arrows on the SIMS images. Reprinted from [63], Copyright (2017), with permission from Elsevier.

conditions. The ^{18}O concentration is represented by $C(^{18}O)$, which is the normalized fraction given by $I(^{18}O^-)/[I(^{16}O^-) + I(^{18}O^-)]$. For Section A, ^{18}O is uniformly distributed within the ScSZ phase with a high concentration of $C(^{18}O)$, ~ 0.25, which is in contrast to that inside the LSM phase, where $C(^{18}O)$ ~ 0.002 or approximately the natural concentration of ^{18}O (figure 3.16(a)). This indicates that the oxygen exchange process mainly occurred on the ScSZ surfaces and that oxygen incorporation was accommodated by the ScSZ bulk under OCV conditions. Here, LSM is considered to be less active under these conditions because of its low oxygen diffusivity. On the other hand, for Section B, which was subjected to biased conditions and located far from the current collector, ^{18}O is limited to the inside of the ScSZ bulk, which is similar to the result obtained for the OCV sample (figure 3.16(b)). This indicates that this specific area is almost inactive, possibly because of its location, and that it does not contribute to the electrochemical reaction. On the other hand, for Section C as shown in figure 3.16(c), a remarkably higher $C(^{18}O)$ ~ 0.55 is obtained for the ScSZ regions, which is accounted for by the applied current which drives the incorporation of oxygen. A closer inspection of several $C(^{18}O)$ profiles across several LSM grains (denoted by lines two to four), reveals that the highest concentrations occur near the cathode/electrolyte interfaces but that the maximum value coincides with the inner regions of the LSM grains. This is interpreted as a possible 'uphill' diffusion behavior induced under cathodically biased conditions, similar to that reported for LSM thin films on YSZ electrolyte. Fitting of the profiles yielded values of $D^* = 1.21 \times 10^{-13}$ cm^2 s^{-1} and $k^* = 1.17 \times 10^{-9}$ cm s^{-1} at the OCV, which are close to those reported for LSM bulk; by contrast, enhancements by factors of about three and four, respectively, are observed under biased conditions. This suggests that the diffusion and surface exchange reactions are both promoted under biased conditions. In addition, there was no significant difference between the values obtained for various locations across the thickness of the cathode.

In a follow-up study, the authors further developed the technique to quantitatively investigate the oxide ion flux incorporated by the air electrode/electrolyte interface [64]. The cell was operated in a mixture of H_2 and Ar on the anode side, and pure O_2 on the air electrode side at 973 K under a constant current density of 0.09 A cm^{-2} for 10 min. The atmosphere was then switched to $^{18}O_2$, which was continued for 70 s. As in the previous study, the diffusion profiles of c_{18O} across the air electrode/electrolyte interfaces were then derived from the SIMS analysis. As shown in figure 3.17, the three profiles obtained at the OCV, Biased #1 (first location), and Biased #2 (second location) were compared and fitted to derive the values of D^* and k^*. At the OCV, $D^* = 9.70 \times 10^{-13}$ m^2 s^{-1} and $k^* = 2.45 \times 10^{-9}$ m s^{-1}, which are approximate to those reported in the scientific literature for polycrystalline YSZ. On the other hand, under biased conditions, D^* showed a marginal enhancement of only 1.2–1.5, whereas k^* was enhanced by one order of magnitude. This enhancement for k^* was attributed to the oxide ion flux incorporated from the cathode/electrolyte interface. Three different kinds of flux were defined, namely, ^{18}O flux via $^{16}O/^{18}O$ exchange at the electrolyte-free surfaces ($J_{18O}^{ex(C/E)}$), $^{18}O^{2-}$ ion flux from the electrochemical reaction at the TPB regions ($J_{18O}^{ion(C/E)}$), and ^{18}O flux arising from the

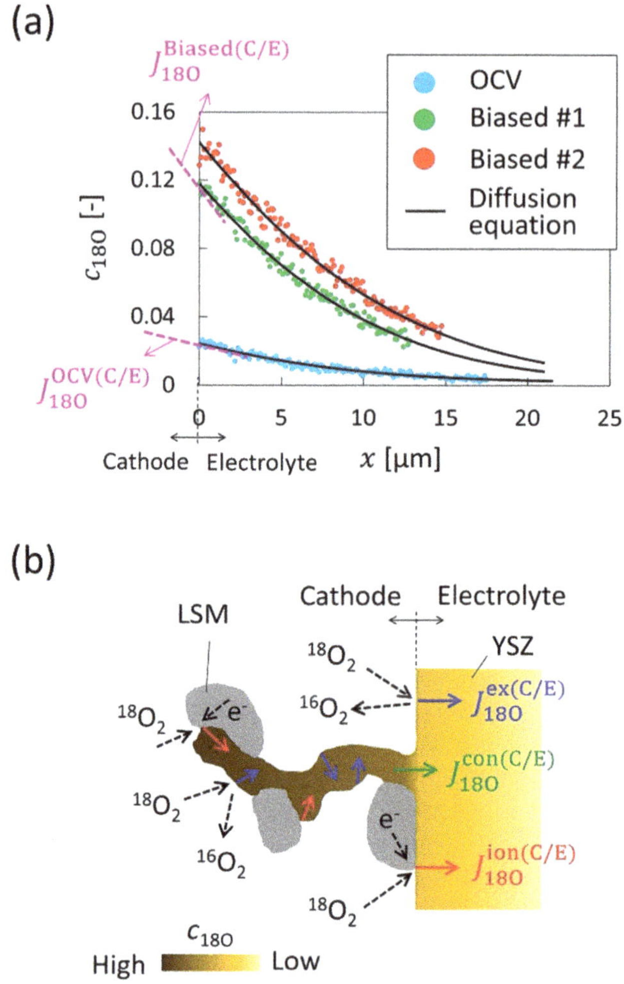

Figure 3.17. Using the SIMS technique to quantitatively investigate the oxide ion flux incorporated from the air electrode/electrolyte interface. The depth profile of the corrected fraction, c_{18O}, is compared for samples tested at the OCV and under biased conditions and fitted using the diffusion equation (a). High concentrations of the isotopic tracer are shown for air electrode/electrolyte areas, including three types of ^{18}O flux in the schematic illustration (b). Reprinted from [64], Copyright (2018), with permission from Elsevier.

difference of c_{18O} between the YSZ in the air electrode region and the YSZ of the electrolyte region ($J_{18O}^{\mathrm{con(C/E)}}$). This is schematically illustrated in figure 3.17(b). Both $J_{18O}^{\mathrm{ex(C/E)}}$ and $J_{18O}^{\mathrm{con(C/E)}}$ are present at the OCV or under biased conditions, but $J_{18O}^{\mathrm{ion(C/E)}}$ is only present under biased conditions. Numerical values for the fluxes were then calculated from the known values of the concentration of oxygen atoms in YSZ, the concentration of ^{18}O in the gas phase, and the experimentally determined c_{18O}. Based on the obtained flux of 1.01–1.43 × 10^{-3} mol m^{-2} s^{-1} at a current density of 0.01 A cm^{-2}, it was inferred that 22%–31% of the overall electrochemical reaction occurred at the air electrode/electrolyte interfaces. The rest of the reaction was assumed to proceed within the porous air electrode.

3.3.3 Scanning probe microscopy

Understanding the fundamental processes that occur at the heterointerfaces of SOCs is crucial to devising ways to address the stability of solid oxide cell performance under realistic operating conditions. One innovative approach capable of locally resolving electrochemical phenomena is that of *operando* direct localized imaging proposed by Nonnenmann *et al* [65]. The authors of this study employed scanning surface potential microscopy (SSPM) (alternatively called Kelvin force microscopy (KFM)) to map local potential variations on the cross-sectional surfaces of symmetrical cells utilizing $La_{0.8}Sr_{0.2}FeO_{3-\delta}$ (LSF) and LSM ($La_{0.8}Sr_{0.2}MnO_{3-\delta}$) electrodes screen printed on YSZ electrolytes. The SSPM system utilized an environmental sample chamber in conjunction with a commercially available scanning probe microscopy (SPM) system. The samples were heated to 600 °C and first imaged in noncontact mode to obtain a topographic image, followed by a measurement using a conductive tip under an AC bias to obtain the corresponding potential distribution across the cross sections. The potential profiles obtained at 600 °C for LSF, an MIEC electrode material, are shown in figure 3.18. The topographic image was used to visualize the electrode–electrolyte interfaces at both sides of the YSZ electrolyte (figures 3.18(b)–(d)). The scanning surface potential images confirmed the existence of local potential variations at various regions of the cross section (figures 3.18(e)–(g)). Here, it

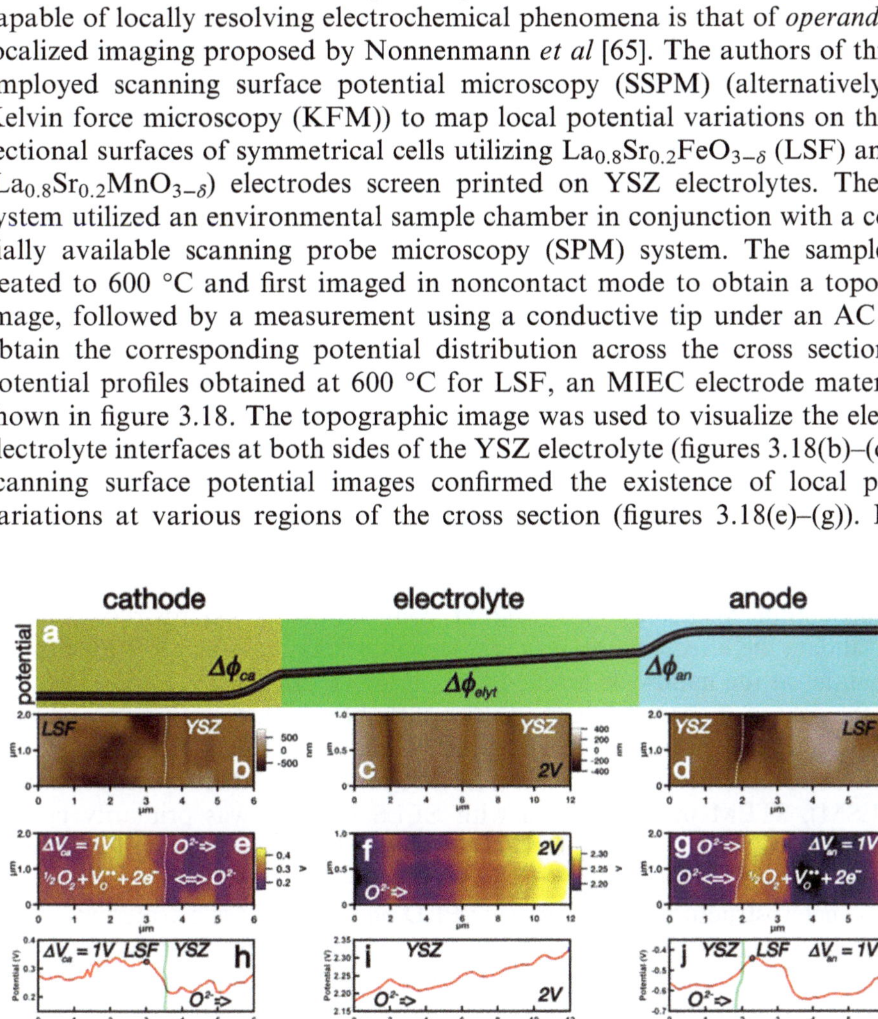

Figure 3.18. Probing localized potential distribution across various layers using the SPM technique: (a) schematic of potential distribution across the air electrode (cathode), the electrolyte, and the fuel electrode (anode). Panels (b)–(d) are topographic images, and (e)–(g) are scanning surface potential images obtained under biased conditions at 600 °C. (h)–(j) Potential profiles across interfaces (indicated by light green lines) obtained from averaging over 15 lines. Reprinted with permission from [65]. Copyright (2013) American Chemical Society.

can be seen that the surface potentials exhibit maxima towards the LSF/YSZ interfaces, whereas a positively sloped linear-type profile is observed across the YSZ electrolyte. It was considered that the variations in potential in the electrode region are indicative of spatial variations in oxygen vacancy concentration due to the oxygen transfer at the interfaces. Specifically, at the cathode region, oxygen transfer to the electrolyte led to an increase in oxygen and an associated decrease in the oxygen vacancy concentration. In a similar manner, at the anode, oxygen introduced into the electrode filled the oxygen vacancies, again resulting in a local decrease in the oxygen vacancy concentration in the vicinity of the interface.

Similar measurements performed for LSM electrode interfaces further confirmed electric potential inhomogeneities in the LSM grain closest to the interface with YSZ, i.e. an increased potential was observed at the grain boundaries in the vicinity of the TPBs in contrast to the potential minima observed within the LSM grain bulk. As LSM inherently contains a low density of oxygen vacancies, the predominant charge transfer mechanism is expected to be determined by the migration of adsorbed molecules across the electrode surface towards the TPB region, where the charge transfer occurs (see figure 3.1). Furthermore, the authors noted a higher activation potential at the air electrode side, indicating a larger barrier at this particular interface compared to that for the fuel electrode side. This is associated with the mechanism responsible for the incorporation of the ionic species into the electrolyte [66]. Although the samples investigated in this study were porous structures, a subsequent study by Zhu *et al* also demonstrated the use of this imaging technique for an *in situ* investigation of vacancy distribution in dense STO/YSZ multilayer films, which form more intimate interfaces [67]. These studies show that oxygen vacancy concentration profiles quantitatively derived from local surface potential perturbations provide important information about the complicated oxygen incorporation processes occurring in complex oxide heterostructures, and further validate the effective use of *in situ* techniques as versatile tools for probing functionalities at the nanoscale level.

With the goal of understanding the nature of the electrochemically active states of electrode materials at the nanoscale level, Kumar *et al* [68] demonstrated the approach of combining high-resolution STEM and electrochemical strain microscopy (ESM). STEM in conjunction with EELS imaging was primarily used to determine the various types of defect (small-angle grain boundaries and antiphase boundaries) on model perovskite surfaces of LSC_{113} and LSC_{214} as well as on $LSC_{214/113}$ heterostructures prepared using PLD on (001) YSZ substrates with GDC buffer layers. However, the ESM technique uses a biased tip which applies a local electric field to the material being probed, which induces ionic redistribution and reversible and irreversible electrochemical processes on the surface. The measurements were performed in the regime for reversible surface deformation, acquiring averaged ESM hysteresis loops for typical bias windows of 8 V and 10 V. Above a critical bias required for the activation of the ORR/OER process, significant hysteresis occurred, resulting in a characteristic (classical) hysteresis loop shape as well as static deformation on the sample surface. Figure 3.19 shows a topographic image (a) acquired for a 15 nm $LSC_{214/113}$ surface and its corresponding ESM

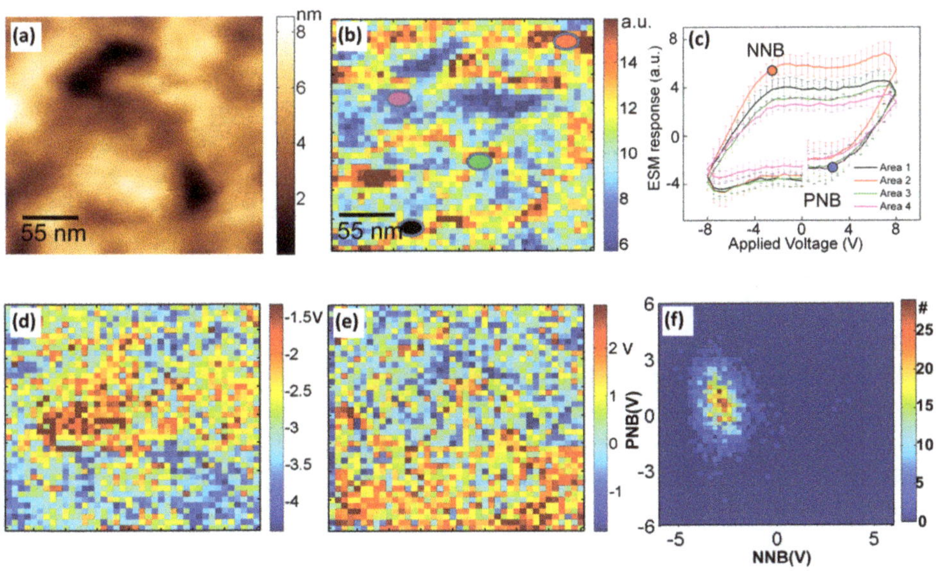

Figure 3.19. Probing the surface of a 15 nm $LSC_{214/113}$ sample. (a) Topographic image, (b) ESM mapping, (c) hysteresis loops obtained from locations indicated by colored ellipses in (b), (d) negative nucleation bias, (e) positive nucleation bias, and (f) a 2D histogram of the negative and positive nucleation biases (denoted by NNB and PNB, respectively). Reprinted with permission from [68]. Copyright (2013) American Chemical Society.

mapping (b) acquired over a 300 nm × 300 nm scan area spatially resolved by a 40 × 40 array. The ESM hysteresis loops for representative areas indicated by colored circles in figure 3.19(b) are further depicted in figure 3.19(c), which show strong variability as a function of area. Furthermore, the negative/positive distributions (referred to as the negative nucleation bias (NNB) and the positive nucleation bias (PNB), respectively) associated with the critical biases required for ORR/OER activation are analyzed and depicted in figures 3.19(d) and (e), respectively. The difference between the NNB and PNB values, i.e. $(PNB-NNB)$ is considered to be the nanoscale analog of the polarization hysteresis. The authors noted that among the samples, $LSC_{214/113}$ had the lowest value, indicating high electrochemical activity and high disorder, while LSC_{113} had the highest value and therefore the least active surface among the samples studied. This also correlates well with the relatively higher ASR obtained for this sample based on electrochemical impedance measurements. Furthermore, the authors noted the strong correlation between the ESM response and the degree of disorder of the material in terms of small-angle grain boundaries, topographic roughness, and the widths of the ESM potential distributions. In this sense, the ESM data can be treated as a good descriptor for electrochemical activity and can therefore be utilized as a local probe. On the other hand, it is still necessary to further correlate the behavior observed in room-temperature ESM data with that obtained under the high-temperature, biased conditions required for cell operation.

3.4 Summary and outlook

In this chapter, we have highlighted several state-of-the-art techniques developed to tune heterointerfaces via a multilayered approach and the appropriate selection of interlayer materials, taking advantage of the versatility of existing thin-film fabrication methods to obtain various types of heterostructure with precise control and reproducibility. Investigations of these structures were facilitated by employing sophisticated analytical tools which allowed analyses to be performed at the nanoscale level, revealing hitherto undiscovered phenomena that would have been difficult to reveal using the conventional structures normally utilized for SOC materials.

Studies of the model heterostructures also provided a fundamental understanding of the behavior frequently observed in practical SOCs, particularly in terms of the factors affecting the ORR process, oxide ion transport, and the long-term degradation/stability of heterointerfaces under operating conditions. Concurrently, it was also shown that the thin-film structures often surpass conventional ones in terms of performance, indicating that there is a great potential to harness the superior properties of these structures for practical applications. Integrating these thin-film structures into conventional cell architecture still remains a challenge, especially with respect to the economic aspect. Research interest in this area is expected to grow within the SOC field in the pursuit of novel materials and structures which may considerably change the course of development.

References

[1] Chroneos A, Yildiz B, Tarancón A, Parfitt D and Kilner J A 2011 Oxygen diffusion in solid oxide fuel cell cathode and electrolyte materials: mechanistic insights from atomistic simulations *Energy Environ. Sci.* **4** 2774–89

[2] Kilner J A and Burriel M 2014 Materials for intermediate-temperature solid-oxide fuel cells *Annu. Rev. Mater. Res.* **44** 365–93

[3] Kawada T *et al* 1999 Oxygen isotope exchange with a dense $La_{0.6}Sr_{0.4}CoO_{3-\delta}$ electrode on a $Ce_{0.9}Ca_{0.1}O_{1.9}$ electrolyte *Solid State Ion.* **121** 271–9

[4] Hayd J, Dieterle L, Guntow U, Gerthsen D and Ivers-Tiffée E 2011 Nanoscaled $La_{0.6}Sr_{0.4}CoO_{3-\delta}$ as intermediate temperature solid oxide fuel cell cathode: microstructure and electrochemical performance *J. Power Sources* **196** 7263–70

[5] Januschewsky J, Ahrens M, Opitz A, Kubel F and Fleig J 2009 Optimized $La_{0.6}Sr_{0.4}CoO_{3-\delta}$ thin-film electrodes with extremely fast oxygen-reduction kinetics *Adv. Funct. Mater.* **19** 3151–6

[6] Tsvetkov N, Lu Q, Sun L, Crumlin E J and Yildiz B 2016 Improved chemical and electrochemical stability of perovskite oxides with less reducible cations at the surface *Nat. Mater.* **15** 1010–6

[7] Jiang S P 2019 Development of lanthanum strontium cobalt ferrite perovskite electrodes of solid oxide fuel cells—a review *Int. J. Hydrog. Energy* **44** 7448–93

[8] Çelikbilek Ö, Jauffrès D, Siebert E, Dessemond L, Burriel M, Martin C L and Djurado E 2016 Rational design of hierarchically nanostructured electrodes for solid oxide fuel cells *J. Power Sources* **333** 72–82

[9] Plonczak P, Bieberle-Huetter A, Sogaard M, Ryll T, Martynczuk J, Hendriksen P V and Gauckler L J 2011 Tailoring of $La_xSr_{1-x}Co_yFe_{1-y}O_{3-\delta}$ nanostructure by pulsed laser deposition *Adv. Funct. Mater.* **21** 2764–75
[10] Ten Elshof J E and Boeijsma J 1996 Influence of iron content on cell parameters of rhombohedral $La_{0.6}Sr_{0.4}Co_{1-y}Fe_yO_3$ *Powder Diffr.* **11** 240–5
[11] Develos-Bagarinao K, Kishimoto H, De Vero J, Yamaji K and Horita T 2016 Effect of $La_{0.6}Sr_{0.4}Co_{0.2}Fe_{0.8}O_{3-\delta}$ microstructure on oxygen surface exchange kinetics *Solid State Ion.* **288** 6–9
[12] Develos-Bagarinao K, De Vero J, Kishimoto H, Ishiyama T, Yamaji K, Horita T and Yokokawa H 2019 Oxygen surface exchange properties and surface segregation behavior of nanostructured $La_{0.6}Sr_{0.4}Co_{0.2}Fe_{0.8}O_{3-\delta}$ thin film cathodes *Phys. Chem. Chem. Phys.* **21** 7183–95
[13] Prestat M, Infortuna A, Korrodi S, Rey-Mermet S, Muralt P and Gauckler L J 2007 Oxygen reduction at thin dense $La_{0.52}Sr_{0.48}Co_{0.18}Fe_{0.82}O_{3-\delta}$ electrodes: part II: experimental assessment of the reaction kinetics *J. Electroceramics* **18** 111–20
[14] Adler S B 2004 Factors governing oxygen reduction in solid oxide fuel cell cathodes *Chem. Rev.* **104** 4791–843
[15] Prestat M, Koenig J F and Gauckler L J 2007 Oxygen reduction at thin dense $La_{0.52}Sr_{0.48}Co_{0.18}Fe_{0.82}O_{3-\delta}$ electrodes: part i: reaction model and faradaic impedance *J. Electroceramics* **18** 87–101
[16] Baumann F S, Fleig J, Habermeier H U and Maier J 2006 Impedance spectroscopic study on well-defined $(La,Sr)(Co,Fe)O_{3-\delta}$ model electrodes *Solid State Ion.* **177** 1071–81
[17] Zhao C, Li Y, Zhang W, Zheng Y, Lou X, Yu B, Chen J, Chen Y, Liu M and Wang J 2020 Heterointerface engineering for enhancing the electrochemical performance of solid oxide cells *Energy Environ. Sci.* **13** 53–85
[18] Kawada T, Sase M, Kudo M, Yashiro K, Sato K, Mizusaki J, Sakai N, Horita T, Yamaji K and Yokokawa H 2006 Microscopic observation of oxygen reaction pathway on high temperature electrode materials *Solid State Ion.* **177** 3081–6
[19] Sase M, Yashiro K, Sato K, Mizusaki J, Kawada T, Sakai N, Yamaji K, Horita T and Yokokawa H 2008 Enhancement of oxygen exchange at the hetero interface of $(La,Sr)CoO_3$/$(La,Sr)_2CoO_4$ in composite ceramics *Solid State Ion.* **178** 1843–52
[20] Crumlin E J, Mutoro E, Ahn S J, La O G J, Leonard D N, Borisevich A, Biegalski M D, Christen H M and Shao-Horn Y 2010 Oxygen reduction kinetics enhancement on a heterostructured oxide surface for solid oxide fuel cells *J. Phys. Chem. Lett.* **1** 3149–55
[21] Zhao C, Liu X, Zhang W, Zheng Y, Li Y, Yu B, Wang J and Chen J 2019 Measurement of oxygen reduction/evolution kinetics enhanced $(La,Sr)CoO_3$/$(La,Sr)_2CoO_4$ hetero-structure oxygen electrode in operating temperature for SOCs *Int. J. Hydrog. Energy* **4** 19102–12
[22] Tsuji Y, Sako S, Nitta K, Yamamoto K, Shao-horn Y, Uchimoto Y and Orikasa Y 2020 Surface analysis of lanthanum strontium cobalt oxides under cathodic polarization at high temperature through operando total-reflection X-ray absorption and X-ray fluorescence spectroscopy *Solid State Ion.* **357** 115502
[23] Orikasa Y, Ina T, Nakao T, Mineshige A, Amezawa K, Oishi M, Arai H, Ogumi Z and Uchimoto Y 2011 X-ray absorption spectroscopic study on $La_{0.6}Sr_{0.4}CoO_{3-\delta}$ cathode materials related with oxygen vacancy formation *J. Phys. Chem.* C **115** 16433–8
[24] Lee W, Han J W, Chen Y, Cai Z and Yildiz B 2013 Cation size mismatch and charge interactions drive dopant segregation at the surfaces of manganite perovskites *J. Am. Chem. Soc.* **135** 7909–25

[25] Zheng Y *et al* 2020 Directly visualizing and exploring local heterointerface with high electro-catalytic activity *Nano Energy* **78** 105236

[26] Xia C, Mi Y, Wang B, Lin B, Chen G and Zhu B 2019 Shaping triple-conducting semiconductor $BaCo_{0.4}Fe_{0.4}Zr_{0.1}Y_{0.1}O_{3-\delta}$ into an electrolyte for low-temperature solid oxide fuel cells *Nat. Commun.* **10** 1–9

[27] Grimaud A, Diaz-Morales O, Han B, Hong W T, Lee Y L, Giordano L, Stoerzinger K A, Koper M T M and Shao-Horn Y 2017 Activating lattice oxygen redox reactions in metal oxides to catalyse oxygen evolution *Nat. Chem.* **9** 457–65

[28] Zhu B, Wang B, Wang Y, Raza R, Tan W, Kim J, Van Aken A and Lund P 2017 Nano energy ceria heterostructure material for new generation fuel cell *Nano Energy* **37** 195–202

[29] Hong T, Zhao M, Brinkman K, Chen F and Xia C 2017 Enhanced oxygen reduction activity on Ruddlesden–Popper phase decorated $La_{0.8}Sr_{0.2}FeO_{3-\delta}$ 3D heterostructured cathode for solid oxide fuel cells *ACS Appl. Mater. Interfaces* **9** 8659–68

[30] Kim Y-M, He J, Biegalski M D, Ambaye H, Lauter V, Christen H M, Pantelides S T, Pennycook S J, Kalinin S V and Borisevich A Y 2012 Probing oxygen vacancy concentration and homogeneity in solid-oxide fuel-cell cathode materials on the subunit-cell level *Nat. Mater.* **11** 888–94

[31] Han B, Stoerzinger K A, Tileli V, Gamalski A D, Stach E A and Shao-horn Y 2017 Nanoscale structural oscillations in perovskite oxides induced by oxygen evolution *Nat. Mater.* **16** 121–7

[32] Kwon J, Choi W S, Kwon Y, Jung R, Zuo J, Lee H N and Kim M 2014 Nanoscale spin-state ordering in $LaCoO_3$ epitaxial thin films *Chem. Mater.* **26** 2496–501

[33] Dusastre V and Kilner J A 1999 Optimisation of composite cathodes for intermediate temperature soild oxide fuel cell applications *Solid State Ion.* **126** 163–74

[34] Leng Y, Chan S H and Liu Q 2008 Development of LSCF-GDC composite cathodes for low-temperature solid oxide fuel cells with thin film GDC electrolyte *Int. J. Hydrog. Energy* **33** 3808–17

[35] Sindirac C, Ahsen A, Ozturk O, Akkurt S, Birss V I and Buyukaksoy A 2020 Fabrication of LSCF and LSCF-GDC nanocomposite thin films using polymeric precursors *Ionics (Kiel).* **26** 913–25

[36] Choi M, Ibrahim I A M, Kim K, Koo J Y, Kim S J, Son J-W, Han J W and Lee W 2020 Engineering of charged defects at perovskite oxide surfaces for exceptionally stable solid oxide fuel cell electrodes *ACS Appl. Mater. Interfaces* **12** 21494–504

[37] Xu W, Lyu F, Bai Y, Gao A, Feng J, Cai Z and Yin Y 2018 Nano energy porous cobalt oxide nanoplates enriched with oxygen vacancies for oxygen evolution reaction *Nano Energy* **43** 110–6

[38] Develos-Bagarinao K, De Vero J, Kishimoto H, Ishiyama T, Yamaji K, Horita T and Yokokawa H 2018 Multilayered LSC and GDC: an approach for designing cathode materials with superior oxygen exchange properties for solid oxide fuel cells *Nano Energy* **52** 369–80

[39] Hayd J, Yokokawa H and Ivers-Tiffee E 2013 Hetero-interfaces at nanoscaled $(La,Sr)CoO_{3-\delta}$ thin-film cathodes enhancing oxygen surface-exchange properties *J. Electrochem. Soc.* **160** F351–9

[40] Berenov A V, Atkinson A, Kilner J A, Bucher E and Sitte W 2010 Oxygen tracer diffusion and surface exchange kinetics in $La_{0.6}Sr_{0.4}CoO_{3-\delta}$ *Solid State Ion.* **181** 819–26

[41] Manning P S, Sirman J D and Kilner J A 1996 Oxygen self-diffusion and surface exchange studies of oxide electrolytes having the fluorite structure *Solid State Ion.* **93** 125–32

[42] Manning P S, Sirman J D, De Souza R A and Kilner J A 1997 The kinetics of oxygen transport in 9.5 mol% single crystal yttria stabilised zirconia *Solid State Ion.* **100** 1–10

[43] Kilner J A, De Souza R A and Fullarton I C 1996 Surface exchange of oxygen in mixed conducting perovskite oxides *Solid State Ion.* **86–88** 703–9

[44] Rodrigo K, Wang H J, Heiroth S, Pryds N, Kuhn L T, Esposito V, Linderoth S, Schou J and Lippert T 2011 Nanostructured PLD-grown gadolinia doped ceria: chemical and structural characterization by transmission electron microscopy techniques *Appl. Surf. Sci.* **257** 5341–6

[45] Develos-Bagarinao K, Budiman R A, Ishiyama T, Yamaji K and Kishimoto H 2020 Leveraging catalytic effects of heterointerfaces through multilayering for superior cathode performance *J. Power Sources* **480** 229094

[46] Sumi H, Yamaguchi T, Hamamoto K, Suzuki T and Fujishiro Y 2013 High performance of $La_{0.6}Sr_{0.4}Co_{0.2}Fe_{0.8}O_3$-$Ce_{0.9}Gd_{0.1}O_{1.95}$ nanoparticulate cathode for intermediate temperature microtubular solid oxide fuel cells *J. Power Sources* **226** 354–8

[47] Leonide A, Sonn V, Weber A and Ivers-Tiffée E 2008 Evaluation and modeling of the cell resistance in anode-supported solid oxide fuel cells *J. Electrochem. Soc.* **155** B36–41

[48] Dos Santos-Gómez L, Sanna S, Norby P, Pryds N, Losilla E R, Marrero-López D and Esposito V 2019 Electrochemical stability of (La,Sr)$CoO_{3-\delta}$ in (La,Sr)$CoO_{3-\delta}$/(Ce, Gd)$O_{2-\delta}$ heterostructures *Nanoscale* **11** 2916–24

[49] Solovyev A A, Kuterbekov K A, Nurkenov S A, Nygymanova A S, Shipilova A V, Smolyanskiy E A, Rabotkin S V and Ionov I V 2021 Anode-supported solid oxide fuel cells with multilayer LSC/CGO/LSC cathode *Fuel Cells* **21** 408–12

[50] Shah M, Voorhees P W and Barnett S A 2011 Time-dependent performance changes in LSCF-infiltrated SOFC cathodes: the role of nano-particle coarsening *Solid State Ion.* **187** 64–7

[51] Hildenbrand N, Boukamp B A, Nammensma P and Blank D H A 2011 Improved cathode/electrolyte interface of SOFC *Solid State Ion.* **192** 12–5

[52] Chrzan A, Karczewski J, Gazda M, Szymczewska D and Jasinski P 2015 Investigation of thin perovskite layers between cathode and doped ceria used as buffer layer in solid oxide fuel cells *J. Solid State Electrochem.* **19** 1807–15

[53] De Vero J C, Develos-Bagarinao K, Kishimoto H, Ishiyama T, Yamaji K, Horita T and Yokokawa H 2018 Enhanced stability of solid oxide fuel cells by employing a modified cathode–interlayer interface with a dense $La_{0.6}Sr_{0.4}Co_{0.2}Fe_{0.8}O_{3-\delta}$ thin film *J. Power Sources* **377** 128–35

[54] Yokokawa H, Sakai N, Horita T, Yamaji K, Brito M E and Kishimoto H 2008 Thermodynamic and kinetic considerations on degradations in solid oxide fuel cell cathodes *J. Alloys Compd.* **452** 41–7

[55] Wang F, Nishi M, Brito M E, Kishimoto H, Yamaji K, Yokokawa H and Horita T 2014 Sr and Zr diffusion in LSCF/10GDC/8YSZ triplets for solid oxide fuel cells (SOFCs) *J. Power Sources* **258** 281–9

[56] Kamecki B, Karczewski J, Jasiński P and Molin S 2021 Improvement of oxygen electrode performance of intermediate temperature solid oxide cells by spray pyrolysis deposited active layers *Adv. Mater. Interfaces* **8** 1–11

[57] Kosacki I, Rouleau C M, Becher P F, Bentley J and Lowndes D H 2005 Nanoscale effects on the ionic conductivity in highly textured YSZ thin films *Solid State Ion.* **176** 1319–26

[58] Garcia-Barriocanal J, Rivera-Calzada A, Varela M, Sefrioui Z, Iborra E, Leon C, Pennycook S J and Santamaria J 2008 Colossal ionic conductivity at interfaces of epitaxial ZrO_2:Y_2O_3/$SrTiO_3$ heterostructures *Science (80-.).* **321** 676–80

[59] Perkins J M *et al* 2010 Anomalous oxidation states in multilayers for fuel cell applications *Adv. Funct. Mater.* **20** 2664–74

[60] De Vero J C, Develos-Bagarinao K, Kishimoto H, Ishiyama T, Yamaji K, Horita T and Yokokawa H 2016 Influence of $La_{0.6}Sr_{0.4}Co_{0.2}Fe_{0.8}O_{3-\delta}$ microstructure on GDC interlayer stability and cation diffusion across the LSCF/GDC/YSZ interfaces *J. Electrochem. Soc.* **163** F1463–70

[61] Develos-Bagarinao K, Yokokawa H, Kishimoto H, Ishiyama T, Yamaji K and Horita T 2017 Elucidating the origin of oxide ion blocking effects at GDC/$SrZr(Y)O_3$/YSZ interfaces *J. Mater. Chem.* A **5** 8733–43

[62] Horita T, Yamaji K and Sakai N 2002 Imaging of oxygen transport at SOFC cathode/electrolyte interface by a novel technique *J. Power Sources* **106** 224–30

[63] Nagasawa T and Hanamura K 2017 Microstructure-scaled active sites imaging of a solid oxide fuel cell composite cathode *J. Power Sources* **367** 57–62

[64] Nagasawa T and Hanamura K 2019 Investigation of oxide ion flux at cathode/electrolyte interface in solid oxide fuel cell *J. Power Sources* **412** 695–700

[65] Nonnenmann S S, Kungas R, Vohs J and Bonnell D A 2013 Direct *in situ* probe of electrochemical processes in operating fuel cells *ACS Nano* **7** 6330–6

[66] Bessler W G, Gewies S and Vogler M 2007 A new framework for physically based modeling of solid oxide fuel cells *Electrochim. Acta* **53** 1782–800

[67] Zhu J, Lee J W, Lee H, Xie L, Pan X, De Souza R A, Eom C B and Nonnenmann S S 2019 Probing vacancy behavior across complex oxide heterointerfaces *Sci. Adv.* **5** 1–10

[68] Kumar A *et al* 2013 Spatially resolved mapping of oxygen reduction/evolution reaction on solid-oxide fuel cell cathodes with sub-10 nm resolution *ACS Nano* **7** 3808–14

Chapter 4

Interface engineering for highly active and stable solid oxide cell electrodes

Mengzhen Zhou and Yan Chen

Solid oxide cells (SOCs) are considered to be one of the cleanest and most efficient energy conversion systems; they are enabled to operate by high temperatures. However, phenomena such as cation enrichment, particle coarsening, contaminant deposition, and secondary phase formation during operation greatly affect the performance of the electrodes, hindering the commercialization of SOCs. In this chapter, we summarize the recent progress in using interface engineering to improve the activity and stability of SOC electrodes. We start with a brief introduction to the operational principles of the SOC electrode. We then outline the degradation mechanisms of both the oxygen electrode and the fuel electrode. The impact of heterointerfaces on the activity and stability of electrode materials is also presented in detail. Finally, the mechanisms by which heterointerfaces influence SOC performance are further discussed.

4.1 An overview of solid oxide cells

SOCs mainly consist of three components: a highly dense electrolyte, a porous air electrode, and a fuel electrode. SOCs are divided into solid oxide fuel cells (SOFCs) and solid oxide electrolysis cells (SOECs) according to their modes of operation. SOFCs can directly convert the energy in chemical fuels into electricity. Using the reverse of the SOFC operational mode, SOECs can store electricity by converting small molecules such as water, CO_2, and N_2 into fuels. The electrolytes of SOCs can be either oxygen ion conductors or proton ion conductors; the corresponding SOCs are referred to as O-SOCs or P-SOCs, respectively. In O-SOFCs, oxygen molecules receive electrons on the cathode surfaces to form oxygen ions, which pass through the dense electrolyte to the fuel electrode, where they react with the fuel (H_2, CO, or C_2H_{2n+2}) to produce H_2O, CO_2, or both. In O-SOECs, H_2O, CO_2, or their mixture are reduced to H_2, CO, or syngas fuel at the fuel electrode though the electrolysis process. During this process, oxygen ions pass through the dense electrolyte and are

doi:10.1088/978-0-7503-4064-9ch4

oxidized into oxygen molecules on the air electrode surface. P-SOCs have a similar operational mechanism, but they use protons as the mobile ion spices. SOCs have attracted widespread attention because of their high energy conversion efficiency, high fuel flexibility, and low emission of contaminants.

The high operating temperatures of SOCs (800 °C–1000 °C for O-SOCs and 400 °C–700 °C for P-SOCs) nevertheless also lead to phenomena such as electrode material agglomeration, poisoning by contaminants, and interactions between the electrode material and the other SOC components, resulting in the degradation of cell performance. The low stability strongly limits the commercial use of SOCs. Reducing the operating temperature can inhibit the degradation of each component of the cell and reduce the cost of using special materials. However, as the operating temperature decreases, the reaction kinetics of the electrodes become slower, contributing to a significant deterioration in cell performance. Therefore, there is a need to design SOC electrode components with novel materials or structures to improve stability and maintain a high level of electrocatalytic activity.

In this chapter, we focus on recent progress in using heterointerface engineering to improve the activity and stability of O-SOC electrodes. All the SOCs mentioned in the following context refer to O-SOCs. In this section, we briefly introduce the principal function of SOC electrodes. In section 4.2, the degradation mechanisms of the SOC electrode are discussed in detail. Section 4.2.1 introduces the degradation mechanisms of air electrodes, including cation segregation, contaminant deposition, and secondary phase generation. In section 4.2.2, the degradation of fuel electrode performance is briefly described, including nickel particle coarsening and migration, coking, and sulfur poisoning. The effect of heterointerfaces on the performance of air electrodes is discussed in section 4.3. In section 4.3.1, we summarize different air electrodes with heterointerfaces, including perovskite oxides with different cation compositions, perovskite oxides with same cation compositions, simple metal oxides and perovskite oxides, and multiphase catalysts and perovskite oxides. In section 4.3.2, we discuss the role of heterointerfaces in determining the performance of air electrodes. In section 4.4, we introduce different fuel electrodes with heterointerfaces, including simple metal-oxide and Ni-based fuel electrodes, perovskite and fluorite oxide/Ni-based fuel electrodes, and perovskite-based fuel electrodes with exsolved nanoparticles. In section 4.4.2, the impacts of heterointerfaces on the activity and stability of fuel electrodes are briefly discussed.

4.1.1 Air electrodes

4.1.1.1 Reaction kinetics in air electrodes

During the operation of an SOFC or an SOEC, the oxygen reduction reaction (ORR) or the oxygen evolution reaction (OER) respectively occurs on the air electrode side. Because of the strong oxygen–oxygen bond, the ORR or OER is normally the main contributor to the energy loss in SOFC or SOEC devices. Here, taking the ORR as an example, we discuss the reaction kinetics on the air electrode. The ORR can be written in Kröger–Vink notation as follows:

$$\frac{1}{2}O_{2,\ gas} + 2e^{-} + V_O^{\cdot\cdot} \rightarrow 2O_O^{\times}. \tag{4.1}$$

The ORR can also be divided into multiple steps that occur on the surfaces and in the bulk of air electrodes. A representative reaction pathway includes the following elemental reaction steps: gas diffusion from the gaseous phase to the gas/electrode interface, surface adsorption (equation (4.2)), dissociation (equation (4.3)), charge transfer (equations (4.4) and (4.5)), and incorporation (equation (4.6)). It should be noted that the dissociated oxygen can either diffuse through the electrode surface or through the bulk of the electrode toward the electrode/electrolyte interface [1–4].

$$O_{2,\ gas} \rightarrow O_{2,\ ad} \tag{4.2}$$

$$O_{2,\ ad} \rightarrow 2O_{ad} \tag{4.3}$$

$$O_{2,\ ad} \xrightarrow{e^-} O^-_{2,\ ad} \xrightarrow{e^-} O^{2-}_{2,\ ad} \xrightarrow{e^-} 2O^-_{ad} \xrightarrow{e^-} O^{2-}_{ad} \tag{4.4}$$

$$O_{ad} \xrightarrow{e^-} O^-_{ad} \xrightarrow{e^-} O^{2-}_{ad} \tag{4.5}$$

$$O^{2-}_{ad} \xrightarrow{V^{\cdot\cdot}_O} 2O^{\times}_O \tag{4.6}$$

The incorporation of O ions into the electrode and the transfer of O ions within the electrode are two key reaction processes during the ORR [5, 6]. The Adler–Lane–Steele (ALS) model [5] has shown that the oxygen surface exchange coefficient (k^*, cm s^{-1}) and the oxygen self-diffusion coefficient (D^*, $cm^2\ s^{-1}$) are two important parameters for describing the reaction kinetics of SOFC air electrodes. The isotopic exchange technique, e.g. an $^{18}O/^{16}O$ exchange experiment, combined with secondary ion mass spectrometry (SIMS), and electrical conductivity relaxation are powerful tools for measuring D^* and k^* [6–12]. The surface exchange of oxygen is found to be affected by the concentration and mobility of oxygen vacancies. Therefore, oxygen vacancies play a critical role in the surface exchange process. Several groups have reported a strong correlation between D^* and k^* [6, 13]. Consequently, many air electrodes with high oxygen ionic conductivity also have good oxygen exchange characteristics [6, 14].

4.1.1.2 Air electrode materials

As discussed above, charge transfer and surface/bulk oxygen diffusion are critical for the ORR [15–18]. Therefore, SOFC air electrodes normally require high ion/electron conductivity and high surface exchange coefficients for the ORR. High ionic conductivity can enable the active region to extend from just the triple phase boundary (TPB) region to the entire air electrode surface through bulk diffusion [19]. Some reports have also stated that high air electrode ionic conductivity not only increased the reaction area but also accelerated the surface reaction rate [1, 20]. In addition, the air electrode needs to have high porosity (to allow the diffusion of oxygen through the air electrode to the air electrode/electrolyte interface), sufficient stability in oxidizing atmospheres, and high chemical compatibility with electrolytes.

Currently, ceramics which are mixed ion–electron conductors (MIECs) are widely used as SOFC air electrode materials due to their high oxygen exchange coefficients and high electronic conductivity [21]. In particular, perovskite-based ($A_mB_nO_x$) MIECs are among the most studied air electrodes, including single perovskites such as $La_{1-x}Sr_xCoO_{3-\delta}$ (LSC), $La_{1-x}Sr_xCo_{1-y}Fe_yO_3$ (LSCF), and $Ba_{1-x}Sr_xCo_{1-y}Fe_yO_{3-\delta}$ (BSCF) [22–24], double perovskites such as $PrBaCoFeO_{5+\delta}$, $Sr_2Co_{1+x}Mo_{1-x}O_6$, and $Ba_2CoMo_{0.5}Nb_{0.5}O_{6-\delta}$ [25–28], as well as Ruddlesden–Popper perovskites such as $La_{2-x}Sr_xCoO_4$, $La_2NiO_{4+\delta}$, and $Pr_2NiO_{4+\delta}$ [29–31]. *A*-site cations in perovskite-based oxides are usually a mixture of rare and alkaline-earth metal ions (such as La, Sr, Ba, Sm, etc.), while the ions in the *B*-site are reducible transition-metal ions (such as Mn, Fe, Co, Ni, etc). Replacing *A*-site cations with a lower-valence metal is a common way of increasing the oxygen non-stoichiometry of perovskite materials. When the *A*-site cations (such as La^{3+}) are replaced by lower-valence metal ions (such as Sr^{2+}, Ba^{2+}, Ca^{2+}, or Mg^{2+}), lattice oxygen atoms are released to maintain charge neutrality, forming oxygen vacancies in the lattice. The oxygen ion migration in most of the single-perovskite oxides is mediated by oxygen vacancies. Therefore, *A*-site doping with low-valence cations usually leads to an improvement in the ionic conductivity of the material [32–36]. The impact of *B*-site cation doping on the ion migration and surface oxygen exchange has been relatively less studied. Recently, Chen *et al* [37] reported an increase in oxygen activity and a decrease in oxygen vacancy formation energy in perovskite ferrite which was obtained by replacing Fe at the *B*-site with Co.

The rate of oxygen diffusion in the bulk of perovskite-based oxides has been reported to be strongly influenced by crystal geometry factors. Due to the mismatch between the radii of *A*-site and *B*-site cations, the perovskite lattice normally exhibits a distortion from cubic symmetry. Goldschmidt's tolerance factor (t) is an important parameter used to describe lattice distortion and is calculated as follows:

$$t = \frac{r_A + r_O}{\sqrt{2}\left(r_B + r_O\right)}, \tag{4.7}$$

where r_A, r_B, and r_O are the radii of the *A* cation, the *B* cation, and O ions, respectively. Cherry *et al* [38] found that when $0.75 < t < 1.0$, the perovskite structure is stable, while at $t = 0.81$, the material has the lowest oxygen migration energies and fast diffusion, leading to high ion conductivities.

4.1.2 The fuel electrode

4.1.2.1 Reaction kinetics in the fuel electrode

The hydrogen oxidation reaction and the hydrogen evolution reaction are the most widely investigated reactions on the fuel electrodes of SOCs. In the SOFC mode, the electrochemical oxidation of H_2 that occurs in a fuel cell can be written in Kröger–Vink notation as follows [39]:

$$H_{2,\,gas} + O^{\times}_{O,\,YSZ} \rightarrow H_2O_{gas} + V^{\cdot\cdot}_{O,\,YSZ} + 2e^-. \tag{4.8}$$

The detailed reaction steps include the adsorption of H_2 molecules on the fuel electrode surface, the dissociation of H_2 molecules into hydrogen atoms, the oxidation reaction that produces protons (H^+), and the reaction of the generated protons with oxygen ions transferred from the air electrode. In the SOEC mode, water vapor from the fuel electrode undergoes a reduction reaction, gaining electrons to produce H_2 [40].

4.1.2.2 Fuel electrode materials

Fuel electrode materials with high activity normally exhibit the following characteristics: high electronic conductivity to allow rapid electron passage through the fuel electrode, sufficient thermal stability to ensure no degradation at high temperatures and in a redox atmosphere, high chemical stability to prevent reactions with contaminants and the formation of insulating phases at the electrode–electrolyte interfaces, chemical compatibility with adjacent components, etc. In addition, they must have a large specific surface area and a continuous porous structure to allow the rapid transport of fuel and reaction by-products [39, 41–43].

Commonly used fuel electrode materials include Ni-based materials, Cu-based materials, and perovskite-based materials. Ni is the most widely used fuel electrode material [39, 41] because of its low cost, good chemical stability, and high electrocatalytic activity. However, when Ni is used as a fuel electrode material on its own, the large mismatch between the coefficient of thermal expansion (CTE) of the electrode and that of the electrolyte leads to performance degradation. Therefore, Ni is often mixed with other materials such as electrolyte materials (e.g. yttria-stabilized zirconia (YSZ), gadolinia-doped ceria (GDC), samarium-doped ceria (SDC), etc.), or metal oxides (e.g. CaO, BaO, SnO_2, TiO_2, etc.) to form composite electrodes in order to improve electrode performance [44–48]. The most widely used composite material is Ni-YSZ, which has high electronic conductivity, the required ionic conductivity, and good electrocatalytic activity. The Ni, which is highly conductive, provides the electron conduction path, and the addition of YSZ provides the conduction path for the diffusion of oxygen ions. Such a combination can extend the three-phase interface, provide a better match for the CTE of the electrolyte, and inhibit the coarsening of Ni [39, 41, 49, 50]. When hydrocarbon fuels are used in SOFCs, Ni-based materials are prone to carbon deposition. Therefore, Cu-based electrodes, which are insensitive to carbon deposition, have attracted much interest. Since Cu has a low melting point (1083 °C) and is not catalytically active, it is often mixed with other materials to make composite electrodes. Representative examples include Cu-SDC, Cu-GDC, and Cu–CeO_2–YSZ [51–55]. Perovskite-based materials, such as $(La_{0.75}Sr_{0.25})Cr_{0.5}Mn_{0.5}O_3$ (LSCM) [56, 57], $(La_{0.75}Sr_{0.25})Cr_{0.5}Ti_{0.5}O_3$ (LSCrT) [58], $SrFeMoO_{6-\delta}$ [59], and $Sr_2MgMoO_{6-\delta}$ [60, 61], are also used as fuel electrode materials in SOCs.

4.2 Electrode degradation

The degradation of electrode performance is affected by the characteristics of the electrode material and the functional environment. The main degradation mechanisms include interactions between the electrodes and the electrolytes, cation segregation, the coarsening of the microscopic structure of the electrodes, mismatches caused by thermal expansion, and interactions between the electrodes and contaminants such as Cr, S, B, CO_2, and H_2O. This section summarizes the main degradation mechanisms proposed in the scientific literature for both the air electrodes and the fuel electrodes.

4.2.1 Air electrode degradation

4.2.1.1 Cation segregation

In perovskite-based air electrodes, certain cations tend to migrate from the lattice to the surface or the interface between the electrode and electrolyte, thereby significantly affecting the catalytic activity and the stability of the electrode [62–65]. In particular, Sr dopant at the *A*-site has been widely observed to segregate to the surfaces of many perovskite-based electrodes during operation, such as $La_{1-x}Sr_xMnO_{3-\delta}$ [66–68], LSC [69–72], and LSCF [23, 73, 74]. For instance, Pan *et al* [74] investigated the degradation mechanism of LSCF and found that the increase in the polarization impedance of the LSCF electrode is linear with the square root of time. As the test temperature increases from 650 °C to 750 °C, the degradation rate increases from 0.0139 to 0.136 under open-circuit voltage (OCV) conditions, which means that the electrode degrades faster at higher test temperatures. Simner *et al* [75] found that LSCF air electrodes showed a 30% power degradation after being tested at 750 °C and 0.7 V for 500 h due to Sr enrichment at the interfaces between the air electrode and the electrolyte and between the air electrode and the current collector. Kubicek *et al* [70] reported that the polarization resistance of LSC samples deposited at 450 °C and 650 °C increased from 0.1 Ω cm^2 to 0.8 Ω cm^2 and from 0.7 Ω cm^2 to 100 Ω cm^2, respectively, after long-term stability tests at 600 °C for 72 h. This difference in degradation rate was due to different amounts of Sr enrichment at the surfaces of electrodes deposited at different temperatures. Cai *et al* [76] found that the surface polarization resistance of LSC thin films increased rapidly during annealing in an oxygen environment. The removal of Sr segregation layers by chemical etching with HCl (0.14 mol L^{-1} for 10 s) could significantly improve the surface exchange kinetics on LSC. Nevertheless, with further annealing after etching, the performance of the LSC cells declined again due to Sr segregation.

The reason that Sr segregation leads to electrode degradation is normally considered to be the formation of non-conductive secondary phases on the surface, such as $Sr(OH)_2$, $SrCO_3$, etc. These secondary phases hinder oxygen exchange and ion transport on the surface, thus reducing the catalytic activity of the electrode for the ORR [76–78]. Chen *et al* [78] revealed the effect of Sr segregation on surface electronic structure and ORR activity of $SrTi_{1-x}Fe_xO_3$ (STF). They found that Sr segregation increased with the increase of Fe content in STF (figure 4.1(a)). Removal of Sr segregation layers by chemical etching can significantly improve surface exchange kinetics on STF. With the combination of scanning tunneling microscopy/spectroscopy (STM/STS) and x-ray photoelectron spectroscopy (XPS), they further discovered that the band gap on the surface of the material increases with the increase of Fe content, due to the generation of a Sr enrichment layer (figure 4.1(b)). The results suggested that the Sr enrichment layer could reduce the oxygen reduction kinetics of STF by hindering

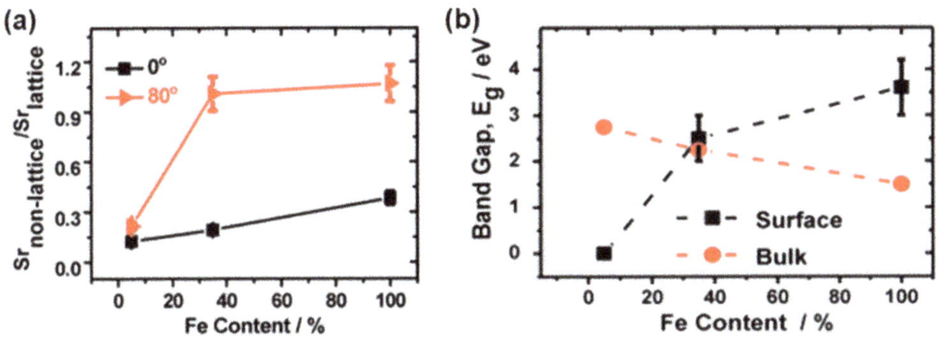

Figure 4.1. (a) The ratio of $Sr_{non\text{-}lattice}/Sr_{lattice}$ on the surface and in the bulk in $SrTi_{1-x}Fe_xO_3$ (STF, x = 5%, 35%, 100%). (b) Comparison of the band gap at the surface and in the bulk for $SrTi_{1-x}Fe_xO_3$ (STF, x = 5%, 35%, 100%). Reproduced from [78] with permission of The Royal Society of Chemistry.

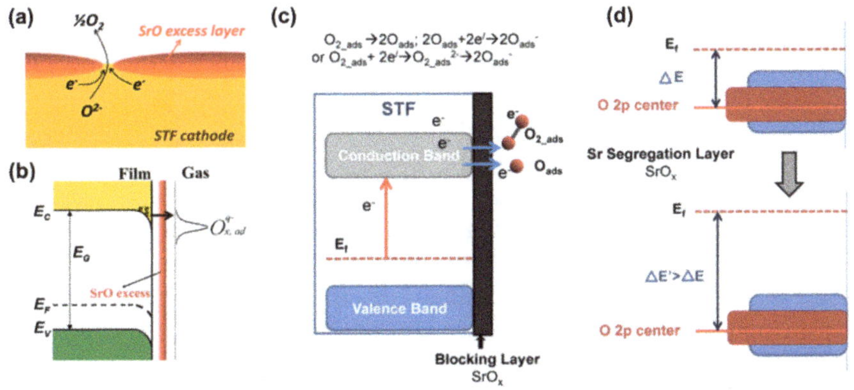

Figure 4.2. (a)–(c) Surface SrO segregation layer hinders charge transfer between the bulk STF and the oxygen species adsorbed on the surface. (d) The down shift of the O 2p center caused by SrO enrichment layer. Reproduced from [77, 78] with permission of The Royal Society of Chemistry.

charge transfer and oxygen exchange process on STF surface (figure 4.2). Similarly, Wang *et al* [73] reported that Sr segregation on porous LSCF electrodes effectively blocked the surface exchange process, leading to decreases in both D^* and k^* by 50% under thermal aging at 800 °C.

4.2.1.2 Deposition of contaminants and secondary phase formation

The interaction between air electrode surfaces and species such as Cr, S, CO_2, and H_2O in the environment leads to contaminant deposition and the formation of insulating secondary phases. Both effects can prevent the oxygen surface exchange process and the diffusion of oxygen within the air electrodes, thus reducing SOC performance.

Volatile chromium is widely observed as a contaminant in SOC systems. The poisoning effect caused by Cr can result in a sharp decline in SOC performance [79–83]. For instance, Jiang *et al* [84, 85] investigated the deposition process of Cr on LSCF and found that Cr was preferentially deposited on the surfaces of the LSCF electrode and chemically reacted with the surface-enriched SrO to form $SrCrO_4$ and Cr_2O_3 phases. These dense $SrCrO_4$ and Cr_2O_3 layers on the surface reduced the electrical conductivity and surface porosity. Huang *et al* [86] quantitatively determined the effect of Cr on the oxygen exchange kinetics of LSCF powder using an *in situ* gas isotope oxygen exchange experiment. The results showed that the apparent activation energy k (E_k) of the aging LSCF was 45.4 kJ mol^{-1}, while the E_k of Cr-poisoned LSCF decreased to 30.3 kJ mol^{-1}. The authors believed that the formation of secondary phases such as $SrCrO_4$ and Cr_2O_3 changed the stoichiometry of LSCF near the surface. This effect not only limited the active sites at which the reaction occurred but also reduced the electrochemical activity of the nearby regions. In addition, the authors found that E_k further decreased to 20.8 kJ mol^{-1} for Cr-poisoned LSCF in the presence of water, suggesting that water could exacerbate Cr poisoning. Ni *et al* [87] found that Cr poisoning led to a significant decrease in conductivity, the oxygen diffusion coefficient, and the oxygen exchange rate of an LSCF electrode by one to two orders of magnitude. They used several techniques, such as focused ion beam scanning electron microscopy (FIB-SEM) and multiple transmission electron microscopy (TEM) to reveal the reaction between Cr and LSCF at the microscale and the nanoscale. The results showed that significant degradation of the air electrode occurred before significant microscale changes took place. The degradation in electrode performance was attributed to the formation of nanosized $SrCrO_4$, the segregation of Cr near the grain boundary, the co-precipitation of Fe and Co, and the reduction of Fe valence due to Cr substitution.

Sulfur dioxide in ambient air is another key contaminant in SOC systems which critically impacts the stability of these devices [88–92]. Wang *et al* [92] found that the polarization resistance of LSCF air electrodes increased with an increase in the SO_2 concentration in the gas environment. Their study confirmed that $SrSO_4$ was generated by the reaction of SO_2 with Sr components and covered the entire grain surface. In addition, the segregation of $CoFe_2O_4$ from the perovskite lattice reduced the oxygen vacancy concentration in the electrode lattice and hindered the surface charge transfer process. Bucher *et al* [89] carried out *in situ* measurements to monitor the changes in the oxygen exchange kinetics of LSC in a SO_2 atmosphere; they found the chemical surface exchange coefficient of oxygen k^* continually decreased due to the formation of $SrSO_4$.

Studies have also shown that the presence of CO_2 and H_2O in the environment can accelerate the degradation of perovskite electrodes containing alkaline-earth elements [93–96]. For example, Yan *et al* [93, 94] studied the effects of CO_2 on the activity and stability of a BSCF electrode. They found that an increase in the concentration of CO_2 exacerbated the toxic effect of CO_2 on BSCF. This effect was attributed to the formation of Sr or Ba carbonate layers on the surface. Chen *et al* [97] reported that the degradation rate of an LSCF electrode noticeably increased as the concentration of CO_2 increased from 1% to 10%. Using *in situ* surface-enhanced Raman spectroscopy (SERS), the researchers found that a $SrCO_3$ peak appeared when pure O_2 was replaced with 10% CO_2 at 500 °C. Sun *et al* [98] found that H_2O could occupy the surface active sites of a BSCF electrode, thereby inhibiting the adsorption, dissociation, and surface diffusion of O_2. Consequently, they observed the appearance of interface resistance in the impedance spectra at medium and low frequencies after introducing H_2O into the system. In addition, Zhao *et al* [99] investigated the degradation mechanism of LSCF electrodes. They found that the degradation of LSCF at 600 °C was mainly due to the strong co-adsorption of CO_2 and H_2O, which reduced oxygen vacancies on the surface. On the other hand, $SrCO_3$ that forms on the surface hinders the activation and surface diffusion of oxygen. In the presence of H_2O, the volatile species of $Sr(OH)_2$ that form could easily migrate to the surface and react with CO_2 to generate $SrCO_3$, thereby intensifying the chemical reaction between CO_2 and the perovskite electrode. Chen *et al* also directly observed the changes in surface chemistry in the presence of H_2O using atmospheric pressure x-ray photoelectron spectroscopy (AP-XPS) [100].

4.2.2 Fuel electrode degradation

4.2.2.1 The coarsening and migration of nickel

Ni-based composites are the most widely used fuel electrode materials in SOCs. Although Ni has high catalytic activity, it undergoes severe coarsening during sintering and operation, resulting in a degradation in cell performance. This coarsening effect is considered to be an important cause of the degradation of SOC performance [101–104] (figures 4.3(a) and (b)). For example, Simwonis *et al* [105] exposed Ni-YSZ cermet in Ar/4% H_2/3% H_2O at 1000 °C for up to 4000 h. The resulting decrease in conductivity was found to be accompanied by an increase in Ni particle size. After 4000 h of testing, the conductivity decreased from 3900 S cm^{-1} to 2600 S cm^{-1}, and the average diameter of the Ni particles increased from 2 to 2.6 μm. Jiao *et al* [106] investigated the degradation mechanism of a Ni/YSZ fuel electrode using the FIB-SEM technique. They found that both the ohmic resistance (R_Ω) and the polarization resistance (R_p) increased with the operational time. After 250 h of testing, the specific surface area decreased by 17.9% and the active TPB decreased by 57%. The authors believed that the increase in resistance was mainly due to the Ni coarsening effect, which resulted in an increase in isolated Ni particles and a decrease in the active TPB. In addition, the reduction of NiO led to a cubic-to-tetragonal monoclinic crystal transition in the YSZ, resulting in the loss of ionic conductivity from the YSZ. Several other studies drew similar conclusions, namely that the coarsening of Ni grains was a major reason for the decrease in the active TPB and the degradation of electrode performance [107–110].

The sintering temperature/time, electrode composition, and ambient atmosphere also have an effect on the coarsening of Ni particles [102, 110, 111]. Jiao *et al* [110] found that fuel electrodes sintered at 500 °C showed the largest initial R_Ω and R_p. The reduced fuel electrode at 800 °C had the highest initial performance, which was attributed to the maximum active TPB density. Here, the average degradation rates of R_Ω with increasing time were 0.30% h^{-1} at 500 °C, 0.24% h^{-1} at 800 °C, and 0.22% h^{-1} at 1000 °C, respectively. The average degradation rates of R_p were 0.44% h^{-1} for 500 °C, 2.09% h^{-1} for 800 °C, and −0.43% h^{-1} for 1000 °C. The enhanced Ni-YSZ interfacial bonding that occurs with increasing temperature is

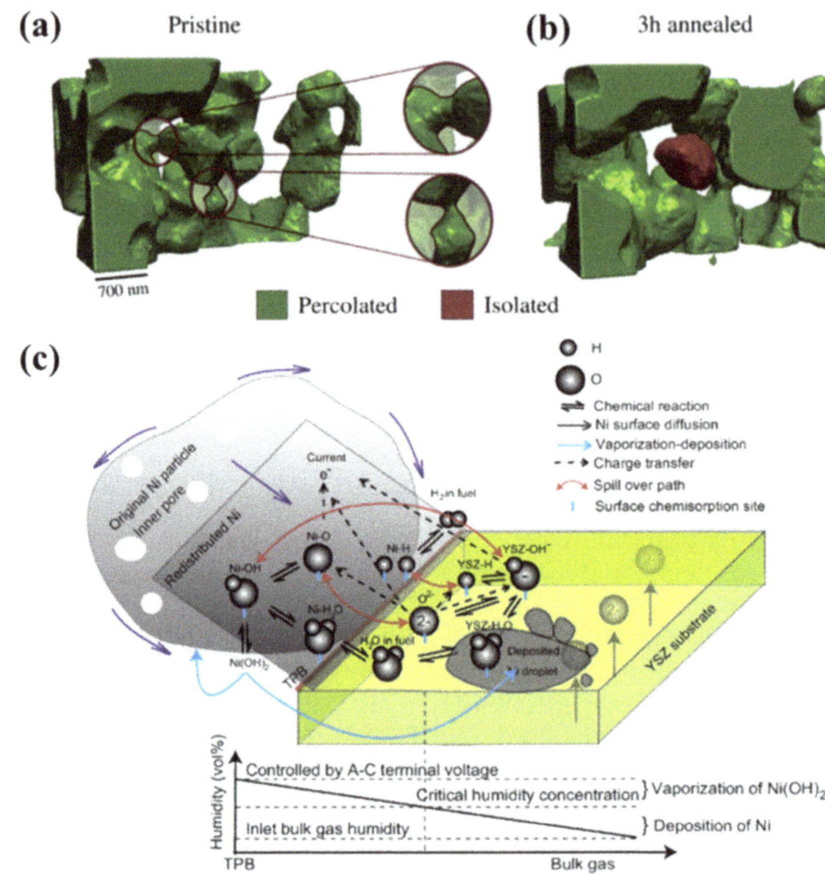

Figure 4.3. (a and b) Morphological evolution of Ni before and after annealing (in dry hydrogen at 850 °C, after 3 h of annealing, newly formed isolated Ni particles were observed). Reprinted from [107], Copyright (2018), with permission from Elsevier. (c) Illustration of Ni densification and vaporization–deposition phenomena in a Ni-YSZ fuel electrode. Reprinted from [112], Copyright (2012), with permission from Elsevier.

beneficial, as it improves the stability of the fuel electrode during long-term operation. The sintering behavior of Ni/YSZ metal–ceramic electrodes was investigated by Holzer *et al* [102]. They found that an increase in the Ni content of Ni/YSZ promoted the agglomeration and grain growth of Ni in the electrode. In addition, it was found that H_2O has a certain beneficial effect on the agglomeration and grain growth of Ni particles.

In addition to Ni coarsening, the Ni migration process has also been widely observed in Ni-based fuel electrodes after long-term operation, which has been attributed to the formation of volatile $Ni(OH)_2$ [112–115]. Jiao *et al* [112] found that the humidity gradient from the TPB outward to the ambient atmosphere drove the surface diffusion of nickel and the vapor deposition of $Ni(OH)_2$. Volatile $Ni(OH)_2$ can be stably present on Ni and NiO when the temperature is below 1000 °C; the critical value of the molar fraction of humidity in H_2 is between 20% and 30% (figure 4.3(c)). Several studies have shown that the impact of Ni migration is more severe than that of Ni coarsening for SOEC devices. During long-term operation, Ni migration leads to significant degradation of the electrode by depleting Ni in the electrode material [114–116]. Wang *et al* [113] tracked the Ni migration process in an SOEC using a phase field model. As the total overpotential increased from 0.153 to 0.234 V, an increase in Ni depletion was clearly observed. As a consequence, the electrochemical reaction region moved away from the electrode–electrolyte interface, extending the oxygen ion transport path and increasing R_Ω. In addition, the Ni migration could intensify the coarsening of the Ni particles and the reduction of active sites.

4.2.2.2 Coking and sulfur poisoning

When carbon-based fuels are used in SOCs, coking or carbon deposition often occurs on the fuel electrode surfaces, leading to degraded cell performance [117–120]. Sumi *et al* [121] found that a large amount of agglomerated amorphous carbon was deposited on the surface of a Ni–YSZ fuel electrode after operating the SOFC in methane fuel at a steam/carbon ratio of 0.1. The increase in polarization resistance was caused by the increased carbon deposition near the fuel electrode/electrolyte interface. In another study, Tao *et al* [122] reported the formation of carbon nanofibers at the Ni-YSZ|YSZ interface during SOEC operation at electrolytic current densities of 2.25 A cm^{-2} with a reactant conversion of 67%. Other studies found that the precipitation and intrusion of carbon into the lattice of Ni can lead to the corrosion of Ni, which results in a decrease in electrode conductivity and catalytic activity [123, 124].

Sulfur poisoning is another main reason for the performance degradation of SOC fuel electrodes. Various sulfur compounds can interact with the fuel electrode and degrade its performance. Shiratori *et al* [125] found that after injecting biogas containing 1 ppm of H_2S into a Ni-ScSZ (scandia-stabilized zirconia) fuel electrode for 1 h, the cell voltage dropped by about 100 mV (i.e. 9% of the initial voltage), and the reaction rate and CO yield decreased by about 40%. Mantilla *et al* [126] carried out durability tests of $La_{1.5}Sr_{1.5}Mn_{1.5}Ni_{0.5}O_{7\pm\delta}$ electrodes at 850 °C under low steam-to-carbon conditions. They observed the formation of harmful sulfides, which resulted in a decrease in the catalytic activity of the electrodes. Wachowski *et al* [127] exposed $SrFe_{0.75}Mo_{0.25}O_{3-\delta}$ (SFM) fuel electrodes to H_2 with 500 ppm of H_2S for 137 h of continuous operation and found that the maximum output power decreased by 50% at 700 °C. In addition, the open-circuit voltage (OCV) of the fuel cell decreased by 33% and 56% when 500 and 5000 ppm H_2S were added, respectively.

The sulfur poisoning was found to strongly depend on the concentration of H_2S, operating temperature, and cell voltage/current [128–131]. For example, Zha *et al* [128] applied a constant cell voltage of 0.6 V to an SOFC with Ni–YSZ fuel electrodes at 800 °C. They observed a sharp drop in cell current (16.67%) within the first few minutes when fuel containing 50 ppm H_2S was introduced. In contrast, when 2 ppm H_2S was introduced, the degradation was much slower and had an initial degradation rate of 12.67%. These results indicated that a higher H_2S

concentration in the fuel led to an increased equilibrium coverage of sulfur adsorption on Ni, which increased the degree of sulfur poisoning. Lee *et al* [129] reported that a high concentration of H_2S resulted in a Ni surface heavily covered by sulfur with few hydroxide sites, which exacerbated cell degradation. Kim *et al* [132] investigated the deleterious effect of sulfur on a Ni-YSZ fuel electrode using *in situ* SERS. They found that the cell current density decreased over time after they replaced pure CH_4 fuel with a fuel containing 0.5 ppm of H_2S at 700 °C. The SERS results showed that the aggregated sulfur, at or near the same location as the Ni-YSZ, deactivated the active sites. Finally, Hauch *et al* [133] found that sulfur-containing impurities led to higher fuel electrode overpotentials (higher cell voltages), thereby exacerbating carbon deposition.

4.3 The influence of heterointerfaces in air electrodes

4.3.1 The performance of SOCs with heterostructured air electrodes

In this section, we summarize the performance of SOCs with different heterostructured air electrodes, including heterostructures consisting of: (a) perovskite oxides with different cation compositions, (b) perovskite oxides with the same cation compositions, (c) simple metal oxides and perovskite oxides, and (d) multiphase catalysts and perovskite oxides.

4.3.1.1 Perovskite oxides with different cation compositions

Lynch *et al* [134] prepared LSCF electrodes with a uniform, continuous, dense $La_{1-x}Sr_xMnO_3$ (LSM) coating by impregnation and found that LSM-coated LSCF air electrodes showed higher performance and durability than those of bare LSCF electrodes. Wang *et al* [135] used $SrCo_{0.9}Nb_{0.1}O_{3-\delta}$ (SCN10) to modify the LSCF electrode and found that the R_p of the cell decreased significantly as the loading of the modified layer increased. When the loading volume of SCN10 was 20 wt%, the electrode exhibited the optimal porosity, and the loaded compound formed a continuous and uniform layer. Compared with that of bare LSCF, the R_p of 20 wt% SCN10-LSCF decreased by 89.4%. The heterostructured electrode also exhibited good stability during a 6000 h test, and the SCN10 modified layer retained its original morphology after the test. Li *et al* [136] introduced a thin layer of *A*-site deficient $La_{0.95}CoO_{3-\delta}$ (LC95) onto a porous LSCF electrode. The LSCF@LC95-based single cell exhibited a peak power of 0.72 W cm^{-2} at 700 °C, which was much higher than the value of 0.53 W cm^{-1} obtained for a cell with a bare LSCF electrode. In a stability test for 120 h at 400 mA cm^{-2}, the degradation rate of an LSCF-based single cell air electrode was approximately 0.13% h^{-1}. No significant performance degradation was observed in the LSCF@LC95-based single cell, and its degradation rate was as low as 0.05% h^{-1}.

Several research groups have found that $BaCoO_{3-x}$ (BCO) decoration can effectively improve the activity and stability of perovskite-based electrodes [137–140]. Zhou *et al* [137] coated an LSC air electrode with BCO nanoparticles to a thickness of about 30 nm (figures 4.4(a) and (b)). Using H_2 as the fuel and air as the oxidizer, the authors constructed reversible protonic ceramic electrochemical cells (R-PCECs). The cell with the decorated electrode obtained maximum power densities of 2.04, 1.60, 1.16, 0.78, and

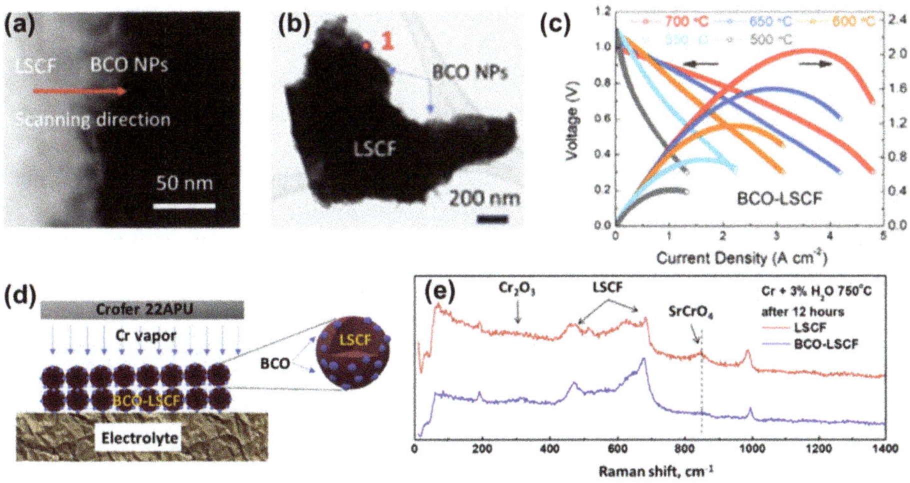

Figure 4.4. (a) A cross-sectional scanning transmission electron microscopy (STEM) image of a BCO-LSCF air electrode. (b) Transmission electron microscopy (TEM) image of the cross-section of an LSCF grain covered by BCO nanoparticles. (c) Typical *I–V–P* curves measured in the SOFC mode. Panels (a)–(c) reproduced from [137] John Wiley & Sons, copyright 2021 Wiley-VCH GmbH]. (d) Schematic of the BCO-LSCF air electrode used for the Cr poisoning test in symmetrical cells; (e) typical Raman spectra for LSCF and BCO-LSCF dense pellets after the Cr poisoning test for 12 h under 3% H_2O at 750 °C. Panels (d)–(e) reprinted from [138], Copyright (2020), with permission from Elsevier.

0.41 W cm^{-2} at 700 °C, 650 °C, 600 °C, 550 °C, and 500 °C, respectively; these values are much higher than the power density of a bare LSCF air electrode (figure 4.4(c)). Pei *et al* [138, 139] further studied the anti-Cr, CO_2 poisoning properties of BCO-LSCF and found that BCO-LSCF exhibited excellent ORR activity and excellent long-term stability in an atmosphere containing Cr, CO_2, and H_2O. The Raman spectra results showed that the $SrCrO_4$ peak (~850 cm^{-1}) intensity on the BCO-LSCF electrode surface was much smaller than that on bare LSCF. A carbonate peak (~1066 cm^{-1}) consistently appeared on the results for the surface of bare LSCF, but not on those for BCO-LSCF after the CO_2 poisoning test. These results suggested that the presence of the surface BCO layer limited the formation of $SrCrO_4$ on the electrode surface (figures 4.4(d) and (e)). Kim *et al* [140] induced BCO nanoparticles on the surface of double-perovskite $PrBa_{0.8}Ca_{0.2}Co_2O_{5+\delta}$ (PBCC). In wet air (containing 3 vol% H_2O), a BCO-PBCC composite electrode showed higher ORR activity and stability than those of a bare PBCC electrode. A cell with the BCO-PBCC electrode exhibited a peak power density of 1.46 W cm^2 at 750 °C, which is much higher than the value of 0.94 W cm^2 obtained for a cell with a bare PBCC electrode. The BCO-PBCC electrode also showed no significant performance degradation during a 500 h durability test.

4.3.1.2 Perovskite oxides with the same cation compositions

Some studies have shown that a perovskite-based heterostructure which consists of two phases with the same cation compositions but different structures exhibits fast oxygen surface exchange kinetics.

Representative examples are heterostructures consisting of single perovskite ($ABO_{3-\delta}$) and a Ruddlesden–Popper phase ($A_2BO_{4+\delta}$). Yashiro *et al* [141] found that the electrode performance of $La_{0.6}Sr_{0.4}CoO_{3-\delta}$ ($LSC64_{113}$)/$LaSrCoO_{4-\delta}$ ($LSC55_{214}$) was much higher than that of LSC_{113} alone. Sase *et al* [142, 143] further carried out ^{18}O isotope exchange experiments and SIMS measurements to determine the oxygen exchange kinetics of LSC_{113}/LSC_{214} heterointerfaces. The results showed that the oxygen exchange process near the heterointerface is several orders of magnitude larger than that of single-phase LSC_{113} or that of LSC_{214}. Crumlin *et al* [144] decorated $La_{0.8}Sr_{0.2}CoO_{3-\delta}$ (LSC_{113}) thin films with an epitaxial growth of $(La_{0.5}Sr_{0.5})_2CoO_{4-\delta}$ (LSC_{214}) by PLD. LSC_{214}-decorated LSC_{113} showed an enhanced k^*, which was three to four orders of magnitude larger than that of single-phase LSC (figure 4.5(a)). Zhao *et al* [145] reported that the ORR/OER activity of a heterostructured LSC_{113}/LSC_{214} thin film was significantly higher than that of LSC_{113}. The k^* of the heterostructure was three to ten times higher than that of single-phase LSC_{113} in the temperature range of 873 K–1123 K. In addition, the k^* of the single-phase LSC_{113} dropped to one sixth of its original value after 8 h of annealing at 873 K, while that of LSC_{113}/LSC_{214} only dropped by half. Ma *et al* [146] prepared vertically aligned nanocomposite (VAN) LSC_{113}/LSC_{214} composite electrodes and tested the temperature dependence of the area specific resistance (ASR) for oxygen exchange. They found that the ORR activity of VAN LSC_{113}/LSC_{214} was significantly improved by about an order of magnitude in comparison with those of single-phase LSC_{113} or LSC_{214} electrodes (figure 4.5(b)). Hong *et al* [147] reported that a three-dimensional (3D)-$La_{0.8}Sr_{0.2}FeO_{3-\delta}$/$(La,Sr)_2FeO_{4-\delta}$ (LSF_{113}/LSF_{214}) heterostructure exhibited a k^* of 4.0×10^{-4} cm s^{-1}, which was much higher than the k^* value of 1.0×10^{-5} cm s^{-1} obtained for single-phase LSF_{113}. Similar results were observed by Demirkal *et al* [148] for the same heterostructure. Zheng *et al* [149] found that the surface oxygen exchange coefficient of an air electrode with a $Nd_{0.5}Sr_{0.5}CoO_{3-\delta}$($NSC_{113}$)/$Nd_{0.8}Sr_{1.2}CoO_{4\pm\delta}$($NSC_{214}$) heterointerface was much higher than that of single-phase NSC_{113} or single-phase NSC_{214}. The ORR kinetics can be further enhanced by increasing the density of the heterointerface.

In addition to the 113/214 heterostructure, Xu *et al* [150] reported a heterostructured air electrode for R-PCECs which consisted of a double-perovskite $PrBaCo_{1.6}Fe_{0.2}Nb_{0.2}O_{5+\delta}$ (PBCFN) backbone covered by $PrBaCo_{1.6}Fe_{0.2}Nb_{0.2-x}O_{5+\delta}$ (Nb-deficient PBCFN) nanoparticles. The heterostructured electrode exhibited high ORR/OER catalytic activity as well as good durability in a reversible cycling test in the fuel cell (FC) and electrolysis cell (EC) modes. In the FC mode, the peak power density of the cell with the heterostructured electrode reached 1.059 W cm^{-2} when wet H_2 (3% H_2O) and wet air (3% H_2O) were fed into the fuel electrode and the air electrode, respectively. In the EC mode, the current density was as high as 2.148 A cm^{-2} at a cell voltage of 1.3 V at 650 °C. Kim *et al* [151] prepared a self-assembled multiphase composite electrode $Ba_{0.5}Sr_{0.5}Co_{0.6}Fe_{0.2}Zr_{0.1}Y_{0.1}O_{3-\delta}$ (BSCFZY) which was composed of Co-rich ($Ba_{0.5}Sr_{0.5}Co_{0.7}Fe_{0.2}Zr_{0.07}Y_{0.03}O_{3-\delta}$) and Zr-rich ($Ba_{0.6}Sr_{0.4}Co_{0.3}Fe_{0.2}Zr_{0.4}Y_{0.1}O_{3-\delta}$) phases. The k^* value of the biphasic composite electrode reached ~7.87×10^{-3} cm s^{-1} at 450 °C, which is much higher than those obtained for bare BSCF (~1×10^{-4} cm s^{-1}) and bare BCFZY (5×10^{-5} cm s^{-1}). At

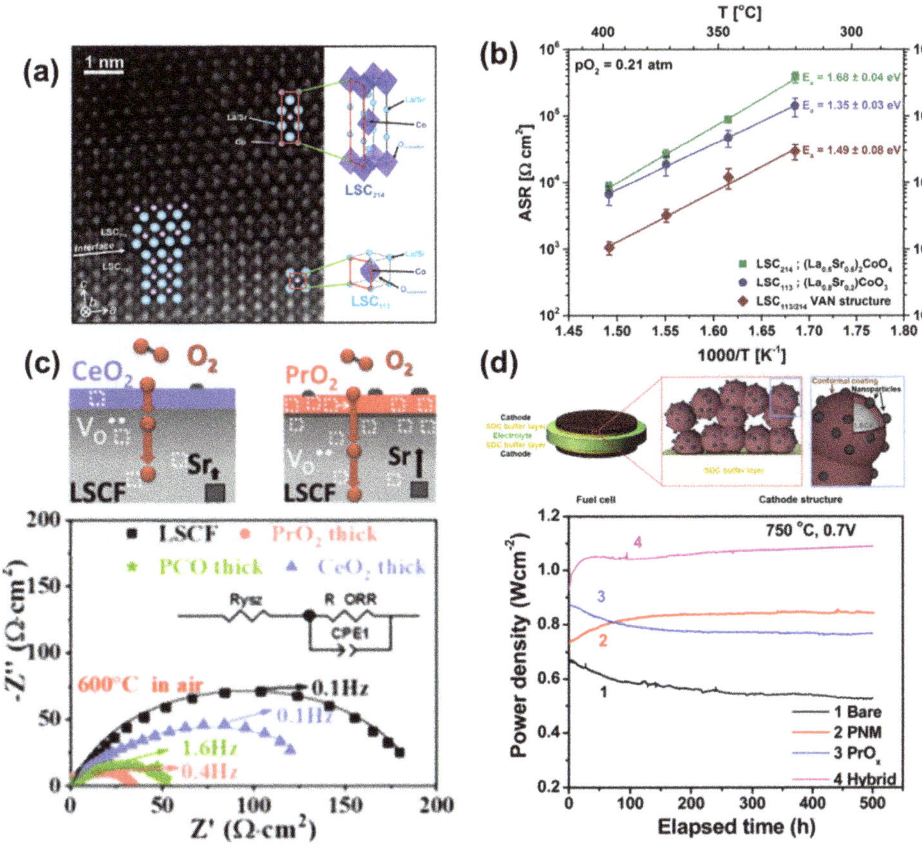

Figure 4.5. (a) High-angle annular dark field scanning transmission electron microscopy (HAADF-STEM) micrograph of an LSC_{113}/LSC_{214} heterointerface. Reprinted with permission from [144]. Copyright (2010) American Chemical Society. (b) The temperature dependence of the ASR of oxygen exchange of VAN LSC_{113}/LSC_{214}. Reproduced from [146] with permission of The Royal Society of Chemistry. (c) (top) The effect of oxygen vacancies in PrO_2-, CeO_2-modified layers on the performance of heterostructured electrodes; (bottom) the electrochemical impedance spectroscopy (EIS) of bare LSCF and PrO_2-, PCO-, and CeO_2-modified LSCF measured in air at 600 °C. Reprinted with permission from [156]. Copyright (2018) American Chemical Society. (d) (top) Schematic diagram of an LSCF electrode skeleton modified by PNM coating and PrO_x nanoparticles; (bottom) stability testing of single cells with bare LSCF, PNM-, PrO_x-, and PNM-PrO_x hybrid catalyst coatings on LSCF air electrodes at a constant cell voltage of 0.7 V at 750 °C using 3% humidified H_2 as the fuel and ambient air as the oxidant. Reprinted with permission from [100]. Copyright (2018) American Chemical Society.

650 °C, the ASR of the composite was 0.013 Ω cm^2, and the output power was ~1.1 W cm^{-2}. The authors observed no obvious performance degradation in a long-term stability test of 310 h.

4.3.1.3 Single-metal oxides and perovskite oxides

Several research groups have found that the introduction of stable, high-valence, and less reducible cations, such as Zr^{4+} [152, 153], Ti^{4+} [154], and Hf^{4+} [155] to the electrode surface can improve the stability of perovskite electrodes. For example, Gong *et al* [152] coated a conformal nanoscale ZrO_2 film onto the surface of LSC by atomic layer deposition (ALD). Compared to those of bare LSC samples, LSC samples with nanoscale ALD-ZrO_2 coatings exhibited higher ORR activity and improved stability. After a stability test lasting for 4000 h, the R_P value of the ZrO_2-LSC electrode was 19 times smaller than that of the bare LSC electrode at 700 °C. Tsvetkov *et al* [154] modified an LSC surface using TiO_2, improving the chemical/electrochemical stability of the LSC surface. The surface oxygen exchange kinetics of an LSC air electrode with a modified layer of 10% Ti content were improved by eight times as a result of preventing Sr segregation. The same group further [155] compared the changes in k^* of LSC modified using different cations (Co, V, Nb, Zr, Ti, Hf, and Al). Studies showed that all samples initially had similar k^* values. With increasing test times, samples with different modification layers showed different degradation rates. The oxygen exchange kinetics of LSC-Hf16 were more than 30 times faster than those of LSC after a stability test lasting 54 h.

Doped cerium dioxide has also been widely used as a modification layer to improve the ORR kinetics of perovskite-based air electrodes. For example, Chen *et al* [156] used pulsed laser deposition (PLD) to prepare a $Pr_xCe_{1-x}O_2$ ($x = 0, 0.2, 1$) -coated LSCF thin-film electrode. They found that PrO_x-LSCF led to the most obvious improvement in ORR kinetics and that the surface oxygen exchange coefficient increased by six times. This improvement was attributed to the high vacancy concentration of the PrO_x surface layer. In addition, CeO_2-LSCF exhibited the best stability and maintained a high oxygen surface exchange coefficient after a long stability test. These effects were attributed to the inhibition of Sr segregation on the surface (figure 4.5c). Zhang *et al* [157] prepared heterostructured nanofibers which were composed of LSCF and CeO_2 nanoparticles using electrospinning technology. The CeO_2/LSCF composite air electrode showed an ASR of 0.154 Ω cm^2 at 600 °C, which was one third of the value obtained for LSCF fibers (0.537 Ω cm^2) and one sixth that of LSCF powder (0.961 Ω cm^2). After a durability test lasting for 200 h, the performance of the composites did not show significant degradation. Similarly, Li *et al* [158] found the same results for a GDC/LSCF composite nanofiber.

Nadeem *et al* [159] deposited NiO on an LSCF electrode using drop coating and sputtering methods. The k^* values of NiO-LSCF prepared by drop coating, NiO-LSCF prepared by sputter coating, and bare LSCF were 3.84×10^{-5} cm s^{-1}, 6.9×10^{-4} cm s^{-1}, and 6.65×10^{-5} cm s^{-1}, respectively. Single cells with a NiO-LSCF composite electrode showed an enhanced peak power density of 1.031 W cm^{-2} at 800 °C, which is much higher than the value of 0.731 W cm^{-2} obtained for a bare LSCF electrode. Yang *et al* [160] found that after decorating the LSCF electrode with MgO, the surface exchange coefficient increased by a factor of 2.4, and the area specific interfacial polarization resistance was reduced by two-fifths at 650 °C. In addition, $PrO_{2-\delta}$ and CuO were also used to improve the ORR activity and stability of a perovskite electrode [161, 162]. It is worth noting that not all the surface modifications led to improved performance. Some modifications using metal oxides such as CoO and Al_2O_3 were shown to have detrimental effects on the oxygen surface exchange process and to accelerate the degradation of electrode performance [163, 164].

4.3.1.4 Multiphase catalysts and perovskite oxides

Chen *et al* [100] coated an LSCF electrode with a composite which consisted of $PrNi_{0.5}Mn_{0.5}O_3$ (PNM) thin films and exsolved PrO_x nanoparticles. The coated PNM-PrO_x hybrid catalyst significantly improved the ORR activity and durability of an LSCF air electrode. At 750 °C, k^* increased from ~5.68×10^{-4} cm s^{-1} for bare LSCF to ~7.5×10^{-4} cm s^{-1} for PNM-PrO_x-modified LSCF. The peak power density increased from 0.79 W cm^{-2} for bare LSCF to 1.21 W cm^{-2} for PNM-PrO_x-modified LSCF. The modified electrode also exhibited excellent working stability for 500 h at a constant voltage of 0.7 V (figure 4.5(d)). Yang *et al* [165] synthesized a multiphase heterostructured air electrode ($Pr_{1.5}Sr_{0.5}Ni_{0.5}Mn_{0.5}O_{4-\delta}$ (S-PNM)), which consisted of PrO_x nanoparticles, $(PrSr)Ni_{0.5}Mn_{0.5}O_{3-\delta}$, and $(PrSr)_2(MnNi)O_{4-\delta}$ formed by self-assembly decomposition. The S-PNM electrode thus obtained showed improved ORR activity and stability in comparison with a single-phase electrode.

Niu *et al* [166] decorated PBCC layers with exsolved BCO nanoparticles on the surfaces of LSCF electrodes. They found that the k^* value of the multiphase-catalyst-coated LSCF was 9.30×10^{-4} cm s^{-1}, which was about two orders of magnitude larger than that of the bare LSCF (1.29×10^{-5} cm s^{-1}) at 650 °C. In the temperature range of 550°C–650°C, the maximum power densities of the cell with the multiphase-catalyst-coated LSCF electrode were 1.6 times higher than those of a cell with a bare LSCF electrode in fuel cell mode. In electrolysis mode, the current density of the cell with the heterostructured electrode was 1.5 times higher than that of a SOEC cell with a bare LSCF electrode.

4.3.2 The role of heterointerfaces in air electrodes

This section discusses the role of heterointerfaces in determining the performance of air electrodes. The proposed mechanisms include the presence of elastic strain near the heterointerface, the extension of the active area, changes in the local electronic structure, suppressed cation segregation, and contaminant deposition.

4.3.2.1 Strain effects

Elastic strain is generally observed near the heterointerface due to the difference in crystal structure. Studies have shown that such elastic strains can influence defect chemistry, charge transfer, cation composition, etc. thereby affecting the oxygen exchange kinetics of the air electrode [167–175].

Using first-principles calculations, Kushima *et al* [167] and Han *et al* [168] demonstrated that tensile strain can promote oxygen vacancy formation and oxygen surface adsorption on $LaCoO_3$. Combining x-ray diffraction (XRD), x-ray absorption spectroscopy (XAS) measurements, and density functional theory (DFT) calculations, Petrie *et al* [172] found that tensile strain can significantly promote the generation of oxygen vacancies in $SrCoO_x$. When a biaxial tensile strain of 2% is applied to $SrCoO_x$, the activation energy barrier (E_a) for oxygen ion migration was reduced by about 30%, and the intercalation enthalpy (H_i) was increased, which facilitated oxygen movement through ordered vacancy channels (OVCs) and reduced the stability of oxygen intercalation. Liu *et al* [176] experimentally observed that an LSC thin film on a $LaAlO_3$ substrate (LSC/LAO) under compressive strain had a lower oxygen vacancy concentration than that of LSC on $SrTiO_3$ (LSC/STO) under tensile strain. These results were consistent with the presence of higher oxygen vacancy formation energy in compressively strained LSC in comparison to the tensilely strained LSC, as revealed by DFT calculations (figure 4.6(a)). Furthermore, Chen *et al* [177] added a conformal Pt-CoO_x layer to an LSCF/GDC composite electrode using ALD. The authors believed that this conformal coating layer imposed a tensile strain on the LSCF, resulting in a decrease in the atomic bonding strength and an increase in the oxygen vacancies of the LSCF/GDC air electrode.

Han *et al* [178] evaluated the effect of the strain near the heterointerface of LSC_{113}/LSC_{214} on the oxygen incorporation and migration kinetics using DFT calculations. They observed a +1.9% in-plane tensile strain in LSC_{113} (001) and a −0.8% in-plane compressive strain in an LSC_{214} (001) layer. Due to the in-plane tensile strain in the LSC_{113} layer, the energy associated with the chemisorption of O_2 on the Co site and the oxygen vacancy formation process decreased, which was

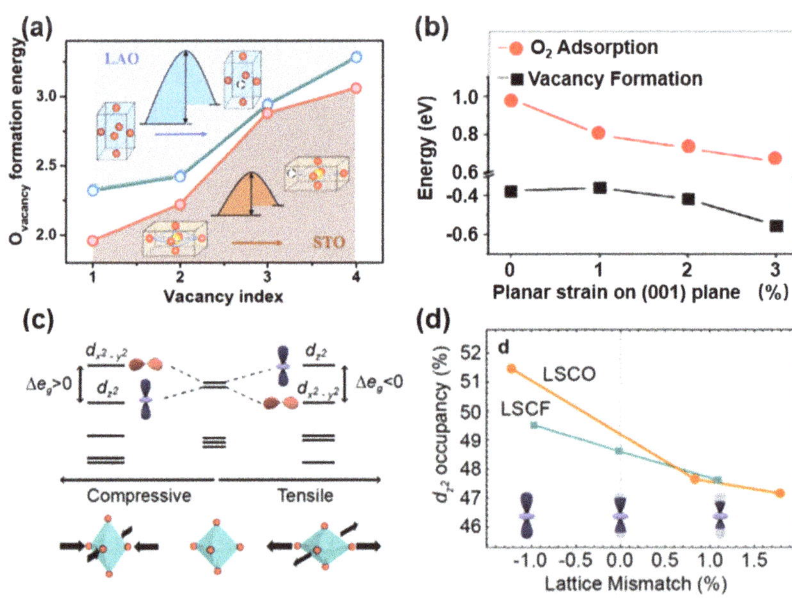

Figure 4.6. (a) Comparison of oxygen vacancy formation energies calculated for LSC/STO and LSC/LAO for different numbers of oxygen vacancies [176] John Wiley & Sons, copyright 2019 The Authors. Published by WILEY-VCH Verlag GmbH & Co. KGaA, Weinheim. (b) Dependence of O_2 adsorption and oxygen vacancy formation energy on planar tensile strain. Reproduced from [178] with permission of The Royal Society of Chemistry. (c) Schematic illustration of the relationship between epitaxial strain and the orbital energy of perovskite oxides [180] John Wiley & Sons, copyright 2021 Wiley-VCH GmbH. (d) Lattice strain versus occupancy of d_{z^2} orbitals [180] John Wiley & Sons, copyright 2021 Wiley-VCH GmbH.

attributed to the weakened hybridization of Co-*d* orbitals with lattice O-*p* orbitals (figure 4.6(b)). These results suggest that the lattice strain near the heterointerface could be favorable for oxygen adsorption and oxygen ion transport. Kang *et al* [179] synthesized a double columnar structure, which consisted of $Sm_{0.5}Sr_{0.5}CoO_{3-\delta}$ (SSC) with tensile strain and $Sm_{0.2}Ce_{0.8}O_{3-\delta}$ (SDC) with compressive strain. The authors believed that the surface oxygen dissociation activity was enhanced by the tensile strain in the SDC, while Sr segregation was suppressed by the compressive strain in SSC. Kubicek *et al* [171] used time-of-flight secondary ion mass spectrometry (ToF-SIMS) and ^{18}O isotope exchange depth profiling (IEDP) to simultaneously measure the surface exchange coefficient k^* and diffusion coefficient D^* of LSC/$SrTiO_3$ (LSC/STO) under tensile strain and LSC/$LaAlO_3$ (LSC/LAO) under compressive strain in the temperature range of 280 °C–475 °C. They showed that LSC/STO under tensile strain had higher k^* and D^* values than those of LSC/LAO under compressive strain over the entire temperature range. Furthermore, a similar strain effect was observed for $La_{1-x}Sr_xCoO_{3-\delta}$ with different compositions ($x = 0.2$ and $x = 0.4$) and for the film after surface chemical etching.

The presence of strain near heterointerfaces was also reported to induce changes in the electronic structure. Cai *et al* [170] studied the strain effects on the surface structure, chemical state, and electronic structure of LSC/LAO and LSC/STO films. Using *in situ* STM/STS measurement, they found that the energy gap of both thin films decreased with increasing temperature. At room temperature, both LSC/LAO and LSC/STO were semiconductors. When the temperature was raised to 300 °C, both films completely converted to metals with an energy gap of zero. The results of DFT calculations further showed that LSC/STO had lower oxygen vacancy formation energy than that of LSC/LAO. As a result, the LSC/STO surface exhibited an enhanced electron state density near the Fermi energy level due to the promotion of surface oxygen vacancy formation. The authors believed that tensile strain in LSC was beneficial for electron transfer between the material's surface and adsorbed oxygen molecules, thereby improving the ORR reaction at elevated temperatures [171].

The impact of strain on the electronic structure of catalysts and ORR/OER activities was also systematically investigated by Fernandez *et al* [180]. They found that the introduction of compressive strain made the energy of the $d_{x^2-y^2}$ orbital higher than that of the d_{z^2} orbital, whereas the introduction of tensile strain did the opposite (figure 4.6(c)). The decrease in the energy of the $d_{x^2-y^2}$ orbital for the tensile films led to an overall decrease in the electron occupancy of the d_{z^2} orbital (figure 4.6(d)). Since the reaction rates for oxygen exchange were accelerated in thin films with lower d_{z^2} occupancy, films under tensile strain exhibited higher ORR/OER performances. Similarly, Pesquera *et al* [181] found that lattice strain in LSM thin films led to the breaking of the perovskite oxygen octahedral symmetry, thereby affecting the ($d_{z^2}/d_{x^2-y^2}$) occupancy. They found that tensile strain favored $d_{x^2-y^2}$ occupancy, whereas compressive strain promoted $d_{x^2-y^2}$ orbital filling.

The presence of lattice strain was also reported to critically impact the cationic composition on the surface. Yu *et al* [182] found that compressive strain inside LSCF film inhibited Sr segregation, while tensile strain enhanced surface Sr segregation. The authors attributed these impacts on Sr segregation to the strain-induced changes in the unit cell volume, which led to changes in the migration energy and diffusion of Sr cations within the LSCF lattice. In contrast, Koo *et al* [183] reported that tensile strain in $SrTi_{0.5}Fe_{0.5}O_{3-\delta}$ (STF) increased the Sr–O bond length, which suppressed Sr enrichment on the surface. In addition, they believed that a certain stretch in the SrO monolayer formed on the STF surface could synergistically reduce Sr enrichment. When a compressive strain was imposed on STF, the vacancy formation energy of Sr ($E_{Sr,vac}$) decreased due to a weakening of the Sr–O bond strength, resulting in surface Sr enrichment.

4.3.2.2 Changes in electronic structure

The introduction of heterointerfaces was reported to directly change the local electronic structure near the interface. Using STM/STS measurement, Kuru *et al* [184] observed that the electronic structure of an STF layer was affected by an adjacent LSM layer; the energy gap reduced from the reported value of ~2 eV to 1 eV in an LSM/STF multilayer structure. In another study, Chen *et al* [185] investigated the changes in the electronic structure of different components in an LSC_{113}/LSC_{214} multilayer structure using a combination of *in situ* STM/STS and grazing-incidence focused ion beam (FIB) milling. The results showed that both phases in the multilayer were semiconductors and that they had a large energy gap difference at room temperature. With an increase in temperature, the energy gaps of both phases decreased or disappeared and transformed into a metal-like state (figures 4.7(a) and (b)). In particular, the authors observed the distinct electronic structure of LSC_{214} in its multilayer form compared to that of single-phase LSC_{214}. These electronic activation effects of LSC_{214} were believed to facilitate the electron transfer process of the LSC_{214} phase with oxygen molecules, thereby strongly promoting the ORR activity of the LSC_{113}/LSC_{214} heterostructure as reported in the literature. Similar electronic activation effects were also reported for an $LSC_{113/214}$ VAN structure by Ma *et al* [146]. They found that when the temperature was increased to more then 250 °C, the electronic structure of the LSC_{214} surface in VAN LSC_{113}/LSC_{214} followed the LSC_{113} phase and became metal-like. In addition, Chen *et al* [186] further probed the distribution of oxygen defects at the LSC_{113}/LSC_{214} and $LSC_{113}/La_2NiO_{4-\delta}$ (LNO_{214}) interfaces using a combination of hard x-ray photoelectron spectroscopy (HAXPES) and high-resolution x-ray diffraction (HRXRD). The experimental results showed that the LSC_{113} in LSC_{113}/LSC_{214} had more oxygen vacancies compared to the LSC_{113} in LSC_{113}/LNO_{214}. In addition, LNO_{214} and LSC_{214}, which form a hetero-structure with LSC_{113}, had less interstitial oxygen or more oxygen vacancies than their single-phase counterparts, respectively. Chen *et al* believed that such changes in the local oxygen defect chemistry were also related to the modified local electronic structure near the interface (figures 4.7(c) and (d)).

Using a combination of ultraviolet–visible absorption spectra (UV–Vis), XPS of Co 2*p* spectra, and valence band offset (VBO) spectra, Zheng *et al* [187] found that the formation of a heterointerface between NSC_{113} and NSC_{214} resulted in a narrowing of the energy gap and a lower energy barrier for electron transition from the valence band (VB) to the conduction band (CB), thus facilitating electron

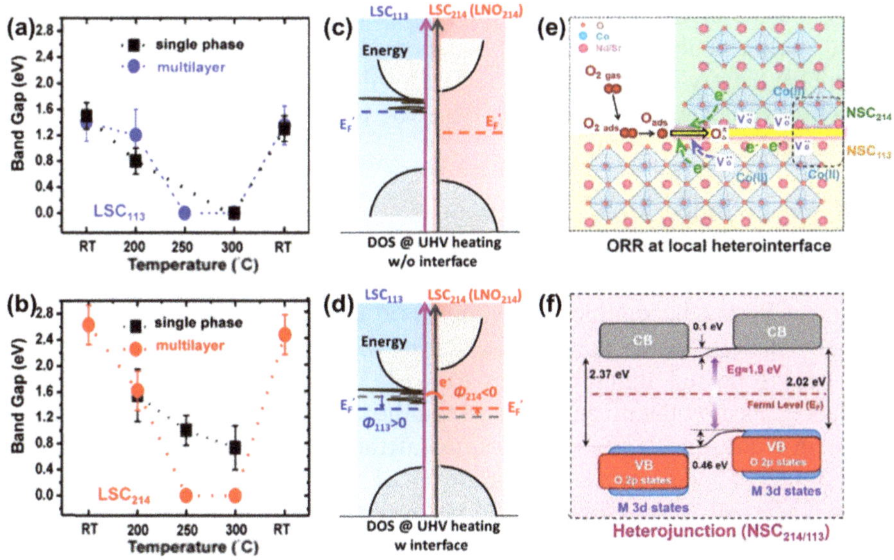

Figure 4.7. (a) and (b) Temperature dependence of the energy gap in single-phase LSC_{113}, single-phase LSC_{214}, and multilayer $LSC_{113/214}$ [185] John Wiley & Sons, copyright 2013 WILEY-VCH Verlag GmbH & Co. KGaA, Weinheim. (c) Misalignment after a loss of oxygen caused by annealing at high temperatures. Reprinted with permission from [186]. Copyright (2018) American Chemical Society. (d) Alignment after annealing due to the creation of local electric fields at the Fermi level in single-phase LSC_{113} and single-phase LSC_{214} (LNO_{214}). Reprinted with permission from [186]. Copyright (2018) American Chemical Society. (e) Schematic diagrams of the ORR process at a local heterointerface. Reprinted from [187], Copyright (2020), with permission from Elsevier. (f) Heterointerface energy band alignment based on valence band offset (VBO) and E_g data. Reprinted from [187], Copyright (2020), with permission from Elsevier.

transfer. In addition, the valence state of Co at the heterointerface was decreased, which was related to an increase in oxygen vacancies. Both the electronic structure and oxygen non-stoichiometry changes near the heterointerface were able to accelerate the ORR kinetics (figures 4.7(e) and (f)).

4.3.2.3 Extending the length of the TPB

For air electrodes with low ionic conductivity, such as LSM, the active site of the ORR or the OER is confined to the TPB region near the electrolyte/electrode interface [19, 144, 188, 189]. Therefore, an effective approach for improving the activity of the electrodes is to extend the triple phase boundary by adding materials with high ionic conductivity. Representative examples are CeO_2-based and Bi_2O_3-based composite air electrodes, such as LSM-GDC [190], LSCF-GDC [191, 192], SSC-SDC [193], LSC-SDC [194], etc [195–198]. Lu *et al* [199] found that when 40 wt% SDC was added to a $SrCo_{0.7}Nb_{0.1}Ni_{0.2}O_{3-\delta}$ air electrode, the area of the electrode TPB was greatly enlarged, thus increasing the k^* and D^* of the electrode. Chen *et al* [200] found that the addition of $Er_{0.4}Bi_{1.6}O_3$ (EBS) to a $Sm_{0.95}CoO_{3-\delta}$ electrode could extend the TPB and enhance the ion conductivity of the electrode. When the mass ratio of ESB was 20 wt%, the cell had the lowest impedance (0.063 Ω cm^2, at 750 °C) and the highest power density (1.66 W cm^{-2} at 750 °C). Baiutti *et al* [194] synthesized a self-assembled thin-film VAN nanocomposite LSM-SDC, which achieved a maximum TPB density of $>10^6$ cm cm^{-2}. The local fast mass transport pathway of the SDC matrix (<10 nm wide) was directly observed by IEDP in combination with atom probe tomography.

4.3.2.4 Suppression of cation segregation

Sr segregation is one of main mechanisms responsible for the degradation of Sr-containing air electrodes [64, 201, 202]. Many reports have stated that the surface enrichment of Sr can be significantly suppressed by the formation of a heterointerface, thereby slowing down the degradation of the air electrode.

It has been reported that Sr moves towards the perovskite surface due to the elastic effect caused by the mismatch between the lattice size of Sr and the host ion at the *A*-site in addition to the electrostatic attraction of surfaces rich in oxygen vacancies [203]. The electrostatic attraction of the abundant positively charged oxygen vacancies ($V_O^{\cdot\cdot}$) on the surface of perovskite oxides to negatively charged *A*-site dopants (such as Sr_{La}') leads to the enrichment of *A*-site dopants on the perovskite surface. It is believed that reducing the surface oxygen vacancy concentration by adding a modified layer to the surfaces of perovskite electrodes can significantly improve the oxygen surface exchange kinetics and stability [153, 155, 204]. For example, using LSC as a model system, Tsvetkov *et al* [155] systematically modified the surface of LSC by adding less reducible metal cations (Me: Nb^{5+}, Ti^{4+}, Zr^{4+}, Hf^{4+}, and Al^{3+}). These metal cations replaced the Co at the *B*-site ($Me_{Co}^{\cdot}$) in the LSC, leading to a decreased surface $V_O^{\cdot\cdot}$ concentration. Consequently, the surface of the doped sample exhibited less positive charge because the $Me_{Co}^{\cdot}$–O bond was stronger than the Co–O bond. The less positively charged surface was able to alleviate the electrostatic attraction of Sr to the surface, thereby stabilizing the cationic composition of the perovskite surface and improving the stability of the air electrode. They further observed a volcano-type dependence of the oxygen exchange kinetic coefficient k_q as a function of the enthalpy of formation of oxygen vacancies in binary oxides (i.e. MeO_x) and the enthalpy of formation of oxygen vacancies in LSC (figure 4.8(a)). Similarly, Bliem *et al* [204] and Wen *et al* [153] found the same inhibitory effect on Sr segregation in a Hf-modified LSM thin film and ZrO-modified LSC, respectively.

Some studies believe that the cations in the modified layer are incorporated into the perovskite surface to form a more stable structure, thereby inhibiting the enrichment of Sr on the surface [134, 205–208]. For example, Lynch *et al* [134] observed the formation of an LSM(C) mixed layer at the interface of an LSM-LSCF air electrode via Co diffusion from the LSCF lattice to the LSM surface layer. Lee *et al* [206] observed the formation of a stable perovskite phase, $La_{1-y}Sr_yCo_{1-x}Mn_xO_{3\pm\delta}$ (LSCMO), at the interface of LSM-modified LSC as a result of the partial substitution of Mn at the *B*-site of LSC. These intermixed phases exhibited lower Sr segregation on the surface and better stability. Li *et al* [205] reported the formation of self-assembled perovskite coatings with high activity by impregnating SrO on the surface of LSC. The authors believed that SrO modification changed the local surface cation composition of

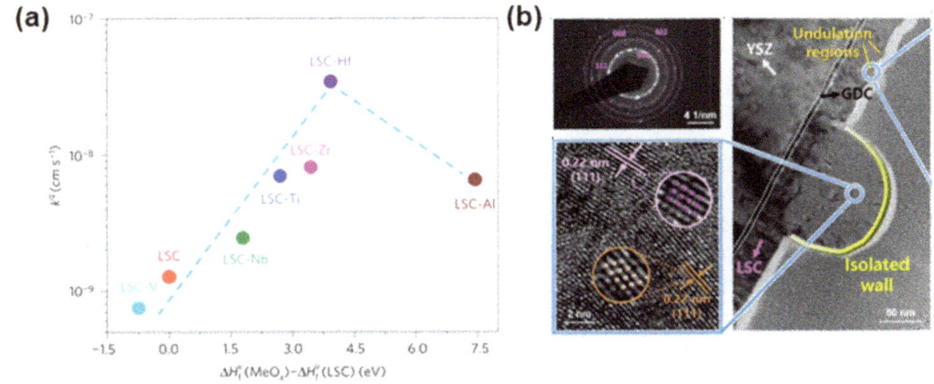

Figure 4.8. (a) Plot of the volcano-type dependence of the oxygen exchange kinetic coefficient k_q versus the difference of the enthalpy of oxygen vacancy formation between dioxides (i.e. MeO_x) and LSC [155] (2016) (Copyright © 2016, Nature Publishing Group). With permission of Springer. (b) TEM images and electron beam diffraction results obtained for a cross-section of SrO-LSC thin film. A diffractogram and a high-resolution TEM image of the surface isolation coating are shown on the left. The overall image of the cross-section of SrO-LSC thin film is shown on the right [205] John Wiley & Sons, copyright 2018 WILEY-VCH Verlag GmbH & Co. KGaA, Weinheim.

the LSC surface and formed an isolated highly catalytic active phase (figure 4.8(b)). Li *et al* [136] reported that after an *A*-deficient $La_{0.95}CoO_{3-\delta}$ (LC95) layer was deposited on an LSCF surface, the surface was stabilized without generating insulating surface Sr species. The authors attributed this stabilization effect to the migration of Sr in the LSCF to the LC95 coating layer, forming a highly active intermediate phase, $La_{0.95}Sr_xCoO_{3-\delta}$, at the heterointerface.

Some studies have shown that the enrichment of Sr at the heterointerface can increase the oxygen vacancy concentration at the interface, thereby enhancing the ORR/OER performance [4, 209–212]. Using coherent Bragg rod analysis (COBRA) and DFT calculations, Fang *et al* [211] found that the thermodynamic stability of Sr_{La} substitution was higher at the LSC_{113}/LSC_{214} interface, which caused Sr segregation near the LSC_{113}/LSC_{214}. This Sr segregation near the interface upshifted the O 2*p* band center to the Fermi level, resulting in a higher local oxygen vacancy concentration and enhanced oxygen mobility. In another study, Zhang *et al* [207] reported that Sr in LSCF perovskites can migrate into a $La_2NiO_{4-\delta}$ (LNO) surface coating layer. This type of Sr doping in LNO increased the electron–hole concentration due to charge compensation. Zheng *et al* [209] reported Sr enrichment at the heterointerface of NSC_{113}/NSC_{214}, resulting in a higher local oxygen vacancy concentration.

4.4 The influence of heterointerfaces in fuel electrodes

4.4.1 The performance of SOCs with heterostructured fuel electrodes

In this section, we provide an overview of different fuel electrodes with heterointerfaces, including heterostructures consisting of: (a) simple metal oxides and Ni, (b) perovskite and fluorite oxide/Ni, and (c) perovskite with exsolved nanoparticles.

4.4.1.1 Fuel electrodes based on simple metal oxides and Ni

Nickel-based fuel electrodes modified by simple metal oxides, such as BaO, CaO, SrO, and MgO, have been extensively investigated as the fuel electrode of SOCs [213–217]. Although these simple metal oxides had almost no electrochemical activity, they were reported to strongly enhance the anti-coking performance of the Ni-based electrode. Takeguchi *et al* [214] and Asamoto *et al* [44] reported that a CaO modification layer effectively suppressed the carbon accumulation at a Ni-YSZ fuel electrode and promoted the steam reforming of methane. Rosa *et al* [215] found that a BaO-modified Ni–Cu/SDC electrode showed no noticeable carbon accumulation after 200 h, whereas the fuel electrode without BaO had significant carbon accumulation on the surface. Itagaki *et al* [218] found that an addition of 0.2 wt% BaO on Ni/SDC was able to significantly improve the performance of the electrode in H_2 and CH_4. The OCV of a cell with a BaO-Ni/SDC fuel electrode was 1.33 V in CH_4, which was 0.06 V higher than that of Ni/SDC. After surface modification, the maximum power density increased from 27.5 mW cm^{-2} to 35 mW cm^{-2} in H_2 fuel and from 61 mW cm^{-2} to 82 mW cm^{-2} in CH_4 fuel. In addition, after the electrode was tested in CH_4 fuel, the BaO modification was found to have inhibited carbon deposition. Wang *et al* [219] deposited TiO_2 on a Ni–Co/Al_2O_3 electrode to form a composite fuel electrode for the partial oxidation of propane for hydrogen production. The TiO_2 phase significantly improved the selectivity of the electrode for H_2 and CO. The highest H_2 concentration was 52.48 vol% for Ni–Co/$Al_2O_3TiO_2$, which was higher than that of 47.92% for Ni–Co/Al_2O_3. At 600 °C, the Ni–Co/Al_2O_3–TiO_2 electrode lasted for more than 20 h and showed a stable hydrogen production rate of about 600 μmol/$g_{cat\cdot s}$. In SOFC mode, the maximum output power density of a Ni–Co/Al_2O_3–TiO_2 electrode was 404 mW cm^{-2} at a current density of 0.8 A cm^{-2}, which was 24.3% higher than that of Ni–Co/Al_2O_3 (325 mW cm^{-2}).

Fluorite-structured cerium dioxide (CeO_2), which has good oxygen ion conductivity, has also been shown to improve the electrocatalytic activity and anti-coking at fuel electrodes [220–223]. For instance, Tiwari *et al* [223] reported that a cell with a CeO_2–NiO–YSZ fuel electrode showed a peak power density of 330 mW cm^{-2} at 800 °C for 5 h of operation, which was much higher than that of a NiO/YSZ fuel electrode (275 mW cm^{-2}). In addition, the CeO_2–NiO/YSZ fuel electrode had good stability, its power density was still 290 mW cm^{-2}, and its polarization impedance only increased by 0.9 Ω cm^2 after operating at 800 °C for 25 h.

Decorating conventional Ni/YSZ fuel electrodes with a doped cerium dioxide layer has been proven to be a powerful strategy for improving the performance and stability of fuel cells. The surface coating layers used include $Zr_{0.35}Ce_{0.65}O_{2-\delta}$ (ZDC) [224], $Mo_{0.1}Ce_{0.9}O_{2+\delta}$ (MDC) [225], SDC [226, 227], GDC, etc [228, 229]. Cimenti *et al* [224] evaluated the performance of Ni/YSZ electrode with ZDC impregnation for the direct utilization of methanol as a fuel. The power density of a cell with a ZDC-Ni/YSZ electrode was 0.252 W cm^{-2}, which was three times higher than that of a cell with a Ni-YSZ electrode. The surface specific resistance was 0.4 Ω cm^2, which was one half of that of Ni/YSZ in methanol at 1073 K at a constant voltage of 0.6 V. After a stability test, the performance of the cell with the ZDC-Ni/YSZ electrode decreased by 9%, which was significantly less than that of the cell with Ni/YSZ electrode (40%). In another work, Lyu *et al* [228] reported that a Ni-YSZ electrode with a GDC catalytic layer had a maximum peak power density of 271 mW cm^{-2} at 600 mA cm^{-2}, which was about 3.3 times higher than that of Ni-YSZ (81 mW cm^{-2}) at 33.3% CH_4 + 66.7% CO_2 in fuel gas at 750 °C.

4.4.1.2 Fuel electrodes based on perovskite oxide and fluorite oxide/Ni

Some perovskite oxides have excellent anti-coking and sulfur tolerance, and have therefore also been used to improve fuel electrode stability [230–233]. For example, Hua *et al* [230] infiltrated $PrBaMn_2O_{5+\delta}$ (PBM) onto the surface of Ni-YSZ fuel electrodes. The peak power densities for Ni/YSZ and PBM-Ni/YSZ electrodes were 1.01 W cm^{-2} and 1.41 W cm^{-2}, respectively, in H_2 containing H_2S at 800 °C. In a long-term stability test, the performance of Ni/YSZ was severely degraded, and its power density decreased from 0.7 W cm^{-2} to 0.54 W cm^{-2} over 24 h. In contrast, the performance of the PBM-modified electrode showed no significant degradation. SEM and temperature programmed oxidation (TPO) analyses revealed that the adsorption of sulfur species and carbon formation resulted in the poor performance and poor stability of the Ni/YSZ fuel electrode in biogas. The PBM modification layer improved the catalytic activity and sulfur tolerance. Xiao *et al* [234] reported that a Ni-based electrode modified using $Sr_2Fe_{1.5}Mo_{0.5}O_6$ significantly improved the cell performance, which achieved peak power densities of 1.134 W cm^{-2} and 0.663 W cm^{-2} in wet H_2 and wet CH_4 at 800 °C, respectively. Kim *et al* [235] reported that SDC modification was able to improve the sulfur tolerance of $Sr_{0.92}Y_{0.08}Ti_{1-x}Ni_xO_{3-\delta}$ (SYTN) fuel electrodes. In pure H_2, the cell performance of an SDC-SYTN electrode did not change significantly after treatments at 200 ppm and 500 ppm of H_2S, while the cell performance of a bare SYTN electrode irreversibly decreased by 35%. Similar effects were observed for $La_{0.4}Sr_{0.5}Ba_{0.1}TiO_3$ and LSM in Ni-YSZ electrodes [231, 232].

4.4.1.3 Fuel electrodes based on perovskite oxides and exsolved nanoparticles

In recent years, perovskite oxides with exsolved metal nanoparticles have attracted a lot of attention as fuel electrode materials for SOCs [236–239] Following thermal annealing under reducing conditions or the application of an appropriate electrical potential, metal nanoparticles can precipitate out from the perovskite lattice to near the surface and become embedded in the oxide matrix in the form of a stable heterointerface [240]. The exsolution generally occurs on the *B*-site of perovskite, and the exsolved ions usually form single-metal nanoparticles such as Fe [241, 242], Co [243–245], Ni [246–248], Cu [249, 250], Re [251] or alloys composed of different *B*-site cations, such as Co–Fe [37, 252–254], Fe–Cu [255, 256], Fe–Ni [257–259], Co–Ni [260, 261], Ni–Co–Fe [262, 263], etc.

Zhou *et al* [245] reported that *in situ* growth of Co nanoparticles on the surface of the layered perovskite $La_{0.8}Sr_{1.2}Fe_{0.9}Co_{0.1}O_{4\pm\delta}$ (LSFC) promoted the hydroxide reaction. The R_p of an LSFC electrode was 1.14 Ω cm^2 in 5% H_2/N_2 at 800 °C, which was much smaller than that of a $La_{0.8}Sr_{1.2}FeO_{4\pm\delta}$ electrode without Co exsolution. SEM and HRTEM showed that the nanoscale Co° particles were uniformly dispersed on the surface of the LSCF electrode. Jo *et al* [250] prepared $La_{0.43}Sr_{0.37}Cu_{0.12}Ti_{0.88}O_{3-\delta}$ (LSCuT) with Cu nanoparticles. The Cu nanoparticles were uniformly distributed on the surface of the perovskite with an average particle size of ~9 nm and a particle density of 111 μm^{-2}. A single cell using the LSCuT electrode exhibited high performance and stability with an R_p value of 0.04 Ω cm^2 and a maximum power density of 1.38 W cm^{-2}, maintaining a stable potential (~0.7 V) over 150 h in 3% H_2O/H_2 fuel at 900 °C. Liu *et al* [252] annealed cubic perovskite $Pr_{0.4}Sr_{0.6}Co_{0.2}Fe_{0.7}Mo_{0.1}O_{3-\delta}$ in a 10% H_2/N_2 reducing atmosphere at 900 °C to obtain double-layered perovskite with *in situ* exsolved Co–Fe alloy nanoparticles $(Pr_{0.4}Sr_{0.6})_3(Fe_{0.85}Mo_{0.15})_2O_7$ (DLP-PSFM). Here, a single cell with a DLP-PSFM electrode had a maximum output power density of 348.84 mW cm^{-2} in C_2H_6 at 750 °C. The same cell also had a high ethylene yield of 41.5%, a high rate of ethane conversion of 45.4%, and a good ethylene selectivity of 91.4%.

The exsolved nanoparticles are embedded in the main body of the perovskite, which enhances its stability, significantly improves sulfur tolerance, and reduces its tendency to suffer from hydrocarbon coking [261, 264–268]. Cui *et al* [265] reported good sulfur tolerance for $La_{0.3}Sr_{0.7}Co_{0.07}Ti_{0.93}O_{3-\delta}$ (LSCoT) with Co nanoparticles. After 48 h, the current density of the LSCoT electrode remained at 400 mA cm^{-2} in humid 5000 ppm H_2S–H_2 at 900 °C. Li *et al* [266] prepared an Fe_3Co_2 alloy nanoparticle-modified Ruddlesden–Popper perovskite oxide electrode Fe_3Co_2–$La_{1.2}Sr_{0.8}Mn_{0.4}Fe_{0.6}O_{4-\delta}$ (LSMF), which exhibited excellent sulfur tolerance at 800 °C in 200 ppm H_2S–H_2 fuel; it achieved a peak power density of 566 mW cm^{-2} and operated stably for 50 h.

Neagu *et al* [264] reported strong adhesion between exsolved Ni nanoparticles and a $La_{0.52}Sr_{0.28}Ni_{0.06}Ti_{0.94}O_3$ matrix, which prevented the particle agglomeration and promoted the anti-coking performance of the electrode. Kwon *et al* [261] synthesized a layered double-perovskite electrode using $PrBaMn_{1.7}Co_{0.1}Ni_{0.2}O_{5+\delta}$ with self-assembled Co–Ni nano-alloys (BMCNO). The current density of the BMCNO fuel electrode did not change much over more than 500 h at a constant pressure at 0.4 V, at 800 °C, in CH_4, compared to that of a PBMO electrode, and no significant coking was observed. This indicated that the BMCNO fuel electrode had good tolerance to carbon coking when used with hydrocarbon fuels.

4.4.2 The role of heterointerfaces in fuel electrodes

In this section, we briefly describe the impacts of heterointerfaces on the activity and stability of fuel electrodes, including changes to their electronic structure, increases in TPB length, and the suppression of carbon deposition.

4.4.2.1 Change of electronic structure

The *d*-band center's position is widely used as a descriptor for the catalytic activity of transition metals [269]. The closer the *d*-band centers are to the Fermi energy level, the easier it is for electrons to transfer from the catalyst to the adsorbate [269]. Several published reports have stated that exsolved nanometal nanoparticles on perovskite oxide matrices exhibited different *d*-band center positions than those of reference nano metals. Joo *et al* [262] prepared Co–Ni–Fe alloy nanoparticles on layered perovskite $PrBaMn_{1.7}Co_{0.1}Ni_{0.2}O_{5+\delta}$ as an SOEC fuel electrode to directly electrolyze CH_4 and CO_2. They found that an electrode with CoNiFe alloy nanoparticles had higher activity than those with Ni monometallic and Co–Ni bimetallic nanoparticles. A comparison of the *d*-band centers revealed that the *d*-band centers of CoNiFe were closer to the Fermi energy level than those of Co–Ni and Ni. As a consequence, the CoNiFe was more likely to provide charge to the adsorbent, weakening the internal binding strength of CH_4 and CO_2, and thus improving catalytic activity (figure 4.9(a)). Lee *et al* [270] reported an increase in electrode catalytic activity as a result of the exsolution of Ni nanoparticles from $La_{0.4}Sr_{0.4}Ti_{0.94}Ni_{0.06}O_{3-\delta}$ (LSTN). DFT showed that hydrogen adsorption and dissociation on positively charged Ni-LSTN catalysts are more favorable than on pure Ni catalysts. The exsolved Ni (−2.92 eV) had a lower *d*-band center than that of pure Ni (−1.28 eV). Therefore, the Ni-LSTN catalyst surface was more ready to accept electrons from H_2, leading to enhanced catalytic activity (figure 4.9(b)).

The use of alkaline-earth metal oxides with strong alkalinity on Ni has been reported to increase the electron density of Ni, thus critically modifying the catalytic activity of Ni-based fuel electrodes [44, 271–273]. For instance, Lay *et al* [271] reported that the use of MgO on Ni led to electron enrichment on the Ni sites, resulting in higher affinity of Ni for the electron attractants C and O. As a result, the steam gasification reaction of water vapor with carbon ($C + H_2O \rightarrow CO + H_2$) was promoted, thus inhibiting the kinetics of coke formation (the Boudouard reaction: $2CO \rightarrow CO_2 + C$) on the heterostructured electrode. Similarly, Matsui *et al* [222] observed that surface modifications using GDC can suppress carbon accumulation caused by CO disproportionation reactions on Ni. They believed that such suppression was due to the local electron transfer process from Ce^{3+} to Ni^0, which led to the high tolerance of the electrode to carbon deposition.

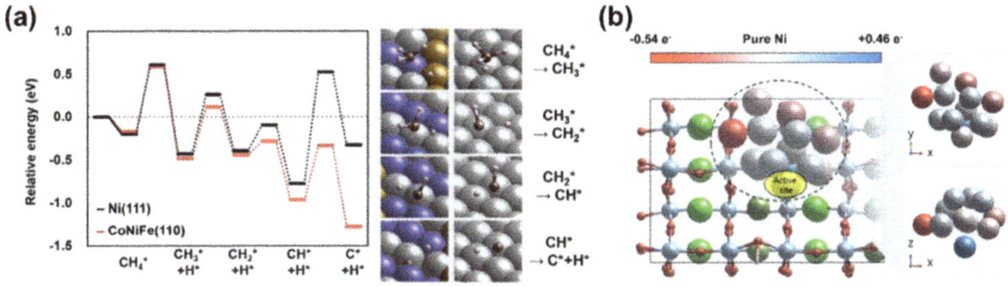

Figure 4.9. (a) Relative energy distribution and transition state images of CH_4 dehydrogenation on Ni (111) and CoNiFe (110) surfaces. The spheres shown in blue, gray, light brown, white, and dark brown are Co, Ni, Fe, H, and C, respectively [262] John Wiley & Sons, copyright 2021 Wiley-VCH GmbH. (b) Atomic charge distribution of Ni in Ni-LSTN (110). Red: relatively large atomic charge state, white: neutral Ni, blue: relatively small atomic charge state. Yellow circles indicate the active sites of H_2 adsorption and dissociation on Ni-LSTN (110). Reprinted from [270], Copyright (2021), with permission from Elsevier.

4.4.2.2 Extending the TPB length

As introduced in the previous section, the agglomeration of metal particles during the preparation and testing of fuel electrodes is one of the main degradation mechanisms of fuel electrodes. Surface modifications using metal oxides have been reported to stabilize the electrode microstructures and prevent their agglomeration, thus expanding the triple phase interface and improving electrocatalytic activity. Representative material systems include MgO–Ni/YSZ [216, 274], SnO_2–Ni/YSZ [275], NbO_5–Ni/YSZ [223], Fe/Cu–$La_{0.5}Sr_{0.6}Fe_{0.8}Cu_{0.15}Nb_{0.05}O_{3-\delta}$ [255], and $FeNi_3$–$Sr_2Fe_xNi_{1-x}MoO_{6-\delta}$ [258].

The modification of fuel electrodes using ion-conducting or electron-conducting phases increased the conductivity and improved the TPB length, thus significantly improving cell performance [222, 223, 225, 226, 229]. For example, Lee *et al* [226] modified Ni/YSZ using SDC, which has a high mixed ionic/electronic conductivity, to increase the TPB length; the SDC coating led to an improvement of about 20% in cell performance under H_2 conditions.

The presence of exsolved metal nanoparticles has also been reported to extend the active region on the surface of perovskite oxide. For instance, Cui *et al* [265] reported that the exsolution of Co nanoparticles with diameters no larger than 10 nm could greatly improve the electron transfer channels in LSCoT electrodes, increasing the concentration of TPB active sites and improving the catalytic activity for fuel oxidation.

4.4.2.3 Anti-coking

Decorating fuel electrodes with materials which can react with water molecules is a widely used strategy for removing the carbon/sulfur deposited on fuel electrodes [48, 213, 218, 276, 277]. A representative example of such a material is CaO. Asamoto *et al* [44] proposed that the following reactions take place on CaO-decorated fuel electrodes, inhibiting carbon accumulation:

$$CaO + C + H_2O \leftrightarrow CaCO_3 + H_2 \tag{4.9}$$

$$CaCO_3 \leftrightarrow CaO + CO_2. \tag{4.10}$$

Mishina *et al* [277] suggested that hydroxyl species form on CaO particles by the dissociative adsorption of water vapor. The hydroxyl species that form at the interface between CaO and Ni may oxidize the deposited carbon by forming CO_x species. A similar mechanism was reported for BaO- [213, 218, 276] and TiO_2- [48] modified Ni-based fuel cells.

It has been reported that carbon dissolves onto Ni particles and then precipitates as fibrous carbon at the interface between the Ni particles and the substrate through a ‘tip growth’ mechanism, leading to severe coking of the electrode [278]. The growth of nanofibrous carbon has been found to be determined by the size of the metal particles and the concentration of fibrous carbon in the metal particles [279, 280]. Therefore, the small size of exsolved particles and their strong interaction with the substrate enables exsolved electrodes to inhibit particle agglomeration and carbon fiber growth [264]. For example, Neagu *et al* [264] reported that the interaction between exsolved particles and $La_{0.4}Sr_{0.4}Ni_{0.03}Ti_{0.97}O_{3-\delta}$ weakened and carbon deposition increased as the size of the exsolved Ni particles increased from 20 nm to 100 nm. Wei *et al* [281] compared the degree of carbon deposition of *in situ* exsolved Ni-modified $La_{0.9}Mn_{0.8}Ni_{0.2}O_3$ (R-NLM; here 'R' refers to the reduced one) catalyst with that of impregnated Ni-modified $LaMnO_3$ catalyst (R-LMN). They found that impregnated Ni particles with larger sizes exhibited weaker binding to Ni-modified $La_{0.9}Mn_{0.8}Ni_{0.2}O_3$; they also exhibited higher carbon deposition and lower activity for the dry reforming reaction. In contrast, the *in situ* exsolved metal nanoparticles showed smaller sizes; these Ni particles were partially embedded in the NLM body by strong interfacial interaction forces. Both their small size and their strong metal-oxide interface inhibited the formation of carbon fibers, resulting in strong anti-coking behavior at 700 °C for 24 h.

4.5 Conclusions

The development of highly active and stable electrodes is critical for the practical application of SOCs. In this chapter, we have summarized the recent progress in using heterointerface engineering to improve the activity and stability of SOC electrodes. In particular, we discussed the principal function of the SOC electrode and highlighted studies of the degradation mechanisms of electrodes. The degradation mechanisms of air electrodes include cation segregation, contaminant deposition, and secondary phase generation. The performance degradation of fuel electrodes, on the other hand, is related to nickel particle coarsening and migration, coking, and sulfur poisoning. Representative heterostructured electrodes and their performances were also summarized. Furthermore, the roles of heterointerfaces in determining the activity of electrodes were introduced in detail, including the effects of strain, changes in electronic structure, increased TPB lengths, the suppression of cation segregation, and anti-coking.

References

[1] Wang L, Merkle R, Mastrikov Y A, Kotomin E A and Maier J 2012 Oxygen exchange kinetics on solid oxide fuel cell cathode materials—general trends and their mechanistic interpretation *J. Mater. Res.* **27** 2000–8

[2] Adler S B 2004 Factors governing oxygen reduction in solid oxide fuel cell cathodes *Chem. Rev.* **104** 4791–843

[3] Chueh W C and Haile S M 2012 Electrochemistry of mixed oxygen ion and electron conducting electrodes in solid electrolyte cells *Annu. Rev. Chem. Biomol. Eng.* **3** 313–41

[4] Mogensen M B, Chatzichristodoulou C, Graves C, Hansen K V, Hansen K K, Hauch A, Jacobsen T and Norrman K 2016 New hypothesis for SOFC ceramic oxygen electrode mechanisms *ECS Trans.* **72** 93–103

[5] Adler S B, Lane J A and Steele B C H 1996 Electrode kinetics of porous mixed-conducting oxygen electrodes *J. Electrochem. Soc.* **143** 3554–64

[6] Kilner J A, De Souza R A and Fullarton I C 1996 Surface exchange of oxygen in mixed conducting perovskite oxides *Solid State Ion. Diffus. React.* **86–88** 703–9

[7] De R A 2006 Souza, A universal empirical expression for the isotope surface exchange coefficients (k*) of acceptor-doped perovskite and fluorite oxides *Phys. Chem. Chem. Phys.* **8** 890–97
[8] Preis W, Bucher E and Sitte W 2002 Oxygen exchange measurements on perovskites as cathode materials for solid oxide fuel cells *J. Power Sources* **106** 116–21
[9] De Souza R and Kilner J 1998 Oxygen transport in $La_{1-x}Sr_xMn_{1-y}Co_yO_{3\pm\delta}$ perovskites: part I. Oxygen tracer diffusion *Solid State Ion.* **106** 175–87
[10] Yoo J, Verma A, Wang S Y and Jacobson A J 2005 Oxygen transport kinetics in $SrFeO_{3-\delta}$, $La_{0.5}Sr_{0.5}FeO_{3-\delta}$, and $La_{0.2}Sr_{0.8}Cr_{0.2}Fe_{0.8}O_{3-\delta}$ measured by electrical conductivity relaxation *J. Electrochem. Soc.* **152** A497–505
[11] Vashook V V, Al Daroukh M and Ullmann H 2001 Oxygen ion diffusion in perovskite-type oxides determined by permeation and by relaxation measurements *Ionics* **7** 59–66
[12] Berenov A V, Atkinson A, Kilner J A, Bucher E and Sitte W 2010 Oxygen tracer diffusion and surface exchange kinetics in $La_{0.6}Sr_{0.4}CoO_{3-\delta}$ *Solid State Ion.* **181** 819–26
[13] Merkle R, Maier J and Bouwmeester H J M 2004 A linear free energy relationship for gas–solid interactions: correlation between surface rate constant and diffusion coefficient of oxygen tracer exchange for electron-rich perovskites *Angew. Chem.* **116** 5179–83
[14] Sahibzada M, Benson S, Rudkin R and Kilner J A 1998 Pd-promoted $La_{0.6}Sr_{0.4}Co_{0.2}Fe_{0.8}O_3$ cathodes *Solid State Ion.* **113** 285–90
[15] Richter J, Holtappels P, Graule T, Nakamura T and Gauckler L J 2009 Materials design for perovskite SOFC cathodes *Monatsh. Chem.—Chem. Monthly* **140** 985–99
[16] Jung W and Tuller H L 2009 Impedance study of $SrTi_{1-x}Fe_xO_{3-\delta}$ ($x = 0.05$ to 0.80) mixed ionic-electronic conducting model cathode *Solid State Ion.* **180** 843–47
[17] Jung W and Tuller H L 2011 A new model describing solid oxide fuel cell cathode kinetics: model thin film $SrTi_{1-x}Fe_xO_{3-\delta}$ mixed conducting oxides-a case study *Adv. Energy Mater.* **1** 1184–91
[18] Colomer M T, Steele B C H and Kilner J A 2002 Structural and electrochemical properties of the $Sr_{0.8}Ce_{0.1}Fe_{0.7}Co_{0.3}O_{3-\delta}$ perovskite as cathode material for itSOFCs *Solid State Ion.* **147** 41–8
[19] Fujimaki Y *et al* 2017 Contribution of triple-phase boundary reaction in cathodic reaction of solid oxide fuel cell *ECS Trans.* **78** 847–53
[20] Fleig J and Maier J 2004 The polarization of mixed conducting SOFC cathodes: effects of surface reaction coefficient, ionic conductivity and geometry *J. Eur. Ceram. Soc.* **24** 1343–47
[21] Kilner J A and Burriel M 2014 Materials for intermediate-temperature solid-oxide fuel cells *Annu. Rev. Mater. Res.* **44** 365–93
[22] Mutoro E, Crumlin E J, Biegalski M D, Christen H M and Shao-Horn Y 2011 Enhanced oxygen reduction activity on surface-decorated perovskite thin films for solid oxide fuel cells *Energy Environ. Sci.* **4** 3689–96
[23] Bucher E, Sitte W, Klauser F and Bertel E 2011 Oxygen exchange kinetics of $La_{0.58}Sr_{0.4}Co_{0.2}Fe_{0.8}O_3$ at 600 °C in dry and humid atmospheres *Solid State Ion.* **191** 61–7
[24] Zhou W, Ran R and Shao Z 2009 Progress in understanding and development of $Ba_{0.5}Sr_{0.5}Co_{0.8}Fe_{0.2}O_{3-\delta}$-based cathodes for intermediate-temperature solid-oxide fuel cells: a review *J. Power Sources* **192** 231–46
[25] Wei T, Zhang Q, Huang Y H and Goodenough J B 2012 Cobalt-based double-perovskite symmetrical electrodes with low thermal expansion for solid oxide fuel cells *J. Mater. Chem.* **22** 225–31

[26] Jin F, Xu H, Long W, Shen Y and He T 2013 Characterization and evaluation of double perovskites $LnBaCoFeO_{5+\delta}$ (Ln = Pr and Nd) as intermediate-temperature solid oxide fuel cell cathodes *J. Power Sources* **243** 10–8
[27] Chen D, Ran R, Zhang K, Wang J and Shao Z 2009 Intermediate-temperature electrochemical performance of a polycrystalline $PrBaCo_2O_{5+\delta}$ cathode on samarium-doped ceria electrolyte *J. Power Sources* **188** 96–105
[28] Deng Z Q, Smit J P, Niu H J, Evans G, Li M R, Xu Z L, Claridge J B and Rosseinsky M J 2009 B cation ordered double perovskite $Ba_2CoMo_{0.5}Nb_{0.5}O_{6-\delta}$ as a potential SOFC cathode *Chem. Mater.* **21** 5154–62
[29] Ghorbani-Moghadam T, Kompany A, Bagheri-Mohagheghi M M and Abrishami M E 2018 High temperature electrical conductivity and electrochemical investigation of $La_{2-x}Sr_xCoO_4$ nanoparticles for IT-SOFC cathode *Ceram. Int.* **44** 21238–48
[30] Montenegro-Hernández A, Vega-Castillo J, Mogni L and Caneiro A 2011 Thermal stability of $Ln_2NiO_{4+\delta}$ (Ln: La, Pr, Nd) and their chemical compatibility with YSZ and CGO solid electrolytes *Int. J. Hydrog. Energy* **36** 15704–14
[31] Kovalevsky A V, Kharton V V, Yaremchenko A A, Pivak Y V, Tsipis E V, Yakovlev S O, Markov A A, Naumovich E N and Frade J R 2007 Oxygen permeability, stability and electrochemical behavior of $Pr_2NiO_{4-\delta}$ based materials *J. Electroceram.* **18** 205–18
[32] Ji Q, Bi L, Zhang J, Cao H and Zhao X S 2020 The role of oxygen vacancies of ABO_3 perovskite oxides in the oxygen reduction reaction *Energy Environ. Sci.* **13** 1408–28
[33] Read M S D, Saiful Islam M, Watson G W, King F and Hancock F E 2000 Defect chemistry and surface properties of $LaCoO_3$ *J. Mater. Chem.* **10** 2298–305
[34] Chen D, Chen C, Baiyee Z M, Shao Z and Ciucci F 2015 Nonstoichiometric oxides as low-cost and highly-efficient oxygen reduction/evolution catalysts for low-temperature electrochemical devices *Chem. Rev.* **115** 9869–921
[35] Teraoka Y, Nobunaga T, Okamoto K, Miura N and Yamazoe N 1991 Influence of constituent metal cations in substituted $LaCoO_3$ on mixed conductivity and oxygen permeability *Solid State Ion. Diffus. React.* **48** 207–12
[36] Stevenson J W, Armstrong T R, Carneim R D, Pederson L R and Weber W J 1996 Electrochemical properties of mixed conducting perovskites $La_{1-x}M_xCo_{1-y}Fe_yO_{3-\delta}$ (M=Sr, Ba, Ca) *J. Electrochem. Soc.* **143** 2722–29
[37] Chen H *et al* 2021 Activating lattice oxygen in perovskite oxide by B-site cation doping for modulated stability and activity at elevated temperatures. dulated stability and activity at elevated temperatures *Adv. Sci.* **8** e2102713
[38] Cherry M, Islam M S and Catlow C R A 1995 Oxygen ion migration in perovskite-type oxides *J. Solid State Chem.* **118** 125–32
[39] Mahato N, Banerjee A, Gupta A, Omar S and Balani K 2015 Progress in material selection for solid oxide fuel cell technology: a review *Prog. Mater Sci.* **72** 141–337
[40] Reisert M, Aphale A and Singh P 2018 Solid oxide electrochemical systems: material degradation processes and novel mitigation approaches *Materials* **11** 2169
[41] Shri Prakash B, Senthil Kumar S and Aruna S T 2014 Properties and development of Ni/YSZ as an anode material in solid oxide fuel cell: a review *Renew. Sustain. Energy Rev.* **36** 149–79
[42] Cowin P I, Petit C T G, Lan R, Irvine J T S and Tao S 2011 Recent progress in the development of anode materials for solid oxide fuel cells *Adv. Energy Mater.* **1** 314–32
[43] Jacobson A J 2009 Materials for solid oxide fuel cells *Chem. Mater.* **22** 660–74

[44] Asamoto M, Miyake S, Sugihara K and Yahiro H 2009 Improvement of Ni/SDC anode by alkaline earth metal oxide addition for direct methane-solid oxide fuel cells *Electrochem. Commun.* **11** 1508–11
[45] Tao Z T, Hou G H, Xu N, Zhang Q F and Ding H P 2014 A mixed proton-oxide ion-electron conducting anode for highly coking-resistant solid oxide fuel cells *Electrochim. Acta* **150** 55–61
[46] Chen Y, Zhang Y X, Lin Y, Yang Z B, Su D, Han M F and Chen F L 2014 Direct-methane solid oxide fuel cells with hierarchically porous Ni-based anode deposited with nanocatalyst layer *Nano Energy* **10** 1–9
[47] Wang F, Wang W, Ran R, Tade M O and Shao Z P 2014 Aluminum oxide as a dual-functional modifier of Ni-based anodes of solid oxide fuel cells for operation on simulated biogas *J. Power Sources* **268** 787–93
[48] Gu D, Zhang G and Zou J 2021 High temperature thermo-photocatalysis driven carbon removal in direct biogas fueled solid oxide fuel cells *Chin. Chem. Lett.* **32** 3548–52
[49] Tao Z, Fu M and Liu Y 2021 A mini-review of carbon-resistant anode materials for solid oxide fuel cells *Sustain. Energy Fuels* **5** 5420–30
[50] Khan M S, Lee S B, Song R H, Lee J W, Lim T H and Park S J 2016 Fundamental mechanisms involved in the degradation of nickel yttria stabilized zirconia (Ni-YSZ) anode during solid oxide fuel cells operation: a review *Ceram. Int.* **42** 35–48
[51] Lu C, Worrell W L, Vohs J M and Gorte R J 2003 A comparison of Cu-Ceria-SDC and Au-Ceria-SDC composites for SOFC anodes *J. Electrochem. Soc.* **150** A1357–59
[52] Gaudillere C, Navarrete L and Serra J M 2014 Syngas production at intermediate temperature through H_2O and CO_2 electrolysis with a Cu-based solid oxide electrolyzer cell *Int. J. Hydrog. Energy* **39** 3047–54
[53] Kumari N, Tiwari P K, Haider M A and Basu S 2017 Electrochemical performance of infiltrated Cu-GDC and Cu-PDC cathode for CO_2 electrolysis in a solid oxide cell *ECS Trans.* **78** 3329–37
[54] He H P, Vohs J M and Gorte R J 2003 Effect of synthesis conditions on the performance of Cu–CeO_2–YSZ anodes in SOFCs *J. Electrochem. Soc.* **150** A1470–75
[55] Costa-Nunes O, Gorte R J and Vohs J M 2005 Comparison of the performance of Cu–CeO_2–YSZ and Ni–YSZ composite SOFC anodes with H_2, CO, and syngas *J. Power Sources* **141** 241–49
[56] Tao S and Irvine J T 2004 Synthesis and characterization of $(La_{0.75}Sr_{0.25})Cr_{0.5}Mn_{0.5}O_{3-\delta}$, a redox-stable, efficient perovskite anode for SOFCs *J. Electrochem. Soc.* **151** A252–59
[57] Zha S, Tsang P, Cheng Z and Liu M 2005 Electrical properties and sulfur tolerance of $La_{0.75}Sr_{0.25}Cr_{1-x}Mn_xO_3$ under anodic conditions *J. Solid State Chem.* **178** 1844–50
[58] Danilovic N, Vincent A, Luo J L, Chuang K T, Hui R and Sanger A R 2009 Correlation of fuel cell anode electrocatalytic and ex situ catalytic activity of perovskites $La_{0.75}Sr_{0.25}Cr_{0.5}X_{0.5}O_{3-\delta}$ (X = Ti, Mn, Fe, Co) *Chem. Mater.* **22** 957–65
[59] Liu Q, Dong X, Xiao G, Zhao F and Chen F 2010 A novel electrode material for symmetrical SOFCs *Adv. Mater.* **22** 5478–82
[60] Gansor P, Xu C, Sabolsky K, Zondlo J W and Sabolsky E M 2012 Phosphine impurity tolerance of $Sr_2MgMoO_{6-\delta}$ composite SOFC anodes *J. Power Sources* **198** 7–13
[61] Marrero-López D, Peña-Martínez J, Ruiz-Morales J C, Pérez-Coll D, Aranda M A G and Núñez P 2008 Synthesis, phase stability and electrical conductivity of $Sr_2MgMoO_{6-\delta}$ anode *Mater. Res. Bull.* **43** 2441–50

[62] Chen K and Jiang S P 2020 Surface segregation in solid oxide cell oxygen electrodes: phenomena, mitigation strategies and electrochemical properties *Electrochem. Energy Rev.* **3** 730–65

[63] Zarabi Golkhatmi S, Asghar M I and Lund P D 2022 A review on solid oxide fuel cell durability: latest progress, mechanisms, and study tools *Renew. Sustain. Energy Rev.* **161** 112339

[64] Koo B, Kim K, Kim J K, Kwon H, Han J W and Jung W 2018 Sr segregation in perovskite oxides: why it happens and how it exists *Joule* **2** 1476–99

[65] Li Y, Zhang W, Zheng Y, Chen J, Yu B, Chen Y and Liu M 2017 Controlling cation segregation in perovskite-based electrodes for high electro-catalytic activity and durability *Chem. Soc. Rev.* **46** 6345–78

[66] Ponce S, Pena M A and Fierro J L G 2000 Surface properties and catalytic performance in methane combustion of Sr-substituted lanthanum manganites *Appl. Catal.* B **24** 193–205

[67] Dulli H, Dowben P A, Liou S H and Plummer E W 2000 Surface segregation and restructuring of colossal-magnetoresistant manganese perovskites $La_{0.65}Sr_{0.35}MnO_3$ *Phys. Rev.* B **62** 14629–32

[68] Bertacco R, Contour J P, Barthelemy A and Olivier J 2002 Evidence for strontium segregation in $La_{0.7}Sr_{0.3}MnO_3$ thin films grown by pulsed laser deposition: consequences for tunnelling junctions *Surf. Sci.* **511** 366–72

[69] Rupp G M, Limbeck A, Kubicek M, Penn A, Stöger-Pollach M, Friedbacher G and Fleig J 2014 Correlating surface cation composition and thin film microstructure with the electrochemical performance of lanthanum strontium cobaltite (LSC) electrodes *J. Mater. Chem.* A **2** 7099–108

[70] Kubicek M, Limbeck A, Frömling T, Hutter H and Fleig J 2011 Relationship between cation segregation and the electrochemical oxygen reduction kinetics of $La_{0.6}Sr_{0.4}CoO_{3-\delta}$ thin film electrodes *J. Electrochem. Soc.* **158** B727–34

[71] Rupp G M, Opitz A K, Nenning A, Limbeck A and Fleig J 2017 Real-time impedance monitoring of oxygen reduction during surface modification of thin film cathodes *Nat. Mater.* **16** 640–46

[72] Rupp G M, Téllez H, Druce J, Limbeck A, Ishihara T, Kilner J and Fleig J 2015 Surface chemistry of $La_{0.6}Sr_{0.4}CoO_{3-\delta}$ thin films and its impact on the oxygen surface exchange resistance *J. Mater. Chem.* A **3** 22759–69

[73] Wang H, Yakal-Kremski K J, Yeh T, Rupp G M, Limbeck A, Fleig J and Barnett S A 2016 Mechanisms of performance degradation of $(La,Sr)(Co,Fe)O_{3-\delta}$ solid oxide fuel cell cathodes *J. Electrochem. Soc.* **163** F581–85

[74] Pan Z, Liu Q, Zhang L, Zhang X and Chan S H 2015 Effect of Sr surface segregation of $La_{0.6}Sr_{0.4}Co_{0.2}Fe_{0.8}O_{3-\delta}$ electrode on its electrochemical performance in SOC *J. Electrochem. Soc.* **162** F1316–23

[75] Simner S P, Anderson M D, Engelhard M H and Stevenson J W 2006 Degradation mechanisms of La–Sr–Co–Fe–O_3 SOFC cathodes *Electrochem. Solid-State Lett.* **9** A478–81

[76] Cai Z, Kubicek M, Fleig J and Yildiz B 2012 Chemical heterogeneities on $La_{0.6}Sr_{0.4}CoO_{3-\delta}$ thin films—correlations to cathode surface activity and stability *Chem. Mater.* **24** 1116–27

[77] Jung W and Tuller H L 2012 Investigation of surface Sr segregation in model thin film solid oxidefuel cell perovskite electrodes *Energy Environ. Sci.* **5** 5370–78

[78] Chen Y, Jung W, Cai Z H, Kim J J, Tuller H L and Yildiz B 2012 Impact of Sr segregation on the electronic structure and oxygen reduction activity of $SrTi_{1-x}Fe_xO_3$ surfaces *Energy Environ. Sci.* **5** 7979–88

[79] Yokokawa H, Yamaji K, Brito M E, Kishimoto H and Horita T 2011 General considerations on degradation of solid oxide fuel cell anodes and cathodes due to impurities in gases *J. Power Sources* **196** 7070–75

[80] Yokokawa H, Horita T, Sakai N, Yamaji K, Brito M, Xiong Y and Kishimoto H 2006 Thermodynamic considerations on Cr poisoning in SOFC cathodes *Solid State Ion.* **177** 3193–98

[81] Fergus J 2007 Effect of cathode and electrolyte transport properties on chromium poisoning in solid oxide fuel cells *Int. J. Hydrog. Energy* **32** 3664–71

[82] Zhou L, Mason J H, Li W and Liu X 2020 Comprehensive review of chromium deposition and poisoning of solid oxide fuel cells (SOFCs) cathode materials *Renew. Sustain. Energy Rev.* **134** 10320

[83] Bucher E, Yang M and Sitte W 2012 In situ investigations of the chromium-induced degradation of the oxygen surface exchange kinetics of IT-SOFC cathode materials $La_{0.6}Sr_{0.4}CoO_{3-\delta}$ and $La_{0.58}Sr_{0.4}Co_{0.2}Fe_{0.8}O_{3-\delta}$ *J. Electrochem. Soc.* **159** B592–96

[84] Jiang S P, Zhang S and Zhen Y D 2006 Deposition of Cr species at (La,Sr)(Co,Fe)O_3 cathodes of solid oxide fuel cells *J. Electrochem. Soc.* **153** A127–34

[85] Jiang S P, Zhang J P and Zheng X G 2002 A comparative investigation of chromium deposition at air electrodes of solid oxide fuel cells *J. Eur. Ceram. Soc.* **22** 361–73

[86] Huang Y L, Hussain A M, Pellegrinelli C, Xiong C and Wachsman E D 2017 Chromium poisoning effects on surface exchange kinetics of $La_{0.6}Sr_{0.4}Co_{0.2}Fe_{0.8}O_{3-\delta}$ *ACS Appl. Mater. Interfaces* **9** 16660–68

[87] Ni N, Cooper S J, Williams R, Kemen N, McComb D W and Skinner S J 2016 Degradation of $(La_{0.6}Sr_{0.4})_{0.95}(Co_{0.2}Fe_{0.8})O_{3-\delta}$ solid oxide fuel cell cathodes at the nanometer scale and below *ACS Appl. Mater. Interfaces* **8** 17360–70

[88] Wang C C, Chen K, Jiang T, Yang Y, Song Y, Meng H, Jiang S P and Lin B 2018 Sulphur poisoning of solid oxide electrolysis cell anodes *Electrochim. Act* A **269** 188–95

[89] Bucher E, Gspan C and Sitte W 2015 Degradation and regeneration of the SOFC cathode material $La_{0.6}Sr_{0.4}CoO_{3-\delta}$ in SO_2-containing atmospheres *Solid State Ion.* **272** 112–20

[90] Liu R R, Taniguchi S, Shiratori Y, Ito K and Sasaki K 2011 Influence of SO_2 on the long-term durability of SOFC cathodes *ECS Trans.* **35** 2255–60

[91] Xiong Y P, Yamaji K, Horita T, Yokokawa H, Akikusa J, Eto H and Inagaki T 2009 Sulfur poisoning of SOFC cathodes *J. Electrochem. Soc.* **156** B588–92

[92] Wang F, Yamaji K, Cho D H, Shimonosono T, Kishimoto H, Brito M E, Horita T and Yokokawa H 2011 Sulfur poisoning on $La_{0.6}Sr_{0.4}Co_{0.2}Fe_{0.8}O_3$ cathode for SOFCs *J. Electrochem. Soc.* **158** B1391–97

[93] Yan A, Cheng M, Dong Y, Yang W, Maragou V, Song S and Tsiakaras P 2006 Investigation of a $Ba_{0.5}Sr_{0.5}Co_{0.8}Fe_{0.2}O_{3-\delta}$ based cathode IT-SOFC *Appl. Catal.* B **66** 64–71

[94] Yan A, Maragou V, Arico A, Cheng M and Tsiakaras P 2007 Investigation of a $Ba_{0.5}Sr_{0.5}Co_{0.8}Fe_{0.2}O_{3-\delta}$ based cathode SOFC *Appl. Catal.* B **76** 320–27

[95] Yan A, Liu B, Dong Y, Tian Z, Wang D and Cheng M 2008 A temperature programmed desorption investigation on the interaction of $Ba_{0.5}Sr_{0.5}Co_{0.8}Fe_{0.2}O_{3-\delta}$ perovskite oxides with CO_2 in the absence and presence of H_2O and O_2 *Appl. Catal.* B **80** 24–31

[96] Hu B, Mahapatra M K, Keane M, Zhang H and Singh P 2014 Effect of CO_2 on the stability of strontium doped lanthanum manganite cathode *J. Power Sources* **268** 404–13

[97] Chen Y *et al* 2018 A highly active, CO_2-tolerant electrode for the oxygen reduction reaction *Energy Environ. Sci.* **11** 2458–66

[98] Sun S and Cheng Z 2016 Effects of H_2O and CO_2 on electrochemical behaviors of BSCF cathode for proton conducting IT-SOFC *J. Electrochem. Soc.* **164** F81–88

[99] Zhao Z, Liu L, Zhang X, Wu W, Tu B, Cui D, Ou D and Cheng M 2013 High- and low-temperature behaviors of $La_{0.6}Sr_{0.4}Co_{0.2}Fe_{0.8}O_{3-\delta}$ cathode operating under CO_2/H_2O-containing atmosphere *Int. J. Hydrog. Energy* **38** 15361–70

[100] Chen Y *et al* 2017 A robust and active hybrid catalyst for facile oxygen reduction in solid oxide fuel cells *Energy Environ. Sci.* **10** 964–71

[101] Iwata T 1996 Characterization of Ni-YSZ anode degradation for substrate-type solid oxide fuel cells *J. Electrochem. Soc.* **143** 1521–25

[102] Holzer L *et al* 2011 Microstructure degradation of cermet anodes for solid oxide fuel cells: quantification of nickel grain growth in dry and in humid atmospheres *J. Power Sources* **196** 1279–94

[103] Jiang S P 2003 Sintering behavior of Ni/Y_2O_3–ZrO_2 cermet electrodes of solid oxide fuel cells *J. Mater. Sci.* **38** 3775–82

[104] Nelson G J *et al* 2012 Three-dimensional microstructural changes in the Ni–YSZ solid oxide fuel cell anode during operation *Acta Mater.* **60** 3491–500

[105] Simwonis D, Tietz F and Stover D 2000 Nickel coarsening in annealed Ni/8YSZ anode substrates for solid oxide fuel cells *Solid State Ion.* **132** 241–51

[106] Jiao Z, Shikazono N and Kasagi N 2012 Quantitative characterization of SOFC nickel-YSZ anode microstructure degradation based on focused-ion-beam 3D-reconstruction technique *J. Electrochem. Soc.* **159** B285–91

[107] De Angelis S, Jørgensen P S, Tsai E H R, Holler M, Kreka K and Bowen J R 2018 Three dimensional characterization of nickel coarsening in solid oxide cells via ex-situ ptychographic nano-tomography *J. Power Sources* **383** 72–9

[108] Trini M, Jørgensen P S, Hauch A, Bentzen J J, Hendriksen P V and Chen M 2019 3D microstructural characterization of Ni/YSZ electrodes exposed to 1 year of electrolysis testing *J. Electrochem. Soc.* **166** F158–67

[109] Monaco F, Hubert M, Vulliet J, Ouweltjes J P, Montinaro D, Cloetens P, Piccardo P, Lefebvre-Joud F and Laurencin J 2019 Degradation of Ni-YSZ electrodes in solid oxide cells: impact of polarization and initial microstructure on the Ni evolution *J. Electrochem. Soc.* **166** F1229–42

[110] Jiao Z and Shikazono N 2015 Quantitative study on the correlation between solid oxide fuel cell Ni-YSZ composite anode performance and reduction temperature based on three-dimensional reconstruction *J. Electrochem. Soc.* **162** F571–78

[111] Kennouche D, Chen-Wiegart Y-c K, Cronin J S, Wang J and Barnett S A 2013 Three-dimensional microstructural evolution of Ni-yttria-stabilized zirconia solid oxide fuel cell anodes at elevated temperatures *J. Electrochem. Soc.* **160** F1293–304

[112] Jiao Z, Takagi N, Shikazono N and Kasagi N 2011 Study on local morphological changes of nickel in solid oxide fuel cell anode using porous Ni pellet electrode *J. Power Sources* **196** 1019–29

[113] Wang Y, Wu C, Zu B, Han M, Du Q, Ni M and Jiao K 2021 Ni migration of Ni-YSZ electrode in solid oxide electrolysis cell: an integrated model study *J. Power Sources* **516** 230660

[114] Liu Q, Zhang Q, Voorhees P W and Barnett S A 2019 Effect of direct-current operation on the electrochemical performance and structural evolution of Ni-YSZ electrodes *J. Phys.: Energy* **2** 014006

[115] Trini M, Hauch A, De Angelis S, Tong X, Hendriksen P V and Chen M 2020 Comparison of microstructural evolution of fuel electrodes in solid oxide fuel cells and electrolysis cells *J. Power Sources* **450** 227599
[116] Hoerlein M P, Riegraf M, Costa R, Schiller G and Friedrich K A 2018 A parameter study of solid oxide electrolysis cell degradation: microstructural changes of the fuel electrode *Electrochim. Act* A **276** 162–75
[117] Watanabe H, Okino R and Hanamura K 2019 Structural evolution of carbon deposition on a Ni/YSZ cermet of a SOFC analyzed by soft X-ray xanes spectroscopy *Int. J. Hydrog. Energy* **44** 24028–35
[118] Skafte T L, Blennow P, Hjelm J and Graves C 2018 Carbon deposition and sulfur poisoning during CO_2 electrolysis in nickel-based solid oxide cell electrodes *J. Power Sources* **373** 54–60
[119] Kim T, Liu G, Boaro M, Lee S I, Vohs J M, Gorte R J, Al-Madhi O H and Dabbousi B O 2006 A study of carbon formation and prevention in hydrocarbon-fueled SOFC *J. Power Sources* **155** 231–38
[120] Macek J, Novosel B, Marinnsek M and SOFC N-Y S Z 2007 Anodes—minimization of carbon deposition *J. Eur. Ceram. Soc.* **27** 487–91
[121] Sumi H, Lee Y-H, Muroyama H, Matsui T, Kamijo M, Mimuro S, Yamanaka M, Nakajima Y and Eguchi K 2011 Effect of carbon deposition by carbon monoxide disproportionation on electrochemical characteristics at low temperature operation for solid oxide fuel cells *J. Power Sources* **196** 4451–57
[122] Tao Y, Ebbesen S D and Mogensen M B 2014 Carbon deposition in solid oxide cells during co-electrolysis of H_2O and CO_2 *J. Electrochem. Soc.* **161** F337–343
[123] Chun C M, Mumford J D and Ramanarayanan T A 2000 Carbon-induced corrosion of nickel anode *J. Electrochem. Soc.* **147** 3680–86
[124] Yue W X *et al* 2019 Enhancing coking resistance of Ni/YSZ electrodes: *in situ* characterization, mechanism research, and surface engineering *Nano Energy* **62** 64–78
[125] Shiratori Y, Oshima T and Sasaki K 2008 Feasibility of direct-biogas SOFC *Int. J. Hydrog. Energy* **33** 6316–21
[126] Vecino-Mantilla S, Simon P, Huvé M, Gauthier G and Gauthier-Maradei P 2020 Methane steam reforming in water-deficient conditions on a new Ni-exsolved Ruddlesden–Popper manganite: coke formation and H_2S poisoning *Int. J. Hydrog. Energy* **45** 27145–59
[127] Wachowski S, Li Z, Polfus J M and Norby T 2018 Performance and stability in H_2S of $SrFe_{0.75}Mo_{0.25}O_{3-\delta}$ as electrode in proton ceramic fuel cells *J. Eur. Ceram. Soc.* **38** 163–71
[128] Zha S W, Cheng Z and Liu M L 2007 Sulfur poisoning and regeneration of Ni-based anodes in solid oxide fuel cells *J. Electrochem. Soc.* **154** B201–6
[129] Lee H S, Lee H M, Park J-Y and Lim H-T 2018 Degradation behavior of Ni-YSZ anode-supported solid oxide fuel cell (SOFC) as a function of H_2S concentration *Int. J. Hydrog. Energy* **43** 22511–18
[130] Pomeroy E D, Maza W A, Steinhurst D A, Owrutsky J C and Walker R A 2020 Electrochemical sulfur oxidation in solid oxide fuel cells studied by near infrared thermal imaging and chronocoulometry *J. Electrochem. Soc.* **167** 164511
[131] Riegraf M, Zekri A, Knipper M, Costa R, Schiller G and Friedrich K A 2018 Sulfur poisoning of Ni/gadolinium-doped ceria anodes: a long-term study outlining stable solid oxide fuel cell operation *J. Power Sources* **380** 26–36

[132] Kim J H, Liu M, Chen Y, Murphy R, Choi Y, Liu Y and Liu M 2021 Understanding the impact of sulfur poisoning on the methane-reforming activity of a solid oxide fuel cell anode *ACS Catal.* **11** 13556–66

[133] Hauch A, Traulsen M L, Küngas R and Skafte T L 2021 CO_2 electrolysis—gas impurities and electrode overpotential causing detrimental carbon deposition *J. Power Sources* **506** 230108

[134] Lynch M E, Yang L, Qin W T, Choi J J, Liu M F, Blinn K and Liu M L 2011 Enhancement of $La_{0.6}Sr_{0.4}Co_{0.2}Fe_{0.8}O_{3-\delta}$ durability and surface electrocatalytic activity by $La_{0.85}Sr_{0.15}MnO_{3\pm\delta}$ investigated using a new test electrode platform *Energy Environ. Sci.* **4** 2249–58

[135] Wang J, Yang T R, Wen Y T, Zhang Y, Sun C W and Huang K 2019 Performance and stability of $SrCo_{0.9}Nb_{0.1}O_{3-\delta}$–$(La_{0.60}Sr_{0.40})_{0.95}(Co_{0.20}Fe_{0.80})O_{3-\delta}$ bilayer cathode for intermediate-temperature solid oxide fuel cells *J. Power Sources* **414** 24–30

[136] Li J, Zhou X, Wu C, Zhao L, Dong B, Wang S and Chi B 2022 Self-stabilized hybrid cathode for solid oxide fuel cell: A-site deficient perovskite coating as solid solution for strontium diffusion *Chem. Eng. J.* **438** 135446

[137] Zhou Y, Zhang W, Kane N, Luo Z, Pei K, Sasaki K, Choi Y, Chen Y, Ding D and Liu M 2021 An efficient bifunctional air electrode for reversible protonic ceramic electrochemical cells *Adv. Funct. Mater.* **31** 2105386

[138] Pei K *et al* 2020 Enhanced Cr-tolerance of an SOFC cathode by an efficient electro-catalyst coating *Nano Energy* **72** 104704

[139] Pei K, Zhou Y, Ding Y, Xu K, Zhang H, Yuan W, Sasaki K, Choi Y, Liu M and Chen Y 2021 An improved oxygen reduction reaction activity and CO_2-tolerance of $La_{0.6}Sr_{0.4}Co_{0.2}Fe_{0.8}O_{3-\delta}$ achieved by a surface modification with barium cobaltite coatings *J. Power Sources* **514** 230573

[140] Kim J H, Yoo S, Murphy R, Chen Y, Ding Y, Pei K, Zhao B, Kim G, Choi Y and Liu M 2021 Promotion of oxygen reduction reaction on a double perovskite electrode by a water-induced surface modification *Energy Environ. Sci.* **14** 1506–16

[141] Yashiro K, Nakamura T, Sase M, Hermes F, Sato K, Kawada T and Mizusaki J 2009 Composite cathode of perovskite-related oxides, $(La,Sr)CoO_{3-\delta}/(La,Sr)_2CoO_{4-\delta}$, for solid oxide fuel cells *Electrochem. Solid-State Lett.* **12** B135–37

[142] Sase M, Hermes F, Yashiro K, Sato K, Mizusaki J, Kawada T, Sakai N and Yokokawa H 2008 Enhancement of oxygen surface exchange at the hetero-interface of $(La,Sr)CoO_{3-\delta}/(La,Sr)_2CoO_{4-\delta}$ with pld-layered films *J. Electrochem. Soc.* **155** B793–97

[143] Sase M, Yashiro K, Sato K, Mizusaki J, Kawada T, Sakai N, Yamaji K, Horita T and Yokokawa H 2008 Enhancement of oxygen exchange at the hetero interface of $(La,Sr)CoO_{3-\delta}/(La,Sr)_2CoO_{4-\delta}$ in composite ceramics *Solid State Ion.* **178** 1843–52

[144] Crumlin E J, Mutoro E, Ahn S J, la O G J, Leonard D N, Borisevich A, Biegalski M D, Christen H M and Shao-Horn Y 2010 Oxygen reduction kinetics enhancement on a heterostructured oxide surface for solid oxide fuel cells *J. Phys. Chem. Lett.* **1** 3149–55

[145] Zhao C H, Liu X G, Zhang W Q, Zheng Y, Li Y F, Yu B, Wang J C and Chen J 2019 Measurement of oxygen reduction/evolution kinetics enhanced $(La,Sr)CoO_{3-\delta}/(La,Sr)_2CoO_{4-\delta}$ hetero-structure oxygen electrode in operating temperature for SOCs *Int. J. Hydrog. Energy* **44** 19102–12

[146] Ma W, Kim J J, Tsvetkov N, Daio T, Kuru Y, Cai Z H, Chen Y, Sasaki K, Tuller H L and Yildiz B 2015 Vertically aligned nanocomposite $La_{0.8}Sr_{0.2}CoO_3/(La_{0.5}Sr_{0.5})_2CoO_4$ cathodes

—electronic structure, surface chemistry and oxygen reduction kinetics *J. Mater. Chem.* A **3** 207–19

[147] Hong T, Zhao M Y, Brinkman K, Chen F L and Xia C R 2017 Enhanced oxygen reduction activity on Ruddlesden–Popper phase decorated $La_{0.8}Sr_{0.2}FeO_{3-\delta}$ 3D heterostructured cathode for solid oxide fuel cells *ACS Appl. Mater. Interfaces* **9** 8659–68

[148] Demirkal E, Ozdemir O, Colakerol Arslan L and Buyukaksoy A 2020 Sources of performance degradation in thin film $La_{0.6}Sr_{0.4}FeO_3$ air electrodes *Solid State Ion.* **355** 115420

[149] Zheng Y, Li Y F, Wu T, Zhang W Q, Zhu J X, Li Z P, Chen J, Yu B, Wang J C and Zhang J J 2018 Oxygen reduction kinetic enhancements of intermediate-temperature SOFC cathodes with novel $Nd_{0.5}Sr_{0.5}CoO_{3-\delta}/Nd_{0.8}Sr_{1.2}CoO_{4\pm\delta}$ heterointerfaces *Nano Energy* **51** 711–20

[150] Xu K *et al* 2022 An efficient steam-induced heterostructured air electrode for protonic ceramic electrochemical cells *Adv. Funct. Mater.* **32** 2110998

[151] Kim J H, Jang K, Lim D-K, Ahn S, Oh D, Kim H, Seo J, Choi P-P and Jung W 2022 Self-assembled nano-composite perovskites as highly efficient and robust hybrid cathodes for solid oxide fuel cells *J. Mater. Chem.* A **10** 2496–508

[152] Gong Y H, Patel R L, Liang X H, Palacio D, Song X Y, Goodenough J B and Huang K 2013 Atomic layer deposition functionalized composite SOFC cathode $La_{0.6}Sr_{0.4}Fe_{0.8}Co_{0.2}O_{3-\delta}$–$Gd_{0.2}Ce_{0.8}O_{1.9}$: enhanced long-term stability *Chem. Mater.* **25** 4224–31

[153] Wen Y T, Yang T R, Lee D, Lee H N, Crumlin E J and Huang K 2018 Temporal and thermal evolutions of surface Sr segregation in pristine and atomic layer deposition modified $La_{0.6}Sr_{0.4}CoO_{3-\delta}$ epitaxial films *J. Mater. Chem.* A **6** 24378–88

[154] Tsvetkov N, Lu Q and Yildiz B 2015 Improved electrochemical stability at the surface of $La_{0.8}Sr_{0.2}CoO_3$ achieved by surface chemical modification *Faraday Discuss.* **182** 257–69

[155] Tsvetkov N, Lu Q Y, Sun L X, Crumlin E J and Yildiz B 2016 Improved chemical and electrochemical stability of perovskite oxides with less reducible cations at the surface *Nat. Mater.* **15** 1010–16

[156] Chen H J *et al* 2018 Improving the electrocatalytic activity and durability of the $La_{0.6}Sr_{0.4}Co_{0.2}Fe_{0.8}O_{3-\delta}$ cathode by surface modification *ACS Appl. Mater. Interfaces* **10** 39785–93

[157] Zhang W W, Wang H C, Guan K, Wei Z Y, Zhang X, Meng J L, Liu X J and Meng J 2019 $La_{0.6}Sr_{0.4}Co_{0.2}Fe_{0.8}O_{3-\delta}/CeO_2$ heterostructured composite nanofibers as a highly active and robust cathode catalyst for solid oxide fuel cells *ACS Appl. Mater. Interfaces* **11** 26830–41

[158] Li D, Jin Y, Liu C, Fu M, Zong X and Xiong Y 2021 Effect of optimal GDC loading on long-term stability of $La_{0.8}Sr_{0.2}Co_{0.2}Fe_{0.8}O_{3-\delta}/Gd_{0.2}Ce_{0.8}O_{1.9}$ composite cathodes *Ceram. Int.* **47** 6591–96

[159] Nadeem M, Hu B B and Xia C R 2018 Effect of NiO addition on oxygen reduction reaction at lanthanum strontium cobalt ferrite cathode for solid oxide fuel cell *Int. J. Hydrog. Energy* **43** 8079–87

[160] Yang Y, Li M, Ren Y Y, Li Y H and Xia C R 2018 Magnesium oxide as synergistic catalyst for oxygen reduction reaction on strontium doped lanthanum cobalt ferrite *Int. J. Hydrog. Energy* **43** 3797–802

[161] Wang H, Zhang W, Guan K, Wei Z, Meng J, Meng J and Liu X 2020 Enhancing activity and durability of A-site-deficient $(La_{0.6}Sr_{0.4})_{0.95}Co_{0.2}Fe_{0.8}O_{3-\delta}$ cathode by surface modification with $PrO_{2-\delta}$ nanoparticles *ACS Sustain. Chem. Eng.* **8** 3367–80

[162] Hong T, Brinkman K and Xia C R 2016 Copper oxide as a synergistic catalyst for the oxygen reduction reaction on $La_{0.6}Sr_{0.4}Co_{0.2}Fe_{0.8}O_{3-\delta}$ perovskite structured electrocatalyst *J. Power Sources* **329** 281–89
[163] Mutoro E, Crumlin E J, Biegalski M D, Christen H M and Shao-Horn Y 2011 Enhanced oxygen reduction activity on surface-decorated perovskite thin films for solid oxide fuel cells *Energy Environ. Sci.* **4** 3689–96
[164] Choi H J, Bae K, Jang D Y, Kim J W and Shim J H 2015 Performance degradation of lanthanum strontium cobaltite after surface modification *J. Electrochem. Soc.* **162** F622–626
[165] Yang Y, Li R, Wu Y, Chu Y, Tian D, Lu X, Guo L, Lin B, Feng P and Ling Y 2020 Highly active self-assembled hybrid catalyst with multiphase heterointerfaces to accelerate cathodic oxygen reduction of intermediate-temperature solid oxide fuel cells *Ceram. Int.* **46** 9661–68
[166] Niu Y *et al* 2022 Highly active and durable air electrodes for reversible protonic ceramic electrochemical cells enabled by an efficient bifunctional catalyst *Adv. Energy Mater.* **2103783**
[167] Kushima A, Yip S and Yildiz B 2010 Competing strain effects in reactivity of $LaCoO_3$ with oxygen *Phys. Rev.* B **82** 115435
[168] Han J W and Yildiz B 2011 Enhanced one dimensional mobility of oxygen on strained $LaCoO_3$(001) surface *J. Mater. Chem.* **21** 18983–90
[169] Mayeshiba T and Morgan D 2017 Strain effects on oxygen vacancy formation energy in perovskites *Solid State Ion.* **311** 105–17
[170] Cai Z H, Kuru Y, Han J W, Chen Y and Yildiz B 2011 Surface electronic structure transitions at high temperature on perovskite oxides: the case of strained $La_{0.8}Sr_{0.2}CoO_3$ thin films *J. Am. Chem. Soc.* **133** 17696–704
[171] Kubicek M, Cai Z, Ma W, Yildiz B, Hutter H and Fleig J 2013 Tensile lattice strain accelerates oxygen surface exchange and diffusion in $La_{1-x}Sr_xCoO_{3-\delta}$ thin films *ACS Nano* **7** 3276–86
[172] Petrie J R, Mitra C, Jeen H, Choi W S, Meyer T L, Reboredo F A, Freeland J W, Eres G and Lee H N 2016 Strain control of oxygen vacancies in epitaxial strontium cobaltite films *Adv. Funct. Mater.* **26** 1564–70
[173] Navickas E *et al* 2017 Dislocations accelerate oxygen ion diffusion in $La_{0.8}Sr_{0.2}MnO_3$ epitaxial thin films *ACS Nano* **11** 11475–87
[174] Hwang J *et al* 2019 Tuning perovskite oxides by strain: electronic structure, properties, and functions in (electro)catalysis and ferroelectricity *Mater. Today* **31** 100–18
[175] Han J W, Jalili H, Kuru Y, Cai Z H and Yildiz B 2011 Strain effects on the surface chemistry of $La_{0.7}Sr_{0.3}MnO_3$ *ECS Trans.* **35** 2097–104
[176] Liu X *et al* 2019 Uncovering the effect of lattice strain and oxygen deficiency on electrocatalytic activity of perovskite cobaltite thin films *Adv. Sci.* **6** 1801898
[177] Chen Y, Romo-De-La-Cruz C O, Paredes-Navia S A, Liang L, Hinerman A, Prucz J, Williams M and Song X 2022 Electrocatalytic surface nanoionics with strained interfaced and colossal conductivity for enhancing durability and performance of solid oxide fuel cell *J. Power Sources* **517** 230715
[178] Han J W and Yildiz B 2012 Mechanism for enhanced oxygen reduction kinetics at the (la, Sr)$CoO_{3-\delta}$/(La,Sr)$_2CoO_{4+\delta}$ hetero-interface *Energy Environ. Sci.* **5** 8598–607
[179] Kang B S, Matsuda J, Ju Y W, Kim H H and Ishihara T 2019 Nano strain induced double columnar oxide as highly active oxygen-dissociation electrode for Ni–Fe metal supported solid oxide fuel cells *Nano Energy* **56** 382–90

[180] Fernandez A, Caretta L, Das S, Klewe C, Lou D, Parsonnet E, Gao R, Luo A, Shafer P and Martin L W 2021 Strain-induced orbital contributions to oxygen electrocatalysis in transition-metal perovskites *Adv. Energy Mater.* **11** 2102175
[181] Pesquera D, Herranz G, Barla A, Pellegrin E, Bondino F, Magnano E, Sanchez F and Fontcuberta J 2012 Surface symmetry-breaking and strain effects on orbital occupancy in transition metal perovskite epitaxial films *Nat. Commun.* **3** 2189
[182] Yu Y, Ludwig K F, Woicik J C, Gopalan S, Pal U B, Kaspar T C and Basu S N 2016 Effect of Sr content and strain on Sr surface segregation of $La_{1-x}Sr_xCo_{0.2}Fe_{0.8}O_{3-\delta}$ as cathode material for solid oxide fuel cells *ACS Appl. Mater. Interfaces* **8** 26704–11
[183] Koo B, Kwon H, Kim Y, Seo H G, Han J W and Jung W 2018 Enhanced oxygen exchange of perovskite oxide surfaces through strain-driven chemical stabilization *Energy Environ. Sci.* **11** 71–7
[184] Kuru Y, Jalili H, Cai Z H, Yildiz B and Tuller H L 2011 Direct probing of nanodimensioned oxide multilayers with the aid of focused ion beam milling *Adv. Mater.* **23** 4543–48
[185] Chen Y, Cai Z H, Kuru Y, Ma W, Tuller H L and Yildiz B 2013 Electronic activation of cathode superlattices at elevated temperatures-source of markedly accelerated oxygen reduction kinetics *Adv. Energy Mater.* **3** 1221–29
[186] Chen Y, Fong D D, Herbert F W, Rault J, Rueff J P, Tsvetkov N and Yildiz B 2018 Modified oxygen defect chemistry at transition metal oxide heterostructures probed by hard x-ray photoelectron spectroscopy and x-ray diffraction *Chem. Mater.* **30** 3359–71
[187] Zheng Y *et al* 2020 Enhanced oxygen reduction kinetics by a porous heterostructured cathode for intermediate temperature solid oxide fuel cells *Energy AI* **2** 100027
[188] Paulson S C and Birss V I 2004 Chromium poisoning of LSM-YSZ SOFC cathodes I. Detailed study of the distribution of chromium species at a porous, single-phase cathode *J. Electrochem. Soc.* **151** A1961–68
[189] Nonnenmann S S, Kungas R, Vohs J and Bonnell D A 2013 Direct *in situ* probe of electrochemical processes in operating fuel cells *ACS Nano* **7** 6330–36
[190] Chen K F, Ai N and Jiang S P 2014 Performance and structural stability of $Gd_{0.2}Ce_{0.8}O_{1.9}$ infiltrated $La_{0.8}Sr_{0.2}MnO_3$ nano-structured oxygen electrodes of solid oxide electrolysis cells *Int. J. Hydrog. Energy* **39** 10349–58
[191] Huang Y-L, Hussain A M and Wachsman E D 2018 Nanoscale cathode modification for high performance and stable low-temperature solid oxide fuel cells (SOFCs) *Nano Energy* **49** 186–92
[192] Kim S *et al* 2021 Microstructure tailoring of solid oxide electrolysis cell air electrode to boost performance and long-term durability *Chem. Eng. J.* **410** 128318
[193] Shimada H, Yamaguchi T, Kishimoto H, Sumi H, Yamaguchi Y, Nomura K and Fujishiro Y 2019 Nanocomposite electrodes for high current density over 3 A cm^{-2} in solid oxide electrolysis cells *Nat. Commun.* **10** 5432
[194] Baiutti F *et al* 2021 A high-entropy manganite in an ordered nanocomposite for long-term application in solid oxide cells *Nat. Commun.* **12** 2660
[195] Aziz A J A, Baharuddin N A, Somalu M R and Muchtar A 2020 Review of composite cathodes for intermediate-temperature solid oxide fuel cell applications *Ceram. Int.* **46** 23314–25
[196] dos Santos-Gomez L, Zamudio-Garcia J, Porras-Vazquez J M, Losilla E R and Marrero-Lopez D 2021 Recent progress in nanostructured electrodes for solid oxide fuel cells deposited by spray pyrolysis *J. Power Sources* **507** 230277

[197] Zhang X, Song Y, Guan F, Zhou Y, Lv H, Wang G and Bao X 2018 Enhancing electrocatalytic CO_2 reduction in solid oxide electrolysis cell with $Ce_{0.9}Mn_{0.1}O_{2-\delta}$ nanoparticles-modified LSCM-GDC cathode *J. Catal.* **359** 8–16

[198] Wan Y H, Hu B B and Xia C R 2017 Oxygen reduction at the three-phase boundary of $PrBaCo_2O_{5+\delta}$-$Sm_{0.2}Ce_{0.8}O_{1.9}$ composite *Electrochim. Act* A **252** 171–79

[199] Lu Y, Ding L, Li M, Li J, Wang X and Ding X 2021 High-performance and CO_2-resistant cathode toward electrocatalytic oxygen reduction for solid oxide fuel cells: doped ceria and $SrCo_{0.7}Nb_{0.1}Ni_{0.2}O_{3-\delta}$ composite *Electrochim. Act* A **398** 139323

[200] Chen Z *et al* 2020 Accelerating effect of polarization on electrode/electrolyte interface generation and electrocatalytic performance of $Er_{0.4}Bi_{1.6}O_3$ decorated $Sm_{0.95}CoO_{3-\delta}$ cathodes *J. Power Sources* **465** 228281

[201] Li Y F, Zhang W Q, Zheng Y, Chen J, Yu B, Chen Y and Liu M L 2017 Controlling cation segregation in perovskite-based electrodes for high electro-catalytic activity and durability *Chem. Soc. Rev.* **46** 6345–78

[202] Zhao C, Li Y, Zhang W, Zheng Y, Lou X, Yu B, Chen J, Chen Y, Liu M and Wang J 2020 Heterointerface engineering for enhancing the electrochemical performance of solid oxide cells *Energy Environ. Sci.* **13** 53–85

[203] Lee W, Han J W, Chen Y, Cai Z and Yildiz B 2013 Cation size mismatch and charge interactions drive dopant segregation at the surfaces of manganite perovskites *J. Am. Chem. Soc.* **135** 7909–25

[204] Bliem R, Kim D, Wang J, Crumlin E J and Yildiz B 2021 Hf deposition stabilizes the surface chemistry of perovskite manganite oxide *J. Phys. Chem.* C **125** 3346–54

[205] Li Y F, Zhang W Q, Wu T, Zheng Y, Chen J, Yu B, Zhu J X and Liu M L 2018 Segregation induced self-assembly of highly active perovskite for rapid oxygen reduction reaction *Adv. Energy Mater.* **8** 1801893

[206] Lee D, Lee Y L, Grimaud A, Hong W T, Biegalski M D, Morgan D and Shao-Horn Y 2014 Enhanced oxygen surface exchange kinetics and stability on epitaxial $La_{0.8}Sr_{0.2}CoO_{3-\delta}$ thin films by $La_{0.8}Sr_{0.2}MnO_{3-\delta}$ decoration *J. Phys. Chem.* C **118** 14326–34

[207] Zhang X X, Zhang H and Liu X B 2014 High performance $La_2NiO_{4+\delta}$-infiltrated $(La_{0.6}Sr_{0.4})_{0.995}Co_{0.2}Fe_{0.8}O_{3-\delta}$ cathode for solid oxide fuel cells *J. Power Sources* **269** 412–17

[208] Ding H P, Virkar A V, Liu M L and Liu F 2013 Suppression of sr surface segregation in $La_{1-x}Sr_xCo_{1-y}Fe_yO_{3-\delta}$: a first principles study *Phys. Chem. Chem. Phys.* **15** 489–96

[209] Zheng Y, Li Y F, Wu T, Zhang W Q, Zhu J X, Li Z P, Chen J, Yu B, Wang J C and Zhang J J 2018 Oxygen reduction kinetic enhancements of intermediate-temperature SOFC cathodes with novel $Nd_{0.5}Sr_{0.5}CoO_{3-\delta}$/$Nd_{0.8}Sr_{1.2}CoO_{4\pm\delta}$ heterointerfaces *Nano Energy* **51** 711–20

[210] Zheng Y *et al* 2020 Directly visualizing and exploring local heterointerface with high electro-catalytic activity *Nano Energy* **78** 105236

[211] Feng Z X *et al* 2014 Anomalous interface and surface strontium segregation in $(La_{1-y}Sr_y)_2CoO_{4\pm\delta}$/$La_{1-x}Sr_xCoO_{3-\delta}$ heterostructured thin films *J. Phys. Chem. Lett.* **5** 1027–34

[212] Zhao L, Zhang J, Becker T and Jiang S P 2014 Raman spectroscopy study of chromium deposition on $La_{0.6}Sr_{0.4}Co_{0.2}Fe_{0.8}O_{3-\delta}$ cathode of solid oxide fuel cells *J. Electrochem. Soc.* **161** F687–693

[213] Islam S and Hill J M 2014 Barium oxide promoted Ni/YSZ solid-oxide fuel cells for direct utilization of methane *J. Mater. Chem.* A **2** 1922–29

[214] Takeguchi T, Kani Y, Yano T, Kikuchi R, Eguchi K, Tsujimoto K, Uchida Y, Ueno A, Omoshiki K and Aizawa M 2002 Study on steam reforming of CH_4 and C_2 hydrocarbons and carbon deposition on Ni-YSZ cermets *J. Power Sources* **112** 588–95

[215] La Rosa D, Sin A, Lo Faro M, Monforte G, Antonucci V and Arico A S 2009 Mitigation of carbon deposits formation in intermediate temperature solid oxide fuel cells fed with dry methane by anode doping with barium *J. Power Sources* **193** 160–64

[216] Qu J F, Wang W, Chen Y B, Deng X and Shao Z P 2016 Stable direct-methane solid oxide fuel cells with calcium-oxide-modified nickel-based anodes operating at reduced temperatures *Appl. Energy* **164** 563–71

[217] Boldrin P, Ruiz-Trejo E, Mermelstein J, Menendez J M B, Reina T R and Brandon N P 2016 Strategies for carbon and sulfur tolerant solid oxide fuel cell materials, incorporating lessons from heterogeneous catalysis *Chem. Rev.* **116** 13633–84

[218] Itagaki Y, Yamaguchi S and Yahiro H 2021 Anodic performance of BaO-added Ni/SDC for solid oxide fuel cell fed with dry CH_4 *Front. Energy Res.* **9** 652239

[219] Wang C, Liao M, Liang B, Jiang Z, Zhong W, Chen Y, Luo X, Shu R, Tian Z and Lei L 2021 Enhancement effect of catalyst support on indirect hydrogen production from propane partial oxidation towards commercial solid oxide fuel cell (SOFC) applications *Appl. Energy* **288** 116362

[220] Chen X, Khor K and Chan S H 2004 Suppression of carbon deposition at CeO_2-modified Ni/YSZ anodes in weakly humidified CH_4 at 850 °C *Electrochem. Solid-State Lett.* **8** A79–82

[221] Sariboğa V and Öksüzömer F 2012 The investigation of active Ni/YSZ interlayer for Cu-based direct-methane solid oxide fuel cells *Appl. Energy* **93** 707–21

[222] Matsui T, Eguchi K, Shirai K, Ozeki T, Okanishi T, Muroyama H and Eguchi K 2017 Redox-induced self-modification of cermet anodes of Ni–CeO_2-based oxide for solid oxide fuel cells *J. Electrochem. Soc.* **164** F1368–74

[223] Tiwari P K and Basu S 2017 CeO_2 and Nb_2O_5 modified Ni-YSZ anode for solid oxide fuel cell *Ionics* **23** 2571–77

[224] Cimenti M, Alzate-Restrepo V and Hill J M 2010 Direct utilization of methanol on impregnated Ni/YSZ and Ni–$Zr_{0.35}Ce_{0.65}O_2$/YSZ anodes for solid oxide fuel cells *J. Power Sources* **195** 4002–12

[225] Chen Y, Bunch J, Jin C, Yang C and Chen F 2012 Performance enhancement of Ni-YSZ electrode by impregnation of $Mo_{0.1}Ce_{0.9}O_{2+\delta}$ *J. Power Sources* **204** 40–5

[226] Lee J M, Kim Y G, Lee S J, Kim H S, Yoon S P, Nam S W, Yoon S D and Yun J W 2014 Carbon-tolerance effects of $Sm_{0.2}Ce_{0.8}O_{2-\delta}$ modified Ni/YSZ anode for solid oxide fuel cells under methane fuel conditions *J. Appl. Electrochem.* **44** 581–88

[227] Choi Y, Brown E C, Haile S M and Jung W 2016 Electrochemically modified, robust solid oxide fuel cell anode for direct-hydrocarbon utilization *Nano Energy* **23** 161–71

[228] Lyu Z, Wang Y, Zhang Y and Han M 2020 Solid oxide fuel cells fueled by simulated biogas: comparison of anode modification by infiltration and reforming catalytic layer *Chem. Eng. J.* **393** 124755

[229] Park B-K, Scipioni R, Cox D and Barnett S A 2020 Enhancement of Ni–$(Y_2O_3)_{0.08}(ZrO_2)_{0.92}$ fuel electrode performance by infiltration of $Ce_{0.8}Gd_{0.2}O_{2-\delta}$ nanoparticles *J. Mater. Chem.* A **8** 4099–106

[230] Hua B, Li M, Sun Y F, Zhang Y Q, Yan N, Li J, Etsell T, Sarkar P and Luo J L 2017 Grafting doped manganite into nickel anode enables efficient and durable energy conversions in biogas solid oxide fuel cells *Appl. Catal.* B **200** 174–81

[231] Vincent A L, Hanifi A R, Luo J L, Chuang K T, Sanger A R, Etsell T H and Sarkar P 2012 Porous YSZ impregnated with $La_{0.4}Sr_{0.5}Ba_{0.1}TiO_3$ as a possible composite anode for SOFCs fueled with sour feeds *J. Power Sources* **215** 301–6

[232] Yoon S P, Han J, Nam S W, Lim T-H, Oh I-H, Hong S-A, Yoo Y-S and Lim H C 2002 Performance of anode-supported solid oxide fuel cell with $La_{0.85}Sr_{0.15}MnO_3$ cathode modified by sol–gel coating technique *J. Power Sources* **106** 160–66

[233] Paige J M, Vu D, Cao T, McIntosh S, Gorte R J and Vohs J M 2021 Investigating the catalytic requirements of perovskite fuel electrodes using ultra-low metal loadings *J. Electrochem. Soc.* **168** 084502

[234] Xiao G and Chen F 2011 Ni modified ceramic anodes for direct-methane solid oxide fuel cells *Electrochem. Commun.* **13** 57–9

[235] Kim K I, Kim H S, Kim H S and Yun J W 2018 H_2S tolerance effects of $Ce_{0.8}Sm_{0.2}O_{2-\delta}$ modification on $Sr_{0.92}Y_{0.08}Ti_{1-x}Ni_xO_{3-\delta}$ anode in solid oxide fuel cells *J. Ind. Eng. Chem.* **68** 187–95

[236] Cao T, Kwon O, Gorte R J and Vohs J M 2020 Metal exsolution to enhance the catalytic activity of electrodes in solid oxide fuel cells *Nanomaterials* **10** 10122445

[237] Hua B, Li M, Sun Y F, Li J H and Luo J L 2017 Enhancing perovskite electrocatalysis of solid oxide cells through controlled exsolution of nanoparticles *ChemSusChem* **10** 3333–41

[238] Sun X, Chen H, Yin Y, Curnan M T, Han J W, Chen Y and Ma Z 2021 Progress of exsolved metal nanoparticles on oxides as high performance (electro)catalysts for the conversion of small molecules *Small* **17** 2005383

[239] Islam Q A, Paydar S, Akbar N, Zhu B and Wu Y 2021 Nanoparticle exsolution in perovskite oxide and its sustainable electrochemical energy systems *J. Power Sources* **492** 229626

[240] Neagu D, Tsekouras G, Miller D N, Ménard H and Irvine J T S 2013 In situ growth of nanoparticles through control of non-stoichiometry *Nat. Chem.* **5** 916–23

[241] Zhu C, Hou S, Hu X, Lu J, Chen F and Xie K 2019 Electrochemical conversion of methane to ethylene in a solid oxide electrolyzer *Nat. Commun.* **10** 1173

[242] Syed K, Wang J, Yildiz B and Bowman W J 2022 Bulk and surface exsolution produces a variety of Fe-rich and Fe-depleted ellipsoidal nanostructures in $La_{0.6}Sr_{0.4}FeO_3$ thin films *Nanoscale* **14** 663–74

[243] Kwon O, Sengodan S, Kim K, Kim G, Jeong H Y, Shin J, Ju Y W, Han J W and Kim G 2017 Exsolution trends and co-segregation aspects of self-grown catalyst nanoparticles in perovskites *Nat. Commun.* **8** 15967

[244] Li Z, Peng M, Zhao Y, Li J and Sun Y 2021 Minimized thermal expansion mismatch of cobalt-based perovskite air electrodes for solid oxide cells *Nanoscale* **13** 20299–308

[245] Zhou J, Shin T H, Ni C S, Chen G, Wu K, Cheng Y H and Irvine J T S 2016 In situ growth of nanoparticles in layered perovskite $La_{0.8}Sr_{1.2}Fe_{0.9}Co_{0.1}O_{4-\delta}$ as an active and stable electrode for symmetrical solid oxide fuel cells *Chem. Mater.* **28** 2981–93

[246] Park B H and Choi G M 2014 Ex-solution of Ni nanoparticles in a $La_{0.2}Sr_{0.8}Ti_{1-x}Ni_xO_{3-\delta}$ alternative anode for solid oxide fuel cell *Solid State Ion.* **262** 345–48

[247] Li Y X, Xie K, Chen S G, Li H X, Zhang Y and Wu Y C 2015 Efficient carbon dioxide electrolysis based on perovskite cathode enhanced with nickel nanocatalyst *Electrochim. Act* A **153** 325–33

[248] Myung J H, Neagu D, Miller D N and Irvine J T S 2016 Switching on electrocatalytic activity in solid oxide cells *Nature* **537** 528–31

[249] Fu L, Zhou J, Yang J, Lian Z, Wang J, Cheng Y and Wu K 2020 Exsolution of Cu nanoparticles in $(LaSr)_{0.9}Fe_{0.9}Cu_{0.1}O_4$ Ruddlesden–Popper oxide as symmetrical electrode for solid oxide cells *Appl. Surf. Sci.* **511** 145525

[250] Jo S, Jeong H G, Kim Y H, Neagu D and Myung J-h 2021 Stability and activity controls of Cu nanoparticles for high-performance solid oxide fuel cells *Appl. Catal.* B **285** 119828

[251] Zubenko D, Singh S and Rosen B A 2017 Exsolution of Re-alloy catalysts with enhanced stability for methane dry reforming *Appl. Catal.* B **209** 711–19

[252] Liu S, Chuang K T and Luo J-L 2015 Double-layered perovskite anode with *in situ* exsolution of a Co–Fe alloy to cogenerate ethylene and electricity in a proton-conducting ethane fuel cell *ACS Catal.* **6** 760–68

[253] Liu S, Liu Q and Luo J-L 2016 CO_2-to-CO conversion on layered perovskite with *in situ* exsolved Co–Fe alloy nanoparticles: an active and stable cathode for solid oxide electrolysis cells *J. Mater. Chem.* A **4** 17521–28

[254] Xi X, Wang X-W, Fan Y, Wang Q, Lu Y, Li J, Shao L, Luo J-L and Fu X-Z 2021 Efficient bifunctional electrocatalysts for solid oxide cells based on the structural evolution of perovskites with abundant defects and exsolved cofe nanoparticles *J. Power Sources* **482** 228981

[255] Song Y F, Yin Y M, Li L S, Chen Z H and Ma Z F 2018 A self-assembled dual-phase composite as a precursor of high-performance anodes for intermediate temperature solid oxide fuel cells *Chem. Commun.* **54** 12341–44

[256] Xi X, Fan Y, Zhang J, Luo J-L and Fu X-Z 2022 In situ construction of hetero-structured perovskite composites with exsolved Fe and Cu metallic nanoparticles as efficient CO_2 reduction electrocatalysts for high performance solid oxide electrolysis cells *J. Mater. Chem.* A **10** 2509–18

[257] Tsekouras G, Neagu D and Irvine J T S 2013 Step-change in high temperature steam electrolysis performance of perovskite oxide cathodes with exsolution of B-site dopants *Energy Environ. Sci.* **6** 256–66

[258] Carrillo A J and Serra J M 2021 Exploring the stability of Fe–Ni alloy nanoparticles exsolved from double-layered perovskites for dry reforming of methane *Catalysts* **11** 741

[259] Wei K, Wang X, Zhu H, Liu H, Wang S, Chen F, Zhou F and Ling Y 2021 Clean and stable conversion of oxygen-bearing low-concentration coal mine gas by solid oxide fuel cells with an additional reforming layer *J. Power Sources* **506** 230208

[260] Park S, Kim Y, Noh Y, Kim T, Han H, Yoon W, Choi J, Yi S-H, Lee W-J and Kim W B 2020 A sulfur-tolerant cathode catalyst fabricated with *in situ* exsolved CoNi alloy nanoparticles anchored on a Ruddlesden–Popper support for CO_2 electrolysis *J. Mater. Chem.* A **8** 138–48

[261] Kwon O, Kim K, Joo S, Jeong H Y, Shin J, Han J W, Sengodan S and Kim G 2018 Self-assembled alloy nanoparticles in a layered double perovskite as a fuel oxidation catalyst for solid oxide fuel cells *J. Mater. Chem.* A **6** 15947–53

[262] Joo S *et al* 2021 Enhancing thermocatalytic activities by upshifting the *d*-band center of exsolved Co–Ni–Fe ternary alloy nanoparticles for the dry reforming of methane *Angew. Chem.* **60** 15912–19

[263] Fan Y, Xi X, Medvedev D, Wang Q, Li J, Luo J-L and Fu X-Z 2021 Emerging anode materials architectured with NiCoFe ternary alloy nanoparticles for ethane-fueled protonic ceramic fuel cells *J. Power Sources* **515** 230634

[264] Neagu D, Oh T S, Miller D N, Menard H, Bukhari S M, Gamble S R, Gorte R J, Vohs J M and Irvine J T S 2015 Nano-socketed nickel particles with enhanced coking resistance grown *in situ* by redox exsolution *Nat. Commun.* **6** 8120

[265] Cui S-H, Li J-H, Zhou X-W, Wang G-Y, Luo J-L, Chuang K T, Bai Y and Qiao L-J 2013 Cobalt doped $LaSrTiO_{3-\delta}$ as an anode catalyst: effect of Co nanoparticle precipitation on SOFCs operating on H_2S-containing hydrogen *J. Mater. Chem.* A **1** 9689–96

[266] Li H, Song Y, Xu M, Wang W, Ran R, Zhou W and Shao Z 2020 Exsolved alloy nanoparticles decorated Ruddlesden–Popper perovskite as sulfur-tolerant anodes for solid oxide fuel cells *Energy Fuels* **34** 11449–57

[267] Sun Y F, Zhou X W, Zeng Y M, Amirkhiz B S, Wang M N, Zhang L Z, Hua B, Li J, Li J H and Luo J L 2015 An ingenious Ni/Ce co-doped titanate based perovskite as a coking-tolerant anode material for direct hydrocarbon solid oxide fuel cells *J. Mater. Chem.* A **3** 22830–38

[268] Song Y, Wang W, Qu J, Zhong Y, Yang G, Zhou W and Shao Z 2018 Rational design of perovskite-based anode with decent activity for hydrogen electro-oxidation and beneficial effect of sulfur for promoting power generation in solid oxide fuel cells *ACS Appl. Mater. Interfaces* **10** 41257–67

[269] Mun Y, Lee S, Kim K, Kim S, Lee S, Han J W and Lee J 2019 Versatile strategy for tuning ORR activity of a single $Fe\text{-}N_4$ site by controlling electron-withdrawing/donating properties of a carbon plane *J. Am. Chem. Soc.* **141** 6254–62

[270] Lee J J, Kim K, Kim K J, Kim H J, Lee Y M, Shin T H, Han J W and Lee K T 2021 In-situ exsolution of Ni nanoparticles to achieve an active and stable solid oxide fuel cell anode catalyst on A-site deficient $La_{0.4}Sr_{0.4}Ti_{0.94}Ni_{0.06}O_{3-\delta}$ *J. Ind. Eng. Chem.* **103** 264–74

[271] Lay E, Metcalfe C and Kesler O 2012 The influence of incorporating mgo into Ni-based cermets by plasma spraying on anode microstructural and chemical stability in dry methane *J. Power Sources* **218** 237–43

[272] Horiuchi T, Sakuma K, Fukui T, Kubo Y, Osaki T and Mori T J A C A G 1996 Suppression of carbon deposition in the CO_2-reforming of CH_4 by adding basic metal oxides to a Ni/Al_2O_3 catalyst *Appl. Catal.* A **144** 111–20

[273] Frusteri F, Arena F, Calogero G, Torre T and Parmaliana A J C C 2001 Potassium-enhanced stability of Ni/MgO catalysts in the dry-reforming of methane *Catal. Commun.* **2** 49–56

[274] Hua B, Zhang W Y, Li M, Wang X, Chi B, Pu J and Li J 2014 Improved microstructure and performance of Ni-based anode for intermediate temperature solid oxide fuel cells *J. Power Sources* **247** 170–77

[275] Yang Q, Chen J, Sun C and Chen L 2016 Direct operation of methane fueled solid oxide fuel cells with Ni cermet anode via Sn modification *Int. J. Hydrog. Energy* **41** 11391–98

[276] Yang L, Choi Y, Qin W T, Chen H Y, Blinn K, Liu M F, Liu P, Bai J M, Tyson T A and Liu M L 2011 Promotion of water-mediated carbon removal by nanostructured barium oxide/nickel interfaces in solid oxide fuel cells *Nat. Commun.* **2** 1359

[277] Mishina T, Fujiwara N, Tada S, Takagaki A, Kikuchi R and Oyama S T 2020 Calcium-modified Ni-SDC anodes in solid oxide fuel cells for direct dry reforming of methane *J. Electrochem. Soc.* **167** 134512

[278] Ferlauto A S, de Florio D Z, Fonseca F C, Esposito V, Muccillo R, Traversa E and Ladeira L O 2006 Chemical vapor deposition of multi-walled carbon nanotubes from nickel/yttria-stabilized zirconia catalysts *Appl. Phys.* A **84** 271–76

[279] Chen D, Christensen K, Ochoafernandez E, Yu Z, Totdal B, Latorre N, Monzon A and Holmen A 2005 Synthesis of carbon nanofibers: effects of Ni crystal size during methane decomposition *J. Catal.* **229** 82–96

[280] Christensen K O, Chen D, Lødeng R and Holmen A 2006 Effect of supports and Ni crystal size on carbon formation and sintering during steam methane reforming *Appl. Catal.* A **314** 9–22

[281] Wei T, Jia L C, Zheng H Y, Chi B, Pu J and Li J 2018 $LaMnO_3$-based perovskite with *in-situ* exsolved Ni nanoparticles: a highly active, performance stable and coking resistant catalyst for CO_2 dry reforming of CH_4 *Appl. Catal.* A **564** 199–207

IOP Publishing

Chapter 5

Nanostructuring the electrodes of solid oxide cells via atomic layer deposition

Kevin Huang

Electrode nanostructuring has been shown to substantially boost the performance of solid oxide cells (SOCs). However, at the high operational temperatures of SOCs, nanostructures are prone to coarsening, resulting in performance degradation. To address this issue, the atomic layer deposition (ALD) technique has recently been investigated as a new way to stabilize SOC performance. This technique can be used to deposit a conformal nanoscale overcoat on the surfaces of nanostructured porous electrodes in SOCs and has been demonstrated to significantly improve performance stability. In this chapter, we review recent progress in the use of ALD to modify nanostructured electrodes in SOCs. The review starts with an introduction describing the need for nanostructured electrodes and why ALD can meet this need. We then briefly present the working principle and basic procedure of the ALD process. The applications of ALD in SOCs are then reviewed for porous oxygen electrodes, the barrier layer, the electrolyte, and fuel electrodes. Finally, the limitations of the ALD technique in SOCs are candidly discussed.

5.1 Introduction

Intermediate-temperature solid oxide cells (IT-SOCs) are currently being actively pursued by commercial SOC developers as a promising technology for clean power generation and green hydrogen production. A key to the success of IT-SOCs is the development of active and stable oxygen electrodes (OEs). Mixed ion and electron conducting (MIEC), oxygen-nonstoichiometric perovskites such as $La_{0.6}Sr_{0.4}Fe_{0.8}Co_{0.2}O_{3-\delta}$ (LSCF) and $La_{0.6}Sr_{0.4}CoO_{3-\delta}$ (LSCo) are often selected for OEs due to their high intrinsic catalytic activity in the oxygen reduction reaction (ORR) and the oxygen evolution reaction (OER). However, the high thermal/chemical expansion coefficients and chemical reactivity with ZrO_2-based electrolytes

doi:10.1088/978-0-7503-4064-9ch5

have prevented these high-performance OEs from being practically applied in IT-SOCs. To maintain the high ORR and OER activities without provoking deleterious reactions and mechanical stress within the cell components, LSCo and LSCF are often used in the form of nanoparticles (NPs) in a prefabricated, porous, ionically conducting, and thermally matched skeleton (e.g. doped CeO_2) at lower temperatures through infiltration processes [1–5]. Nanostructured OEs have high surface areas and high catalytic activities, which allows OEs to support fast ORRs and OERs, thus considerably lowering the cell resistance, while the ionically conducting skeleton provides an ionic pathway and structural support for the ORR-/OER-active NPs.

However, the central issue for such nanostructured electrodes is the instability of their morphology due to the clustering of NPs under the high-temperature operational conditions used for SOCs; this results in the loss of reactive sites and catalytic activity over time [1, 6–11]. The stability of previously tested nanostructured OEs was less than 1000 h, even in the temperature range of 650 °C–700 °C [4, 5], and it can be significantly shorter at higher temperatures. Therefore, the preservation of the catalytic activity of ORR-/OER-active NPs under SOC operational conditions has become the central requisite for the development of nanostructured OEs. Several mitigation approaches have been described in the literature; representative examples include: (1) the formation of a 'core–shell' structure [7], (2) alloying the active metal with more refractory metals [12], and (3) co-impregnating the active metal with a ceramic phase [8, 13]. However, so far, none of these methods has proven effective in extending OE lifetimes to the level required by commercial applications.

The performance degradation also derives from the LSCo or LSCF itself. In many early lab-scale investigations in which an LSCo or LSCF bulk layer was used as an IT-OE, noticeable performance degradation was seen [14–18]. In-situ and ex-situ spectroscopic techniques have revealed ample evidence that this performance degradation is associated with Sr segregation into SrO(s) in those Sr-containing OEs [19–23], which is a chemical process strongly dependent on temperature, the partial pressure of oxygen (pO_2), and the electrical potential [18–23]. Since the ORR/OER activity of OEs varies critically with their surface atomic structure (e.g. electronic configuration) and composition (e.g. cation concentration and oxygen nonstoichiometry) [14–28], full or partial coverage of the OE by a passive $SrO_{(s)}$ layer blocks the ORR-/OER-active sites. While the phenomenon of surface Sr segregation has been well recognized and studied in the literature, mitigation methods that can reduce this detrimental effect are surprisingly lacking.

The ALD method has recently been increasingly used as an effective way to stabilize nanostructures against high-temperature coarsening. Successful examples include the use of nanoscale conformal coatings to stabilize OE NPs for SOCs and the use of catalytic NPs for chemical conversions [29–32]. In this chapter, we provide a brief overview of the efforts to apply ALD to deposit nanoscale thin films over SOC electrode nanostructures in order to stabilize their high performance. The reader is also referred to some excellent reviews on this topic [33–36].

5.2 The atomic layer deposition process

ALD is an emerging leading-edge thin-film technology capable of depositing highly conformal and uniform nanoscale films on the surfaces of almost any open bulk geometries [37–45]. As a type of chemical vapor deposition (CVD) technique, the process uses self-limiting surface reactions to deposit a film onto a substrate surface one atomic layer at a time [46]. To achieve monatomic layer deposition, the process proceeds in an ABAB… binary sequence of chemical reactions between gas-phase precursor molecules and a solid surface [46, 47]. The digital nature of the thin-film growth permits the ALD process to control the film thickness with great precision and excellent uniformity. Due to its CVD nature, ALD is also able to produce conformal films as an exact copy of the outer surfaces of a substrate, an attribute that is unobtainable from solution and line-of-sight-dependent methods (e.g. sol-gels, physical vapor deposition, etc). Although it was originally developed to coat silicon wafers in the semiconductor industry, ALD has since been widely extended to deposit nanoscale films of metals and oxides on a range of porous media including powders [48–50], gas separation membranes [51–53], silicon [54], and ZrO_2 [55]. A representative microstructure of an ALD-synthesized, conformal nanofilm of Fe_2O_3 on ZrO_2 nanoparticles is illustrated in figure 5.1.

A complete repeating ALD cycle consists of four distinct steps: (1) adsorption of the gaseous precursor molecules onto the surface of a substrate; (2) removal of the excess precursor and its byproducts from the sample's surface by inert gas purging; (3) introduction of a gaseous reactant (e.g. an oxidizer) to remove ligands and restore sites; and (4) the removal of excess reactants and their byproducts by inert gas purging. These steps are illustrated in figure 5.2 using a popular ALD TiO_2 process as an example, in which $TiCl_4$ and water vapor are used as the precursor and oxidizer, respectively, to remove the ligands [33].

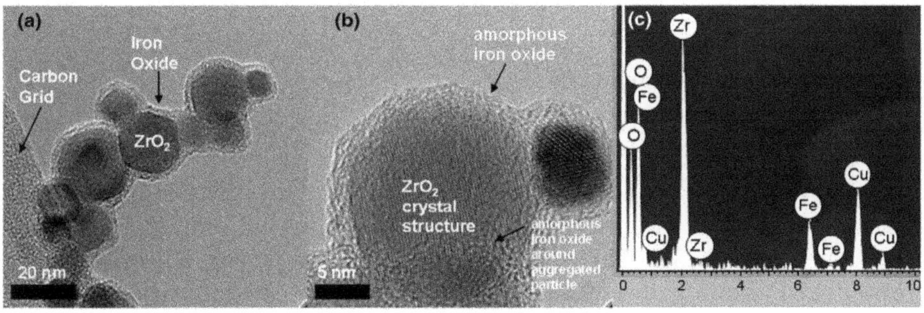

Figure 5.1. (a) and (b) High-resolution transmission electron microscopy (HRTEM) images of an ALD-Fe_2O_3 nanoscale overcoat on ZrO_2 particles after 200 coating cycles. (c) Confirmation of the presence of iron obtained using energy dispersive x-ray spectroscopy (EDS) [55]. Reprinted from [55], Copyright (2021), with permission from Elsevier.

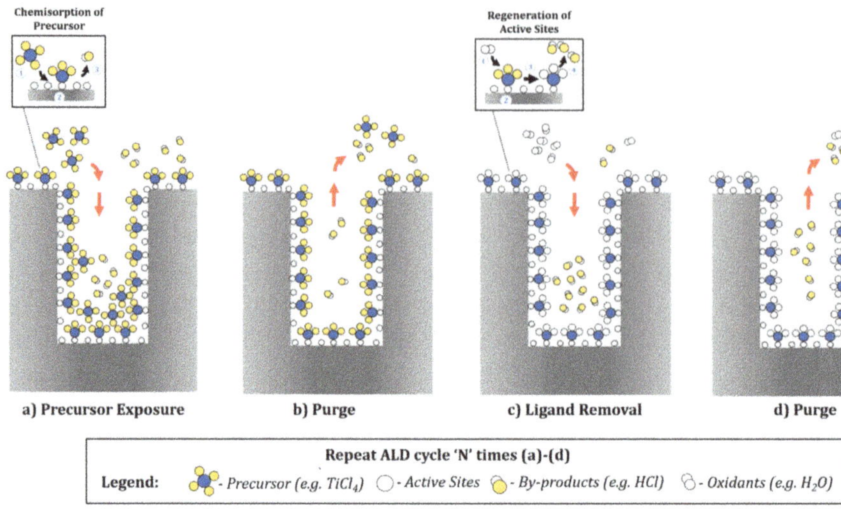

Figure 5.2. Schematic of one ALD cycle in a simplified porous structure. (a) The introduction of precursor molecules and subsequent precursor chemisorption on the surface; (b) removal of the unreacted precursor; (c) the introduction of reactants that react with the chemisorbed precursor molecules; and (d) removal of the excess reactants and products. This is an example of the ALD process for TiO_2 deposition, which uses $TiCl_4$ as the precursor and H_2O as the reactant.

5.3 The applications of atomic layer deposition in the porous electrodes of solid oxide cells

The use of ALD in SOCs has primarily been concentrated on surface modifications of electrodes by ALD-derived nanoscale conformal coatings to achieve stability and activity through coarsening confinement, Sr-segregation suppression, and active site retention. The deposition of ultrathin electrolytes or OE barrier layers by ALD has also been attempted for low-temperature SOCs [56–62]. In the following, we review these ALD applications in SOCs.

5.3.1 Surface modification of OEs

Since high-performance OEs containing NPs are unstable at high temperatures due to the coarsening and loss of their active surface areas, their practical use in SOCs is limited. To address this problem, Huang's group at the University of South Carolina first demonstrated the use of ALD to deposit a conformal, nanoscale, protective coating over OEs [29, 30]. The initial intention was to confine NP growth using a conformal ALD coating of a refractory material, which led to the first trial of ZrO_2. However, the concern was that ZrO_2 is not ORR/OER active; thus, a layer of ZrO_2 on the surface of OEs would block the ORR/OER reactions and increase polarization resistance. However, the results obtained by testing an ALD-ZrO_2 overcoated nanostructured OE were surprising. Figure 5.3 shows that 5 nm ZrO_2-overcoated LSCo NPs exhibited a much more stable performance than that of an uncoated baseline OE over 4000 h, exhibiting only 1/18 of the degradation rate of

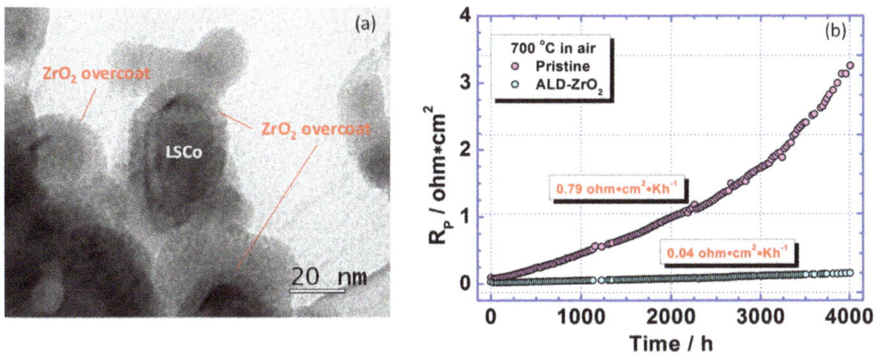

Figure 5.3. (a) ALD-ZrO_2 overcoated with LSCo NPs; (b) long-term stability of nanostructured LSCo OEs, coated vs. uncoated. Reprinted with permission from [29]. Copyright (2021) American Chemical Society.

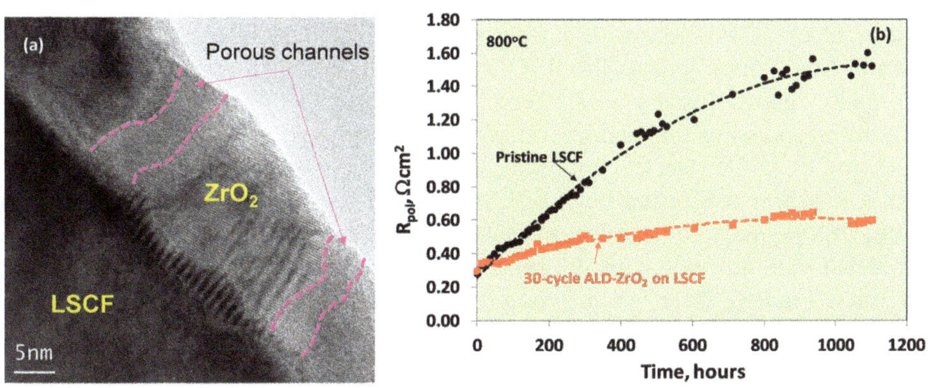

Figure 5.4. (a) A nanoscale ALD-ZrO_2 overcoat on an LSCF+GDC OE; (b) the long-term stability of nanostructured LSCF+GDC OEs, coated vs. uncoated. Reprinted with permission from [30]. Copyright (2021) American Chemical Society.

the uncoated OE. While the degradation of the baseline OE is straightforward, understanding the performance improvement in the ZrO_2-overcoated OE is not. Huang *et al* eventually concluded through experimental and theoretical analyses that the performance enhancement was derived from the multifunctionality of the ALD-ZrO_2: (1) confinement of NP growth; (2) micropores in the ZrO_2-overcoat, which allow O_2 to diffuse to the underlying ORR-/OER-active LSCo surface; (3) suppression of Sr segregation from the LSCo bulk to the surface by electrostatic repulsion between the negatively charged surface defect Co_{Zr}' and negatively charged bulk Sr-defect Sr_{La}'. The latter was indirectly confirmed by Nicholas' group at Michigan State University in a recent study which showed that ZrO_2 can serve as a Sr getter [63]. Huang's group showed that the ALD-ZrO_2 overcoat also worked on the benchmark SOC OE, i.e. LSCF+GDC, appreciably improving its stability by suppressing Sr segregation (see figure 5.4) [30]. Using an epitaxial LSCo film produced via pulsed laser deposition (PLD) and supported on a yttria-stabilized zirconia (YSZ) single-crystal substrate, Huang's group further provided direct evidence of Sr segregation and determined the kinetics of the latter by *in situ* ambient pressure x-ray photoemission spectroscopy (APXPS). The group used electrochemical impedance spectroscopy (EIS) to show that a layer of ALD-ZrO_2 0.8 nm thick helped to reduce Sr segregation and thus polarization resistance [64]. Similarly, Song *et al* reported that performance enhancement by a factor of 1.5–1.7 and long-term stability were obtained at 0.8 V and 750 °C by overcoating ALD-ZrO_2 with Pt NPs on a porous OE consisting of $La_{1-x}Sr_xMnO_3$ (LSM) and YSZ [65]. In a similar study, the same group replaced ZrO_2 with CoO_x and achieved a similar power enhancement and stability [66].

Chang *et al* also observed the beneficial effect of an ALD-YSZ coating on a functional Pt OE [58]. In this case, the thin YSZ overcoat prevented the Pt from sintering, retaining the porosity of the Pt during operation at 500 °C. Interestingly, the YSZ overcoat also increased the ORR activity by a factor of 2.5. The authors argued that the improvement was attributed to an increased three phase boundary density, especially near the Pt–electrolyte interface, which benefited from the addition of an ion-conducting component to the Pt OE. In similar works, Karimaghaloo *et al* and Liu *et al* also reported that a YSZ overcoat on a porous Pt OE helped preserve ORR activity by suppressing the agglomeration of the porous Pt [67, 68].

With an ORR-/OER-active material, Choi *et al* demonstrated the use of ALD to deposit a 1–12 nm conformal layer of LSC over an LSCF porous skeleton, resulting in significantly enhanced performance [69]. In this case, the nanoscale LSC provided high ORR/OER activity without raising a concern about coarsening due to its conformity. Finally, there is a growing interest in directly modifying the surface of OE powders by ALD before they are made into inks for screen printing for commercial SOC systems [70].

However, the beneficial effect of ALD-ZrO_2 has not always been observed. Gorte's group reported that the deposition of a 1 nm ALD-ZrO_2 overcoat significantly decreased the performance of an OE made from $La_{0.8}Sr_{0.2}FeO_{3-\delta}$ (LSF) [71]. The study attributed the discrepancy to the different ALD reactor systems used in the studies, which resulted in very different growth rates and different ALD processes. Using an LSF-YSZ skeleton, the same group observed the beneficial effect of sub-monolayers (<1 nm) of ALD-La_2O_3 [71]. One explanation is that LSF phase may be *A*-site deficient at the surface, such that a small amount of La provided by ALD could complete the perovskite structure, thus improving performance. Results obtained following the deposition of La_2O_3–Fe_2O_3 mixtures also showed improved cathode performance, even for relatively thicker films. One important conclusion of this study was that ALD may be suitable for enhancing O_2 adsorption sites during the ORR process.

5.3.2 Modifications to SOC fuel electrodes

The benchmark material used for SOC fuel electrodes (FEs) is a mechanical mixture of Ni and YSZ, which is also known as Ni–YSZ cermet (i.e. a ceramic–metallic composite) [72]. The Ni provides the electrical conductivity and catalytic activity for fuel oxidation reactions (FORs) or fuel production reactions (FPRs), while the YSZ provides oxide-ion conductivity, helps to prevent Ni sintering, and retains porosity.

In contrast to the wide availability of reports for OEs, the application of ALD-derived nanoscale conformal coatings on FEs has not been widely reported. In general, two strategies have taken advantage of ALD coatings: (1) mitigating Ni NP coarsening at high temperatures; (2) providing coking and sulfur poisoning resistances. It is well known that the electrochemical performance of Ni–YSZ FEs strongly depends on the size of the Ni catalyst; the smaller the Ni particle, the higher the catalytic performance. However, at high operating temperatures, nanoscale Ni particles suffer from severe coarsening, causing a loss of electrical conductivity and catalytic activity. Cui *et al* demonstrated the beneficial effect of ALD-YSZ in limiting Ni coarsening at high temperature on a Si wafer substrate [73]. Similarly, McIntosh *et al* reported an ALD-derived, sub-monolayer conformal coating of mixed metal oxide films that was used to enhance the thermal stability and coking tolerance of the Ni–YSZ cermet [74].

5.4 Application in dense electrolytes

Since practical electrolytes in SOCs are too thick to deposit by ALD, most of the work in this area involves surface modifications of the electrolyte [56] or the implantation of a barrier layer between the OE and the electrolyte to prevent deleterious reactions or block electronic current [57–61]. Oh *et al* reported a high-performance 450 °C solid oxide fuel cell (SOFC) that consisted of a nanoporous anodized aluminum oxide (AAO) substrate, a porous Pt OE, and a FE that sandwiched a sputtered samarium-doped ceria (SDC) and ALD-YSZ bilayer electrolyte [61] (see figure 5.5). The YSZ layer was expected to block electronic current from the SDC under reducing atmospheres. Similarly, Wu *et al* reported an AAO-supported low-temperature SOFC with an ALD-derived ultrathin blocking layer between PVD-derived YSZ and GDC layers. The results showed improvements in the open circuit voltage, power density, and reproducibility [62].

5.5 The limitations of ALD in overcoating porous media

While the potential benefit of ALD for SOC performance is well recognized, much of the past ALD work on porous SOC electrodes has used the same equipment and procedures developed by the semiconductor industry for deposition on flat and dense surfaces, where the flow resistance and surface area are relatively low. To deposit the surfaces of porous structures such as SOC electrodes, the ALD process parameters need to be carefully adjusted to allow enough time for the gas-phase precursor, the oxidizer, and the purge gas to penetrate the pores and fulfill their duties. Commercial ALD systems generally have two operational modes: flow and exposure. The former is suited to flat and dense surfaces, while the latter is designed for porous media. When the exposure mode is used, all reactants are allowed to stay in contact with the porous substrates for a longer time before being evacuated. However, several precautions are suggested when overcoating porous media by ALD.

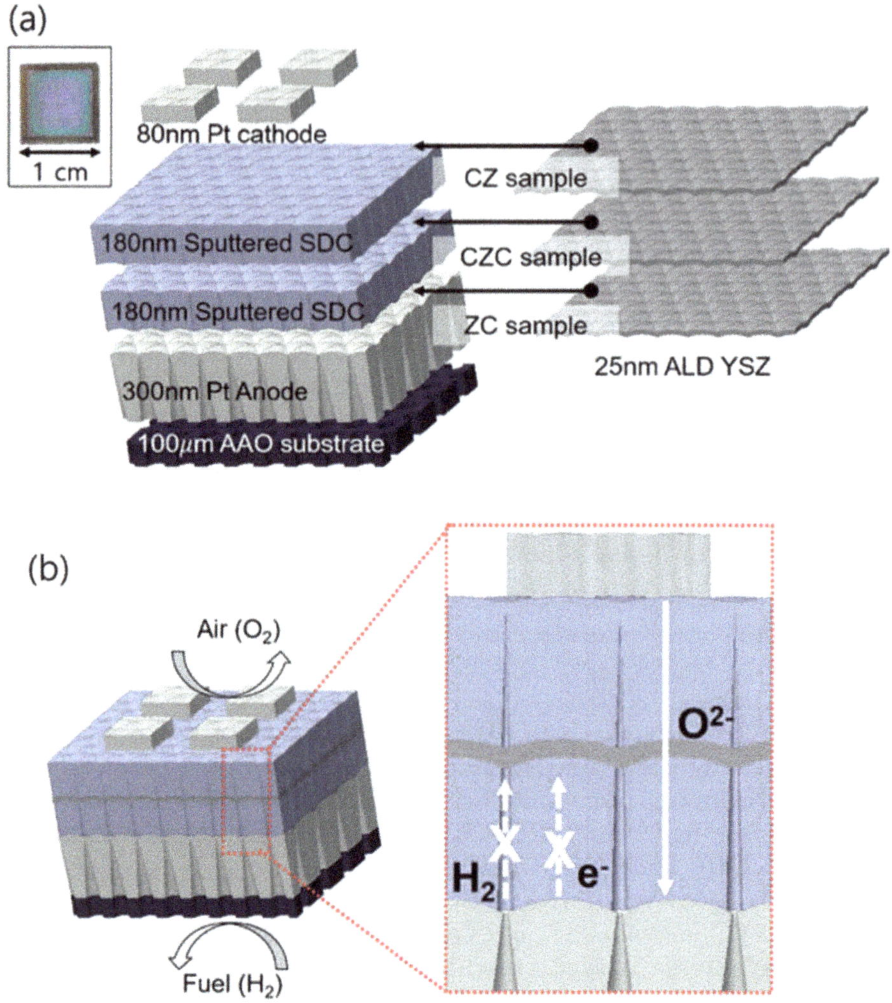

Figure 5.5. Illustration of (a) the structures of CZ, CZC, and ZC samples (a photo of the actual cell is shown in the inset image) and (b) the operational principle of the CZC sample [61]. (CZ: YSZ/SDC bilayer CZC: YSZ/SDC/YSZ trilayer ZC: SDC/YSZ bilayer) Reproduced from [61] with permission of The Royal Society of Chemistry.

(1) *Determination of the correct growth rate.*

For a sample with a flat and dense surface, for which the diffusion of the precursor molecules on the surface is rapid, the growth rate can be rather accurately determined by an *in situ* quartz crystal microbalance (QCM) sensor as a part of the ALD system, or by the x-ray reflectivity (XRR) method. For a porous material with a high surface area, however, the QCM readings are likely to deviate from the actual coating thickness, and so should only be used as a reference. The XRR method is limited to thin-film samples. A good practice is always to use transmission electron microscopy

(TEM) to visually determine the coating thickness and correlate it with the ALD cycle number under specific coating conditions.

(2) *Recognition of the diffusional resistance of the precursor/oxidizer/purge gas through the porous media.*

The aspect ratio of porous SOC electrodes is relatively high. To minimize diffusion limitations in powders, a rotating or fluidized bed can be used [75, 76]. Some commercial ALD systems offer a powder tumbler for this purpose. However, a diffusional limitation still exists if the individual particles remain porous [77]. For prefabricated porous SOC electrode skeletons, a common practice is to develop specially tailored recipes that include the exposure mode, exposure time, pulse time, etc. to maximize the penetration of the reactants into the porous structure.

(3) *Recognition of the low precursor utilization during the ALD process.*

Organic precursors designed for ALD are usually expensive chemicals in order to meet the narrow temperature window requirements of ALD reactors. Therefore, temperatures must be high enough for the precursor to react with the surface and low enough to avoid chemical vapor deposition. For high-value semiconductor devices, this might not be a problem, as long as the precursors are inexpensive. For SOC applications, this cost issue could be prohibitive if the utilization is too low. However, due to the nature of the large surface areas involved, ALD requires a smaller number of cycles for the same mass loading than the number required for flat and dense surfaces. Therefore, a best practice that can be used to alleviate the cost issue is to minimize the number of ALD cycles required to achieve the needed thickness.

(4) *Recognition of the effect of low vapor pressures of precursors on the growth rate.*

Many ALD precursors are solids, the vapor pressures of which sensitively vary with the temperature of the precursor holder. Therefore, identification of the right holder temperature and the corresponding thermal insulation to ensure a uniform temperature distribution are critically important in order to avoid clogging during deposition. This issue can be solved by flowing a carrier gas over the solid precursors instead of relying on the self-vaporization of the precursors.

5.6 Summary and outlook

ALD is a powerful tool capable of controlling the surface structure and composition of porous SOC electrodes at the atomic scale. Early studies have provided strong evidence that supports the effectiveness of ALD surface modifications in preserving the electrochemical activity of nanostructured electrodes. These promising results were primarily demonstrated in nanostructured SOC OEs. The enhancement mechanisms of ALD conformal coatings are understood to be multifunctional: (1) they inhibit the coarsening of NPs; (2) they suppress Sr segregation; and (3) they enhance nanoporosity. In contrast, the applications of ALD coatings in the FEs and electrolytes of SOCs have been rather limited, partly due to the relatively high maturity level of these two functional layers in SOCs and partly due to the reduced effectiveness of ALD in making these two layers. We have also discussed the precautions and considerations to be taken into account when ALD coating is performed on porous media. The factors to be considered include an estimation of the growth rate/thickness, a recognition of diffusional resistance, the vapor pressures of precursors, and the utilization/cost of the precursors.

References

[1] Vohs J M and Gorte R J 2009 High-performance SOFC cathodes prepared by infiltration *Adv. Mater.* **21** 943–56

[2] Kurokawa H, Yang L M, Jacobson C P, De Jonghe L C and Visco S J 2007 Y-doped $SrTiO_3$ based sulfur tolerant anode for solid oxide fuel cells *J. Power Sources* **164** 510–8

[3] Shah M and Barnett B A 2008 Solid oxide fuel cell cathodes by infiltration of $La_{0.6}Sr_{0.4}Co_{0.2}Fe_{0.8}O_{3-\delta}$ into Gd-doped ceria *Solid State Ion.* **179** 2059–64

[4] Sholklapper T Z, Kurokawa H, Jacobson C P, Visco S J and De Jonghe L C 2007 Nanostructured solid oxide fuel cell electrodes *Nano Lett.* **7** 2136–41

[5] Sholklapper T Z, Radmilovic V, Jacobson C P, Visco S J and De Jonghe L C 2007 Synthesis and stability of a nanoparticle-infiltrated solid oxide fuel cell electrode *Electrochem. Solid-State Lett.* **10** B74–76

[6] Tucker M C, Lau G Y, Jacobson C P, DeJonghe L C and Visco S J 2007 Performance of metal-supported SOFCs with infiltrated electrodes *J. Power Sources* **171** 477–82

[7] Kim J S, Wieder N L, Abraham A J, Cargnello M, Fornasiero P, Gorte R J and Vohs J M 2011 Highly active and thermally stable core-shell catalysts for solid oxide fuel cells *J. Electrochem. Soc.* **158** B596–600

[8] Klemenso T, Thyden K, Chen M and Wang H J 2010 Stability of Ni-yttria stabilized zirconia anodes based on Ni-impregnation *J. Power Sources* **195** 7295–301

[9] Wang W S, Gross M D, Vohs J M and Gorte R J 2007 The stability of LSF-YSZ electrodes prepared by infiltration *J. Electrochem. Soc.* **154** B439–45

[10] Shah M, Voorhees P W and Barnett S A 2011 Time-dependent performance changes in LSCF-infiltrated SOFC cathodes: The role of nano-particle coarsening *Solid State Ion.* **187** 64–7

[11] Lee S, Miller N and Gerdes K 2012 Long-term stability of SOFC composite cathode activated by electrocatalyst infiltration *J. Electrochem. Soc.* **159** F301–8

[12] Liang F L, Chen J, Jiang S P, Wang F Z, Chi B, Pu J and Li J 2009 Mn-Stabilised microstructure and performance of Pd-impregnated YSZ cathode for intermediate temperature solid oxide fuel cells *Fuel Cells* **9** 636–42

[13] Imanishi N, Marata K, Hirano A, Takeda Y, Yamamoto O and Yamahara K 2009 LSM-YSZ cathode with infiltrated cobalt oxide and cerium oxide nanoparticles *Fuel Cells* **9** 215–1

[14] Simner S P, Anderson M D, Engelhard M H and Stevenson J W 2006 Degradation mechanisms of La–Sr–Co–Fe–O_3 SOFC cathodes *Electrochem. Solid-State Lett.* **9** A478–81

[15] Mai A, Becker M, Aseenmacher W, Tietz F, Hathiramani D, Ivers-Tiffée E, Stöver D and Mader W 2006 Time-dependent performance of mixed conducting SOFC cathodes *Solid State Ion.* **177** 1965–8

[16] Hjalmarsson P, Søgarrd M and Mogensen M 2008 Electrochemical performance and degradation of $(La_{0.6}Sr_{0.4})_{0.99}CoO_{3-\delta}$ as porous SOFC-cathode *Solid State Ion.* **179** 1422–6

[17] Hagen A, Liu Y L, Barfod R and Hendriksen P V 2008 Assessment of the cathode contribution to the degradation of anode-supported solid oxide fuel cells *J. Electrochem. Soc.* **155** B1047–52

[18] Oh D, Gostovic D and Wachsman E D 2012 Mechanism of $La_{0.6}Sr_{0.4}Co_{0.2}Fe_{0.8}O_3$ cathode degradation *J. Mater. Res.* **27** 1992–9

[19] Finsterbusch M, Lussier A, Schaefer J A and Idzerda Y U 2012 Electrochemically driven cation segregation in the mixed conductor $La_{0.6}Sr_{0.4}Co_{0.2}Fe_{0.8}O_{3-\delta}$ *Solid State Ion.* **212** 77–80

[20] Crumlin E J, Mutoro E, Liu Z, Grass M E, Biegalski M D, Lee Y L, Morgan D, Christen H M, Bluhm H and Yang S H 2012 Surface strontium enrichment on highly active perovskite for oxygen electrocatalysis in solid oxide fuel cells *Energy Environ. Sci.* **5** 6081–8

[21] Cai Z, Kubicek M, Fleig J and Yildiz B 2012 Chemical heterogeneities on $La_{0.6}Sr_{0.4}CoO_{3-\delta}$ thin films—correlations to cathode surface activity and stability *Chem. Mater.* **24** 1116–27

[22] Caillol N, Pijolat M and Siebert E 2007 Investigation of chemisorbed oxygen, surface, segregation and effect of post treatments on $La_{0.8}Sr_{0.2}MnO_3$ powder and screen-printed layers for solid oxide fuel cell cathodes *Appl. Surf. Sci.* **253** 4641–8

[23] Fisher T T, Fong D D, Eastman J A, Baldo P M, Highland M J, Fuoss P H, Balasubramaniam K R, Meador J C and Salvador P A 2008 In-situ characterization of strontium surface segregation in epitaxial $La_{0.7}Sr_{0.3}MnO_3$ thin films as a function of partial pressure *Appl. Phys. Lett.* **93** 151904

[24] Tanaka H and Misono M 2001 Advances in designing perovskite catalysts *Curr. Opin. Solid State Mater. Sci.* **5** 381–7

[25] Serra J M, Vert V B, Betz M, Haanappel V A C, Meulenberg W A and Tietz F 2008 Screening of A-substitution in the system $A_{0.68}Sr_{0.3}Fe_{0.8}Co_{0.2}O_{3-\delta}$ for SOFC cathodes *J. Electrochem. Soc.* **155** B207–14

[26] Pena M A and Fierro J L G 2001 Chemical structures and performance of perovskite oxides *Chem. Rev.* **101** 1981–2018

[27] Shimizu T 1980 Effect of electronic structure and tolerance factor on Co oxidation activity of perovskite oxides *Chem. Lett.* **1** 1–4

[28] Henrich V E 1985 The surfaces of metal oxides *Rep. Prog. Phys.* **48** 1481–541

[29] Gong Y, Palacio D, Song X, Patel P L, Liang X, Goodenough J B and Huang K 2013 Stabilizing nanostructured solid oxide fuel cell cathode with atomic layer deposition *Nano Lett.* **13** 4340–5

[30] Gong Y, Song X, Liang X, Goodenough J B and Huang K 2013 Atomic layer deposition functionalized composite SOFC cathode LSCF-GDC: enhanced long-term stability *Chem. Mater.* **25** 4224–31

[31] Lu J, Fu B, Kung M C, Xiao G, Elam J W, Kung H H and Stair P C 2012 Coking- and sintering-resistant palladium catalysts achieved through atomic layer deposition *Science* **335** 1205–8

[32] Yan H, He K, Samek I A, Jing D, Nanda M G, Stair P C and Notestein J M 2021 Tandem In_2O_3–Pt/Al_2O_3 catalyst for coupling of propane dehydrogenation to selective H_2 combustion *Science* **371** 1257–60

[33] Onn T M, Küngas R, Fornasiero P, Huang K and Gorte R J 2018 Atomic layer deposition on porous materials: problems with conventional approaches to catalyst and fuel cell electrode preparation *Inorganics* **6** 34

[34] Cassir M, Ringuedé A and Niinistö L 2010 Input of atomic layer deposition for solid oxide fuel cell applications *J. Mater. Chem.* **20** 8987–93
[35] Detavernier C, Dendooven J, Sree S P, Ludwig K F and Martens J A 2011 Tailoring nanoporous materials by atomic layer deposition *Chem. Soc. Rev.* **40** 5242–53
[36] Marichy C, Bechelany M and Pinna N 2012 Atomic layer deposition of nanostructured materials for energy and environmental applications *Adv. Mater.* **24** 1017–32
[37] He Y, Yu X, Wang Y, Li H and Huang X 2011 Alumina-coated patterned amorphous silicon as the anode for a lithium-ion battery with high coulombic efficiency *Adv. Mater.* **23** 4938–41
[38] Lee J H, Hon M H, Chung Y W and Leu I C 2011 The effect of TiO_2 coating on the electrochemical performance of ZnO nanorod as the anode material for lithium-ion battery *Appl. Phys.* A **102** 545–50
[39] Snyder M Q, Trebukhova S A, Ravdel B, Wheeler M C, DiCarlo J, Tripp C P and DeSisto W J 2007 Synthesis and characterization of atomic layer deposited titanium nitride thin films on lithium titanate spinel powder as a lithium-ion battery anode *J. Power Sources* **165** 379–85
[40] Fu L J, Liu H, Li C, Wu Y P, Rahm E, Holze R and Wu H Q 2006 Surface modifications of electrode materials for lithium-ion batteries *Solid State Sci.* **8** 113–28
[41] Verma P, Maire P and Novák P 2010 A review of the features and analyses of the solid electrolyte interphase in Li-ion batteries *Electrochim. Acta* **55** 6332–41
[42] Xu K 2010 Electrolytes and interphasial chemistry in Li-ion devices *Energies* **3** 135–54
[43] Winter M 2009 The solid electrolyte interphase—the most important and the least understood solid electrolyte in rechargeable Li batteries *Z. Phys. Chem.* **223** 1395–406
[44] Jung Y S, Cavanagh A S, Dillon A C, Groner M D, George S M and Lee S H 2010 Enhanced stability of $LiCoO_2$ cathodes in lithium-ion batteries using surface modification by atomic layer deposition *J. Electrochem. Soc.* **157** A75–81
[45] Lee J T, Wang F M, Cheng C S, Li C C and Lin C H 2010 Low-temperature atomic layer deposited Al_2O_3 thin film on layer structure cathode for enhanced cycle ability in lithium-ion batteries *Electrochim. Acta* **55** 4002–6
[46] George S M, Ott A W and Klaus J W 1996 Surface chemistry for atomic layer growth *J. Phys. Chem.* **100** 13121–31
[47] Ritala M and Leskelä M 2001 Atomic layer deposition *Handbook of Thin Films* 1 (San Diego, CA: Academic) 2
[48] Ferguson J D, Weimer A W and George S M 2000 Atomic layer deposition of ultrathin and conformal Al_2O_3 films on BN particles *Thin Solid Films* **371** 95–104
[49] Ferguson J D, Weimer A W and George S M 2000 Atomic layer deposition of SiO_2 films on BN particles using sequential surface reactions *Chem. Mater.* **12** 3472–80
[50] Ferguson J D, Weimer A W and George S M 2002 Atomic layer deposition of boron nitride using sequential exposures of BCl_3 and NH_3 *Thin Solid Films* **413** 16–25
[51] Ott A W, Klaus J W, Johnson J M, George S M, McCarley K C and Way J D 1997 Modification of porous alumina membranes using Al_2O_3 atomic layer-controlled deposition *Chem. Mater.* **9** 707–14
[52] Berland B S, Gartland I P, Ott A W and George S M 1998 In situ monitoring of atomic layer controlled pore reduction in alumina tubular membranes using sequential surface reactions *Chem. Mater.* **10** 3941–50
[53] Cameron M A, Gartland I P, Smith J A, Diaz S F and George S M 2000 Atomic layer deposition of SiO_2 and TiO_2 in alumina tubular membranes: pore reduction and effect of surface species on gas transport *Langmuir* **16** 7435–44

[54] Ducso C, Khanh N Q, Horvath Z, Barsony I, Utriainen M, Lehto S, Nieminen M and Niinisto L 1996 Deposition of tin oxide into porous silicon by atomic layer epitaxy *J. Electrochem. Soc.* **143** 683–7
[55] Scheffe J R, Francésa A, King D M, Liang X, Branch B A, Cavanagh A S, George S M and Weimer A W 2009 Atomic layer deposition of iron(III) oxide on zirconia nanoparticles in a fluidized bed reactor using ferrocene and oxygen *Thin Solid Films* **517** 1874–9
[56] Kim Y B, Holme T P, Gür T M and Prinz F B 2011 Surface-modified low-temperature solid oxide fuel cell *Adv. Funct. Mater.* **21** 4684–90
[57] Nieminen M, Lehto S and Niinistö L 2001 Atomic layer epitaxy growth of $LaGaO_3$ thin films *J. Mater. Chem.* **11** 3148–53
[58] Gourba E, Ringuede A, Cassir M, Billard A, Päiväsaari J, Niinistö J, Putkonen M and Niinistö L 2003 Characterisation of thin films of ceria-based electrolytes for intermediate temperature—solid oxide fuel cells (IT-SOFC) *Ionics* **9** 15–20
[59] Fan Z, Chao C C, Hossein-Babaei F and Prinz F B 2011 Improving solid oxide fuel cells with yttria-doped ceria interlayers by atomic layer deposition *J. Mater. Chem.* **21** 10903–6
[60] Chang I, Ji S, Park J, Lee M H and Cha S W 2015 Ultrathin YSZ coating on Pt cathode for high thermal stability and enhanced oxygen reduction reaction activity *Adv. Energy Mater.* **5** 1402251
[61] Oh S, Park J, Shin J W, Yang B C, Zhang J, Jang D Y and An J 2018 High performance low-temperature solid oxide fuel cells with atomic layer deposited-yttria stabilized zirconia embedded thin film electrolyte *J. Mater. Chem.* A **6** 7401–8
[62] Yu W, Ji S, Cho G Y, Noh S, Tanveer W H, An J and Cha S W 2015 Atomic layer deposition of ultrathin blocking layer for low-temperature solid oxide fuel cell on nano-porous substrate *J. Vac. Sci. Technol.,* A **33** 01A145
[63] Zhang Y, Wen Y, Huang K and Nicholas J 2020 Zirconia overcoats as on-board strontium getters for improved solid oxide fuel cell nano-composite cathode durability *ACS Appl. Energy Mater.* **3** 4057–67
[64] Wen Y, Yang T, Lee D, Lee H N, Crumlin E J and Huang K 2018 Temporal and thermal evolutions of surface Sr-Segregation in pristine and atomic layer deposition modified $La_{0.6}Sr_{0.4}CoO_{3-\delta}$ epitaxial films *J. Mater. Chem.* A **6** 24378–88
[65] Chen Y, Gerdes K and Song X 2016 Nanoionics and nanocatalysts: conformal mesoporous surface scaffold for cathode of solid oxide fuel cells *Nature* **6** 32997
[66] Chen Y, Liang L, Navia S A P, Hinerman A, Gerdes K and Song X 2019 Synergetic interaction of additive dual nanocatalysts to accelerate oxygen reduction reaction in fuel cell cathodes *ACS Catal.* **9** 6664–71
[67] Karimaghaloo A, Andrade A M, Grewal S, Shim J H and Lee M H 2017 Mechanism of cathodic performance enhancement by a few-nanometer-thick oxide overcoat on porous Pt cathodes of solid oxide fuel cells *ACS Omega* **2** 806–13
[68] Liu K Y, Fan L, Yu C C and Su P C 2015 Thermal stability and performance enhancement of nano-porous platinum cathode in solid oxide fuel cells by nanoscale ZrO_2 capping *Electrochem. Commun.* **56** 65–9
[69] Choi H J *et al* 2018 Surface tuning of solid oxide fuel cell cathode by atomic layer deposition *Adv. Energy Mater.* **8** 1802506
[70] Sbrockey N, Aindow M, Deljoo B, Ghezel-Ayagh H, Torabi A and Tompa G 2017 Fluidized bed production of surface functionalized powders for solid oxide fuel cell cathodes *ECS Trans.* **78** 817–25

[71] Rahmanipour M, Cheng Y, Onn T M, Donazzi A, Vohs J M and Gorte R J 2017 Modification of LSF-YSZ composite cathodes by atomic layer deposition *J. Electrochem. Soc.* **164** F879–84

[72] Koide H, Someya Y, Yoshida T and Maruyama T 2000 Properties of Ni/YSZ cermet as anode for SOFC *Solid State Ion.* **132** 253–60

[73] Cui X, Zdunek A D, Jursich G and Takoudis C G 2015 Nanostructured Ni–YSZ by atomic layer deposition *ECS J. Solid-State Sci. Technol.* **4** P429–35

[74] McIntosh S, Gorte R J, Vohs J M and Curran C 2019 Enhancing SOFC anode stability through atomic layer deposition *ECS Meet. Abstract.* **MA2019-01** 1725

[75] Longrie D, Deduytsche D and Detavernier C 2014 Reactor concepts for atomic layer deposition on agitated particles: a review *J. Vac. Sci. Technol.* **A32** 010802

[76] Grillo F, Kreutzer M T and van Ommen J R 2015 Modeling the precursor utilization in atomic layer deposition on nanostructured materials in fluidized bed reactors *Chem. Eng. J.* **268** 384–98

[77] Haukka S, Kytökivi A, Lakomaa E-L, Lehtovirta U, Lindblad M, Lujala V and Suntola T 1995 The utilization of saturated gas–solid reactions in the preparation of heterogeneous catalysts *Stud. Surf. Sci. Catal.* **91** 957–66

Chapter 6

Novel electrode materials and nano-oxide composite electrolytes

Gang Chen

The ionic conductivities of traditional oxide ceramic electrolytes such as yttria-stabilized zirconia (YSZ), strontium/magnesium-doped lanthanum gallate (LSGM), gadolinia-doped ceria (GDC), and the high-temperature proton conductor yttria-doped barium zirconate (BZY) are 10^{-2} S·cm^{-1} or less at 550 °C. Recent studies have found that the ionic conductivity of electrolytes in ceramic fuel cells with electrodes made from lithium compounds such as $Ni_{0.8}Co_{0.15}Al_{0.05}LiO_2$ (NCAL) can easily exceed 0.1 S·cm^{-1} at 500 °C. It has been found that the reduction of the NCAL anode in H_2 produces a $LiOH/Li_2CO_3$ mixture, which diffuses into the electrolyte and forms an 'oxide–molten salt' composite electrolyte with a core–shell structure. The 'oxide–molten salt' composite electrolyte has very high ionic conductivity, which enables cells to achieve a maximum power density of more than 500 mW·cm^{-2} at 500 °C. In this chapter, the research progress, material composition, working mechanism, and current problems of this new ceramic fuel cell are comprehensively reviewed.

6.1 Introduction

Solid oxide fuel cells (SOFCs) are energy conversion devices that operate at intermediate and high temperatures (>400 °C) and can use energy carriers such as H_2, NH_3, CH_4 and C as their fuels. They have the advantages of fuel flexibility, high energy conversion efficiency (power generation efficiency 40%–65%, comprehensive energy efficiency ≥ 90%), and are the key technology required to achieve the clean utilization of fossil energy and carbon neutrality [1, 2]. Traditional SOFCs usually operate at a high temperatures between 700 °C and 900 °C. The high operating temperatures cause problems related to the cost and long-term stability of SOFC stacks, which seriously limit their commercialization [3–5]. Reducing the operating temperature of SOFCs to less than 600 °C while ensuring that the cell has good

doi:10.1088/978-0-7503-4064-9ch6

power generation performance and a low cost of manufacture is one of the most promising technological ways to realize the large-scale commercialization of SOFCs [6]. However, when the operating temperatures of SOFCs are reduced, the electrochemical performance of the cell is significantly reduced due to a significant decrease in the ionic conductivity of the electrolyte and the poor catalytic activity of the electrodes [7–10]. The ionic conductivities of the traditional electrolytes, i.e. YSZ, GDC, and LSGM, decrease significantly with as the operational temperature decreases. For instance, the ionic conductivity of a YSZ electrolyte is 0.1 $S{\cdot}cm^{-1}$ at 1000 °C, which decreases to 0.03 $S{\cdot}cm^{-1}$ and 0.002 $S{\cdot}cm^{-1}$ at 800 °C and 600 °C, respectively. On the other hand, the ionic conductivity of LSGM is 0.17 $S{\cdot}cm^{-1}$ at 800 °C, which decreases to 0.0011 $S{\cdot}cm^{-1}$ at 600 °C. The ionic conductivity of GDC at 600 °C is only 0.019 $S{\cdot}cm^{-1}$ [11–14]. The catalytic activity of traditional SOFC cathode materials such as $La_{1-x}Sr_xCo_{1-x}Fe_xO_{3-\delta}$ (LSCF) and $La_{1-x}Sr_xMnO_{3-\delta}$ (LSM) for the oxygen reduction reaction (ORR) also deteriorates significantly with a decrease in the operating temperature [15]. Therefore, the development of ceramic fuel cells with new electrolyte and electrode material systems not only allows the operating temperature of the cell to be reduced and ensures excellent electrochemical performance but also provides a new idea that can be used to solve the bottleneck of ceramic fuel cell commercialization.

6.2 Lithium compound ceramic fuel cells

6.2.1 Introduction to lithium compound ceramic fuel cells

In recent years, a new type of symmetrical electrode ceramic fuel cell with a lithium compound such as $LiNiO_2$ or NCAL as its electrodes and a ceramic oxide such as GDC, $SrTiO_3$ (STO), or BZY as its electrolyte has attracted wide attention because its maximum power density (MPD) exceeds 500 $mW{\cdot}cm^{-2}$ at 550 °C, and the ionic conductivity of its electrolyte can reach 0.221–0.476 $S{\cdot}cm^{-1}$ [16–21]. In 2002, Zhu *et al* reported that adding an alkaline carbonate containing Li, Na, or K to an oxide electrolyte could significantly improve its ionic conductivity [22]. Zhu *et al* studied a ceramic fuel cell with a Sm-doped CeO_2 (SDC)–Na/Li_2CO_3 composite as its electrolyte and obtained an MPD of 250 $mW{\cdot}cm^{-2}$ at 400 °C [23]. Raza *et al* prepared an SDC-Na_2CO_3 electrolyte by a carbonate coprecipitation method and prepared symmetrical electrode fuel cells with ZnO/NiO nanocomposites as the cathode and the anode. The cell achieved excellent MPDs of 1257, 1107, and 840 $mW{\cdot}cm^{-2}$ at 520 °C, 500 °C, and 480 °C, respectively [24]. It was later found that a new type of lithium compound electrode ceramic fuel cell (LCCFC) could be prepared by a co-pressing method with lithium compounds as the electrodes and an oxide material as the electrolyte instead of adding carbonate into the electrolyte. This new LCCFC also achieved very good electrochemical performance at low temperature [25–37]. Chen *et al* used foamed Ni-NCAL as a symmetrical electrode and nanocrystalline $Ce_{0.9}Gd_{0.1}O_{2-\delta}$ (GDC) as an electrolyte to prepare a ceramic fuel cell with an MPD of 592 $mW{\cdot}cm^{-2}$ at 550 °C, and the ionic conductivity of the electrolyte was about 0.37 $S{\cdot}cm^{-1}$ [20]. Chen *et al* found that the reason why this LCCFC has very high MPDs at low temperatures and its electrolyte has very high ionic conductivity is directly related to the use of a lithium compound electrode [38]. It was found that lithium compound plays an important role in the significant increase in the electrolyte's conductivity and in reducing the polarization impedance (R_p) of the hydrogen oxidation reaction (HOR) on the anode and the ORR on the cathode. In this chapter, the research progress in the field of LCCFCs in recent years is summarized from the perspectives of cell composition, working mechanism, factors influencing electrochemical performance, and current problems.

6.2.2 The structural composition of lithium compound ceramic fuel cells

As the name suggests, lithium compounds are used as the electrodes of lithium compound electrode ceramic fuel cells. Figure 6.1(a) shows a schematic diagram of the material composition of the LCCFC components. The dark gray parts on both sides represent the lithium compound electrode in the LCCFC. The electrode material used in LCCFCs can be one of a variety of Li-containing compounds, such as $LiNi_{0.8}Co_{0.2}O_2$, $LiNiO_2$, $Li_xAl_yCo_{1-y}O_2$ or NCAL [33, 39–42]. The orange part in the middle of figure 6.1(a) represents the electrolyte of the cell, which can be a mixture of oxide and carbonate, such as SDC+Na_2CO_3, SDC+ $(Li/K)_2CO_3$, or GDC+$(Li/K)_2CO_3$ [43–45] or a single oxide powder material such as STO, La-doped

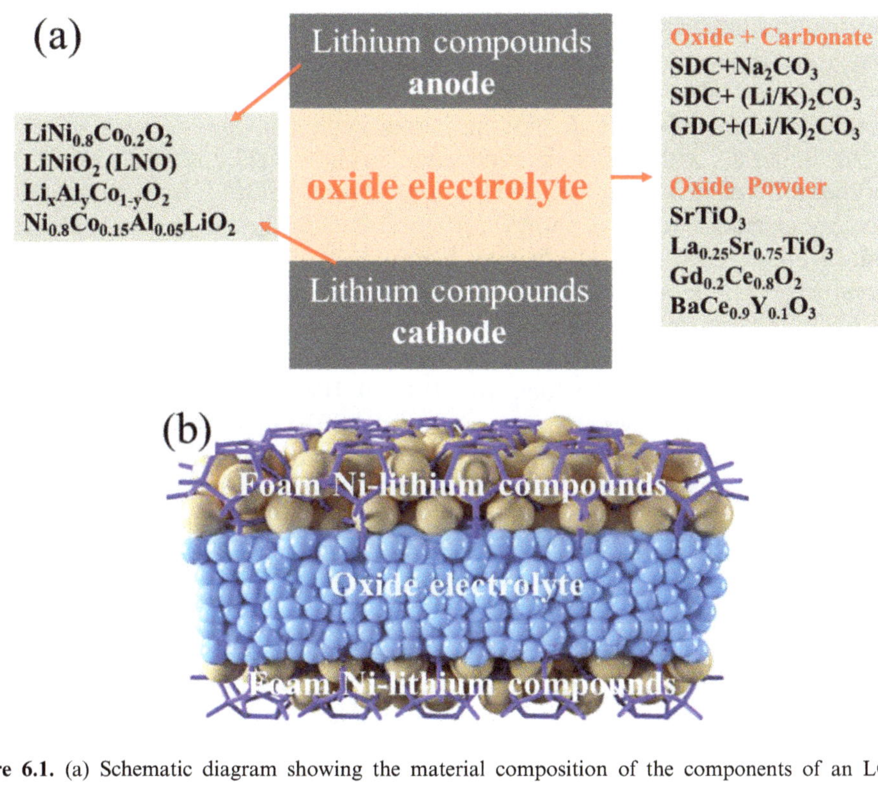

Figure 6.1. (a) Schematic diagram showing the material composition of the components of an LCCFC, (b) schematic diagram of the cross-sectional structure of an LCCFC . Reprinted from [42], Copyright (2021), with permission from Elsevier.

$SrTiO_3$ (LST), GDC, or yttria-doped barium cerate (BCY) [20, 21, 32, 34]. Figure 6.1(b) shows a schematic diagram of the cross-sectional structure of an LCCFC, in which the upper and lower parts are foamed Ni–lithium compound electrodes, the blue mesh part represents foamed Ni, the light-yellow spheres represent lithium compound electrodes, and the middle blue spheres represent the oxide electrolyte. The characteristic of this new LCCFC is that the preparation process is relatively simple: only the foamed nickel coated with the lithium compound electrode and the oxide electrolyte needs to be co-pressed, and neither the electrolyte nor the electrode needs to undergo the high-temperature sintering process used for traditional SOFCs [20, 21, 32]. As shown in figure 6.1(b), the electrolyte in the fuel cell with this new structure is formed by the close packing of oxide particles exposed to high-pressure pressing. The electrode and electrolyte are closely connected, and the foamed nickel plays the role of skeletal support and current collector [20, 21, 32].

6.2.3 The electrochemical performance of lithium compound electrode ceramic fuel cells

As can be seen from the schematic diagram of the cell structure in figure 6.1, the biggest difference between LCCFCs and traditional SOFCs is that the materials used for the electrolyte differ significantly. The electrolytes of traditional SOFCs are basically made of classical oxygen ion conductors, proton conductors, or mixed conductors of oxygen ions and protons. The electrolyte materials of LCCFCs can chosen from many semiconductor materials that do not have ionic conductivity, or even some semiconductor materials with electronic conduction, such as STO and LST [21, 32].

Another difference between LCCFCs and traditional SOFCs is that LCCFCs can still achieve excellent power generation performance in the range of 400 °C–550 °C without using a very thin electrolyte membrane. Zhu *et al* mixed the *p*-type semiconductor LSCF and the ionic conductor samarium with calcium co-doped ceria (SCDC) in a mass ratio of 60:40 as an electrolyte and used foamed Ni-NCAL as an electrode to prepare a cell with the structure Ni-NCAL|LSCF-SCDC|NCAL-Ni. When H_2 was used as the fuel, an MPD of 1080 $mW{\cdot}cm^{-2}$ was obtained at 550 °C [26]. Figure 6.2 shows some cross-sectional SEM images and *IV–IP* curves of LCCFCs with different oxides as the electrolyte and foamed Ni-NCAL as a symmetrical electrode [20, 21, 32]. Figures 6.2(b) and (a) show a cross-sectional SEM image and the corresponding *IV–IP* curves of a cell with LST as the electrolyte, respectively. Before a fuel cell performance test at 550 °C, the ionic conductivity of the LST electrolyte was very low, but when hydrogen was introduced into the cell, the ionic conductivity of the LST electrolyte reached 0.221 $S{\cdot}cm^{-1}$, and the MPD of the cell reached 908.2 $mW{\cdot}cm^{-2}$ [32]. Figures 6.2(d) and (c) show the cross-sectional SEM image and corresponding *IV–IP* curves of a cell with the structure Ni-NCAL|STO|NCAL-Ni under different testing temperatures, respectively [21]. As can be seen from figure 6.2(c), the MPD values of the cell with the STO electrolyte were 310 $mW{\cdot}cm^{-2}$, 440 $mW{\cdot}cm^{-2}$, and 620 $mW{\cdot}cm^{-2}$ at 450 °C, 500 °C, and 550 °C, respectively. The thickness of the STO electrolyte was about

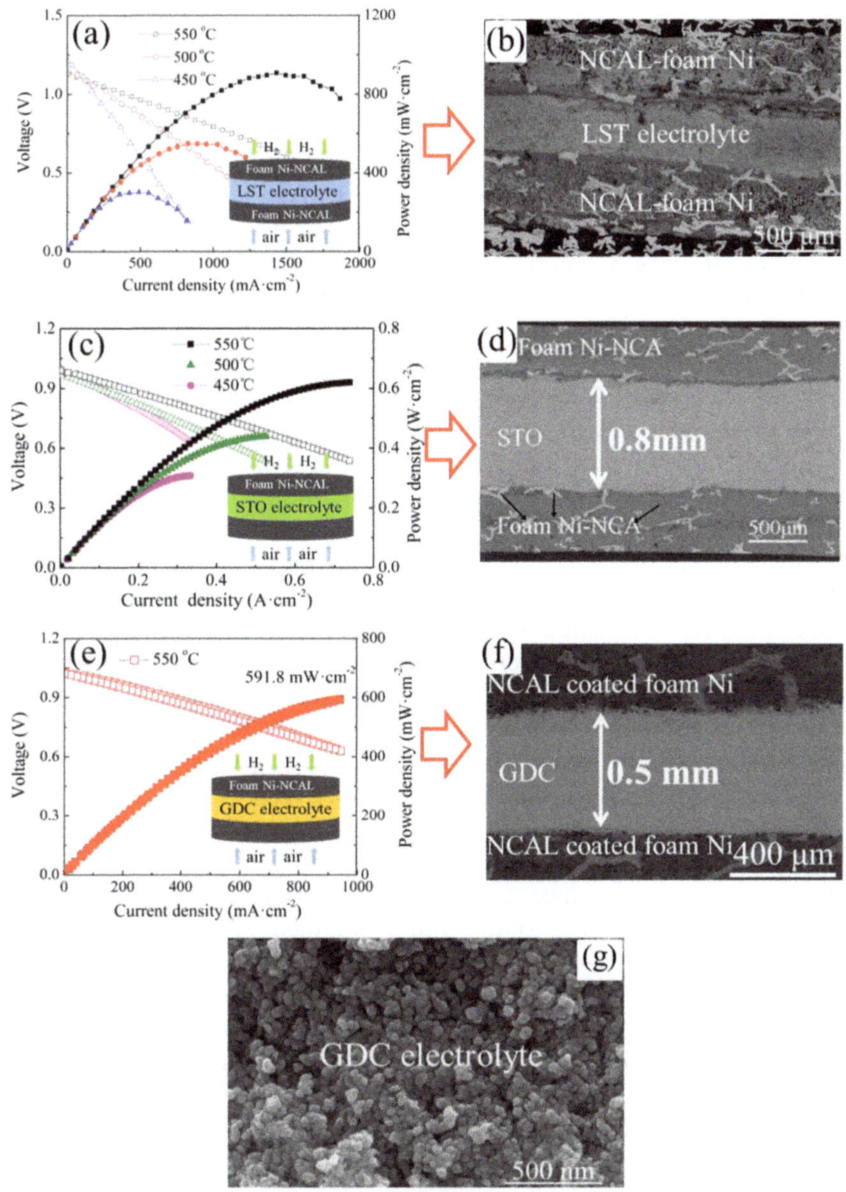

Figure 6.2. (a) *IV–IP* curves of the cell with an LST electrolyte. (b) SEM images of a cross section of the symmetrical cell. Reprinted with permission from [32]. Copyright (2018) American Chemical Society. (c) *IV–IP* curves of the cell with an STO electrolyte. (d) SEM images of a cross section of the symmetrical cell. Reproduced from [21] with permission of The Royal Society of Chemistry. (e) *IV–IP* curves of the cell with a GDC electrolyte. (f)–(g) SEM images of cross sections of the symmetrical cell. Reprinted with permission from [20]. Copyright (2019) American Chemical Society.

0.8 mm. According to the electrochemical impedance spectroscopy (EIS) results, the ionic conductivities of the STO electrolyte at 450 °C, 500 °C, and 550 °C were 0.16 $S{\cdot}cm^{-1}$, 0.21 $S{\cdot}cm^{-1}$, and 0.24 $S{\cdot}cm^{-1}$, respectively. It was found that when the semiconductor materials LST and STO powder were used as electrolytes, there was no short circuit in the cell and the ionic conductivity of the electrolyte was one order of magnitude higher than that of a traditional SOFC electrolyte at the same temperature [21, 32]. Interestingly, it was found that regardless of the composite electrolyte used, such as one produced by following the method described by Zhu *et al* (adding electronic conductors or semiconductor materials to the electrolyte) or an electrolyte made of semiconductor materials such as LST or STO, ceramic fuel cells prepared using lithium compounds as the electrode did not exhibit any obvious internal short circuit under H_2 fuel cell test conditions, and the open-circuit voltages (OCVs) of the cells were 1 V or even higher [21, 25–32]. Zhu *et al* believed that the reason why there was no short circuit in the electrolyte in this new type of ceramic fuel cell is that Schottky junctions formed at the interface between semiconductor materials inside the electrolyte, generating a semiconductor junction effect and thus preventing the occurrence of internal electron short circuits [26].

Studies by Chen *et al* found that in LCCFCs with semiconductors such as LST and STO as electrolytes and foamed Ni-NCAL as symmetrical electrodes, the ohmic resistances of the cells measured when air was introduced into both sides of the cell before the performance test was very large, that is to say, the ionic conductivity of the electrolyte was very low [21, 32]. For example, the total impedance resistance of a cell with LST as the electrolyte, according to the EIS spectra, under conditions in which air was introduced into both sides of the cell, was about 27 580 $\Omega{\cdot}cm^{-2}$. According to a rough analysis, the ohmic resistance (R_o) of the LST electrolyte was about 10 030 $\Omega{\cdot}cm^{-2}$. The cell with the STO electrolyte had a total resistance of about 160 000 $\Omega{\cdot}cm^{-2}$ when both sides were filled with air, and the cell with the BZY electrolyte had a total impedance resistance of about 16 800 $\Omega{\cdot}cm^{-2}$ in air/air at 550 °C; both of these are similar to the value obtained for the cell with the LST electrolyte. Such a high resistance indicates that the ionic conductivities of the LST, STO, and BZY electrolytes are very low before H_2 fuel is introduced into the cell. In other words, there is no electron conduction in the electrolyte in this case, so there is no possibility of a short circuit [19, 21, 32]. Figures 6.2(e)–(g) show the *IV–IP* curves, cross-sectional SEM image, and magnified SEM image of the GDC electrolyte of the cell with nanocrystalline GDC as the electrolyte and foamed Ni-NCAL as the symmetrical electrode, respectively. As can be seen from figure 6.2(e), the MPD of the cell with nanocrystalline GDC as the electrolyte was 591.8 $mW{\cdot}cm^{-2}$ at 550 °C [20]. The ionic conductivity of the nanocrystalline GDC electrolyte was about 0.37 $S{\cdot}cm^{-1}$, which was an order of magnitude higher than that of a dense GDC electrolyte at this temperature [11–14, 20]. It can be seen from figure 6.2(g) that the nanocrystalline GDC electrolyte in the cell showed a dense powder compact packing structure, which is different from the microstructure of the GDC electrolyte sintered at a high temperature of more than 1400 °C. Due to the nonuse of a high-temperature sintering process, no large grains or growing grain boundaries were observed in the nanocrystalline GDC electrolyte, which can be seen in figure 6.2(g).

6.3 The influence of lithium compound electrodes

6.3.1 Reasons for the high ionic conductivity of the electrolytes in lithium compound electrode ceramic fuel cells

It should be noted that some nontraditional ionic conductor materials, such as LST, STO, and LSCF, can also be used as electrolytes in LCCFCs, and show ionic conductivities higher than 0.1 $S{\cdot}cm^{-1}$ at about 500 °C [20, 21, 32]. Chen *et al* used foamed Ni-NCAL as a symmetrical electrode and LST, STO, BCY, BZY, and GDC as electrolytes in the preparation of single cells [20, 21, 32]. It was found that before a performance test of the cells under H_2/air fuel cell test conditions, the electrical conductivity of the electrolytes was not very high, and even some electrolytes were almost insulators. However, the ionic conductivities of the electrolytes measured in an H_2/air atmosphere reached 0.221–0.476 $S{\cdot}cm^{-1}$ at 550 °C [20, 21, 32]. It was found that when H_2 was introduced into the cells, the structure or phase composition of the electrolyte of this LCCFC was greatly changed after the electrochemical performance test. Chen *et al* used the EIS results of a cell with LST as the electrolyte before and after testing in H_2 to draw a schematic diagram of the change in the ionic conductivity of the LST electrolyte before and after a power generation performance test, which can be observed in figure 6.3(a) [32]. Figure 6.3(a) shows the ionic conductivity of the LST electrolyte at 550 °C when air was introduced into both sides of the cell before the cell performance test, which was very small, i.e. the cell was almost an insulator. However, the ionic conductivity of the LST electrolyte rapidly increased to 0.221 $S{\cdot}cm^{-1}$ a few minutes after H_2 fuel was introduced into the anode of the cell [32]. Figures 6.3(b) and (c) show high-resolution transmission electron microscope (HRTEM) images of the LST powder before the performance test and HRTEM images of LST powder scraped from the cell after the performance test, respectively [32]. It can be seen from figures 6.3(b) and (c) that a layer of amorphous compound with a thickness of 2–3 nm had appeared on the surface of the LST electrolyte after the fuel cell performance test [32]. The same results were found for the cells with the STO electrolyte and those with the GDC electrolyte [20, 21]. Figures 6.3(d) and (e) show HRTEM images of the STO powder before the performance test and STO powder scraped from the cell after the performance test [21]. Finally, figures 6.3(f) and (g) show HRTEM images of the GDC powder before the performance test and HRTEM images of GDC powder scraped from the cell after the performance test [20]. It can be seen from the HRTEM images that an amorphous layer with a thickness of about 1–4 nm was formed on the surface of the electrolyte particles after the performance test. The formation of this amorphous layer is directly related to the introduction of H_2 into the anode of the LCCFC, because there is no obvious amorphous layer on the surface of these oxide materials before the fuel cell performance tests.

In order to find out why the ionic conductivity of the electrolyte in LCCFC with NCAL as electrodes increased rapidly after H_2 was introduced into the cell, He *et al* used a GDC electrolyte with a thickness of 0.5 mm to assemble cells in the form of NCAL|GDC|Pt and Pt|GDC|Pt structures with Pt and NCAL as the anode, respectively. It should be noted that GDC electrolytes sintered at 1550 °C only conduct oxygen ions. Figure 6.4(a) shows the *IV–IP* curves with NCAL and Pt as

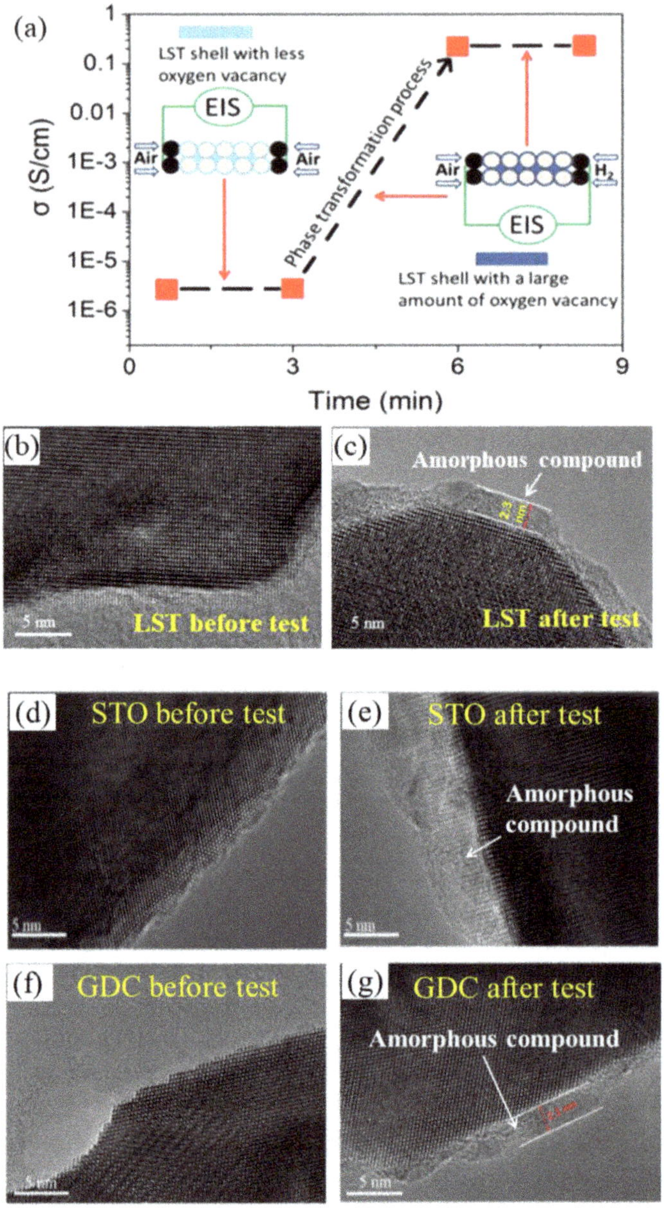

Figure 6.3. (a) Total electrical conductivity of the cell with the LST electrolyte in air/air and in air/H_2. Reprinted with permission from [32]. Copyright (2018) American Chemical Society. (b) and (c) HRTEM images of LST powder before and after the performance test. Reprinted with permission from [32]. Copyright (2018) American Chemical Society. (d) and (e) HRTEM images of STO powder before and after the performance test. Reproduced from [21] with permission of The Royal Society of Chemistry. (f) and (g) HRTEM images of GDC powder before and after the performance test. Reprinted with permission from [20]. Copyright (2019) American Chemical Society.

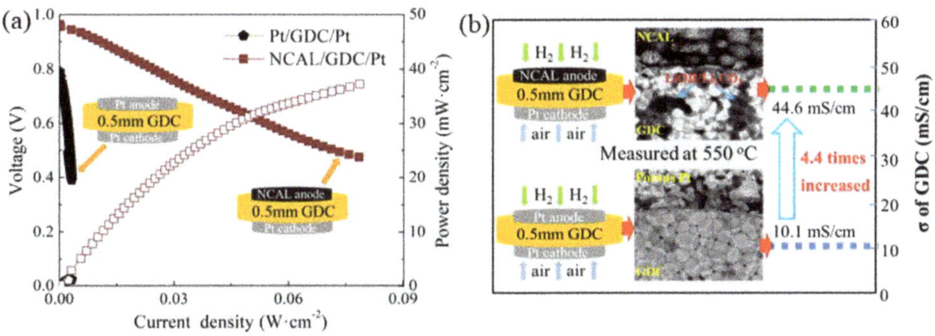

Figure 6.4. (a) *IV–IP* curves of fuel cells with the structures of NCAL/GDC/Pt and Pt/GDC/Pt in H_2/air at 550 °C, (b) schematic diagram of the ionic conductivity changes of the GDC electrolyte of the cells with NCAL and Pt as the anode, respectively. Reprinted with permission from [38]. Copyright (2020) American Chemical Society..

anodes in H_2/Air at 550 °C. It can be seen from figure 6.4(a) that the MPD of the cell with an NCAL anode reached 37 mW·cm^{-2}, which was 32 times higher than that of the cell with a Pt anode [38]. Figure 6.4(b) shows a schematic diagram of the ionic conductivity of the GDC electrolytes from the two cells with different anodes calculated from the EIS results and the SEM images after a performance test. It was found that the ionic conductivity of the GDC electrolyte improved from 10.1 mS·cm^{-1} to 44.6 mS·cm^{-1} after the anode was changed from Pt to NCAL [38]. It is well known that the ionic conductivity of a GDC electrolyte measured with a Pt electrode is its intrinsic conductivity. Therefore, when NCAL is used as the anode, the reason for the significant increase in the ionic conductivity of the GDC electrolyte is directly related to NCAL. From the SEM image in figure 6.4(b), it can be seen that a large number of gray phases appeared inside the GDC electrolyte in the cell with NCAL as the anode, which were distributed around the grain boundaries of GDC particles. In some places, the dark gray phase in the electrolyte extended further, almost covering the whole grain surface. The generation of this dark gray phase in the GDC electrolyte should be related to the introduction of H_2 into the NCAL anode, and it should also be the main reason for the improvement in the conductivity of the GDC electrolyte. He *et al* analyzed the characterization results of x-ray photoelectron spectroscopy (XPS), Fourier transform infrared spectroscopy (FT-IR), thermogravimetric analyzer–differential scanning calorimetry (TGA-DSC), and inductively coupled plasma emission spectrometry (ICP), and found that this dark gray phase and the amorphous substances on the surfaces of the electrolyte particles in the HRTEM image in figure 6.3 are similar substances, the main component of which is a mixture of LiOH and Li_2CO_3 [38]. It has been found that after reducing NCAL at 550 °C in H_2 for a period of time, NCAL is reduced to Ni, LiOH, Li_2CO_3 and other oxides containing Al [36, 38]. There are two main reasons why Li_2CO_3 emerged in the NCAL electrode during the fuel cell performance test. One is that the NCAL electrode reacted with CO_2 due to the long-term exposure to the air. The other is that the NCAL electrode reacted with the CO_2 that was generated by the high-temperature decomposition of organic solvents, such as terpineol present in the electrode slurry and the binder during the preparation process of the cells [20, 21, 36–38]. Driven by the chemical potential difference, the mixture of LiOH and Li_2CO_3 diffuses into the oxide electrolyte and forms an 'oxide–LiOH/Li_2CO_3' composite electrolyte [36, 38]. The formation of an oxide–LiOH/Li_2CO_3 composite electrolyte should be the real reason why the electrolytes of ceramic fuel cells with NCAL as the electrode have very high ionic conductivities at 450 °C–550 °C [36, 38].

In order to further determine which substance plays a leading role in improving the electrolyte conductivity after LiOH and Li_2CO_3 diffuse into the electrolyte, Zhang *et al* prepared composite electrolyte materials with different proportions of Li_2CO_3 and LiOH by mechanical blending in BCY and prepared cells with the structures Pt|BCY–x%Li_2CO_3 and y%LiOH|Pt ($x = 0, 5, 10, 30$; $y = 0, 2, 5, 8$) with porous platinum as a symmetric electrode [37]. Using EIS tests and analyses of the BCY–y%LiOH ($y = 2, 5, 8$) and BCY–x% Li_2CO_3 ($x = 5, 10, 30$) composite electrolytes containing different proportions of LiOH and Li_2CO_3, they found that the ionic conductivity of the BCY electrolyte at 550 °C was about 0.001 $S{\cdot}cm^{-1}$, while the ionic conductivity of the composite electrolyte with added Li_2CO_3 was significantly increased [37]. After mixing Li_2CO_3 with the BCY electrolyte, the ionic conductivity of the composite electrolyte increased by two orders of magnitude, and with the further addition of Li_2CO_3, the ionic conductivity of the composite electrolyte also gradually increased [37]. However, the ionic conductivity of the BCY–y%LiOH composite electrolyte with added LiOH first increased and then decreased with an increase in the amount of LiOH. Compared with a pure BCY electrolyte, the ionic conductivity of the BCY–LiOH composite electrolyte was only two times higher, but this value was significantly lower than that of the BCY–Li_2CO_3 composite electrolyte. The ionic conductivity of the composite electrolyte composed of BCY–5%Li_2CO_3–5%LiOH was lower than that of the composite electrolyte composed of BCY–5%Li_2CO_3 but one order of magnitude higher than that of the composite electrolyte composed of BCY–5%LiOH. This indicates that Li_2CO_3 plays a decisive role in improving the ionic conductivity of the BCY composite electrolytes [37]. It also reveals that the formation of Li_2CO_3 may play a more important role in improving the ionic conductivity of the electrolyte in the 'oxide–LiOH/Li_2CO_3' composite electrolyte formed during the performance test of the LCCFC with NCAL as the electrode.

Figure 6.5 shows a schematic diagram of the working mechanism of the composite electrolyte ceramic fuel cell with foamed Ni-NCAL as the symmetrical electrode, drawn according to the experimental results [46]. It can be seen that in the performance test of a ceramic fuel cell with the structure 'foamed Ni-NCAL electrode|oxide electrolyte|foamed Ni-NCAL electrode,' when hydrogen was introduced into the anode, the NCAL anode was reduced to form the LiOH/Li_2CO_3 mixture. At the cell performance test temperature of 450 °C–550 °C, the LiOH/Li_2CO_3 mixture was in a molten or partially molten state. Driven by the chemical potential, it migrated into the oxide electrolyte through the electrode/electrolyte interface and was coated on the surface of oxide particles to form an 'oxide–molten

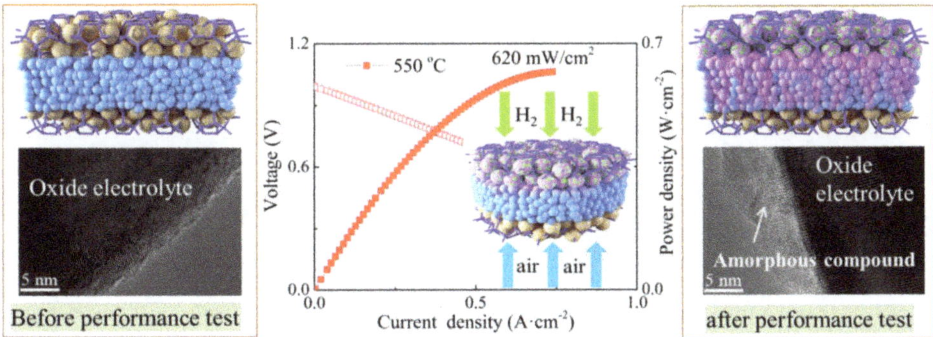

Figure 6.5. Schematic diagram of the working mechanism of a fuel cell with NCAL as the symmetrical electrode and oxide as the electrolyte in H_2. Reprinted from [46], Copyright (2021), with permission from Elsevier.

salt' composite electrolyte with a core–shell structure [46]. This core–shell 'oxide–molten salt' composite electrolyte had a very high ionic conductivity, which enabled the cell to achieve a MPD of more than 500 mW·cm^{-2} at around 500 °C [46]. The characteristics of high ionic conductivity and high power output make it a very competitive choice for the development of low-temperature oxide ceramic fuel cell technology. The presence of the molten salt inside the anode and electrolyte also solves the problem of gas leakage inside the cell, so that the OCV of the cell can reach the theoretical value calculated by the Nernst equation for hydrogen fuel cells [46].

6.3.2 The influence of $Ni_{0.8}Co_{0.15}Al_{0.05}LiO_2$ electrodes on cell performance

From the above research results, it can be seen that the main reason why LCCFCs have excellent power generation performance at low temperatures is related to the use of lithium compound electrodes. Imran *et al* used a composite of lithium nickel zinc oxide (LNZ) and GDC as an electrolyte with NCAL-coated foamed Ni and Au as symmetrical electrodes and current collectors, respectively, to study the effect of NCAL electrodes on cell performance [47]. Figures 6.6(a) and (b) show schematic diagrams of ceramic fuel cells with Au- and NCAL-coated foamed Ni as the symmetrical electrodes, respectively, and *IV–IP* curves tested at 550 °C [47]. Figure 6.6(b) shows the test cell with Au as the electrode, which achieved an MPD of 357 mW·cm^{-2} at 550 °C. However, when the NCAL-coated foamed Ni was used as the electrode, the MPD reached 801 mW·cm^{-2}, proving that NCAL has better catalytic performance for electrode reactions [47].

Chen *et al* used GDC sintered at 1550 °C with a thickness of 0.5 mm as an electrolyte support and NCAL coated with foamed nickel as a symmetrical electrode to prepare a single cell with the structure Ni-NCAL/GDC/Ni-NCAL [33]. The catalytic performances of the foamed Ni-NCAL electrode for the oxygen reduction reaction (ORR) at the cathode and the hydrogen oxidation reaction (HOR) at the anode of the cell were tested and studied using the three-electrode method [33]. Figure 6.7(a) shows a schematic diagram of the three-electrode cell performance test method, in which Pt wire wound around the GDC electrolyte was used as the

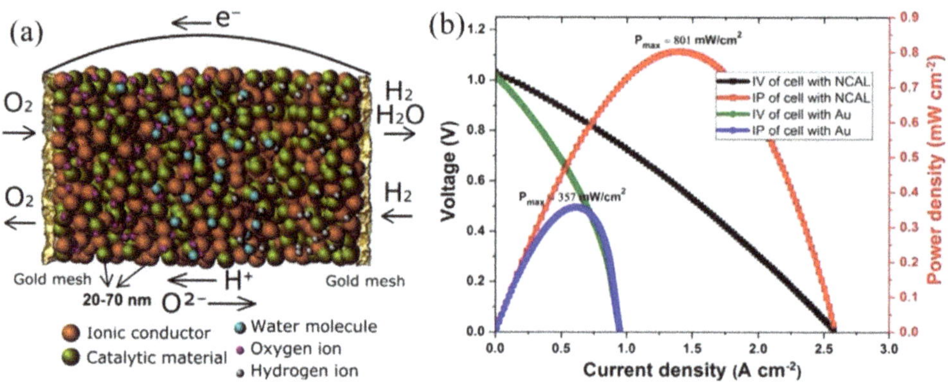

Figure 6.6. (a) Schematic structure of a single-layer ceramic fuel cell, (b) *IP–IV* curves of single-component fuel cells utilizing Ni foamed coated with NCAL as the current collector on both sides of the cell and Au as the current collector on both sides of the cell measured at 550 °C. Reprinted from [47], Copyright (2018), with permission from Elsevier.

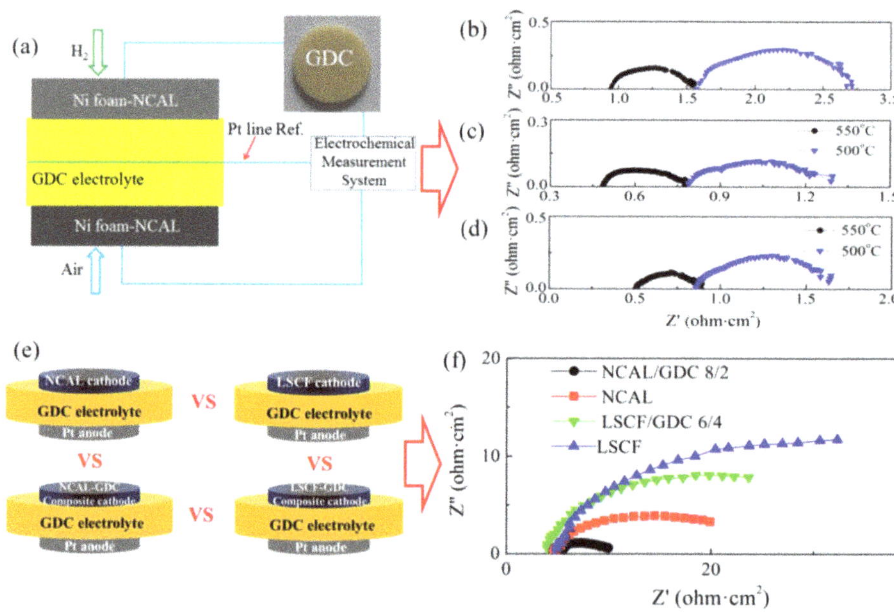

Figure 6.7. (a) Schematic of the cell performance test. Reprinted from [33], Copyright (2018), with permission from Elsevier. (b)–(d) EIS spectra of cells tested at 500 °C and 550 °C: (b) full cell, (c) reference electrode and half-cell of anode, (d) reference electrode and half-cell of cathode. Reprinted from [33], Copyright (2018), with permission from Elsevier. (e) Schematic diagram of the structure of a single cell using 0.8 mm thick GDC sintered at 1550 °C as the electrolyte and four different materials as the cathode. Reprinted from [50], Copyright (2021), with permission from Elsevier. (f) EIS spectra of single cells with NCAL, NCAL/GDC (at a mass ratio of 8:2), LSCF, and LSCF/GDC (at a mass ratio of 6:4) as the cathode at 550 °C under H_2/air conditions. Reprinted from [50], Copyright (2021), with permission from Elsevier.

reference electrode [33]. The impedance spectra of the full cell, a half-cell consisting of the cathode and a reference electrode, and a half-cell consisting of the anode and a reference electrode were measured at 550 °C and 500 °C under H_2/air fuel cell test conditions [33]. In the impedance spectra, the difference between the intersection of the high-frequency region and the low-frequency region of the impedance spectrum arc and the *x*-axis represents the polarization resistance (R_p) of the cell. The R_p values of the HOR and the ORR are denoted by $R_{(RE–H2)}$ and $R_{(RE–air)}$, respectively. Figures 6.7(b)–(d) show the impedance spectra of the full cell, the half-cell of an anode and a reference electrode, and the half-cell of a cathode and a reference electrode, respectively. From the results, it can be seen that at 550 °C and 500 °C, the $R_{(RE–H2)}$ values measured for the half-cell of the anode and a reference electrode were 0.31 $\Omega \cdot cm^{-2}$ and 0.469 $\Omega \cdot cm^{-2}$, respectively, indicating that the foamed Ni-NCAL anode had excellent HOR catalytic activity [33]. The $R_{(RE–air)}$ values measured for the half-cell of the cathode and a reference electrode were 0.393 $\Omega \cdot cm^{-2}$ and 0.729 $\Omega \cdot cm^{-2}$ respectively, which were far lower than those of a traditional LSCF/GDC composite cathode, indicating that NCAL has good ORR catalytic activity [48, 49].

Yang *et al* prepared four single cells with GDC electrolyte pellets as the support, porous Pt as the anode, and NCAL, an NCAL/GDC composite, LSCF, and an LSCF/GDC composite as the cathodes [50]. The low-temperature ORR catalytic performances of the NCAL cathode and the composite cathode composed of NCAL and GDC were studied by testing the power generation performance of the cells and also using EIS. It is well known that the LSCF/GDC composite cathode is a classic SOFC cathode material with very good ORR catalytic performance in the intermediate temperature range [48, 49]. Figure 6.7(e) shows a schematic structural diagram of the four single cells with different cathodes. Figure 6.7(f) shows the EIS spectra of the single cells with NCAL, NCAL/GDC (at a mass ratio of 8:2), LSCF, and LSCF/GDC (at a mass ratio of 6:4) as cathodes at 550 °C under H_2/air conditions [50]. It was found that the performance of the NCAL/GDC composite cathode was better than that of the LSCF/GDC composite cathode, and that its R_p as a cathode was smaller than that of LSCF as a cathode [50]. The results of x-ray photoelectron spectroscopy (XPS) showed that after the mixture of NCAL and binder was sintered, an NCAL-Li_2CO_3 composite phase with a core–shell structure containing lithium carbonate and a large number of oxygen vacancies was formed on the surface of the NCAL [50]. The formation of oxygen vacancies was closely related to the space charge layer formed at the interface between the NCAL and the Li_2CO_3, which should be the reason why NCAL has high ionic conductivity [50]. Yang *et al* proposed that the increase in the oxygen vacancy concentration caused by the space charge layer in the NCAL/GDC composite cathode was the reason for its excellent ORR performance, although the specific mechanism needs further study [50].

6.3.3 The effects of different lithium compound electrodes

In addition to the lithium compound NCAL, which contains Ni, other lithium compounds include $LiNi_{0.83}Co_{0.11}Mn_{0.06}O_2$ (LNCM-811), $LiNi_{0.6}Co_{0.2}Mn_{0.2}O_2$

(LNCM-622), $LiNi_{0.5}Co_{0.2}Mn_{0.3}O_2$ (LNCM-523), $LiMnO_2$ (LMO) and $LiCoO_2$ (LCO). In order to study the influence of different lithium compound electrodes on the performance of ceramic fuel cells, Zhou *et al* prepared single cells with LNCM-811, LNCM-622, LNCM-523, LCO, and LMO as symmetrical electrodes using GDC as electrolyte and tested the electrochemical performance of the cells at 550 °C [51]. Figure 6.8(a) shows the *IV–IP* curves of the cells with LNCM-811, LNCM-622, LNCM-523, LCO, and LMO as symmetrical electrodes at 550 °C in H_2. It can be seen that the OCV of the cell with LNCM-811 as the electrode rises rapidly, reaches

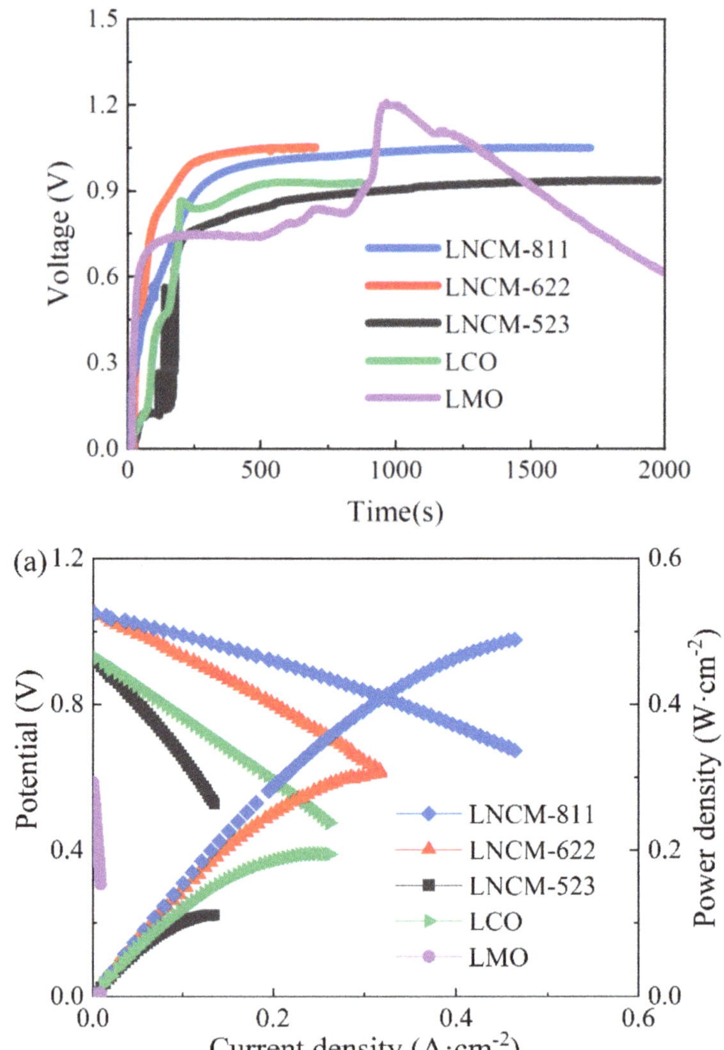

Figure 6.8. OCV curves and *IV–IP* curves of the five cells with different symmetrical electrodes at 550 °C under H_2/air fuel cell test conditions. (a) OCV curves of the cells over time, (b) *IV–IP* curves of the cells. Reprinted from [51], Copyright (2022), with permission from Elsevier.

more than 1 V in about 8 min, and stabilizes at 1.05 V after 15 min. The OCV of the cell with LNCM-523 as the electrode rose rapidly at the beginning, began to rise more slowly after about 4 min, and stabilized at 0.97 V after 40 min. The OCV of the cell with LCO as the electrode increased rapidly at the beginning and increased to 0.988 V in about 9 min, but then slowly decreased. The OCV of the cell with LMO as the electrode increased rapidly at the beginning, its highest OCV reaching 1.16 V, and then began to decline rapidly, dropping to 0.6 V after 30 min. Figure 6.8(b) shows the *IV–IP* curves of the cells with LNCM-811, LNCM-622, LNCM-523, LCO, and LMO as symmetrical electrodes in H_2 at 550 °C, respectively. It can be seen that the MPDs of the cells with LNCM-811, LNCM-622, and LNCM-523 as electrodes were 488.71 $mW{\cdot}cm^{-2}$, 306.76 $mW{\cdot}cm^{-2}$ and 111.73 $mW{\cdot}cm^{-2}$, respectively. The MPD of the cell decreased significantly with decreasing nickel content in the LNCM electrodes. The MPD of the cell with LCO as the electrode was 196.9 $mW{\cdot}cm^{-2}$, while that of the cell with the LMO electrode was only 4.24 $mW{\cdot}cm^{-2}$ [51]. Among the LCCFCs with five different electrodes, the cell with the LNCM-811 electrode had the largest MPD. The OCVs and MPDs of the cells with LCO and LMO electrodes were lower than those the cells with LNCM-811 as electrodes, and the OCV and MPD of the cell with LMO electrodes were the lowest. As the amount of $LiOH/Li_2CO_3$ molten salt produced by the anode of the cell with the LMO electrode is small after it is reduced in hydrogen, its OCV rises rapidly at the beginning. In the later stages, because the amount of molten salt is insufficient and there is molten salt diffusion in the electrolyte, the internal seal of the cell changes, so the OCV of the cell with the LMO electrode exhibits a rapid decline [51]. The performance of the cell also deteriorates significantly. Through a series of XRD, SEM, and XPS characterizations of the powders reduced in hydrogen, such as LNCM-811, LNCM-622, LNCM-523, LCO, and LMO, it was found that the amount of LiOH molten salt produced by the reduction of the five different lithium compounds by hydrogen differed significantly. From the OCV and *IV–IP* curves of the cells, it can be seen that the catalytic activities of Ni, Co, and Mn in the lithium compound electrode for H_2 oxidation were also different. Among them, Ni and Co had excellent catalytic activities for the HOR, whereas Mn had the worst catalytic activity for the HOR. Therefore, the reduction of the lithium compound electrode in hydrogen to produce enough molten salt is the key to the excellent electrochemical performance of the cell. The presence of enough of the metal elements in the electrode materials that can be used as catalysts for the HOR is also a necessary condition for the cell to have excellent performance [51].

6.4 The effects of the operational temperature on the electrochemical performance of lithium compound electrode ceramic fuel cells

6.4.1 A sudden drop in the electrolyte ionic conductivity at low temperatures

Our previous study found that the performance of ceramic fuel cells with NCAL electrodes decreases significantly as the temperature continues to decrease below 450 °C [19, 21]. The composite electrolyte formed by oxide–$LiOH/Li_2CO_3$ has a very high ionic conductivity, which is directly related to the melting state of

$LiOH/Li_2CO_3$ at the operating temperature of the cell [37, 38]. With a decrease in the cell's operating temperature, the melting state of $LiOH/Li_2CO_3$ also changes, which definitely has a huge impact on the ionic conductivity of the electrolyte. To understand the change in the electrolyte's ionic conductivity in a ceramic fuel cell with a lithium compound electrode when the operational temperature of the cell is less than 450 °C, Zhang *et al* selected GDC as the electrolyte and foamed Ni-NCAL as the symmetrical electrode and prepared a cell with the structure Ni-NCAL|GDC|Ni-NCAL [42]. Figures 6.9(a) and (b) show the *IP–IV* curves of the as-prepared cell tested at 350 °C–550 °C and the variation of the GDC electrolyte's ionic conductivity with the operational temperature [42]. Zhang *et al* found that the MPD of the cell made from Ni-NCAL|GDC|Ni-NCAL with GDC as the electrolyte was 643.63 mW·cm^{-2} at 550 °C. They found that the MPD of this cell decreased linearly with decreasing operational temperatures at temperatures greater than 400 °C. However, when the operating temperature dropped to 400 °C, the MPD of the cell decreased significantly [51]. EIS test results showed that the ionic conductivity of the GDC electrolyte was 0.574 S·cm^{-1} at 550 °C and it decreased linearly with decreasing cell operating temperatures. When the temperature dropped to about 400 °C, the ionic conductivity of the electrolyte decreased significantly from 0.251 S·cm^{-1}at 425 °C to 0.026 S·cm^{-1} at 400 °C [51]. Therefore, the significant decrease in the electrolyte's ionic conductivity was the main reason for the sudden drop in the MPD [42].

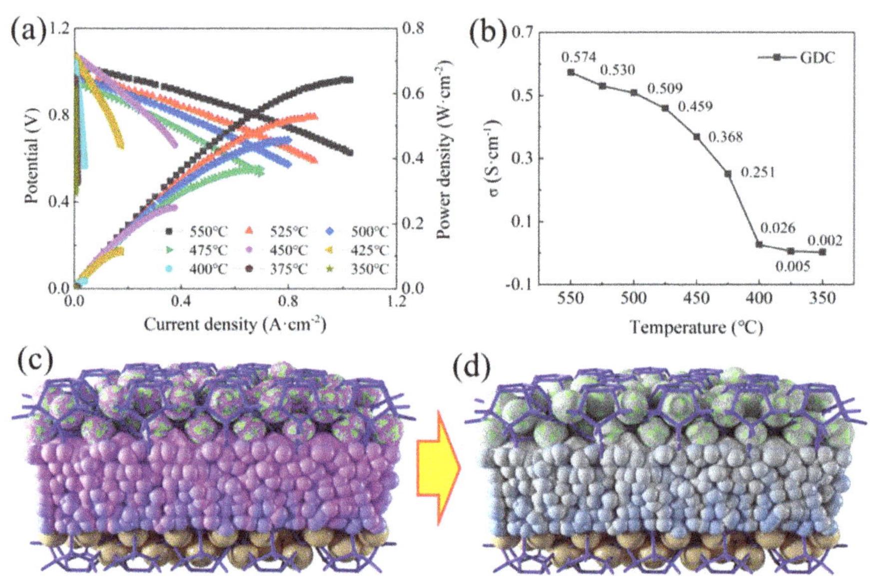

Figure 6.9. (a) The *IP–IV* curves of the cell with GDC as the electrolyte and foamed Ni-NCAL as the symmetrical electrode at 350 °C–550 °C, (b) the variation of the MPD with the operational temperature. (c) (d) Proposed schematic diagram of the molten salt state and structural change of the NCAL electrode ceramic fuel cell at different stages, (c) during the performance test in H_2 at 550 °C–419 °C, and (d) during a performance test in H_2 at a temperature less than the melting temperature of the salt in the cell. Reprinted from [42], Copyright (2021), with permission from Elsevier.

According to results obtained using XPS, FT-IR, and TGA-DSC, the NCAL anode reacts with H_2 to generate molten $LiOH/Li_2CO_3$ salt during the cell test, and the melting point of the salt is 419 °C [42]. Figures 6.9(c) and (d) show schematic diagrams of the changes in the internal material composition and state of the ceramic fuel cell with NCAL as the symmetrical electrode in the H_2 fuel performance test [42]. Figure 6.9(c) shows a structural diagram of the cell during a performance test at 550 °C–425 °C. The purple molten salt in the cell electrolyte is in the molten state at 550 °C–419 °C. When the working temperature drops below 419 °C, i.e. less than the melting point of LiOH, the mixture of $LiOH/Li_2CO_3$ changes from the molten state to the solid state; in other words, the molten salt becomes the solid gray substance illustrated in figure 6.9(d). Therefore, the reason why the electrolyte has very high ionic conductivity is that the $LiOH/Li_2CO_3$ generated in the NCAL anode diffuses to the electrolyte during the operation of the cell and appears in a molten state at 419 °C or above [42]. The phenomenon of molten salt solidification of $LiOH/Li_2CO_3$ and other compounds below 419 °C is the reason why the ionic conductivity of the electrolyte decreases significantly.

6.4.2 The effect of the initial operating temperature on the electrochemical performance of lithium compound electrode ceramic fuel cells

From the results shown in figure 6.9, it is not difficult to realize that the performance of the cell is greatly affected by the melting state of the lithium compounds in the electrolyte. Since the lithium compound produced by the reduction of the NCAL anode is in a molten state, its melting state, migration rate, and total amount must be greatly affected by temperature. Therefore, the influence of the initial reduction temperature on the ionic conductivity of the electrolyte in an NCAL anode ceramic fuel cell and the electrochemical performance of the cell was studied [52]. In order to study the effect of the initial reduction temperature on the power generation performance of the prepared cell, Chen *et al* prepared cells and tested their electrochemical performances at 550 °C, 500 °C, and 450 °C, respectively [52]. The electrochemical performances of the cells reduced at 550 °C and 500 °C were tested at the reduction temperature, and then gradually cooled at intervals of 50 °C. The electrochemical performances of the cells were evaluated at 500 °C and 450 °C, and the cells tested were labeled C550 and C500, respectively. The electrochemical performance of the cell reduced at 450 °C was only tested at 450 °C [52].

As can be seen from figure 6.10(a), MPD of the cell with an initial test temperature of 550 °C at 450 °C was 222.1mW·cm^{-2}, and the MPD of the cell with an initial test temperature of 500 °C at 450 °C was 124.9 mW·cm^{-2} [42]. However, the MPD of the cell with an initial test temperature of 450 °C was only 17.5 mW·cm^{-2} at 450 °C. At the same test temperature, the MPDs of the cells decreased significantly with the decrease in the initial test temperature. According to the rough calculation of the R_o of the cell in figure 6.10(b), the ionic conductivities of the electrolytes of the three cells with initial test temperatures of 550 °C, 500 °C, and 450 °C at 450 °C were 0.288, 0.165, and 0.011S·cm^{-1}, respectively. Therefore, it can

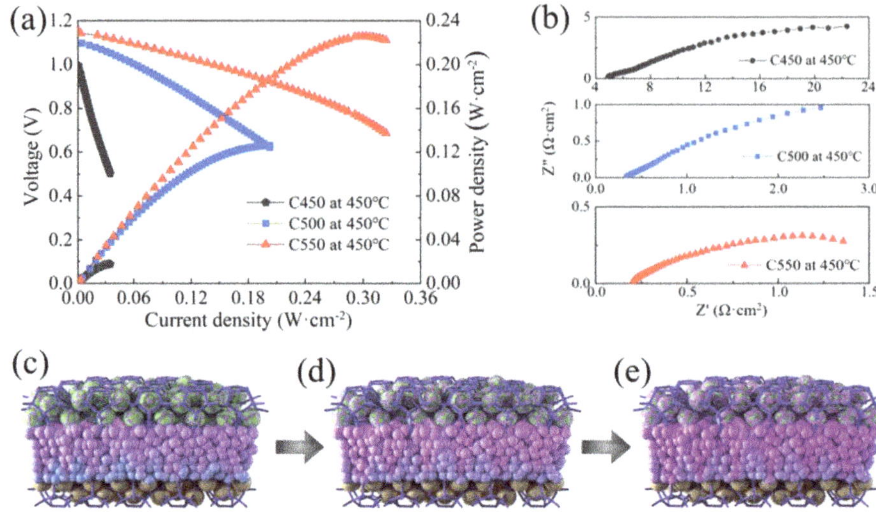

Figure 6.10. Electrochemical performances of three cells with initial test temperatures of 550 °C, 500 °C, and 450 °C at 450 °C. (a) *IV–IP* curves; (b) impedance spectra measured under open-circuit conditions; (c)–(e) schematic diagrams of cross sections of the cells after testing at different temperatures for (c) cell C450, (d) cell C500, and (e) cell C550. Reprinted with permission from [52]. Copyright (2021) American Chemical Society.

be observed that the initial test temperature of the cell has a significant impact on the change in the ionic conductivity of the electrolyte [52].

Previous studies found that in a cell with GDC sintered at different temperatures as the electrolyte and foamed Ni-NCAL as the symmetrical electrode, the ionic conductivity of the GDC electrolyte decreased significantly with an increase in the sintering temperature [36, 38]. Through a quantitative analysis of the LiOH/Li_2CO_3 entering the electrolyte, it was found that the ionic conductivity of the GDC electrolyte was directly proportional to the amount of LiOH/Li_2CO_3 entering it [36, 38]. In order to verify whether the amount of LiOH/Li_2CO_3 produced at different NCAL anode reduction temperatures and reduction degrees of NCAL were different, NCAL/GDC half cells were reduced in a hydrogen atmosphere at 450 °C, 500 °C, and 550 °C. According to the SEM of a cross section of the GDC pellet, with an increase in reduction temperature, the amount of amorphous molten salt at the NCAL/GDC interface was greater, and the NCAL particles were reduced more fully, indicating that higher reduction temperatures produce more molten lithium compounds [42]. Combined with XRD, XPS, and FT-IR characterizations, it was found that the higher the initial test temperature, the faster and greater the amount of LiOH/Li_2CO_3 mixture produced in the NCAL anode which enters the electrolyte [52]. Therefore, during the electrochemical performance test, the electrochemical performance of the cell can be improved at a lower temperature by selecting the method of first reducing the cell at a higher temperature and then cooling it. The above research results are summarized in the schematic diagram shown in figures 6.10(c)–(e). According to this diagram, when the initial test temperature is higher, the amount of molten lithium compounds (shown in purple) produced is higher, and these lithium compounds have a significant impact on the performance of the cell.

6.5 Conducted carriers and the corresponding transference numbers in electrolytes

As mentioned above, during the performance test of the cell with NCAL as the electrode, NCAL was reduced in an H_2 atmosphere to produce the $LiOH/Li_2CO_3$ mixture. The molten $LiOH/Li_2CO_3$ complex covers the oxide electrolyte surfaces of STO, LST, BCY, and GDC, which is the key to improving their ionic conductivities. He *et al* found that the ionic conductivity of the composite electrolyte formed by GDC and lithium compounds including $LiOH/Li_2CO_3$ was 4.4 times higher than the ionic conductivity of a pure GDC electrolyte [38]. However, the carrier types in the composite electrolyte and the transference numbers of each carrier are still unclear. To clarify this, a symmetrical electrode cell was prepared using GDC pellets sintered at 1550 °C with thickness of 0.5 mm as the electrolyte and NCAL as the electrode. The possibility of proton conduction in the electrolyte of ceramic fuel cells using NCAL as electrode was investigated using a hydrogen concentration cell test, and the carrier type and ion conduction mechanism in the composite electrolyte were analyzed [53].

Figure 6.11 shows the variation curve of the OCV over time for a cell with the structure Pt/GDC/Pt and a cell with the structure NCAL/GDC/NCAL in the hydrogen concentration cell test using pure hydrogen and low-concentration hydrogen (3% H_2 + 97% Ar) at 550 °C. The proton transference numbers in the concentration cell can be calculated according to equation (6.2), where E_{obs} represents the cell voltage observed during the concentration cell test.

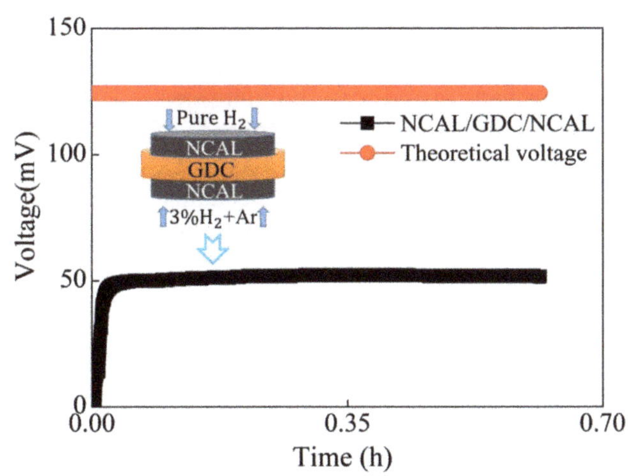

Figure 6.11. OCV versus time curve for the cell with the structure NCAL|GDC|NCAL in the hydrogen concentration cell test at 550 °C.

$$E_{\mathrm{Nernst}} = \frac{RT}{nF}\mathrm{In}\left(\frac{p_1}{p_2}\right) \tag{6.1}$$

$$t_{\mathrm{ion}} = \frac{E_{\mathrm{obs}}}{E_{\mathrm{Nernst}}} \tag{6.2}$$

According to equation (6.2), the proton transference number of the GDC electrolyte in the cell is about 42%. It is a very interesting phenomenon that GDC, which is a classical oxygen ion conductor, shows obvious proton conduction in the cell with NCAL as the electrode [53]. He *et al* speculated that the dark gray phase displayed within the GDC electrolyte was a $LiOH/Li_2CO_3$ complex on the basis of the results of XPS, FT-IR, and ICP characterizations [38]. In other words, the electrolyte of the cell with NCAL as the electrode changed into a $LiOH + Li_2CO_3 + Li_2O$/GDC composite electrolyte during the hydrogen concentration test [53]. The H_2 concentration cell test results indicate that the carriers in the GDC/$LiOH + Li_2CO_3$+Li_2O composite electrolyte include protons, and the proton transference number is about 45%. GDC is an oxygen ion conductor, so the GDC/$LiOH + Li_2CO_3 + Li_2O$ composite electrolyte may also have the ability to conduct oxygen ions. Thus, the carriers of this composite electrolyte may include oxygen ions. In addition, it should be noted that the lithium hydroxide contained in the composite electrolyte is in a molten state during the hydrogen concentration testing. Therefore, the OH^- in the molten LiOH may also be one of the carriers. It should be noted that, unlike conventional molten carbonate fuel cells, which require CO_2 on the cathode side as fuel, this composite electrolyte ceramic fuel cell was tested with a foamed Ni-NCAL symmetric electrode, and the oxidant on the air side was 'synthetic air' consisting of N_2 and O_2 (which does not contain CO_2), so that we can exclude the possibility of CO_3^{2-} migration as one of the internal carriers within the electrolyte [20, 21, 32]. In summary, we can speculate that the carriers in the $LiOH + Li_2CO_3 + Li_2O$/GDC composite electrolyte formed in the hydrogen concentration cell test atmosphere in the cell with NCAL as the electrode should be H^+, OH^-, and/or O^{2-} [53]. In the fuel cell with the GDC/$LiOH + Li_2CO_3 + Li_2O$ composite as the electrolyte, the conduction of O^{2-} should mainly rely on the GDC phase. Its conduction mechanism and corresponding electrode reaction are the same as that of an SOFC with a pure oxygen ion conductor as the electrolyte [54, 55]. The proton conduction medium should be the $LiOH + Li_2CO_3 + Li_2O$ composite phase or at the interface with GDC, although the specific mechanism still needs further study. The anodic H_2 oxidation reaction associated with proton conduction is shown in equation (6.3):

$$H_2 \rightarrow 2H^+ + 2e^- \tag{6.3}$$

After the proton is conducted to the air electrode through the GDC/$LiOH + Li_2CO_3 + Li_2O$ composite electrolyte, the ORR takes place, as shown in equation (6.4):

$$O_2 + 4H^+ + 4e^- \rightarrow 2H_2O \tag{6.4}$$

Therefore, we speculate that the H_2O generated by the ORR (which involves protons) on the air side represents a prerequisite for the conduction of OH^-.

The H_2O generated by the reaction (equation 6.4) further reacts with O_2 on the air side to generate OH^- (as shown in equation 6.5). OH^- is then conducted to the anode through the molten LiOH in the composite electrolyte and reacts with H_2 to release electrons (as shown in equation 6.6) [51].

$$O_2 + 2H_2O + 4e^- \rightarrow 4OH^- \tag{6.5}$$

$$2H_2 + 4OH^- \rightarrow 4H_2O + 4e^- \tag{6.6}$$

In brief, the carriers of the GDC/LiOH+Li_2CO_3+Li_2O composite electrolyte formed by the LCCFC using an NCAL electrode under H_2/air fuel cell test conditions should be H^+, OH^-, and/or O^{2-}. The transference numbers of the carriers are closely related to the change in temperature. However, the specific conduction mechanism of protons in the composite electrolyte and the transference numbers of OH^- need to be further studied.

6.6 The durability of lithium compound electrode ceramic fuel cells

At present, research into these new LCCFCs mainly focuses on their working mechanisms and the development of new cell materials. Most of this research is performed with the objective of pursuing high output power. There are only few reports of their long-term stability. Some reports have provided durability data for some cells over dozens of hours, but has been no in-depth study of the mechanisms responsible for the degradation in cell performance [20, 54–56]. Previous research has found that a cell with nanocrystalline GDC as the electrolyte and foamed Ni-NCAL as the symmetrical electrode could operate stably for 9 h at a current density of 0.076 $mA\cdot cm^{-2}$, and the output voltage was maintained at about 1 V. From the 9th hour, the cell began to degrade [20]. In order to study the effect of the chemical reaction between the electrolyte and the lithium compound electrode on cell performance, Dai *et al* used CeO_2, TiO_2, ZrO_2, and YSZ as electrolytes and foamed Ni-NCAL as the symmetrical electrode in the preparation of Ni-NCAL|oxide electrolyte|NCAL-Ni single cells [35]. They found that the chemical reaction between the oxide electrolyte material and LiOH/Li_2CO_3 greatly affected the OCV and the electrochemical performance of the cell [35]. Figure 6.12(a) shows the variation curves of the OCV over time of the cells with CeO_2, TiO_2, ZrO_2, and YSZ electrolytes at 550 °C in H_2 [35]. As can be seen from figure 6.12(a), the OCVs of the four cells with different electrolytes increased rapidly from zero in the first few minutes. The maximum OCVs of the four cells with CeO_2, TiO_2, ZrO_2, and YSZ electrolytes reached 1.04 V, 1.12 V, 1.01 V, and 0.99 V, respectively. However, the OCV of the cells with TiO_2 and ZrO_2 electrolytes began to decrease rapidly after reaching the maximum value for more than 10 min. When tested for 60 min, the OCVs of the cells with TiO_2 and ZrO_2 electrolytes decreased to 0.59 V and 0.27 V, respectively. The OCV of the cell with the YSZ electrolyte slowly decreased after reaching a maximum value of 0.99 V and dropped to 0.91 V after 60 min. It should be noted that the OCV of the cell with the CeO_2 electrolyte remained relatively stable after reaching the maximum value. Figure 6.12(b) shows the *IP–IV* curves of

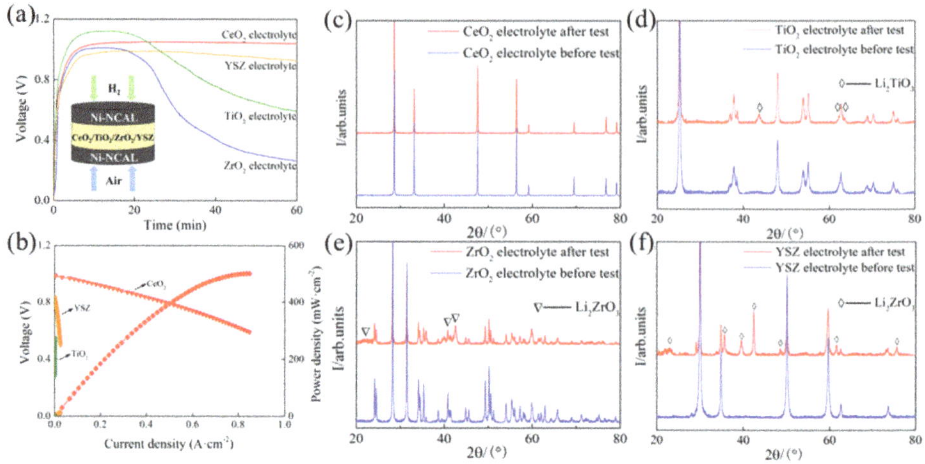

Figure 6.12. (a) OCV versus time curves of the cells using different oxide electrolytes after H_2 was introduced into the cells, (b) *IP–IV* curves of the cells with CeO_2, TiO_2, and YSZ electrolytes at 550 °C, (c)-(f) XRD patterns of the CeO_2, TiO_2, ZrO_2, and YSZ electrolytes before and after the performance test: (c) CeO_2, (d) TiO_2, (e) ZrO_2, and (f) YSZ. Reprinted from [35], Copyright (2022), with permission from Elsevier.

the cells with CeO_2, TiO_2, ZrO_2, and YSZ electrolytes at 550 °C in H_2. As can be seen from figure 6.12(b), the MPDs of the cells with CeO_2, TiO_2, and YSZ electrolytes are 599.6, 0.09, and 12.5 mW·cm^{-2}, respectively. The MPD values of the cells using TiO_2 and YSZ as electrolytes are much lower than that of the cell using the CeO_2 electrolyte.

Figures 6.12(c)–(f) show the XRD patterns of the electrolyte powders of CeO_2, TiO_2, ZrO_2, and YSZ in the cells before and after the electrochemical performance test in H_2 [35]. According to the XRD results, it was found that there was no obvious phase structural change and no new phase appeared in the CeO_2 electrolyte after the electrochemical performance test. Compared with the XRD patterns of the electrolyte powders before the electrochemical performance test in H_2, Li_2TiO_3 and Li_2ZrO_3 peaks appeared in the XRD patterns of TiO_2, ZrO_2, and YSZ electrolyte powders scraped from the cells after the electrochemical performance test [35]. Combined with the results of SEM-EDS and FT-IR, it was found that during the electrochemical performance test of the cells, the $LiOH/Li_2CO_3$ mixture produced by the reduction of the NCAL anode by H_2 diffuses into the electrolyte [35]. During the electrochemical performance test of the cells with TiO_2, ZrO_2, and YSZ electrolytes, the electrolytes react with the $LiOH/Li_2CO_3$ mixture to produce Li_2TiO_3 and Li_2ZrO_3, so that the molten salt of $LiOH/Li_2CO_3$ is consumed, resulting in a significant reduction in the OCV and MPD of the cells [35]. Therefore, the chemical reaction between the electrolyte and $LiOH/Li_2CO_3$ in the cell at 550 °C is the main reason for the OCV and the performance degradation of the cell [35]. Since the CeO_2 electrolyte does not react with $LiOH/Li_2CO_3$ compounds, the cell has a relatively stable OCV and a high MPD. However, TiO_2, ZrO_2, and YSZ electrolytes react chemically with $LiOH/Li_2CO_3$ compounds during the electrochemical performance test, so that an 'oxide–molten salt' composite electrolyte cannot be formed. There are few or no channels for rapid transmission of the carriers in the electrolyte, resulting in a very low cell MPD. Therefore, the chemical reactivity of the electrolyte materials with the NCAL electrode and/or $LiOH/Li_2CO_3$ compounds at high temperature has a very important impact on the electrochemical performance and durability of LCCFCs.

6.7 Current issues and future research directions for lithium compound electrode ceramic fuel cells

It can be seen from the current research results that for LCCFCs, when H_2 is introduced into the lithium compound anode, the $LiOH/Li_2CO_3$ mixture produced by the reduction of the anode is partially molten at the test temperature of 450 °C–550 °C, which then diffuses into the oxide electrolyte under the effect of chemical potential and forms an 'oxide–molten salt' composite electrolyte with a special core–shell structure. The lithium compound molten salt produced by the reduction of the anode is the key to obtaining excellent power generation performance from LCCFCs at low temperatures. Therefore, the amount and state of the lithium compound molten salt in the cell play a crucial role in determining the power generation performance and the durability of the cell.

From the long-term stability data of the LCCFCs reported so far, it can be seen that there are still some problems to be solved in their long-term stability. The key to improving the durability of LCCFCs is to understand the working mechanism of the cell and reveal the relevant rules and mechanisms of the microstructural and phase changes of the various components of the cell during long-term operation. The specific problems should focus on material migration inside the LCCFC, the factors affecting the diffusion and migration of lithium compounds molten salts in the cell, the conduction mechanism of protons in the composite electrolyte, the transference number of OH^-, and the factors that have important impacts on the long-term stability of the cell. In addition, new electrolyte materials need to be further developed to improve the power generation performance and durability of LCCFCs and reduce the cell cost.

References

[1] McIntosh S and Gorte R J 2004 Direct hydrocarbon solid oxide fuel cells *Chem. Rev.* **104** 4845–65

[2] Singhal S C and Kendall K 2002 High-temperature solid oxide fuel cells: fundamentals. Design and applications *Mater. Today* **5** 55

[3] Zhao H L, Zheng Y, Yang C Y, Shen Y N, Du Z H and Swierczek K 2013 Electrochemical performance of $Pr_{1-x}Y_xBaCo_2O_{5+\delta}$ layered perovskites as cathode materials for intermediate-temperature solid oxide fuel cells *Int. J. Hydrog. Energy* **38** 16365

[4] Jiang L, Li F S and Wei T 2014 Evaluation of $Pr_{1+x}Ba_{1-x}Co_2O_{5+\delta}$ ($x = 0$–0.30) as cathode materials for solid-oxide fuel cells *Electrochem. Acta* **133** 364

[5] Chen D J, Chen C, Dong F F, Shao Z P and Ciucci F 2014 Cobalt-free polycrystalline $Ba_{0.95}La_{0.05}FeO_{3-\delta}$ thin films as cathodes for intermediate-temperature solid oxide fuel cells *J. Power Sources* **250** 188
[6] Wachsman E D and Lee K T 2011 Lowering the temperature of solid oxide fuel cells *Science* **334** 935
[7] Smeacetto F, Salvo M, Ajitdoss L C, Perero S, Moskalewicz T and Boldrini S *et al* 2010 Yttria-stabilized zirconia thin film electrolyte produced by RF sputtering for solid oxide fuel cell applications *Mater. Lett.* **64** 2450
[8] Ding D, Li X, Lai S Y, Gerdes K and Liu M 2014 Enhancing SOFC cathode performance by surface modification through infiltration *Energy Environ. Sci.* **7** 552
[9] Suzuki T, Hasan Z, Funahashi Y, Yamaguchi T, Fujishiro Y and Awano M 2009 Impact of anode microstructure on solid oxide fuel cells *Science* **325** 852
[10] Waldbillig D, Wood A and Ivey D G 2005 Electrochemical and microstructural characterization of the redox tolerance of solid oxide fuel cell anodes *J. Power Sources* **145** 206
[11] Fuentes R O and Baker R T 2009 Structural, morphological and electrical properties of $Gd_{0.1}Ce_{0.9}O_{1.95}$ prepared by a citrate complexation method *J. Power Sources* **186** 268–77
[12] Yamamoto O, Arati Y, Takeda Y, Imanishi N, Mizutani Y, Kawai M and Nakamura Y 1995 Electrical conductivity of stabilized Zirconia with Ytterbia and Scandia *Solid State Ion.* **79** 137–42
[13] Smeacetto F, Salvo M L, Ajitdoss C, Perero S, Moskalewicz T, Boldrini S, Doubova L and Ferraris M 2010 Yttria-stabilized zirconia thin film electrolyte produced by RF sputtering for solid oxide fuel cell applications *Mater. Lett.* **64** 2450–3
[14] Huang K Q, Tichy R and Goodenough J B 1998 Superior perovskite oxide-ion conductor; strontium- and magnesium-doped $LaGaO_3$: III, performance tests of single ceramic fuel cells *J. Am. Ceram. Soc.* **81** 2581–5
[15] Ding D, Li X, Lai S Y, Gerdes K and Liu M 2014 Enhancing SOFC cathode performance by surface modification through infiltration *Energy Environ. Sci.* **7** 552–75
[16] Zhu B, Raza R and Qin H *et al* 2011 Single-component and three-component fuel cells *J. Power Sources* **196** 6362–5
[17] Hu H Q, Lin Q Z and Zhu Z G *et al* 2014 Fabrication of electrolyte-free fuel cell with $Mg_{0.4}Zn_{0.6}O/Ce_{0.8}Sm_{0.2}O_{2-\delta}$–$Li_{0.3}Ni_{0.6}Cu_{0.07}Sr_{0.03}O_{2-\delta}$ layer *J. Power Sources* **248** 577–81
[18] Mi Y, Zhang W and Deng H *et al* 2017 Rare-earth oxide $Li_{0.3}Ni_{0.9}Cu_{0.07}Sr_{0.03}O_{2-\delta}$ composites for advanced fuel cells *Int. J. Hydrog. Energy* **42** 22214–21
[19] Chen G, Luo Y D and Sun W K *et al* 2018 Electrochemical performance of a new structured low temperature SOFC with BZY electrolyte *Int. J. Hydrog. Energy* **43** 12765–73
[20] Chen G, Sun W K and Luo Y D *et al* 2019 Advanced fuel cell based on new nanocrystalline structure $Gd_{0.1}Ce_{0.9}O_2$ electrolyte *ACS Appl. Mater. Interfaces* **11** 10642
[21] Chen G, Liu H L and He Y *et al* 2019 Electrochemical mechanisms of an advanced low-temperature fuel cell with a $SrTiO_3$ electrolyte *J. Mater. Chem.* A **7** 9638–45
[22] Zhu B, Liu X G and Zhou P *et al* 2002 Innovative solid carbonate-ceria composite electrolyte fuel cells *Fuel Cells Bull.* **1** 8–12
[23] Zhu B, Yang X T and Xu J *et al* 2003 Innovative low temperature SOFCs and advanced materials *J. Power Sources* **118** 47–53
[24] Raza R, Liu Q H and Nisar J *et al* 2011 ZnO/NiO nanocomposite electrodes for low-temperature solid oxide fuel cells *Electrochem. Commun.* **13** 917–20

[25] Wu J S, Zhu B and Mi Y Q *et al* 2012 A novel core-shell nanocomposite electrolyte for low temperature fuel cells *J. Power Sources* **201** 164–8

[26] Zhu B, Huang Y Z and Fan L D *et al* 2016 Novel fuel cell with nanocomposite functional layer designed by perovskite solar cell principle *Nano Energy* **19** 156–64

[27] Qiao Z, Xia C and Cai Y X *et al* 2018 Electrochemical and electrical properties of doped CeO_2–ZnO composite for low-temperature solid oxide fuel cell applications *J. Power Sources* **392** 33–40

[28] Zhu B, Ma Y and Wang X D *et al* 2011 A fuel cell with a single component functioning simultaneously as the electrodes and electrolyte *Electrochem. Commun.* **13** 225–7

[29] Zhu B, Raza R, Abbas G and Singh M 2011 An electrolyte-free fuel cell constructed from one homogenous layer with mixed conductivity *Adv. Funct. Mater.* **21** 2465–9

[30] Meng Y J, Wang X Y and Xia C *et al* 2018 High-performance SOFC based on a novel semiconductor-ionic $SrFeO_{3-\delta}$–$Ce_{0.8}Sm_{0.2}O_{2-\delta}$ membrane *Int. J. Hydrog. Energy* **43** 12697–703

[31] Khan M A, Raza R and Lima R B *et al* 2014 Effect of titania concentration on the grain boundary conductivity of calcium-doped ceria electrolyte *Ceram. Int.* **40** 9775–81

[32] Chen G, Zhu B and Deng H *et al* 2018 Advanced fuel cell based on perovskite La-$SrTiO_3$ semiconductor as the electrolyte with superoxide-ion conduction *ACS Appl. Mater. Interfaces* **10** 33179–86

[33] Chen G, Sun W K and Luo Y D *et al* 2018 Investigation of layered $Ni_{0.8}Co_{0.15}Al_{0.05}LiO_2$ in electrode for low-temperature solid oxide fuel cells *Int. J. Hydrog. Energy* **43** 417

[34] Chen G, Zhang X B and Luo Y D *et al* 2020 Ionic conduction mechanism of a nanostructured BCY electrolyte for low-temperature SOFC *Int. J. Hydrog. Energy* **45** 24108–15

[35] Dai R, Chen G and Wei K *et al* 2022 Effect of chemical reactions between electrolyte and lithium compounds on the electrochemical performance of the ceramic fuel cells *Carbon Resour. Convers.* **5** 131–8

[36] He Y, Chen G and Zhang X B *et al* 2020 Influence of sintering temperature on ceramic fuel cell electrolyte conductivity with lithium-compound electrode *Ceram. Int.* **46** 17545–52

[37] Zhang X B, Chen G and He Y *et al* 2021 Effect of Li_2CO_3 and LiOH on the ionic conductivity of $BaCe_{0.9}Y_{0.1}O_3$ electrolyte in SOFCs with a lithium compound electrode *Int. J. Hydrog. Energy* **46** 9948–56

[38] He Y, Chen G and Zhang X B *et al* 2020 Mechanism for major improvement in SOFC electrolyte conductivity when using lithium compounds as anode *ACS Appl. Energy Mater.* **3** 4134–8

[39] Fan L D, Su P C and Layer-structured 2016 $LiNi_{0.8}Co_{0.2}O_2$: a new triple ($H^+/O^{2-}/e^-$) conducting cathode for low temperature proton conducting solid oxide fuel cells *J. Power Sources* **306** 369

[40] Zhang L M, Yang J and Li J 2014 A novel composite cathode for intermediate temperature solid oxide fuel cell *J. Power Sources* **269** 723

[41] Lan R *et al* 2014 Novel proton conductors in the layered oxide material $Li_xAl_{0.5}Co_{0.5}O_2$ *Adv. Energy Mater.* **4** 1301683

[42] Zhang R, Chen G, Chen Z, Dai R, Lv X, Lou T, Li Y and Geng S 2021 Investigation of the sudden drop of electrolyte conductivity at low temperature in ceramic fuel cell with $Ni_{0.8}Co_{0.15}Al_{0.05}LiO_2$ electrode *Int. J. Hydrog. Energy* **46** 27793–800

[43] Zhu B *et al* 2012 Nanocomposite electrode materials for low temperature solid oxide fuel cells using the ceria-carbonate composite electrolytes *Int. J. Hydrog. Energy* **37** 19351

[44] Huang J, Mao Z Q, Liu Z X and Wang C 2007 Development of novel low-temperature SOFCs with co-ionic conducting SDC-carbonate composite electrolytes *Electrochem. Commun.* **9** 2601–5

[45] Benamira M, Ringuede A, Hildebrandt L, Lagergren C, Vannier R-N and Cassir M 2012 Gadolinia-doped ceria mixed with alkali carbonates for SOFC applications: II—an electrochemical insight *Int. J. Hydrog. Energy* **37** 19371

[46] Yang D, Chen G and Zhang L *et al* 2021 Low temperature ceramic fuel cells employing lithium compounds: a review *J. Power Sources* **503** 230070

[47] Imran A M, Sami J and Riina J *et al* 2018 Wide bandgap oxides for low-temperature single-layered nanocomposite fuel cell *Nano Energy* **53** 391–7

[48] Zhao E Q, Ma C, Yang W, Xiong Y P, Li J Q and Sun C W 2013 Electrospinning $La_{0.8}Sr_{0.2}Co_{0.2}Fe_{0.8}O_{3-\delta}$ tubes impregnated with $Ce_{0.8}Gd_{0.2}O_{1.9}$ nanoparticles for an intermediate temperature solid oxide fuel cell cathode *Int. J. Hydrog. Energy* **38** 6821

[49] Jiang S P 2019 Development of lanthanum strontium cobalt ferrite perovskite electrodes of solid oxide fuel cells—a review *Int. J. Hydrog. Energy* **44** 7448

[50] Yang D, Chen G and Liu H L *et al* 2021 Electrochemical performance of a $Ni_{0.8}Co_{0.15}Al_{0.05}LiO_2$ cathode for a low temperature solid oxide fuel cell *Int. J. Hydrog. Energy* **46** 10438–47

[51] Zhou J, Chen G, Chen Z, Wei K, Guan G, Geng S and Abudula A 2022 Effects of different lithium compound electrodes on the electrochemical performance of the ceramic fuel cells *Int. J. Hydrog. Energy* **47** 36640–9

[52] Chen Z, Chen G and Zhang R *et al* 2021 Toward understanding of temperature dependence of an advanced ceramic fuel cell with $Ni_{0.8}Co_{0.15}Al_{0.05}LiO_2$ as an electrode *ACS Appl. Energy Mater.* **4** 8386–94

[53] Wei K, Chen G, Chen Z, Dai R, Yu K and Geng S 2022 Ionic conduction of ceramic/molten salt composite electrolyte in fuel cell with lithium compound as electrode *J. Electrochem. Soc.* **169** 094503

[54] Abbas G, Raza R and Ashfaq M *et al* 2014 Electrochemical study of nanostructured electrode for low-temperature solid oxide fuel cell (LTSOFC) *Int. J. Energy Res.* **38** 518–23

[55] Khan I, Asghar M I, Lund P D and Basu S 2017 High conductive $(LiNaK)(2)CO_3$–$Ce_{0.85}Sm_{0.15}O_2$ electrolyte compositions for IT-SOFC applications *Int. J. Hydrog. Energy* **42** 20904–9

[56] Gao J, Xu S and Akbar M *et al* 2021 Single layer low-temperature SOFC based on $Ce_{0.8}Sm_{0.2}O_{2-\delta}$–$La_{0.25}Sr_{0.75}Ti_1O_{3-\delta}$–$Ni_{0.8}Co_{0.15}Al_{0.05}LiO_{2-\delta}$ composite material *Int. J. Hydrog. Energy* **46** 9775–81

IOP Publishing

Nanoengineered Materials for Solid Oxide Cells

Katherine Develos-Bagarinao

Chapter 7

Tailoring nanostructured fuel electrode materials for solid oxide cells through redox exsolution

Xiangling Yue and Shuai He

Fabricating solid oxide cell (SOC) electrodes (particularly their key components, such as electrocatalysts) at the nanoscale has generally been acknowledged to be advantageous for greatly extending their triple phase boundaries (TPBs), thus enhancing their performance and stability, diminishing interfacial reactions, allowing more diverse material choices, etc. It also has the potential to save materials and reduce operational costs. Various techniques used to introduce nanostructures have been addressed in other chapters of this book; these include atomic layer deposition (see chapter 5) and infiltration (see chapter 8). In this chapter, we focus on redox exsolution, which is a relatively new technique for creating nanomaterials in SOCs. We explain this concept, the fundamentals of its design, control of the evolving nanomaterials' microstructure and properties, and its applications in SOCs, followed by a summary of the advances in this strategy and its outlook for SOC development.

7.1 The concept of exsolution and its development

Exsolution is typically employed in materials that have an ABO_3 perovskite structure, which is well known for its flexibility in accommodating various dopants/valences within its lattice and its consequent versatility in modifying materials' structures, properties, and functionalities.

Exsolution refers to a method in which a catalytically active metal and/or metals, predominantly transitional metals, are doped into the crystal lattice of a parent oxide, forming a single-phase solid solution under oxidizing conditions (e.g. in air) and released as nanoclusters or particles on the surface of the oxide following extrinsic reduction, as shown in equations (7.1) and (7.2) and figure 7.1(a) [1], for example as a result of exposure to a H_2-containing atmosphere or the application of a cathodic potential.

doi:10.1088/978-0-7503-4064-9ch7

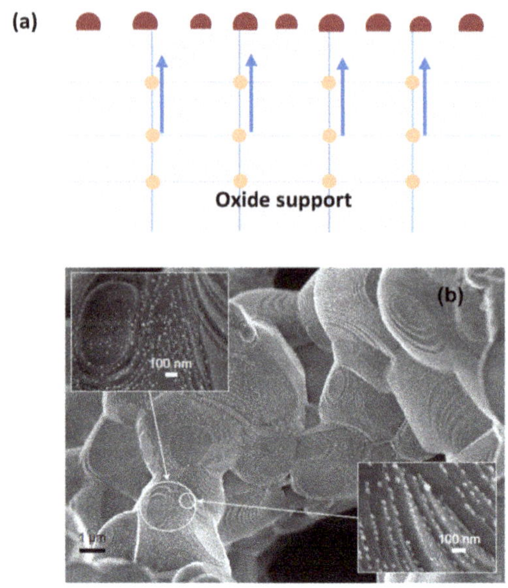

Figure 7.1. (a) Schematic of the exsolution phenomenon and (b) the exsolution-induced microstructure of a (La, Ce)(Ni, Ti)O_3 oxide material with prominent nanoparticles decorating the surface after reduction, whereas the surface was clean and smooth (free of nanofeatures) prior to reduction. Reproduced with permission from [2]. © University of St. Andrews and Dr Dragos Neagu.

$$A_{1-\alpha}BO_{3-\delta} \xrightarrow{\text{Exsolution}} (1-\alpha)ABO_{3-\delta'} + \alpha B \tag{7.1}$$

$$ABO_{3-\delta} \xrightarrow{\text{Exsolution}} (1-\alpha)ABO_{3-\delta''} + \alpha AO + \alpha B \tag{7.2}$$

Here, α stands for the extent of the A-site deficiency, which is an important factor that impacts the exsolution behavior of nanoparticles from a perovskite host (see section 7.3 for a detailed explanation).

The first step of the exsolution method is similar to the typical doping that is often used with perovskites to tune their properties. However, the choices of metal dopants as well as their reducibility in the following reduction treatment need to be considered during the materials design to ensure the formation of a single solid solution as well as the subsequent effective release of nanoparticles following reduction. From a microstructural point of view, these materials can be observed to have clean and nanofeature-free surfaces after the first step (i.e. before reduction); however, following reduction by either a reducing agent (such as H_2 or CO) or cathodic polarization, nanoparticles (or other types of nanostructure) can be seen decorating the surface of the supporting oxide, while the latter retains its perovskite structure. A schematic of the exsolution phenomenon and a typical microstructure generated by exsolution are displayed in figure 7.1. The numerous Ni nanoparticles in figure 7.1(b) were reported to be homogeneously distributed on the surface of $La_{0.8}Ce_{0.1}Ni_{0.4}Ti_{0.6}O_3$ after reduction in 5% H_2/Ar for 20 h at 930 °C [2].

The early concept of exsolution was reported in 2002 by Nishihata *et al*, who introduced the idea of the self-regeneration of precious catalyst particles, which helped to suppress particle growth in a three-way catalyst system for automotive emission control [3]. Pd particle aggregation and the resulting catalyst deactivation were severe problems in traditional Pd-impregnated γ-Al_2O_3 (Pd/alumina) catalyst systems. In Nishihata *et al*'s work, Pd was doped onto the *B*-site of a perovskite, forming a single-phase oxide, $LaFe_{0.57}Co_{0.38}Pd_{0.05}O_3$. When such a perovskite oxide was exposed to alternating oxidative and reductive environments at high temperature (900 °C)—conditions similar to the fluctuations in the redox characteristics of the automotive emission exhaust—it was discovered that Pd could reversibly move in and out of the perovskite structure, retaining its particle size at 1–3 nm, whereas impregnated Pd particles on the alumina support increased to 120 nm after 100 h. As a result, the Pd–perovskite catalyst maintained its high activity, but the Pd–alumina catalyst suffered from a performance deterioration of about 10% after an identical stability test. Due to the excellent lifetime performance demonstrated by the said self-regeneration process, $LaFe_{0.95}Pd_{0.05}O_3$ has been used as a commercial three-way catalyst for automotive emission control in gasoline engines ever since.

The approach was later extended to the incorporation of other precious-metal catalysts, such as Rh and Pt, within a wide range of perovskite compositions, including $CaTiO_3$, $SrTiO_3$, $LaAlO_3$, etc [4]. Although the extent of self-regeneration varied in different perovskite systems, it has generally been accepted that precious-metal-containing perovskite catalysts, also known as 'intelligent catalysts,' in which metallic clusters or nanoparticles emerge from the perovskite oxide upon reduction and dissolve back into the perovskite after oxidation, can successfully maintain a high level activity in prolonged cycled operations even at high temperatures. This technology, based on self-regeneration in a redox cycling environment, later known as redox exsolution, is an effective way of manufacturing highly active materials by careful design and selection of supporting oxide and metal catalyst, reducing the amount of precious-metal catalysts needed. It may thus solve the socioeconomic issues associated with the supply and demand of precious metals.

The success of these regenerative materials in catalysis attracted some attempts to apply the same methodology to the fabrication of solid oxide fuel cell (SOFC) electrodes. Using the exsolution concept, Barnett *et al* examined the capability of Ru, Pd, and Ni to dissolve into and precipitate out of an (La, Sr)CrO_3 oxide matrix and evaluated their performance as SOFC anodes [5–8]. It was reported that Ru could be incorporated into the chosen perovskite material and nucleated as metallic nanoclusters on the surfaces of oxide anodes at the initial reduction stage, which contributed to a continuous improvement in cell performance due to decreased anode polarization resistance for the first 300 h [5]. The precipitated Ru nanoclusters exhibited stable particle sizes of $\leq$5 nm for ~300 h and coarsened slightly over time up to 1000 h at 800 °C [7]. In addition, the outgrowth of bulk Ru from the perovskite and the corresponding anode performance were studied as functions of the dopant concentration and the reduction temperature [7]. The particle density increased following increased Ru concentration in the bulk at a given temperature, while increasing the reduction temperature enlarged the size and decreased the population of the precipitated Ru nanoclusters.

It is worth noting that a nanoscale Ni catalyst was also produced through exsolution from a $La_{0.8}Sr_{0.2}Cr_{0.82}Ni_{0.18}O_{3-\delta}$ perovskite, yielding enhanced SOFC performance, which extended the applicability of the exsolution approach from precious-metal to Earth-abundant transition metals. However, the performance improvement was less pronounced compared to that of its Ru analog, which might coincide with the finding that the segregated Ni nanoclusters were much larger than the Ru (20 vs. 5 nm) variant and that the Ni nanocatalyst coarsened markedly, growing from 20 to 50 nm after being exposed to humidified H_2 for 311 h at 800 °C, whereas the Ru nanoclusters retained their size after an identical test [5].

At that time, the materials employing exsolution methodology investigated in the mainstream were confined to *A*-site stoichiometric perovskite compositions, which was concomitant with a small degree of nanoparticle exsolution, which gave rise to a limited population of nanocatalysts and hence insufficient fuel electro-oxidation activity at the SOFC anode.

The culmination of the exsolution concept has occurred in the last decade, when Irvine and Neagu *et al* extensively investigated titanate perovskites with the ability to grow numerous nanoparticles offering compositional diversity, including metals, oxides, and/or a mixture of the two on the surfaces of perovskites [1, 9–11]. By tuning the degree of non-stoichiometry (i.e. the deviation from the ideal ABO_3 stoichiometry) and varying compositions to tailor the surface characteristics of the parent oxide, dynamic exsolution of nanoparticles could be facilitated significantly, affording unprecedented control over particle size, distribution, and surface anchorage, particularly when *A*-site deficiency was adopted. Furthermore, they discovered the socketed nature of the exsolved particles with respect to their underlying perovskite support; in other words, the particles are deeply submerged into the perovskite with an epitaxial orientation. This strong interfacial interaction between the *in situ* grown nanoparticles and the oxide was believed to lead to a tremendously enhanced resistance against particle coarsening at elevated temperatures and coking in carbonaceous atmospheres and other functionalities to be explored.

Unlike the early studies of 'intelligent' catalysts, depending on the compositions of both the perovskite and the nanophase to be produced, the exsolved particles do not necessarily dissolve back into the perovskite matrix upon reoxidation, but are partly oxidized, retaining their anchorage to their supports at their initial location. This is particularly the case when *A*-site deficiency is adopted to boost the exsolution phenomena. The reason for this is the synergy between the *A*-site oxide and the segregation of the *B*-site metal. When an *A*-site deficiency is present, the precipitation of the *B*-site element is not accompanied by the formation of the *A*-site oxide on the surface, as occurs in the stoichiometric analog; thus, it is less likely for the *B*-site metal to revert back to the perovskite lattice with the dissolution of the *A*-site oxide and is more likely to lead to metal oxidation. Nevertheless, the oxidized nanophase can still possess superior catalytic activity or even provide new functionalities, as exemplified by NiCo alloy nanoparticles precipitated from $La_{0.7}Ce_{0.1}Ni_{0.3}\,Co_{0.1}Ti_{0.6}O_3$ [10]. In a CO oxidation reactor, the exsolved NiCo alloy oxidized following the light off reaction conditions and reorganized its particles into tilted cube-like structures, which resulted in a nearly threefold increase in catalytic activity compared to the as-prepared NiCo metallic catalysts. Additionally, these NiCo cubes (with a particle size of 10 nm) exhibited doubled CO oxidation rates at high temperatures (>450 °C) and fivefold higher NO oxidation rates across the entire temperature range compared to the metallic state of the catalysts. The oxidized NiCo alloy cubes were demonstrated to rival a commercial Pt catalyst due to their remarkable catalytic activity and stability, which resulted from synergistic effects taking place at the cube–perovskite interface (which showed a semi-enclosed nanostructure), offering new pathways for the base metal oxide nanostructure to achieve Pt-like functionality.

Since the publication of this series of key findings, exsolution materials have been in the spotlight for catalysis and energy applications, including SOCs, and many researchers have joined this interesting area. The low P_{O2} environment in the fuel electrode compartment of SOCs is a good match for the conditions required to trigger the *in situ* exsolution of nanomaterials; therefore, SOCs can favorably take advantage of the operating conditions required to produce nanoscale electrode materials without any pretreatment, offering simplicity in manufacture together with excellent material activity and stability. Efforts have been made to validate the concept of exsolution in various fuel electrodes, covering a wide range of oxide materials, in order to enhance cell performance in fuel cell and/or electrolysis cell applications (see section 7.4). Success has been manifested in a wide array of perovskite systems, and the exsolved phases have offered compositional diversity as well. In contrast to the early studies, the vast majority of reports have addressed Earth-abundant base metal exsolutions, which are economically favorable considering the high cost and limited availability of noble metals.

7.2 Exsolution vs. infiltration

As an alternative technology for producing nanostructured SOC electrodes, exsolution has often been compared with infiltration, with respect to their surface morphologies, their interactions between the nanophase and the substrate, as well as their efficiencies in boosting the performance and stability of SOCs. As has been mentioned, it is well known that both technologies can produce diverse compositions and functionalities with nanofeatures, extending TPBs and consequently enhancing SOC performance compared to the traditional ceramic mechanical mixing approach. However, due to the striking difference in methodology, some outstanding phenomena have been generally acknowledged for each of these technologies, as will be discussed in the following.

As indicated in figure 7.1(a), exsolution is an inside-out method. The catalytically active component is first dissolved in the lattice of a host material, forming a solid solution—often under oxidizing conditions—and then grows on the surface as nanoparticles or nanoclusters; the latter typically takes place in a reducing atmosphere. In this way, the exsolved catalyst (regardless of its form, i.e. particles, cubes, rods, etc.) is coherently linked to the host perovskite at the interface, which is known as a socketed structure between the segregated nanophase and the parent

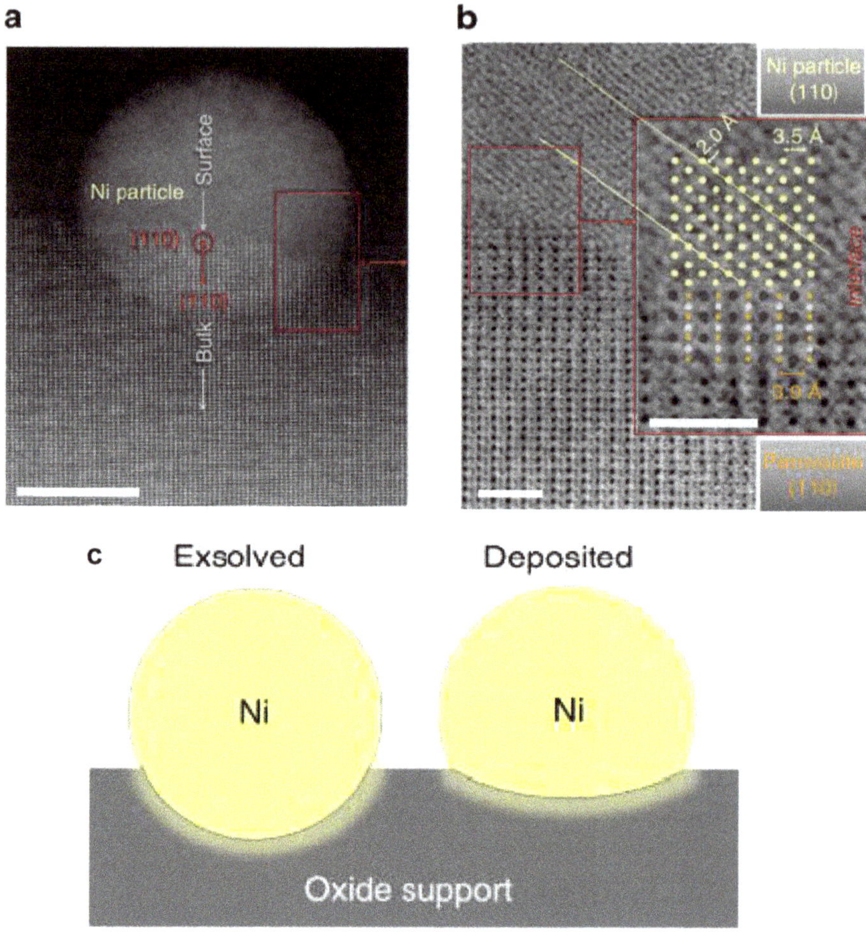

Figure 7.2. Transmission electron microscopy (TEM) micrograph of (a) a Ni particle exsolved from a (110) native surface facet of $La_{0.52}Sr_{0.28}Ni_{0.06}Ti_{0.94}O_3$ after ageing (3% H_2O 5% H_2/Ar, 900 °C, 60 h) and (b) the metal-perovskite interface highlighting the atomic planes and orientations after reduction (5%H_2/Ar, 900 °C, 12 h). (c) A schematic illustration of the interactions of exsolved and deposited Ni particles with the underlying support [9] Copyright © 2015, The Author(s). With permission of Springer.

oxide, as shown in figures 7.2(a) and (b). Infiltration is a fundamentally different method: nanoparticles are introduced by loading catalyst precursor solutions into a preformed porous solid support with the aid of capillary force and subsequent thermal treatment and reduction. When this approach is used, both the liquid–solid interactions between the catalyst precursor solution and the supporting scaffold as well as the subsequent solid–solid interactions after thermal treatment and reduction affect the interactions between the nanoparticles and their support.

Because of the intrinsically different procedures used for these two approaches, there are two crucial differences in the microstructure and morphology of the resultant nanofeatures. First, the distribution of nanoparticles and the particle size

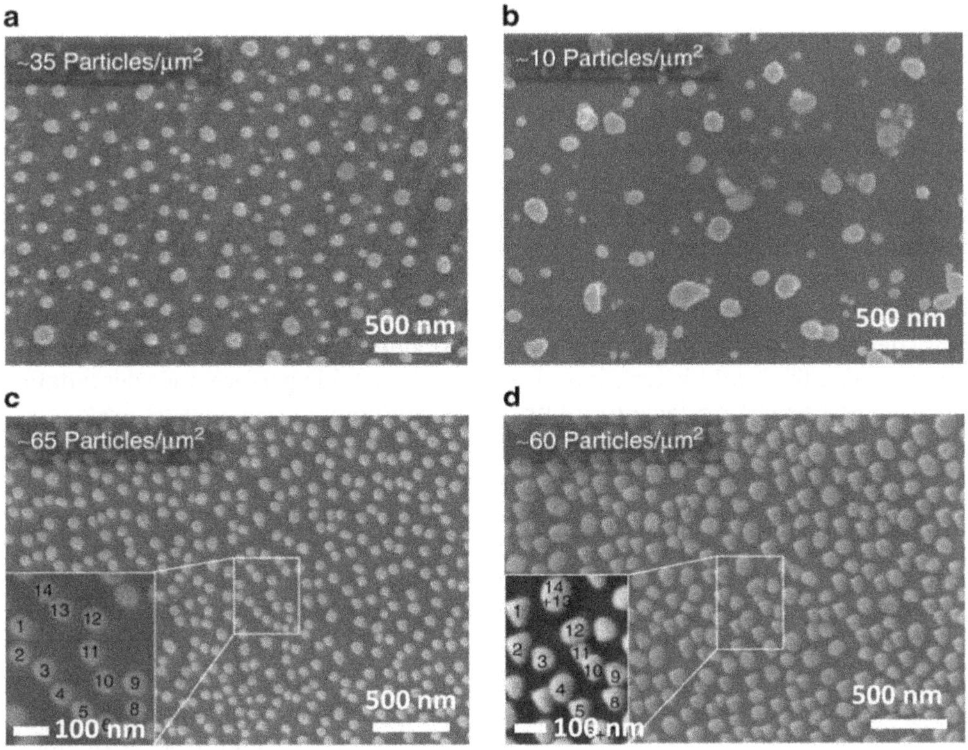

Figure 7.3. Comparison of deposited and exsolved Ni nanoparticles and their thermal stability. (a and b) SEM images of vapor-deposited Ni nanoparticles on $La_{0.4}Sr_{0.4}TiO_3$ (a) before and (b) after ageing (H_2, 650 °C, 24 h and 800 °C, 6 h); (c and d) SEM images of Ni particles exsolved from a $La_{0.52}Sr_{0.28}Ni_{0.06}Ti_{0.94}O_3$ (5% H_2/Ar, 900 °C, 12 h) cleaved surface (c) before and (d) after ageing (5% H_2/Ar, 900 °C, 70 h) [9] Copyright © 2015, The Author(s). With permission of Springer.

can be vastly contrasting. When infiltration is used, the distribution of the nanophase is largely dependent on the structure of the porous scaffold/substrate and the properties associated with the precursor solution, such as its concentration, pH value, solvent type, chelating agent used, etc. which makes it complex to manipulate the distribution and the size of the nanoparticles [12]. When exsolution is used, the exsolved phase initially forms a single solid solution within the framework of the support; when the environment is altered, metal cations diffuse from the bulk to the surface of the support and become reduced, decorating the surface with a metallic or oxide nanophase (or even a mixture), leading to a more homogeneous coverage of nanofeatures, as demonstrated in figure 7.3 where the microstructures of deposited and exsolved Ni nanoparticles are compared. Note that not only the coverage but also the particle density, size, and shape can also be conveniently controlled by altering the initial doping concentration, reduction temperature, P_{O2}, or non-stoichiometry in the parent oxide composition in exsolution (see section 7.3).

Second, the attachment of the nanoparticles (or other forms) to their support is tremendously distinct in the exsolution compared to that in infiltration processing.

This attachment is a key factor impacting the morphology and microstructural evolution during ageing and long-term operation. It is well known that the infiltrated nanoparticles are weakly attached to their support, which was validated in an experiment in which shallow pits remained in an infiltrated system after the catalyst was etched away by diluted nitric acid [9]. Due to this weak interaction, the infiltrated nanoparticles tend to agglomerate easily at high temperatures and in prolonged operations. Figures 7.3(a) and (b) aptly display the microstructural evolution one could expect from infiltrated/impregnated nanoparticles subjected to an ageing test, although these nanoparticles were obtained by vapor deposition, which has similar issues. The growth and coalescence of nanoparticles would negate the benefits of the nanostructure due to a reduction in surface area and the active sites available for catalytic reactions, leading to inevitable performance degradation, which is the greatest obstacle to infiltration technology.

In contrast, exsolved nanoparticles are deeply embedded into the substrate, as they grow from the perovskite lattice (figures 7.1(a) and 7.2) by diffusion through the bulk to the surface and nucleation on the surface following reduction. In fact, an epitaxial growth of exsolved Ni particles along the interface was observed by electron diffraction, as exhibited previously in figure 7.2(b). These strong interactions endow the exsolved nanoparticles with excellent thermal stability against ageing. As demonstrated in figures 7.3(c) and (d), the Ni nanoparticles precipitated via the exsolution route grew only slightly during a 70 h ageing test at 900 °C, and their particle density was almost unchanged; their deposited Ni counterparts, on the contrary, experienced a considerable growth in particle size and hence a significant loss of particle density in a similar ageing test at an even lower temperature, i.e. 800 °C (figures 7.3(a) and (b)). The stronger attachment of the exsolved particles to the support was also evidenced by the much deeper pits in the exsolved materials than those of the infiltrated materials after the nanoparticles were removed in a similar etching experiment [9]. This contrast in the extent of the interactions between the nanoparticles and their substrates in the two systems, i.e. exsolution and infiltration, is also schematically illustrated in figure 7.2(c), which makes it easier to understand/interpret the striking difference in the performance and stability of SOC electrodes fabricated by infiltration and exsolution. The stronger interactions between nanoparticle and the substrate are clearly beneficial for increasing the robustness of the system as well as maintaining high performance, reducing the long-term operational cost of SOCs.

Furthermore, it has been found that the stronger interfacial interactions in the exsolution materials disrupted the carbon fiber growth mechanism, leading to completely different coking behavior compared to the infiltrated or deposited nanocatalyst system. As depicted in figure 7.4, the growth of carbon fiber in the infiltrated or deposited Ni catalyst system was believed to follow a tip-growth mechanism, in which the Ni catalyst was lifted by carbon fiber forming underneath it due to the weak interaction between the Ni catalyst and its support, which prompted the growth and accumulation of lengthy fibers, as seen in figure 7.5(a). However, in the exsolved system, the growth of fiber was massively inhibited because the Ni nanocatalyst was deeply embedded in the support,

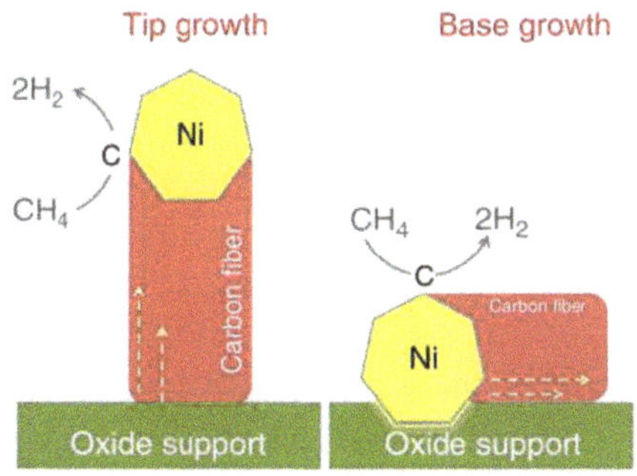

Figure 7.4. Illustration of the possible carbon fiber growth mechanisms of infiltrated (left) and exsolved (right) Ni nanoparticles [9] Copyright © 2015, The Author(s). With permission of Springer.

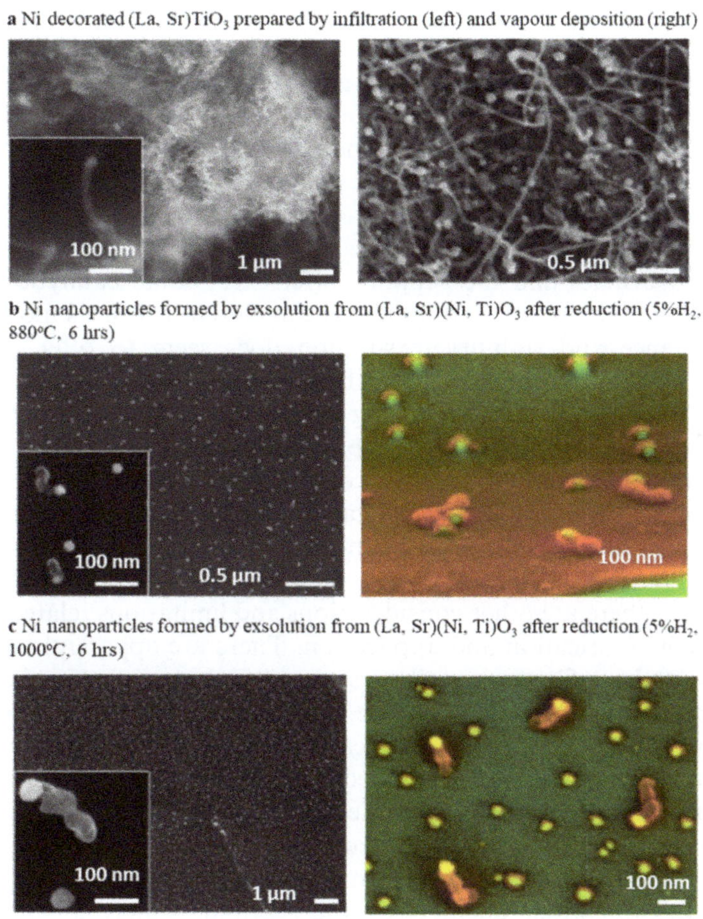

Figure 7.5. SEM micrographs of (La, $Sr)_{0.8}TiO_3$ with Ni nanoparticles prepared via different routes and conditions after identical coking tests (20% CH_4/H_2 at 800 °C, 4 h) (red denotes carbon and yellow represents Ni in the false-color images in b and c; adapted) [9] Copyright © 2015, The Author(s). With permission of Springer.

rendering it hard to be lifted by carbon fiber grown underneath, hence a base growth mechanism was favored. In the base growth scenario, very short carbon fibers were found beside the Ni nanoparticles, as can be observed in figures 7.5(b) and (c) where negligible coking occurred on the surface of the (La, Sr)(Ni, Ti)O_3 material after a coking test in 20% CH_4/H_2 at 800 °C for 4 h. In conclusion, the exsolved catalyst displayed remarkable coking resistance and hence experienced only minor coking due to the strong particle–support interaction, while the impregnated catalyst suffered from severe carbon formation by virtue of the weak interaction between the catalyst and the support.

Another characteristic of exsolution materials is their superior sulfur tolerance, which results from the strong interactions between the exsolved particles and their support. When compared with a commercial Ni/Al_2O_3–SiO_2 catalyst, both the exsolved Ni/$La_{0.8}Ce_{0.1}Ni_{0.4}Ti_{0.6}O_3$ and Ni–Fe/$La_{0.5}Sr_{0.4}Ni_{0.1}Fe_{0.1}Ti_{0.8}O_3$ materials showed better tolerance to sulfur poisoning after sequential exposure to 50 ppm of SO_2 gas for 15, 30, and 60 min; the Ni–Fe decorated system demonstrated no change in catalytic activity for CO oxidation after the sulfur poisoning experiment [13]. High H_2S tolerance and sustained stability was also demonstrated for *in situ* exsolved NiCo alloy nanoparticles anchored on a Ruddlesden–Popper support when up to 100 ppm of H_2S was introduced into the CO/CO_2 fuel mixture used for CO_2 electrolysis [14].

In summary, exsolution materials, due to their unique interactions between the exsolved nanoparticles and the support, offer exceptional catalytic activity and stability, new functionalities, and more to be explored. From the point of view of SOC performance and stability, exsolution does seem to offer a number of advantages over infiltration, such as higher resistance to particle coalescence, carbon deposition, and sulfur attacking. From the perspective of manufacture, infiltration can be troublesome, as the process needs to be repeated to meet the required mass loading and it offers less control over particle size and distribution, while exsolution can rather simply take advantage of the SOC operating conditions to complete the reduction necessary to generate nanoparticles and can manipulate the particle size and population by tailoring the reduction conditions. Yet, there are other considerations and limitations related to materials design and SOC fabrication and application. There are apparently more options for catalysts in the infiltration method so long as the catalyst is chemically and thermochemically compatible with the support, whereas the options for the exsolution approach are confined to catalysts that can be dissolved into the support lattice structure and easily reduced and grown on the surface upon reduction. In addition, the bulk properties of the support, such as its conductivity, can be dramatically altered by the dopant, doping level, and perovskite support used in the exsolution strategy, while in the case of the infiltration strategy, the bulk properties are often dictated by the scaffold, which is not significantly impacted by the impregnates and their concentrations. Given the differences between infiltration and exsolution, there are possible synergies between these two approaches that may combine their respective advantages [15]. However, this has not been investigated extensively.

7.3 Tailoring material morphologies and functionalities through exsolution

It is essential to understand the mechanisms of the exsolution phenomenon, such as the fundamentals and driving forces of such an intriguing process. Such an understanding provides the insights and general guidance required to tailor materials' properties, morphologies, and functionalities for practical applications through exsolution. In the following sections, the exsolution thermodynamics and driving forces that have been generally accepted will be discussed first, followed by an introduction to the factors that have been proven effective for tuning the exsolution of nanoparticles.

7.3.1 The thermodynamics and driving forces of exsolution

From a thermodynamic point of view, the exsolution of a metal (M) is governed by its Gibbs free energy change (ΔG) from M^{n+} to M^0 during reduction, which acts as the main driving force for exsolution (see equation 7.3). Figure 7.6 shows the $\Delta G^{900}_{\text{red.}}$ of reduction of oxides to either metals or selected oxides at 900 °C in H_2 [2]. According to the figure, certain metals that are known to have the exsolution capability exhibit $\Delta G^{900}_{\text{red.}} < 0$, e.g. Fe, Co, Ni, Cu, and noble metals. This means that from a theoretical perspective, the reduction of these metal cations is thermodynamically favored and spontaneous at 900 °C. Conversely, in the cases of Ti, Mn, and Zn, positive $\Delta G^{900}_{\text{red.}}$ is seen, indicating unfavorable reduction or exsolution.

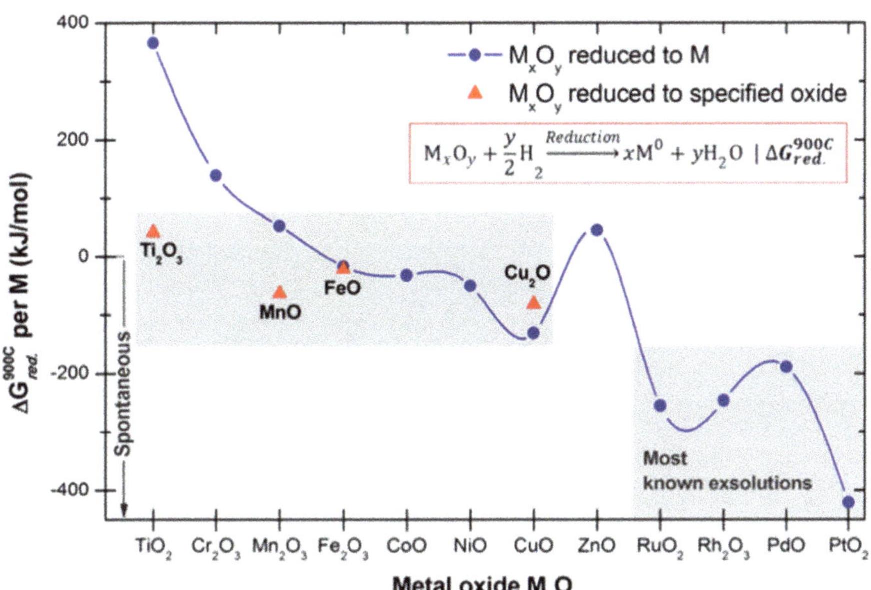

Figure 7.6. The Gibbs free energy of the reduction of oxides to either metals or selected oxides at 900 °C in H_2. Reproduced with permission from [2]. © University of St. Andrews and Dr Dragos Neagu.

$$M_xO_y + \frac{y}{2}H_2 \xrightarrow{\text{Reduction}} xM^0 + yH_2O \quad |\Delta G^T_{\text{red.}} \tag{7.3}$$

The Gibbs free energy of reduction $\Delta G^T_{\text{red.}}$ of a metal only serves as a primitive guide for predicting the phenomenon of exsolution and comparing the exsolution capabilities of different metals. In realistic exsolution systems, the ability of a metal to exsolve is complicated and can be influenced by different factors. As mentioned earlier, it has been accepted that exsolution comprises several processes, namely cation diffusion, reduction, nucleation, and growth, as indicated in figure 7.7 [16]. Figure 7.8 describes the Gibbs free energy change during nucleation and growth. According to the classical isothermal nucleation theory, nucleation is related to the ΔG of the nucleus at the critical particle size (r^*), i.e. the critical free energy (ΔG^*), which can be expressed as follows:

$$\Delta G^* = \Delta G^*_{\text{bulk}} + \Delta G^*_{\text{interface}} + \Delta G^*_{\text{surface}}, \tag{7.4}$$

where the terms on the right-hand side of the equation denote the ΔG^* of bulk reduction, that of the newly formed metal/perovskite interface, and that of the metal surface, respectively. These values are system- and temperature-dependent, and the latter two terms are highly relevant to the surface morphology. For example, in a Ni exsolution system $La_{0.43}Sr_{0.37}Ni_{0.06}Ti_{0.94}O_{3-\delta}$ (LSNT), Ni^{2+} cations diffuse from the bulk to the surface of the material and are subsequently reduced to Ni^0 metal. The reduced nuclei assemble into small nanoparticles, which grow in size with reduction time and temperature [16]. From the exsolving material's perspective, the extent of its exsolution is determined by the surface/bulk mechanical energy (the strain effect)

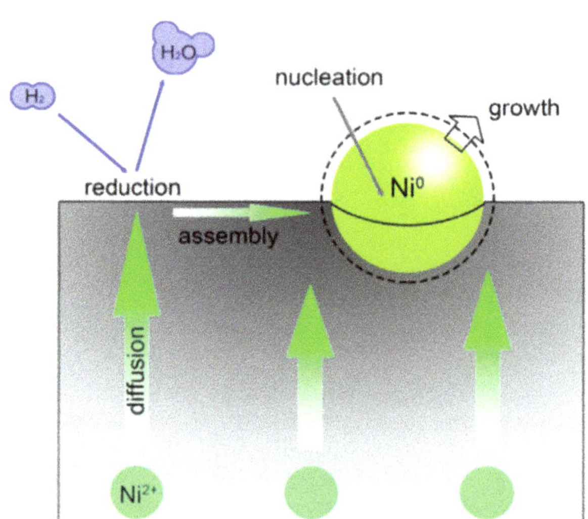

Figure 7.7. A scheme of the exsolution process: Ni cations in the perovskite lattice first diffuse to the surface and are subsequently reduced to Ni^0 metal; the particle size increases as a function of time and temperature. Reprinted from [16], Copyright (2016), with permission from Elsevier.

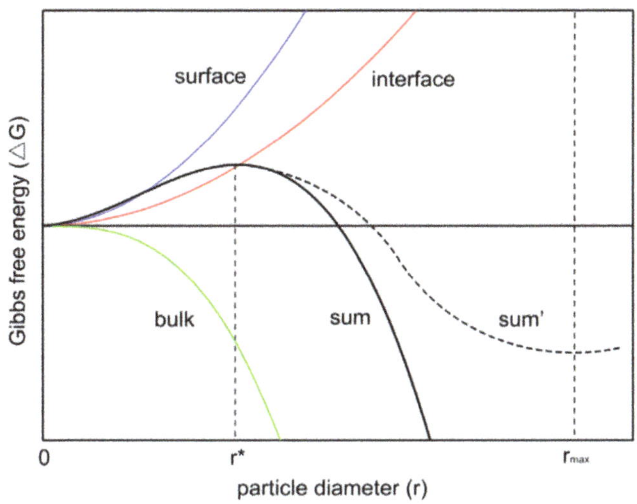

Figure 7.8. Gibbs free energy change during nucleation and growth. Reprinted from [16], Copyright (2016), with permission from Elsevier.

and the metal cation supply and cation diffusion rate, which can be influenced by many intrinsic and extrinsic factors (see below for a detailed discussion).

In principle, the exsolution of nanoparticles is driven by the gradient of the exsolving species across the material bulk to the surface, which can be controlled and promoted in various ways, for instance by material composition, defect chemistry (e.g. non-stoichiometry and lattice defects), surface features, oxygen partial pressure (P_{O2}), temperature/time, electrochemical bias, strain, light, etc. Neagu *et al* introduced *A*-site deficiency (α = 0.2) into a $La_{0.52}Sr_{0.28}Ni_{0.06}Ti_{0.94}O_3$ titanate and observed a significant promotion of nanoparticle exsolution compared to its *A*-site stoichiometric counterpart [1]. Han *et al* grew $La_{0.2}Sr_{0.7}Ni_{0.1}Ti_{0.9}O_{3-\delta}$ thin films on various lattice-mismatched substrates, which induced different degrees of lattice misfit strain [17]. Compressive-strained films showed a larger number of exsolved particles than tensile-strained films, eventually reaching more than 1100 particles/ μm^2 with a particle size as small as ~5 nm via strain control. Myung *et al* applied an electrolysis voltage (~2 V) to a SOC electrode made from $La_{0.43}Ca_{0.37}Ni_{0.06}Ti_{0.94}O_{3-\delta}$ and found that such an electrical reduction remarkably facilitated Ni exsolution (~a few minutes) compared to chemical reduction (~tens of hours) [18].

One fascinating aspect of exsolution is that the final exsolved structure can be controlled and adjusted by sophisticated material design and the subsequent treatment. In the following sections, the typical controlling factors used to drive exsolution and determine materials' morphologies will be introduced.

7.3.2 Controlling factors for the exsolution process

Since the exsolution strategy facilitates a broad range of functional applications in the field of catalysis and electrocatalysis, it is important to identify the respective factors that control and determine the properties of exsolved nanoparticles. As mentioned earlier, factors such as stoichiometry, the surface characteristics of the parent material, oxygen partial pressure, electrochemical bias, etc. have been found very effective in influencing exsolution behavior. In this section, we discuss the mechanisms responsible for the methods commonly utilized to tune the exsolution process, and typical examples are given where necessary.

7.3.2.1 Stoichiometry

The stoichiometry of a perovskite plays a crucial role as an intrinsic factor in controlling and tailoring the exsolution. The general stoichiometry of perovskite oxides is ABO_3; this conventionally possesses a large cation at the *A*-site and a smaller cation at the *B*-site. By adjusting a stoichiometric perovskite to be a non-stoichiometric one, its exsolution behavior can be vastly altered. In the ABO_3 framework, non-stoichiometry or defects can take the form of *A*-site deficiency/excess, *B*-site deficiency/excess, or oxygen deficiency/excess. Among these, *A*-site deficiency has been most frequently used to promote exsolution and has been regarded highly effective.

Unlike the exsolution that occurs in near-stoichiometric perovskites, the pioneering study by Neagu *et al* demonstrated that *A*-site deficiency ($A/B = (1-x)/1$) plays a crucial role in boosting *B*-site exsolution, thus facilitating the production of large quantities of embedded nanoparticles [1]. Figure 7.9 shows that significant numbers of Ni nanoparticles were exsolved from an *A*-site deficient $La_{0.52}Sr_{0.28}Ni_{0.06}Ti_{0.94}O_3$ backbone after reduction, while no exsolution was observed from *A*-site stoichiometric $La_{0.3}Sr_{0.7}Ni_{0.06}Ti_{0.94}O_{3.09}$.

One possible mechanism for the beneficial effect of *A*-site deficiency is that upon creating *A*-site deficiency in a perovskite ($A_{1-x}BO_3$), a large number of oxygen vacancies are generated. To retain the balance of charges and equilibrium, the perovskite lattice tends to release the local 'excess' of *B*-site cations. Upon reduction, the excess *B*-site cations are driven across the perovskite lattice by its gradient and diffuse to the surface. This process is also accelerated by the oxygen partial pressure

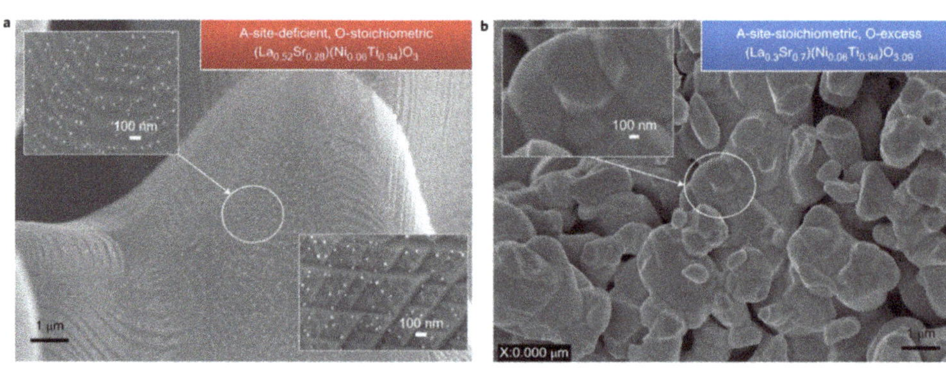

Figure 7.9. SEM micrographs showing (a) exsolution from initially *A*-site deficient $La_{0.52}Sr_{0.28}Ni_{0.06}Ti_{0.94}O_3$ and (b) *A*-site stoichiometric $La_{0.3}Sr_{0.7}Ni_{0.06}Ti_{0.94}O_{3.09}$ after reduction at 930 °C for 20 h in 5% H_2/Ar [2]. Reproduced with permission from [2]. © University of St. Andrews and Dr Dragos Neagu.

gradient. The diffused BO_x is reduced to metallic B and grows as exsolved nanoparticles. Therefore, the stoichiometry of the A-site deficient perovskite after exsolution approaches $ABO_{3-\delta'}$, which is more thermodynamically stable (see equation (7.1)). Since the perovskite is already A-site deficient, the formation of AO (oxide) upon exsolution is unlikely to occur.

To date, introducing A-site deficiency into the material structure has been extensively employed for various perovskite oxide systems. For example, the A-site deficient transition-metal-doped titanates (e.g. $(La, Sr)_{1-\alpha}Ti_{1-x}M_xO_{3-\delta}$, M = Fe, Co, Ni, Cu, Mn, etc.) clearly benefit in terms of the degree of exsolution, compared to the A-site stoichiometric design [19–23].

7.3.2.2 Surfaces

Solid materials interact with their surroundings through surfaces. The physical and chemical compositions of these surfaces determine the nature of the interactions. Therefore, surfaces influence many crucial properties of the material. *In situ* exsolution is practically a surface phenomenon; the first layers of atoms at the top surface interact with the reducing media, providing a driving force for cation diffusion, nucleation, and growth. The attributes of exsolved nanoparticles are largely influenced by the surface properties; therefore, this section discusses the impact of different surface characteristics on nanoparticles' exsolution behavior.

The surfaces of a material right after its fabrication (e.g. by calcination), without further treatments, are regarded as native surfaces. Reports suggest that the stoichiometry of a perovskite's native surface is not necessarily identical to that of the bulk. Surface A-site enrichment in perovskites has been reported on numerous occasions for various systems [24–27] and has also been predicted by calculations [28]. Neagu *et al* studied the surfaces of a $La_{0.52}Sr_{0.28}Ni_{0.06}Ti_{0.94}O_3$ sample by x-ray photoelectron spectroscopy (XPS), as shown in figure 7.10 [2]. Instead of the expected $A_{0.8}BO_3$ stoichiometry, the native surface displayed considerable A-site

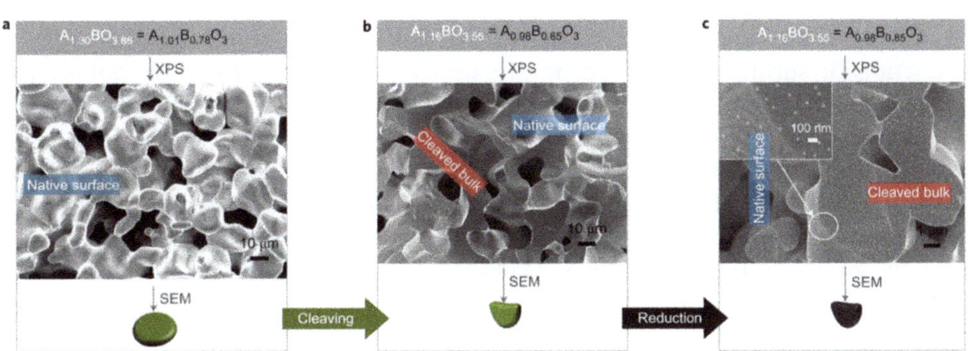

Figure 7.10. Microstructure (obtained via scanning electron microscopy (SEM)) and the corresponding surface stoichiometry (obtained via XPS) of a $La_{0.52}Sr_{0.28}Ni_{0.06}Ti_{0.94}O_3$ porous sample: as prepared (~1400 °C, air), that is, the native surface (a); after cleaving, which exposes the bulk-type surfaces with nominal stoichiometry (b); after reducing the cleaved sample in the XPS in 5%H_2/Ar (30 min isothermal steps at 600 °C, 700 °C, 800 °C, and 900 °C) (c) [2]. Reproduced with permission from [2]. © University of St. Andrews and Dr Dragos Neagu.

and O-site excess, $A_{1.30}BO_{3.86}$ (figure 7.10(a)). Szot and co-workers showed that the A-site- and O-site-rich surface layer is usually, as the stoichiometry implies, an A-site- and O-site-excess perovskite variant inherently formed during annealing in oxidizing conditions at high temperatures [29, 30]. Therefore, in A-site deficient perovskites, the native surface layer acts to locally suppress B-site exsolution, as it belongs to the excess perovskite non-stoichiometric class, which does not naturally promote B-site exsolution. Due to the unfavorable conditions for exsolution, the resulting nanoparticle precipitates tend to be very large in dimensions and low in population.

In comparison to native surfaces, cleaved surfaces (i.e. fractured surfaces) display the nominal stoichiometry of the material. During cleaving, most of the fracture occurs through the grains and thus exposes several 'bulk surfaces' of nominal $A_{0.8}BO_3$. As shown in figure 7.10(b), where the averaged compositions of the cleaved and native surfaces are measured, the surface A-site excess decreases from $A_{1.3}BO_{3.86}$ to $A_{1.16}BO_{3.55}$. The importance of studying a cleaved surface is that it represents the exsolution conditions as they were designed. When reduced under the same conditions, the cleaved surfaces are far superior in terms of producing smaller and more uniformly distributed nanoparticles than the native surfaces. As presented in figure 7.10(c), the nanoparticles exsolved on the cleaved surfaces are more abundant and much better distributed than those of the native surfaces. Thus, by investigating cleaved surfaces, one may obtain a more representative perspective of the capability of different non-stoichiometries to produce exsolution [17].

In addition to native and cleaved surfaces, exsolution from a material's surface (which is typically polycrystalline) can differ drastically from area to area of the same material. Such a phenomenon is not rare in exsolution studies, as surfaces separated by grain boundaries also have varying crystallographic orientations. Since the nanoparticles grow epitaxially, surface orientation plays an essential role in determining the crystal structure and the exsolution behavior of the nanoparticles. It has been reported that the exsolved nanoparticle size and population vary dramatically between the (111), (110), and (100) surface orientations, as shown in figure 7.11 [31]. Such a phenomenon could be linked to the differences in interfacial energies as a function of surface orientation; it can be used to control the morphology and nucleation behavior of exsolved nanoparticles. It should also be pointed out that in thin-film exsolution studies, the orientation might be controlled by the strain between the film and the support, thus impacting the exsolution process. Nevertheless, the electrode materials used in SOCs are generally random in orientation, and it is common for the surface orientations to deviate (or tilt) far from the major lattice planes (e.g. (111), (100), (200), etc). Therefore, it is very challenging to use the surface orientations to control nanoparticle exsolution in many practical applications.

Moreover, although less reported, the exsolution behavior of particulate materials varies significantly with their grain size. The grain size reflects the surface area of a particle and correlates closely with the particle's surface energy. In many catalytic applications, e.g. SOC electrodes, catalysts are utilized in particle form, and smaller particles are generally favored since they possess a larger specific surface area, hence

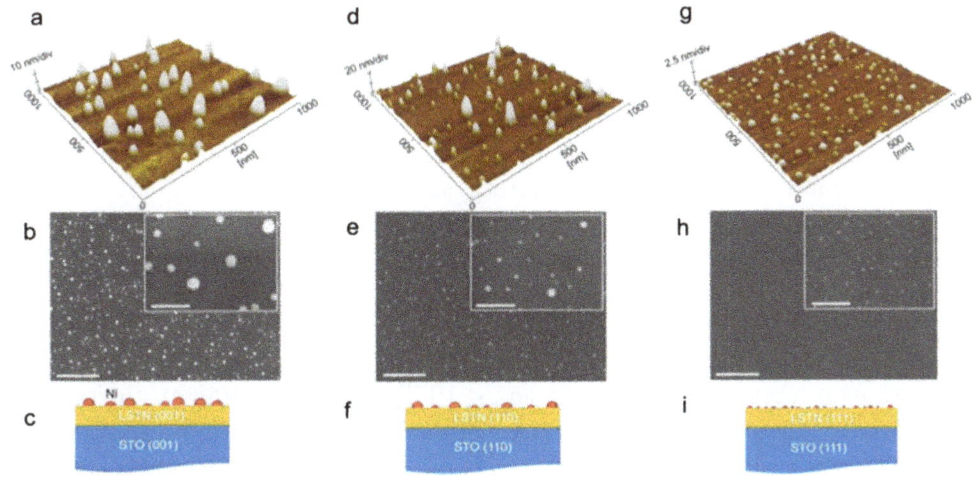

Figure 7.11. Exsolution from $SrTiO_3$ (STO)-supported $La_{0.2}Sr_{0.7}Ti_{0.9}Ni_{0.1}O_{3-\delta}$ films with surface orientations of (a), (b), (c) (001); (d), (e), (f) (110); and (g), (h), (i) (111). Reprinted with permission from [31]. Copyright (2019) American Chemical Society.

better activity. However, in exsolution systems, a 'critical surface area' or 'critical grain size' may exist. As the name suggests, the critical surface area refers to the surface area below which exsolution becomes very difficult or no longer occurs. For example, when the SOC electrode particle size is reduced to ~1 μm, the number of exsolved NPs is very low compared to the number exsolved by electrode particles of ~4 μm. Therefore, the adjustment of particle size could be an effective way of tailoring exsolution. However, there might be trade-offs in terms of SOC performance.

Apart from a material's intrinsic surfaces as discussed above, surfaces that are externally processed or treated can show notably different exsolution properties to those of intrinsic surfaces (either improved or worsened). For example, in some scenarios, the as-prepared material needs to be polished to obtain a large flat area with the desired surface roughness. During polishing, the *A*-site- and O-site-rich surfaces are removed, thus exposing the bulk surfaces. The polished surface is able to generate tremendous amounts of nanoparticles over a large area and can therefore be conveniently used to study the influence of stoichiometry and doping on exsolution. However, it should also be noted that polished surfaces cannot fully represent the intrinsic properties of the materials studied. During polishing, the material's surfaces are highly strained, which may cause the accumulation of certain lattice defects, making the surface energy high. Such modifications would promote the exsolution process in a similar way to the lattice tensile strain effect reported by researchers [17].

In a word, since the surface characteristics affect the exsolution behavior dramatically, it is of paramount importance to understand and study the interrelations between the surface features and the exsolution process. Of all the kinds of surface discussed above, the fractured/cleaved and polished ones should be the best options for studying the aptitude of different non-stoichiometries to produce exsolution, since they represent identical or nearly identical stoichiometries to that of the material as it was designed. Moreover, using the surface orientation to tune exsolution may be most suitable for thin-film applications, as the catalysts are likely polycrystalline in most catalytic scenarios, thus posing a significant challenge to the qualitative control of their surface lattice structures. In contrast, adjusting the particle size of a catalyst might be the most practical way to tune the exsolution and nanoparticles' traits, especially for SOC applications.

7.3.2.3 Oxygen partial pressure

The exsolution of metal nanoparticles mainly occurs in a reducing atmosphere, i.e. at a low oxygen partial pressure ($p_{O_2} < 10^{-16}$ atm, depending on the system). Therefore, p_{O_2} is probably one of the most important extrinsic factors that critically governs the exsolution process. It has also been widely used to promote and tune the scale of exsolved nanoparticles. The biggest advantage of utilizing p_{O_2} is that it can be conveniently changed to obtain a wide spectrum of either reducing or oxidizing atmospheres, thus providing a handy way to study and control exsolution. For example, p_{O_2} can be changed by mixing H_2 and an inert gas (e.g. N_2, Ar) at varying ratios or by changing the atmospheric temperature or humidity of the gas, etc.

A sufficiently reductive atmosphere creates a gradient of p_{O_2} from the bulk to the surface of the studied material. Such a gradient of p_{O_2} acts as a driving force that induces cation diffusion and subsequent exsolution, and this process starts by stripping away the oxygen atoms from the very top surface layers of the material. The reducible *B*-site metal cation is then reduced to keep the system charge neutral. Exsolution continues to occur as long as there is a sufficiently high p_{O_2} gradient, and the above process naturally slows down over time due to the limited cation supply.

By carefully adjusting the p_{O_2} of the atmosphere, the composition and properties of exsolved nanoparticles can be tuned. For example, using *A*-site deficient $Sr_{0.95}Ti_{0.3}Fe_{0.63}Ni_{0.07}O_{3-\delta}$, Barnett *et al* decreased the p_{O_2} of the atmosphere from 4.0×10^{-20} to 2.6×10^{-20} atm by switching the gas composition from 10% H_2/3% H_2O/87% Ar to 30% H_2/3% H_2O/67% Ar at 850 °C, and observed that the composition of exsolved alloy nanoparticles was altered from $Fe_{0.3}Ni_{0.7}$ to $Fe_{0.5}Ni_{0.5}$. It was predicted that the Fe content should decrease as the p_{O_2} increases [32]. In addition, Yue *et al* successfully induced the co-exsolution of CeO_2 and Ni nanoparticles by treating $La_{0.8}Ce_{0.1}Ti_{0.6}Ni_{0.4}O_3$ material in air (oxidative, ~1100 °C–1300 °C) and then in 5% H_2/95% Ar (reductive, 900 °C) [33]. This suggests an intriguing use of vastly different p_{O_2} environments to trigger the consecutive exsolution of two functional phases. Moreover, by repetitively switching between reducing/oxidative atmospheres, i.e. applying redox cycling, exsolution could be significantly enhanced. For example, after a $Sr_2Fe_{1.4}Ru_{0.1}Mo_{0.5}O_{6-\delta}$ perovskite was exposed to four sequential redox cycles, the population of exsolved RuFe alloy nanoparticles reached ~21 000 particles μm^{-2}, which was ~3.6 times higher than that of a sample after the first round of reduction treatment [34].

7.3.2.4 Potential bias

The exsolution of finely distributed nanoparticles generally requires a lengthy reduction process (e.g. ~20 h) in either a pure or diluted H_2 atmosphere. The sluggish cation diffusion and exsolution kinetics make the entire process time-consuming, thus chemically induced exsolution is impractical or expensive in many applications. In 2016, a striking electrochemically induced exsolution phenomenon was reported by Irvine's group [18]. In contrast to lengthy chemical reduction, electrochemical reduction allows the exsolution of fine nanoparticles to take place within a couple of minutes. This technique is particularly useful in SOC applications, as the potential bias (i.e. the high electrolysis voltage) can conveniently be applied to the fuel electrode, 'instantly' boosting the performance of the cell due to accelerated exsolution. At 900 °C, the population of the Ni nanoparticles from $La_{0.43}Ca_{0.37}Ni_{0.06}Ti_{0.94}O_{3-\delta}$ induced by potential bias (2 V, 150 s) is significantly higher than that obtained via chemical reduction (5% H_2, 20 h), i.e. 370 versus 90 particles μm^{-2}. The details of the cell performance improvement are discussed in section 7.4.

The slow kinetics of H_2-induced exsolution is probably related to the much lower p_{O_2} experienced by the sample (typically ~10^{-19} atm), compared with the p_{O_2} obtained through an electrolysis voltage (~10^{-35} atm at 2 V). When the voltage is applied, the sudden flux of electrons flowing through the perovskite lattice greatly facilitates particle nucleation, leading to the formation of smaller yet significantly more nanoparticles.

Since the report of electrochemical reduction and its beneficial effects, its application in exsolution systems has been gaining increasing attention. Myung *et al* employed an electrochemical reduction at 2.3 V to induce the exsolution of Cu nanoparticles from a $La_{0.43}Sr_{0.37}Cu_{0.12}Ti_{0.88}O_{3-\delta}$ (LSCuT) titanate fuel electrode [35]. Compared to its Cu-infiltrated $La_{0.4}Sr_{0.4}TiO_{3-\delta}$ (LST) counterpart, the electrically exsolved Cu–LSCuT displayed much better thermal stability and finer nanoparticle distribution. A cell that used exsolved LSCuT as its fuel electrode exhibited an excellent maximum power density (MPD) of 1.38 W cm^{-2} at 900 °C in 3% H_2O/97% H_2, and superior operational stability was demonstrated at 1.5 A cm^{-2} for ~150 h. Similarly, Fan *et al* utilized an electrochemical treatment to trigger Co–Fe alloy nanoparticle exsolution from $La_{0.43}Ca_{0.37}Ti_{0.8}Co_{0.1}Fe_{0.1}O_{3-\delta}$ (LCTCF) and demonstrated that the electrocatalytic activity of LCTCF could be manipulated by adjusting the externally applied voltage [36]. The electrically boosted LCTCF cell produced a much higher MPD (896 mW cm^{-2} in 3% H_2O/97% H_2, 3 V for several minutes) than that of its chemically reduced counterpart (210 mW cm^{-2} in 3% H_2O/97% H_2, 20 h) at 900 °C.

Using potential bias to trigger exsolution is effective and particularly useful in SOC applications, as it enables the convenient *in situ* tuning of the electrode's microstructure and activity to be completed rapidly. Nevertheless, it is worth mentioning that potential bias is not a universal technique and that excessive electrical reduction may lead to material decomposition and thus performance degradation. The underlying mechanism responsible for the characteristics of electrochemical reduction needs further investigation.

In summary, the exsolution phenomenon is fascinating, as it can be optimized through sophisticated material design and post-treatment. In addition to the controlling factors discussed above (i.e. stoichiometry, surface characteristics, oxygen partial pressure and potential bias), other factors, e.g. light, strain, phase transition, etc. have also been reported to influence the exsolution process to some extent [17, 37, 38]; since these factors are not common practice and are relatively less studied in the SOC field, their discussion is not included in this chapter.

7.4 Examples of the use of exsolution in solid oxide cells

Exsolution-derived nanostructures have been utilized for a wide range of applications, especially in SOCs. Compared to other applications, the use of exsolution materials as electrodes (mainly fuel electrodes) in SOCs offers a couple of unique advantages. First, the fuel atmosphere (e.g. pure or diluted H_2) and high operational temperature (600 °C–900 °C) of SOCs provide ideal conditions for the convenient reduction of the fuel electrode material, triggering the *in situ* exsolution of nanoparticles. Second, the exsolved transition metals (e.g. Ni) are frequently used as anode components in SOFC and exhibit high catalytic activity for fuel oxidation. Third, the exsolved nanocatalysts show superior resistance to agglomeration and coking due to their inherent socketed interface structure. This is particularly useful for the fuel electrodes of SOCs, as these cells often need to be operated with hydrocarbon fuels in the long term. Last but not least, as SOCs are electrochemical devices, a large electrical bias can be applied to the electrode materials to trigger abundant nanoparticle exsolution in a short space of time.

Therefore, the study and application of exsolved materials in SOCs has been increasing exponentially in recent years. In addition to the discussion above, this growth has been motivated by the fact that exsolution can be tuned by factors such as stoichiometry, the surface characteristics of the parent material, oxygen partial pressure, etc. enabling delicate control of the material's functionality to address varying application scenarios in SOCs. Specifically, introducing *A*-site deficiency into a perovskite fuel electrode promotes *B*-site cation exsolution and hence improves SOC performance; implementing external potential bias can facilitate the exsolution of large quantities of fine NPs and thus notably enhance the catalytic activity of the material. In this context, this section introduces the ways in which exsolution can yield material functionality and advancement in SOCs, focusing in particular on work published since 2013. The progressive development and optimization of exsolution systems in order to improve SOC performance is summarized and discussed. The following sections are categorized on the basis of the exsolved species, with a focus on the performance improvement resulting from exsolution.

7.4.1 Metal and metal alloy exsolution

To date, Ni has undoubtedly been the most investigated transition metal for exsolution, due to the ease with which it exsolves and its superior catalytic activity as a fuel electrode for SOCs. The extensive application of exsolved Ni systems has been advocated, as it delivers higher activity and durability than the conventional deposition process. Evidence suggests that the exsolution of large quantities of Ni nanoparticles facilitates the catalysis of reactant molecules, e.g. H_2 absorption and dissociation, thus improving Ni's catalytic activity. In SOFCs, the H_2 fuel dissociative adsorption that takes place on exsolved Ni is followed by the adsorbed atomic hydrogen spilling over onto the electrode oxide surface for electrochemical oxidation. Tsekouras *et al* examined the performance of Ni-exsolved $La_{0.4}Sr_{0.4}Ni_{0.06}Ti_{0.94}O_{3-\delta}$ as a cathode for solid oxide electrolyzer cell (SOEC) steam electrolysis [39]. The presence of electrocatalytically active Ni nanoparticles and higher oxygen vacancy ($V_{O}^{\cdot\cdot}$, introduced by exsolution) concentrations drastically lowered the activation barrier of steam electrolysis as compared to the undoped counterpart, i.e. $La_{0.4}Sr_{0.4}TiO_{3-\delta}$ (LST). Therefore, compared to an LST cell, the use of Ni exsolution led to a step change in steam electrolysis performance, in which the current density (j) increased from 0.09 A cm^{-2} to 0.19 A cm^{-2} at 1.6 V and at 900 °C in 47% H_2O/53% N_2. Sun *et al* prepared Ni-doped $(La_{0.7}Sr_{0.3})(Cr_{0.85}Ni_{0.15})O_{3-x}$ (LSCNi) for SOFC anodes, and the exsolution of abundant Ni nanoparticles remarkably increased the conductivity and catalytic activity of this material [19]. The MPD of the exsolved cell was 460 mW cm^{-2} in 5000 ppm of H_2S–H_2, compared to only 135 mW cm^{-2} for cells with stoichiometric LSCNi, which had little Ni exsolution. The performance of these exsolved materials is still poor compared to the performance of state-of-the-art Ni–YSZ, mainly due to the low activity of the fuel electrode materials themselves as well as the thick electrolyte support. Continuing efforts have been devoted to material design and cell structure optimization in order to improve SOC performance.

Later, in 2016, an innovative and exciting method for triggering exsolution was reported by Irvine *et al* [18]. In contrast to the lengthy chemical reduction (by H_2, generally taking 10–30 h), Ni nanoparticle exsolution from a $La_{0.43}Ca_{0.37}Ni_{0.06}Ti_{0.94}O_{3-\gamma}$ (LCNT) fuel electrode was achieved at 2 V in 50% H_2O/N_2 for a few seconds (see section 7.3 for a detailed mechanistic discussion). Moreover, as compared to H_2 reduction, the electrochemical reduction led to a higher degree of exsolution, a higher particle population (370 versus 90 particles μm^{-2}), and a smaller average particle size (15 versus 20 nm). The electrochemical process not only promotes Ni exsolution, but also effectively enhances the electronic and ionic conductivities of the titanate electrode. Therefore, the performance of an LCNT (~10 μm)|$Zr_{0.89}Sc_{0.1}Ce_{0.01}O_{2-\gamma}$ (~80 μm)|$Zr_{0.89}Sc_{0.1}Ce_{0.01}O_{2-\gamma}$–$(La_{0.8}Sr_{0.2})_{0.95}MnO_{3-\gamma}$ (~20 μm) cell was 'switched on' drastically. The MPD of the cell in SOFC mode reached ~2 W cm^{-2} at 900 °C in 3% H_2O/H_2, while the current density in SOEC mode soared to 2.75 A cm^{-2} at 1.3 V and 900 °C in 50% H_2O/N_2.

In addition to the predominant Ni exsolution studies, other transition metals such as Fe, Co, Cu, and their alloys are actively being investigated for SOC applications. As discussed in section 7.3, the exsolution of Fe is more difficult than that of Ni. Tsekouras *et al* revealed that the performance of Fe-exsolved $La_{0.4}Sr_{0.4}Fe_{0.06}Ti_{0.94}O_{3-\delta}$ (LSFT) is much lower than that of its Ni doped and exsolved counterpart when used as a SOEC steam electrolysis cathode, with significantly lower nanoparticle coverage [39]. However, Fe alloyed with other exsolved metals is very effective in promoting the catalytic activity of these materials. Sun *et al* studied the effect of Fe doping on the electrocatalytic activity of a $(La_{0.7}Sr_{0.3})(Cr_{0.85}Ni_{0.15-x}Fe_x)O_{3-\delta}$ (LSCNi–Fe,

$x = 0.0375$)–YSZ composite fuel electrode. The exsolved bimetallic Ni–Fe nanoparticles promoted the SOFC performance in 5000 ppm of H_2S–H_2 fuel, exhibiting a MPD of ~600 mW cm^{-2} at 800 °C (compared to ~500 mW cm^{-2} for LSCNi, $x = 0$). It has been suggested that the Fe doping not only altered the material's reduction behavior but also increased the number of oxygen vacancies [20]. Ni–Fe alloy exsolution has also been instrumental in improving SOEC performance. Liu *et al* investigated the activity of an Fe–Ni exsolved $La_{0.6}Sr_{0.4}Fe_{0.8}Ni_{0.2}O_{3-\delta}$ (LSFN)–GDC composite cathode for CO_2 electrolysis [40]. The results indicated that Fe–Ni exsolution significantly improved the cathode kinetics for CO_2 electrolysis, with a high j of 1.78 A cm^{-2} and a high faradaic efficiency (98.8%) at 1.6 V and 850 °C. The exsolved cathode material also displayed superior tolerance to carbon deposition. Doping a small amount of Fe into a Ni-containing material enhances its activity; similarly, adding a small quantity of Ni to a ferrite enhances its performance. Barnett *et al* [32] evaluated the effectiveness of doping Ni into a ferrite, $Sr_{0.95}(Ti_{0.3}Fe_{0.7-x}Ni_x)O_{3-\delta}$ (STFN, $x = 0.07$), and discovered exsolved $Ni_{0.5}Fe_{0.5}$ nanoparticles uniformly dispersed on the perovskite surface, whereas no nanoparticles were seen on $SrTi_{0.3}Fe_{0.7}O_{3-\delta}$ (STF), as seen in figure 7.12(a). The addition of Ni to STFN, i.e. the exsolution of $Ni_{0.5}Fe_{0.5}$ particles, dramatically reduced the anode polarization by as much as four times at 800 °C in 3% H_2O/H_2, resulting in an MPD of ~1 W cm^{-2} (see figure 7.12(b)).

Compared to Ni exsolution systems, Co metal can also readily exsolve due to its similar Gibbs free energy of reduction (see section 7.3.1 for details). The application of Co exsolution systems in SOCs has also been widely reported. An interesting work published by Sun *et al* found that the crystal structure of a Co-doped perovskite, $Pr_{0.5}Ba_{0.5}Co_{0.1}Mn_{0.9}O_x$ (PBCM), was altered to a layered structure, $PrBaMn_{1.8}Co_{0.2}O_5$, by reduction, which triggered the exsolution of massive quantities of finely dispersed Co nanoparticles [38]. Density functional theory (DFT)

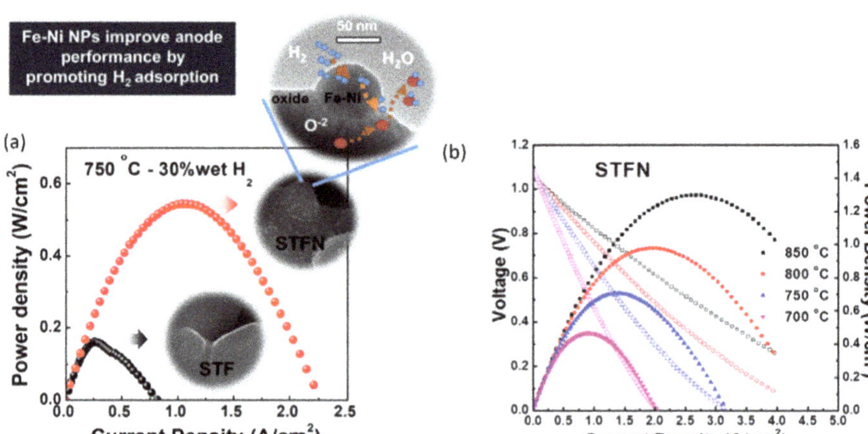

Figure 7.12. (a) A comparison of the SOFC performances of reduced STF and STFN anode cells at 750 °C in 30% H_2/67% Ar/3% H_2O and (b) the I–V curves and power output of the STFN cell at different temperatures in 3% H_2O/H_2. Reprinted from [32], Copyright (2018), with permission from Elsevier.

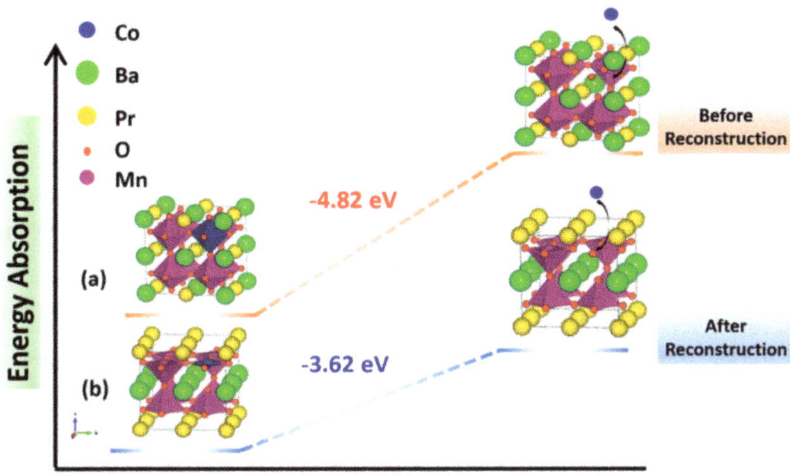

Figure 7.13. A schematic diagram of PBCM (a) before and (b) after crystal reconstruction, showing the ordering of each element, the formation of oxygen vacancies, and the exsolved Co. Reproduced with permission from [38]. Copyright (2016) American Chemical Society.

calculations suggested that the crystal reconstruction (CR, i.e. phase change or lattice rearrangement) of the PBCM double perovskite induced the creation of large numbers of ordered oxygen vacancies and thus the loss of coordinated oxygen surrounding Co cations, which served as the driving force for fast Co nanoparticle growth. Figure 7.13 shows a schematic diagram of the simulated PBCM exsolution process before and after reconstruction, which indicates the relatively kinetically unfavorable exsolution of Co from perovskite without CR (the Co formation energy is −4.82 eV without CR vs −3.62 eV with CR) due to its full coordination properties and the absence of consecutive diffusion paths. At 900 °C, a fuel cell using an exsolved PBCM–GDC composite anode showed a good MPD of 900 mW cm^{-2} with syngas; it also displayed superior CO_2 electrolysis performance with a high *j* of 2.5 A cm^{-2} at 1.5 V in 70 vol% CO_2/30 vol% CO.

In addition to the exsolution of Co alone, alloys of Co with other metals such as Fe have also been investigated by many research groups. In a direct hydrocarbon-fed SOFC study, exsolved Co–Fe nanoparticles were reported to boost the conductivity, stability, and activity of the perovskite $Pr_{0.4}Sr_{0.6}Co_{0.2}Fe_{0.7}Nb_{0.1}O_{3-\delta}$ (PSCFN) [41]. An exsolved Co–Fe–PSCFN anode cell displayed a decent power output with wet C_3H_8 (3% H_2O) fuel, achieving an MPD of 0.92 W cm^{-2} at 850 °C; the single cell was able to run stably in wet C_3H_8 at a *j* of 0.4 A cm^{-2} at 800 °C for ~600 h. Moreover, some studies found that exsolved Co–Fe alloy particles were capable of redissolving into and regenerating from the substrate lattice under certain conditions. Lai *et al* presented a perovskite SOFC anode material $La_{0.3}Sr_{0.7}Cr_{0.3}Fe_{0.6}Co_{0.1}O_{3-\delta}$ (LSCrFeCo), which allowed the exsolution (upon reduction) and dissolution (upon oxidation) of Co–Fe alloy nanoparticles to take place at 700 °C. The Co–Fe nanocatalysts improved the MPD of an LSCrFeCo–GDC composite anode by more than 75% in comparison to that of its cobalt-free

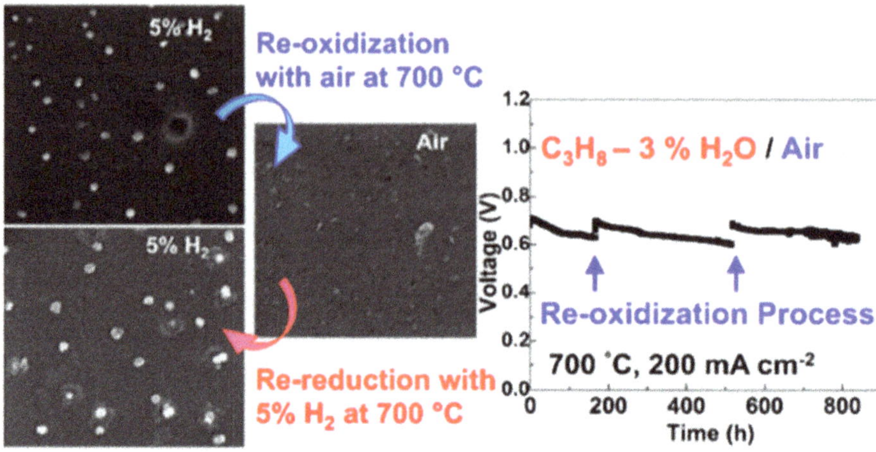

Figure 7.14. A self-regenerating LSCrFeCo anode drives off carbon deposition in C_3H_8 fuel and reactivates its catalyst during redox cycles. Reprinted with permission from [42]. Copyright (2018) American Chemical Society.

counterpart; the self-regenerating anode (i.e. dissolving and regenerating Co–Fe nanoparticles) was able drive off carbon deposition in propane fuel and reactivate its catalyst during redox cycles [42], as presented in figure 7.14.

The exsolution of alloy nanoparticles can also be achieved by combining infiltration and *in situ* exsolution techniques. An interesting study published by Kim *et al* reported the use of a smart selective exsolution process to produce alloyed Co–Fe nanoparticles through topotactic ion exchange (see figure 7.15(b)), in which infiltrated Fe guest cations were exchanged with Co host cations in $PrBaMn_{1.7}Co_{0.3}O_{5+\delta}$ (PBMCo) [15]. This process changed the host composition to $PrBaMn_{1.7}Fe_{0.3}O_{5+\delta}$ (PBMF), liberating all the Co cations from the host lattice, thus leading to an increase in the number of exsolved nanoparticles. According to the DFT calculation in figure 7.15(a), the topotactic cation exchange between Co and Fe occurred spontaneously due to the favorable incorporation energy (−0.41 eV) and exchange energy (−0.34 eV) of the infiltrated guest Fe on the host material PBMCo. The MPD of the PBMCo–Fe cell reached 1.83 W cm^{-2} in humidified H_2 (3% H_2O) at 800 °C, as displayed in figure 7.15(c).

In comparison to the Ni, Co, and Fe exsolution systems, the study of Cu exsolution has been relatively less reported, partly due to the difficulty in doping an adequate amount of Cu into the perovskite lattice and its subsequent exsolution. Cu is a good catalyst for SOC fuel electrodes, and it also improves their resistance to coking and sulfur poisoning. Ma *et al* managed to dope 15 mol.% Cu into the *B*-site of $La_{0.5}Sr_{0.5}Fe_{0.8}Cu_{0.15}Nb_{0.05}O_{3-\delta}$ (LSFCNb), and the exsolution of Cu was realized after reduction at 800 °C [43]. A symmetrical cell that used LSFCNb as the fuel and air electrodes showed good performance in H_2, syngas, and H_2S-containing H_2 fuels, with MPDs of 0.76, 0.64, and 0.71 W cm^{-2} at 800 °C, respectively. However, the LSFCNb oxide decomposed after reduction, forming a composite consisting of a K_2NiF_4-type $SrFeLaO_4$ layered perovskite, $LaFeO_3$, and Cu. It appears that the

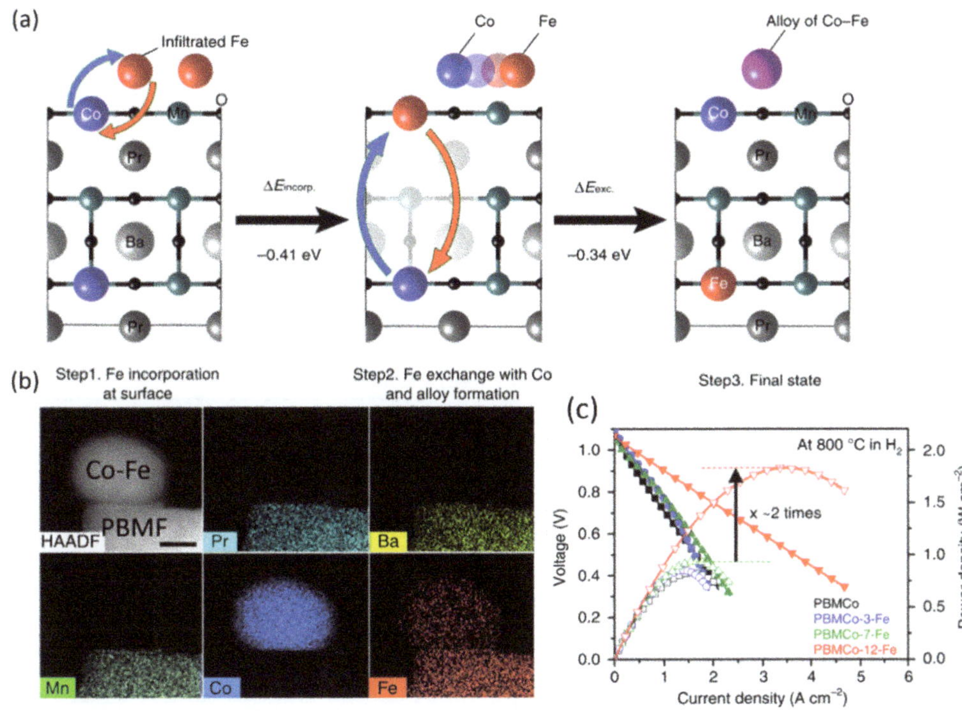

Figure 7.15. (a) Topotactic ion exchange energetics for the mechanism of particle exsolution via Fe infiltration on a PBMCo surface. (b) Scanning transmission electron microscopy–energy dispersive x-ray spectroscopy (STEM-EDS) element map of Pr, Ba, Mn, Co, and Fe; scale bar 20 nm. (c) *I–V* curves and the MPDs of PBMCo–*x*-Fe samples at 800 °C in humidified H_2 (3% H_2O); *x* stands for the amount (wt.%) of Fe infiltrated. Reproduced with permission from reference [15] Copyright © 2019, The Author(s). With permission of Springer.

parent perovskite was not robust enough to retain its structure upon reduction and Cu exsolution. A similar phenomenon was also mentioned in a report in which Ni-doped spinel oxide $CuFe_2O_4$ decomposed after reduction, forming Cu–Fe–Ni alloy nanoparticles [44]. A recent Cu exsolution study was reported by Myung *et al* [35]. In their study, 12 mol.% Cu was successfully doped into $La_{0.43}Sr_{0.37}Cu_{0.12}Ti_{0.88}O_{3-\delta}$ (LSCuT) via a solid-state synthesis method. The exsolution of Cu was achieved by electrochemical reduction (~a few seconds) at 2.3 V and at 900 °C, as shown in figures 7.16(a) and (b). The exsolved LSCuT cell exhibited an MPD of 1.38 W cm^{-2} at 900 °C in wet H_2 and excellent operational stability for ~150 h (in figures 7.16(c) and (d)). The exsolution of Cu nanoparticles did not decompose the perovskite structure of the LSCuT titanate.

In brief, metal exsolution imparts unique, additional, and enhanced properties to the support material, and depending on the metal species, the functionality of the material can be tuned precisely. Apart from the transition metals discussed above, many other metal exsolution systems are employed for SOC applications, e.g. noble metals (Ag, Ir, Ru, Pd, etc.) and ternary alloy exsolution systems, which further broadens the applicability of exsolution materials to SOCs. For detailed

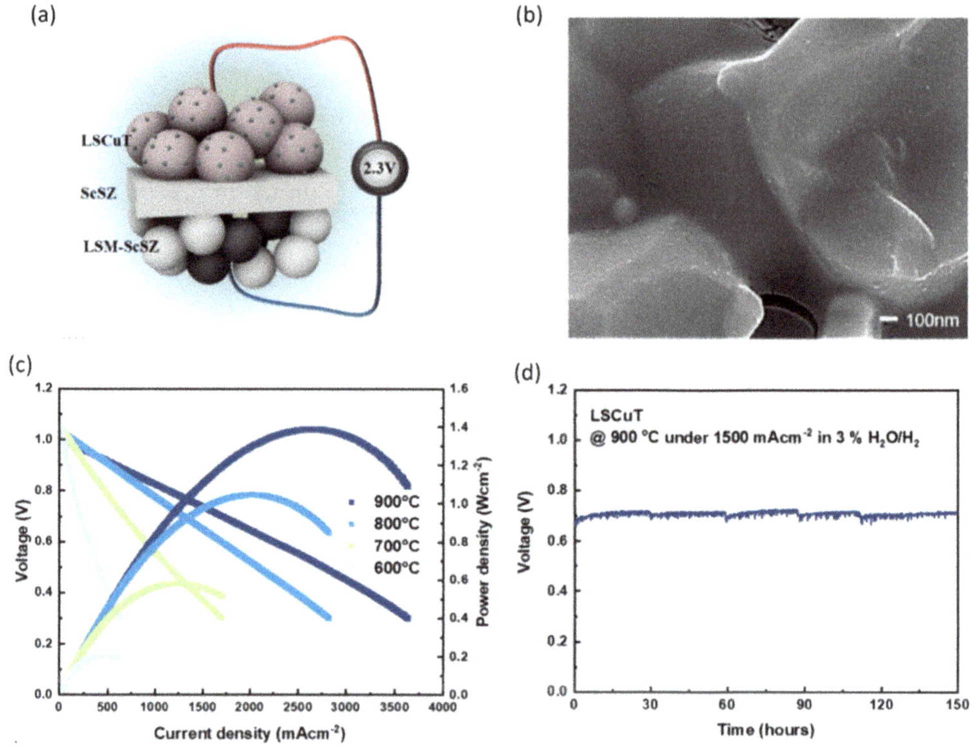

Figure 7.16. (a) A schematic of the electrochemical reduction of the LSCuT cell at 2.3 V. (b) An SEM image of the electrochemically reduced LSCuT electrode. (c) *I–V* curves and power density of the reduced LSCuT cell at different temperatures and (d) the results of a test of its stability at 900 °C in 3% H_2O/H_2. Reprinted from [35], Copyright (2020), with permission from Elsevier.

information, the reader is referred to two recent review papers published by Kousi *et al* [45] and Tang *et al* [46].

7.4.2 Metal and metal oxide exsolution

In addition to *B*-site transition-metal exsolution, oxide segregation/exsolution from the perovskite support may also occur along with the *B*-site metal exsolution. The co-exsolved oxides can have many forms and compositions, depending on the structure of the parent perovskite structure. As the segregated oxide can also influence the properties of the base material, it is equally important to discuss and explore the potential of oxide co-exsolution in energy applications.

Kim *et al* studied an *A*-site layered oxygen-deficient perovskite, $NdBaMn_2O_{5+\delta}$ (NBMO), and discovered that the exsolution of MnO occurred during a reduction process [47]. The perovskite experienced a phase transition, forming an *A*-site layered perovskite structure, as shown in figure 7.17. As an anode, the MnO-nanoparticle-decorated NBMO exhibited excellent SOFC performance, showing a MPD of ~0.6 W cm^{-2} at 800 °C in wet H_2. However, the stability of this exsolved MnO anode under SOFC operating conditions was not studied; thus it is unclear whether

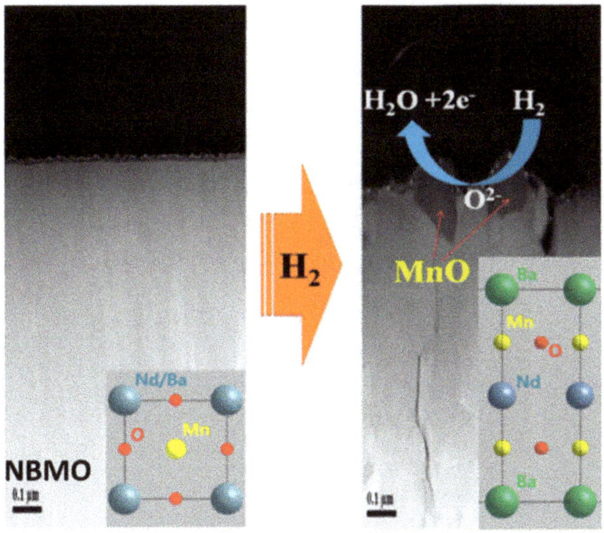

Figure 7.17. High-angle annular dark-field scanning TEM image of $Nd_{0.5}Ba_{0.5}MnO_3$ film (a) before and (b) after reduction. The precipitation of MnO after reduction is highlighted. Reprinted with permission from [47]. Copyright (2017) American Chemical Society.

this material can run stably over a long time period. Guo *et al* [48] reported an *A*-site and *B*-site co-exsolution phenomenon that occurred in a ferrite perovskite, $Ba_{0.3}Sr_{0.7}Fe_{0.9}Mn_{0.1}O_{3-\delta}$ (BSFM), in which Fe nanoparticles and SrO/BaO whiskers were exsolved after reduction at 800 °C for 5 h in 50% H_2/50%Ar gas. While the perovskite structure was retained, the exact mechanism of co-exsolution was not discussed and thus remains unclear. The exsolved BSFM anode exhibited an MPD of 538 mW cm^{-2} at 800 °C in H_2, and decent stability was demonstrated for 60 h. The exsolved SrO/BaO whiskers in the anode were speculated to function synergistically with Fe nanoparticles in improving the oxygen dissociation/dissolution kinetics.

Moreover, the co-exsolution of well-known active oxides such as CeO_2 along with transition metals can be an effective way to reinforce the functionality of the base material. Li *et al* doped 5 at.% Ru into the perovskite $Sr_{0.8}Ce_{0.2}FeO_3$ (SCF), and the Ru-doped SCF was used as the electrodes of a symmetrical SOFC (SSOFC) [49]. Surface segregation of nanoscale SrO, CeO_2, and Ru^0 was observed under fuel conditions. The symmetrical cell showed MPDs of 846 mW cm^{-2} and 310 mW cm^{-2} at 800 °C in H_2 and C_3H_8, respectively. It was suggested that the exsolution of Ru^0 facilitated the H_2 oxidation reaction and that the CeO_2 reinforced the absorption of propane molecules, while the simultaneous superficial exsolution of SrO was found to be important for coke resistance under carbonaceous fuel. Nevertheless, the segregation/exsolution of SrO–CeO_2–Ru^0 was due to the surface decomposition of the support, and the distribution of ceria was difficult to identify. In another study, the co-exsolution of clearly visible ceria cubes and Ni nanoparticles was realized from a $La_{0.8}Ce_{0.1}Ni_{0.4}Ti_{0.6}O_3$ (LCeNT) host material [33]. The exsolution of ceria cubes was achieved by annealing LCeNT in air at high temperature (1000–1300 °C),

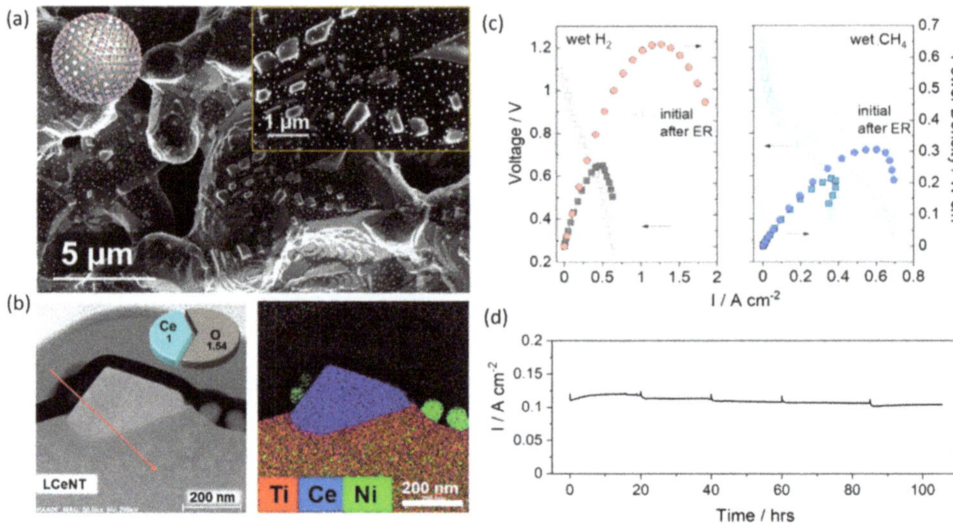

Figure 7.18. (a) An SEM image of the exsolved LCeNT with CeO_2 cubes and Ni nanoparticles. (b) STEM-EDS element map at the exsolved interface, showing the composition of the ceria, the Ni, and the substrate. (c) *I–V* curves and power densities of the exsolved LCeNT cell at 900 °C fed with wet H_2 and wet CH_4 (3% H_2O). (d) Stability test of the exsolved LCeNT cell in wet CH_4 at 850 °C and at 0.7 V. Reproduced from [33], Copyright (2021), with permission from Elsevier.

while the subsequent Ni exsolution was triggered by a reduction treatment; see figures 7.18(a) and (b). The perovskite structure of LCeNT was preserved throughout the treatment. While the exsolved Ni nanoparticles favor H–H and C–H bond breaking, CeO_2 can improve the material's coke resistance. The co-exsolved LCeNT cell showed good performance in H_2 and CH_4 fuels, with MPDs of 642 and 306 mW cm^{-2} at 900 °C, respectively (in figure 7.18(c)); superior stability was demonstrated by an extended period of operation (~105 h) in 3% H_2O/CH_4 at 850 °C and at 0.7 V, as shown in figure 7.18(d).

In a word, a remarkable improvement in the performance of SOCs due to the use of exsolution has been widely reported in the past decade, and the exsolution technique has been perceived as an effective and versatile way to reinforce the functionality of the electrode material. While keeping the structure of the base material stable is important, it is intriguing that the exsolved species and composition can be adjusted through sophisticated material design and post-treatment.

7.5 Advances and outlook

As a result of vigorous development in materials engineering and the mechanistic study of exsolution, substantial advances have been made with increasingly high performance reported, much higher than that achieved using conventional Ni–YSZ cermet electrodes in SOCs. Some efforts have been made to evaluate the stability of emerging exsolution materials in SOCs, though the duration of the stability tests needs to be significantly extended to validate the superior durability and cost-effectiveness of these materials in SOCs for commercial applications. Continuous efforts have been made to improve the depth of understanding of the mechanisms controlling the exsolution process and the particle–substrate interface, as well as to extend the concept to a wide range of structures, such as fluorite structures, spinel structures, etc.

Exsolution has predominantly been applied in oxide materials with perovskite-type structures, including single perovskites (ABO_3), double perovskites ($A_2BB'O_{6-\delta}$ and $AA'B_2O_{5+\delta}$), and Ruddlesden–Popper (RP) phases (such as $A_2BO_{4+\delta}$), as already explained in previous sections. In addition, there have been a few reports of the use of exsolution in compounds other then perovskite-derived structures in SOC applications. Kui *et al* explored $NiMn_2O_4$ mixed with Ni or Ni–YSZ, (Ni, Cu)-doped ceria ($CeO_{2-\delta}$), and $Nb_{1.33}(Ti_{0.8}M_{0.2})_{0.67}O_4$ (M = Mn, Cr) as candidate cathode materials for CO_2 electrolysis through exsolution manipulation [50]. In the mixtures of $NiMn_2O_4$–Ni or –Ni/YSZ, the spinel-type $NiMn_2O_4$ decomposed to metallic Ni and MnO_x during reduction, resulting in nanosized MnO_x anchored on a metallic Ni or a Ni–YSZ composite cathode. The resultant MnO_x–Ni interfaces were found to be critical and advantageous for promoting the chemisorption of CO_2 molecules compared to a Ni-only cathode. Similarly, in the case of Ni/Cu-doped ceria and $Nb_{1.33}(TiM)_{0.67}O_4$, favorable oxygen non-stoichiometry was introduced by doping, which was beneficial for CO_2 accommodation on the surface according to DFT calculations. The coupling of an *in situ* exsolved interface and the oxygen vacancies created by doping and reduction greatly facilitated the oxygen transfer processes and improved electrode activity for CO_2 electrolysis. These results prove that exsolution is a very useful and intriguing platform for creating functionally active interfaces in a wide range of structures, which can be exploited to guide the development of materials and material design in SOCs.

In addition to an expansion in the number of structures that accommodate the exsolution concept, we are seeing the emergence of new phenomena discovered in exsolution materials that have been applied in SOCs to enhance electrode activity, such as topotactic exsolution, redox cycling-promoted exsolution, etc (see section 7.4). These new discoveries help to tackle the issues associated with exsolution, namely, that it is generally a slow and time-consuming process and often generates insufficient nanoparticles. The topotactic exsolution reported in the $PrBaMn_{1.7}Co_{0.3}O_{5+\delta}$ system (see the explanation near figure 7.15 in section 7.4) showed that cation exchange due to the distinct segregation energies of metals in the host lattice can be utilized to speed up the production of sufficient nanoparticles for various applications. Although there are only a few examples in double-perovskite materials, this approach can potentially be extended to other perovskite-related structures, using a better understanding of the fundamentals controlling metal diffusion kinetics and their stability in the host structure. The topotactic exsolution concept, in which the guest ions (introduced by a deposition technique, e.g. impregnation) are exchanged with host ions, also demonstrates a possible synergy between the exsolution and deposition techniques that can be used to generate advanced nanomaterials; however, this perspective needs to be validated by further development.

It is important to point out that the fast development of *in situ* characterization techniques, especially those that can work at high temperatures (600 °C–900 °C), has contributed significantly to the advances made in exsolution materials; such techniques have been shown to be extremely helpful tools for elucidating the intricate processes and reaction mechanisms involved in exsolution and its applications. Real-time recording of individual nanoparticle growth at the surface of perovskite oxides became possible using next-generation environmental TEM with ultrahigh spatial and temporal resolutions [51]. An analysis of this data disclosed the atomic-scale processes that underpin the formation of socketed, highly strained interfaces that give exsolved nanoparticles their outstanding stability and reactivity. Using operando photoelectron spectroscopy at SOEC operating temperatures, Opitz and co-workers revealed the formation of a carbonate intermediate that developed on the oxide surface upon cathodic polarization; its amount increased with the concentration of oxygen vacancies in the thin-film electrode material, which suggested that vacant oxygen lattice sites were the predominant adsorption sites for CO_2, in agreement with the proposals of other researchers [34, 50, 52]. It is not our intention to give an exhaustive list, but there are many examples of the use of *in situ*/operando techniques to study exsolution materials as well as their applications in SOCs, including steam electrolysis and CO_2 electrolysis, which are instrumental in gaining insights into mechanistic understanding as well as materials design and device operation [53–55]. However, it is worth noting that the further development of these techniques, especially those that can accommodate realistic devices and operating conditions, is critical. Today, many of these *in situ*/operando measurements are performed using a simplified model cell configuration under compromised conditions due to instrumental limitations, which might limit the relevance of the results to real cell operations in SOCs.

In conclusion, exsolution provides a flexible and versatile platform for growing and tuning nanomaterials that are highly desirable for use in SOCs in order to maximize their performance and enhance their stability. Compared to the infiltration technique, exsolution processing offers great control over particle size and population as well as stronger interfacial interactions between particles and their support due to the socketed nature of the nanoparticles, which leads to their outstanding thermal stability, excellent resistance to coking, and tolerance of sulfur poisoning. Exsolution has been subjected to rapid development, and great advances have been made in materials design, property tuning, mechanistic understanding, and the exploration of functionality, taking advantage of the increasing effort being invested in *in situ*/operando characterization techniques and computational modeling development. Emerging new phenomena have been discovered, and there is a vast space for the further development of these materials (e.g. new compositions and/or advanced processing to speed up the generation of sufficient nanoparticles) and for an improved understanding of the structure–property–performance relationship, as well as the continued optimization of material applications in SOCs to provide enhanced performance and prolonged operation at reduced cost, eventually leading to accelerated commercialization.

References

[1] Neagu D, Tsekouras G, Miller D N, Menard H and Irvine J T S 2013 In situ growth of nanoparticles through control of non-stoichiometry *Nat. Chem.* **5** 916–23

[2] Neagu D 2013 Materials and microstructures for high temperature electrochemical devices through control of perovskite defect chemistry *PhD Thesis* (University of St Andrews) https://research-repository.st-andrews.ac.uk/handle/10023/3606

[3] Nishihata Y, Mizuki J, Akao T, Tanaka H, Uenishi M, Kimura M, Okamoto T and Hamada N 2002 Self-regeneration of a Pd-perovskite catalyst for automotive emissions control *Nature* **418** 164–7
[4] Tanaka H, Taniguchi M, Uenishi M, Kajita N, Tan I, Nishihata Y, Mizuki J, Narita K, Kimura M and Kaneko K 2006 Self-regenerating Rh- and Pt-based perovskite catalysts for automotive-emissions control *Angew. Chem. Int. Ed. Engl.* **45** 5998–6002
[5] Madsen B D, Kobsiriphat W, Wang Y, Marks L D and Barnett S A 2007 Nucleation of nanometer-scale electrocatalyst particles in solid oxide fuel cell anodes *J. Power Sources* **166** 64–7
[6] Madsen B D, Kobsiriphat W, Wang Y, Marks L D and Barnett S 2007 SOFC Anode performance enhancement through precipitation of nanoscale catalysts *ECS Trans.* **7** 1339–48
[7] Kobsiriphat W, Madsen B D, Wang Y, Marks L D and Barnett S A 2009 $La_{0.8}Sr_{0.2}Cr_{1-x}Ru_xO_{3-\Delta}$–$Gd_{0.1}Ce_{0.9}O_{1.95}$ solid oxide fuel cell anodes: Ru precipitation and electrochemical performance *Solid State Ion.* **180** 257–64
[8] Bierschenk D M, Potter-Nelson E, Hoel C, Liao Y, Marks L, Poeppelmeier K R and Barnett S A 2011 Pd-substituted $(La,Sr)CrO_{3-\Delta}$–$Ce_{0.9}Gd_{0.1}O_{2-\Delta}$ solid oxide fuel cell anodes exhibiting regenerative behavior *J. Power Sources* **196** 3089–94
[9] Neagu D, Oh T S, Miller D N, Menard H, Bukhari S M, Gamble S R, Gorte R J, Vohs J M and Irvine J T S 2015 Nano-socketed nickel particles with enhanced coking resistance grown *in situ* by redox exsolution *Nat. Commun.* **6** 8120–7
[10] Neagu D *et al* 2017 Demonstration of chemistry at a point through restructuring and catalytic activation at anchored nanoparticles *Nat. Commun.* **8** 1855
[11] Irvine J T S, Neagu D, Verbraeken M C, Chatzichristodoulou C, Graves C and Mogensen M B 2016 Evolution of the electrochemical interface in high-temperature fuel cells and electrolysers *Nat. Energy* **1** 15014–26
[12] Connor P A *et al* 2018 Tailoring SOFC electrode microstructures for Improved performance *Adv. Energy Mater.* **8** 1800120
[13] Papaioannou E I, Neagu D, Ramli W K W, Irvine J T S and Metcalfe I S 2018 Sulfur-tolerant, exsolved Fe–Ni alloy nanoparticles for CO oxidation *Top. Catal.* **62** 1149–56
[14] Park S, Kim Y, Noh Y, Kim T, Han H, Yoon W, Choi J, Yi S-H, Lee W-J and Kim W B 2020 A sulfur-tolerant cathode catalyst fabricated with *in situ* exsolved CoNi alloy nanoparticles anchored on a Ruddlesden–Popper support for CO_2 electrolysis *J. Mater. Chem.* A **8** 138–48
[15] Joo S, Kwon O, Kim K, Kim S, Kim H, Shin J, Jeong H Y, Sengodan S, Han J W and Kim G 2019 Cation-swapped homogeneous nanoparticles in perovskite oxides for high power density *Nat. Commun.* **10** 697
[16] Gao Y, Chen D, Saccoccio M, Lu Z and Ciucci F 2016 From material design to mechanism study: nanoscale Ni exsolution on a highly active *A*-site deficient anode material for solid oxide fuel cells *Nano Energy* **27** 499–508
[17] Han H, Park J, Nam S Y, Kim K J, Choi G M, Parkin S S P, Jang H M and Irvine J T S 2019 Lattice strain-enhanced exsolution of nanoparticles in thin films *Nat. Commun.* **10** 1471
[18] Myung J H, Neagu D, Miller D N and Irvine J T S 2016 Switching on electrocatalytic activity in solid oxide cells *Nature* **537** 528–31
[19] Sun Y, Li J, Zeng Y, Amirkhiz B S, Wang M, Behnamian Y and Luo J 2015 A-site deficient perovskite: the parent for *in situ* exsolution of highly active, regenerable nano-particles as SOFC anodes *J. Mater. Chem.* A **3** 11048–56

[20] Sun Y F, Li J H, Cui L, Hua B, Cui S H, Li J and Luo J L 2015 A-site-deficiency facilitated *in situ* growth of bimetallic Ni–Fe nano-alloys: a novel coking-tolerant fuel cell anode catalyst *Nanoscale* **7** 11173–81
[21] Ding S, Li M, Pang W, Hua B, Duan N, Zhang Y-Q, Zhang S-N, Jin Z and Luo J-L 2020 A-site deficient perovskite with nano-socketed Ni–Fe alloy particles as highly active and durable catalyst for high-temperature CO_2 electrolysis *Electrochim. Acta* **335** 135683
[22] Hu S, Zhang L, Cai L, Cao Z, Jiang Q, Yu W, Wu Y, Zhu X and Yang W 2020 Iron stabilized 1/3 A-site deficient La–Ti–O perovskite cathodes for efficient CO_2 electroreduction *J. Mater. Chem.* A **8** 21053–61
[23] Yu N, Jiang G, Liu T, Chen X, Miao M, Zhang Y and Wang Y 2020 Understanding the A-site non-stoichiometry in perovskites: promotion of exsolution of metallic nanoparticles and the hydrogen oxidation reaction in solid oxide fuel cells *Sustain. Energy Fuels* **5** 401–11
[24] Horvath G, Gerblinger J, Meixner H and Giber J 1996 Segregation driving forces in perovskite titanates *Sens. Actuators* B **32** 93–9
[25] Deak D S 2013 Strontium titanate surfaces *Mater. Sci. Technol.* **23** 127–36
[26] Bonnell D A and Garra J 2008 Scanning probe microscopy of oxide surfaces: atomic structure and properties *Rep. Prog. Phys.* **71** 044501
[27] Szot K, Speier W, Carius R, Zastrow U and Beyer W 2002 Localized metallic conductivity and self-healing during thermal reduction of $SrTiO_3$ *Phys. Rev. Lett.* **88** 075508
[28] Islam M 2002 Computer modelling of defects and transport in perovskite oxides *Solid State Ion.* **154-55** 75–85
[29] Szot K and Speier W 1999 Surfaces of reduced and oxidized $SrTiO_3$ from atomic force microscopy *Phys. Rev.* B **60** 5909–26
[30] Szot K, Pawelczyk M, Herion J, Freiburg C, Albers J, Waser R, Hulliger J, Kwapulinski J and Dec J 1996 Nature of the surface layer in ABO_3-type perovskites at elevated temperatures *Appl. Phys.* A **62** 335–43
[31] Kim K J *et al* 2019 Facet-dependent *in situ* growth of nanoparticles in epitaxial thin films: the role of interfacial energy *J. Am. Chem. Soc.* **141** 7509–17
[32] Zhu T, Troiani H E, Mogni L V, Han M and Barnett S A 2018 Ni-substituted Sr(Ti,Fe)O_3 SOFC anodes: achieving high performance via metal alloy nanoparticle exsolution *Joule* **2** 478–96
[33] He S, Li M, Hui J and Yue X 2021 In-situ construction of ceria-metal/titanate heterostructure with controllable architectures for efficient fuel electrochemical conversion *Appl. Catal.* B **298** 120588–99
[34] Lv H *et al* 2021 Promoting exsolution of RuFe alloy nanoparticles on $Sr_2Fe_{1.4}Ru_{0.1}Mo_{0.5}O_{6-\Delta}$ via repeated redox manipulations for CO_2 electrolysis *Nat. Commun.* **12** 5665
[35] Jo S, Jeong H G, Kim Y H, Neagu D and Myung J-h 2021 Stability and activity controls of Cu nanoparticles for high-performance solid oxide fuel cells *Appl. Catal.* B **285** 119828
[36] Fan W, Sun Z and Bai Y 2022 Manipulating electrocatalytic activity of perovskite oxide through electrochemical treatment *Small* **18** e2107131
[37] Chen Z *et al* 2020 Organic photochemistry-assisted nanoparticle segregation on perovskites *Cell Rep. Phys. Sci.* **1** 4618
[38] Sun Y F, Zhang Y Q, Chen J, Li J H, Zhu Y T, Zeng Y M, Amirkhiz B S, Li J, Hua B and Luo J L 2016 New opportunity for *in situ* exsolution of metallic nanoparticles on perovskite parent *Nano Lett.* **16** 5303–9

[39] Tsekouras G, Neagu D and Irvine J T S 2013 Step-change in high temperature steam electrolysis performance of perovskite oxide cathodes with exsolution of *B*-site dopants *Energy Environ. Sci.* **6** 256–66

[40] Liu S, Liu Q and Luo J-L 2016 Highly stable and efficient catalyst with *in situ* exsolved Fe–Ni alloy nanospheres socketed on an oxygen deficient perovskite for direct CO_2 electrolysis *ACS Catal.* **6** 6219–28

[41] Yang C, Li J, Lin Y, Liu J, Chen F and Liu M 2015 In situ fabrication of CoFe alloy nanoparticles structured $(Pr_{0.4}Sr_{0.6})_3(Fe_{0.85}Nb_{0.15})_2O_7$ ceramic anode for direct hydrocarbon solid oxide fuel cells *Nano Energy* **11** 704–10

[42] Lai K-Y and Manthiram A 2018 Self-regenerating Co–Fe nanoparticles on perovskite oxides as a hydrocarbon fuel oxidation catalyst in solid oxide fuel cells *Chem. Mater.* **30** 2515–25

[43] Zhou N, Yin Y-M, Chen Z, Song Y, Yin J, Zhou D and Ma Z-F 2018 A Regenerative coking and sulfur resistant composite anode with cu exsolution for intermediate temperature solid oxide fuel cells *J. Electrochem. Soc.* **165** F629–34

[44] Kang B, Matsuda J and Ishihara T 2019 Cu–Fe–Ni nano alloy particles obtained by exsolution from $Cu(Ni)Fe_2O_4$ as active anode for SOFCs *J. Mater. Chem.* A **7** 26105–15

[45] Kousi K, Tang C, Metcalfe I S and Neagu D 2021 Emergence and future of exsolved materials *Small* **17** 2006479–505

[46] Tang C, Kousi K, Neagu D and Metcalfe I S 2021 Trends and prospects of bimetallic exsolution *Chemistry* **27** 6666–75

[47] Sengodan S, Ju Y-W, Kwon O, Jun A, Jeong H Y, Ishihara T, Shin J and Kim G 2017 Self-decorated MnO nanoparticles on double perovskite solid oxide fuel cell anode by *in situ* exsolution *ACS Sustain. Chem. Eng.* **5** 9207–13

[48] Zhao J, Pu Y, Li L, Zhou W and Guo Y 2020 Efficient ferrite-based perovskite anode for solid oxide fuel cells with *A*-site and *B*-site co-exsolution *Energy Fuels* **34** 10100–8

[49] Li B, He S, Li J, Yue X, Irvine J T S, Xie D, Ni J and Ni C 2020 A Ce/Ru co-doped $SrFeO_{3-\Delta}$ perovskite for a coke-resistant anode of a symmetrical solid oxide fuel cell *ACS Catal.* **10** 14398–409

[50] Wang W, Gan L, Lemmon J P, Chen F, Irvine J T S and Xie K 2019 Enhanced carbon dioxide electrolysis at redox manipulated interfaces *Nat. Commun.* **10** 1550

[51] Neagu D *et al* 2019 In situ observation of nanoparticle exsolution from perovskite oxides: from atomic scale mechanistic insight to nanostructure tailoring *ACS Nano* **13** 12996–3005

[52] Opitz A K *et al* 2017 Surface chemistry of perovskite-type electrodes during high temperature co_2 electrolysis investigated by operando photoelectron spectroscopy *ACS Appl. Mater. Interfaces* **9** 35847–60

[53] Opitz A K *et al* 2020 Understanding electrochemical switchability of perovskite-type exsolution catalysts *Nat. Commun.* **11** 4801

[54] Yu Y, Mao B, Geller A, Chang R, Gaskell K, Liu Z and Eichhorn B W 2014 CO_2 activation and carbonate intermediates: an operando AP-XPS study of CO_2 electrolysis reactions on solid oxide electrochemical cells *Phys. Chem. Chem. Phys.* **16** 11633–9

[55] Kim K, Joo S, Huang R, Kim H J, Kim G and Han J W 2021 Mechanistic insights into the phase transition and metal ex-solution phenomena of $Pr_{0.5}Ba_{0.5}Mn_{0.85}Co_{0.15}O_{3-\Delta}$ from simple to layered perovskite under reducing conditions and enhanced catalytic activity *Energy Environ. Sci.* **14** 873–82

Chapter 8

Engineering nanostructured fuel electrodes for high-performance, durable SOCs through infiltration

Miao Yu, Bhaskar Reddy Sudireddy and Ming Chen

This chapter presents an assessment of the various types of infiltrated solid oxide cell (SOC) fuel electrode together with the details of the infiltration technique. First, we outline the fundamental principles of the SOC and the state-of-the-art fuel electrode materials and their limitations. The infiltration technique has been applied to further improve the electrochemical performance and durability of SOCs; it is an effective method for modifying the microstructure of the porous fuel electrode backbone that can result in extended electrocatalytic active sites. The infiltration procedure, different infiltration methods, and key process parameters, such as the precursor chemistry, the heat treatment temperature and the number of infiltration cycles, are briefly discussed, and a summary of the process parameters based on experimental data from previous reports is presented. The performance and/or durability of SOCs with electrocatalyst-infiltrated fuel electrodes are reviewed according to the type of fuel electrode, e.g. cermet-based and oxide-based fuel electrodes. For instance, infiltrated Ni-based cermet fuel electrodes are discussed from various perspectives, such as improving sulfur tolerance or the resistance of the electrode to carbon deposition for hydrocarbon fuel utilization, decreasing the polarization resistance (R_p) or area specific resistance (ASR), and improving the cell performance for steam, CO_2, or co-electrolysis operations. In addition, the modification of perovskite and fluorite-based fuel electrodes through the introduction of active metals and/or metal oxides is also briefly discussed. Finally, the ongoing challenges and opportunities of the infiltration method in terms of scalability and the industrialization of infiltrated SOCs are outlined in the outlook.

8.1 Introduction

Solid oxide cells are electrochemical energy conversion devices that are operated at high temperatures (600 °C–1000 °C). When operated in the fuel cell mode, chemical

doi:10.1088/978-0-7503-4064-9ch8

energy in the fuel is converted into electricity and heat, while in the electrolysis mode, electricity is provided to the cell to generate fuel gases. The relevant fuel gases include hydrogen, CO, and syngas (CO+H_2). SOCs are considered to be much more efficient than low-temperature energy conversion technologies such as proton exchange membrane cells or alkaline cells, due to the rapid reaction kinetics that benefit from their high operating temperatures [1, 2].

An SOC typically consists of two porous electrodes separated by a dense layer of electrolyte (see figure 8.1(a)). In the fuel cell mode, gaseous oxygen (or air) is supplied to the oxygen electrode and is reduced to oxide ions (O^{2-}). The oxide ions are transported through the electrolyte and react with the fuel gases from the fuel electrode, producing electrons, either H_2O or CO_2 (depending on the fuel gas type), and heat through the oxidation reaction. The electrons thus produced are externally transported back to the oxygen electrode. In the electrolysis mode, electricity is supplied to the SOC. Steam or CO_2 fed to the fuel electrode is reduced to hydrogen or CO and oxygen anions. In the opposite case, i.e. electrolysis mode, oxygen anions migrate through the electrolyte to the oxygen electrode and evolve as oxygen gas. The efficiency of an SOC is always less than 100% due to the irreversible losses caused by ionic transportation through the electrolyte, mass transport losses in the electrodes, and charge transport losses through the electrodes [1]. An SOC stack is composed of a number of SOC cells connected in series, separated by metallic interconnects that also function as separators and gas distributors for the two adjacent electrodes (see figure 8.1(b)).

In order to minimize the losses in an SOC system, several requirements for different cell components need to be fulfilled. For the electrolyte, sufficient ionic conductivity, chemical stability, mechanical strength, and electronic insulation are necessary [3]. Yttria-stabilized zirconia (YSZ), scandia-stabilized zirconia (ScSZ), and gadolinia-doped ceria (CGO) are the most common electrolyte materials. The materials used for the fuel electrode must meet the requirements of high catalytic activity, good chemical stability in humid reducing atmospheres, enough porosity to allow sufficient gas transport and triple phase boundary (TPB) sites, a thermal expansion coefficient (TEC) that matches those of the other cell components, and sufficient mechanical strength for the fuel electrode support cell configuration. TPBs

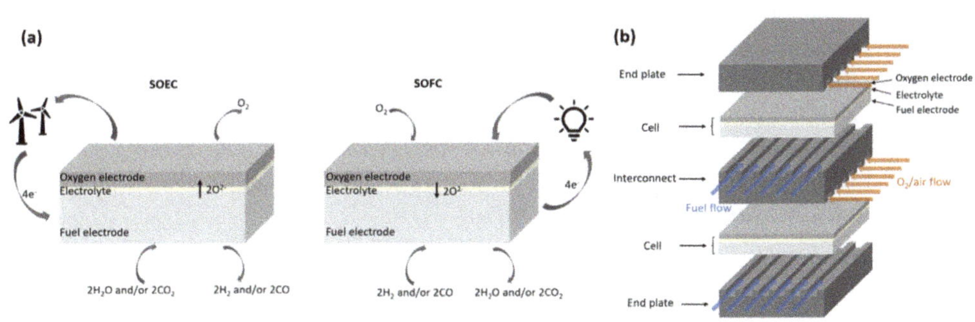

Figure 8.1. (a) The working principles of SOCs operated in SOEC mode and SOFC mode. (b) Schematic illustration of a planar-type SOC stack.

are the sites where the electron-conducting phase, the ion-conducting phase, and the gas phase meet and where electrochemical reactions involving electrons, oxygen ions, and gas molecules take place. If a hydrocarbon is used as the fuel gas, tolerance to carbon deposition and sulfur poisoning is also a requirement. Ni/YSZ cermet is the most widely used fuel electrode material due to the high catalytic activity of Ni and its flexible/abundant supply. In addition, Ni/CGO, and Ni/samaria-doped ceria (SDC) are also commonly used. For the oxygen electrode, similar requirements need to be fulfilled including high catalytic activity, high electronic conductivity, high oxygen ion conductivity, chemical and mechanical stability, sufficient porosity, and a matching TEC. A wide range of oxygen electrode materials have been studied in the past few decades, with a strong focus on doped perovskites including Sr-doped lanthanum manganite (LSM), strontium and cobalt co-doped lanthanum ferrite (LSCF), etc. Among all the cell/stack components, the requirements for the interconnect (IC) material are the most severe, since the interconnect is in direct contact with harsh environments on both sides. These requirements include high electronic conductivity, chemical stability in both oxidizing and reducing atmospheres, gas tightness, mechanical reliability, and a TEC that matches those of other cell/stack components. To promote the commercialization of SOC technology, cost-effectiveness is, of course, a common requirement for all the component materials in the SOC system.

Various SOC cell architectures or configurations have been developed. The most common cell configurations (in terms of the cell-supporting layer) include fuel- or oxygen-electrode-supported, electrolyte-supported, or metal-supported cells. With regard to the cell/stack architecture, planar, tubular, and flat-tubular cells/stacks have been developed for different applications. Planar SOFCs are usually applied in stationary power plants, transportation, and military applications [4]. Tubular and flat-tubular SOFCs have been widely used in small-scale combined heat and power (micro-CHP) systems and auxiliary power units (APUs) [5]. For SOEC applications, the planar architecture is the most popular. The international SOFC/SOEC technology providers are Bloom Energy (US), Sunfire (Germany), Ceres (UK), Elcogen (Finland), Topsoe (Denmark), Kyocera (Japan), SOLIDpower (Italy), SOFCMAN (China), etc. Further details about the the status of the development of SOC technology can be found in [6].

This chapter focuses on the fuel electrodes of SOCs, in particular, infiltrated fuel electrodes. The properties of the electrode backbone and the state-of-the-art fuel electrode materials are given in section 8.2. Section 8.3 introduces the infiltration technique, including the general procedure, the different types of infiltration method, and the processing parameters. Section 8.4 provides examples of SOC fuel electrodes infiltrated with electrocatalysts to enhance performance and durability. Finally, section 8.5 discusses the challenges and prospects of the infiltration technique related to the industrialization of SOC technology.

8.2 State-of-the-art SOC fuel electrodes

If steam or hydrogen is used as the reactant gas, the hydrogen evolution reaction (HER) and the hydrogen oxidation reaction (HOR) occur in the fuel electrode in the electrolysis mode and the fuel cell mode, respectively. The representative half-cell reactions are presented below:

$$H_2O + 2e^- \rightarrow H_2 + O^{2-} \text{ (HER)} \quad (8.1)$$

$$H_2 + O^{2-} \rightarrow H_2O + 2e^- \text{ (HOR)} \quad (8.2)$$

As mentioned in section 8.1, the requirements for the fuel electrodes include adequate electronic and ionic conductivity, high electrocatalytic activity, sufficient TPB zones, and a matching TEC. Ni/YSZ cermet has become the dominant material for the SOC fuel electrode due to its high electrochemical activity for the HER and HOR reactions at high temperatures and its good thermal and structural stability [7, 8]. In Ni/YSZ cermet, Ni provides catalytic activity for the electrochemical reactions, while YSZ offers sufficient oxygen ion conductivity as well as a TEC which matches that of the YSZ electrolyte. In addition, Ni/YSZ cermet has a high electronic conductivity of 1100–1200 S cm^{-1} at 600 °C [9]. The abundant supply of raw materials also makes Ni/YSZ the primary fuel electrode material [2]. Therefore, the fabrication and optimization of Ni/YSZ electrode microstructure has been extensively studied in order to further improve electrodes' electrochemical activity and enhance their durability [10–13]. For example, Hauch *et al* [13] optimized the microstructure of a Ni/YSZ electrode to increase its electrolytic performance at high steam partial pressures ($p(H_2O)/p(H_2)$ = 90/10) and high current densities (above −1 A cm^{-2}). By analyzing the electrode/cell performance and microstructural changes after long-term electrolytic operation, they concluded that the optimum microstructure of the Ni/YSZ electrode should have a relatively dense structure, a fine particle size distribution (PSD), and matching PSDs for the pore, Ni, and YSZ phases. Their optimized electrode had a Ni/8YSZ volume ratio of 40/60 and a porosity of 21.8% after reduction. In general, a Ni/YSZ electrode with a fine and uniform structure, a large surface area, and a sufficient porosity is highly desirable.

Although the state-of-the-art Ni/YSZ cermet electrode has many advantages, it also suffers from degradation under various test conditions. A number of different degradation phenomena have been reported, including Ni migration [14, 15], carbon deposition due to the use of hydrocarbon fuels [16–20], impurity (e.g. sulfur) poisoning [21–24], secondary phase formation [25, 26], insufficient low-temperature performance [27–30], and limited lifetime at high current densities [31–33]. Tailoring the microstructure of the Ni-based cermet electrodes and/or optimizing the operating conditions help to reduce such degradation and extend the lifetime of the electrode.

In addition to Ni/YSZ, alternative fuel electrode materials have also been extensively studied in an attempt to overcome the limitations of conventional Ni/YSZ electrodes. In recent decades, oxide-based materials (including perovskite and fluorite-structured materials) have attracted great interest as fuel electrode materials. Mixed ionic and electronic conducting (MIEC) oxides are of great interest, as their reaction sites are extended from the traditional TPBs to the entire surface, i.e. two phase boundaries. Gadolinia-doped ceria has been widely studied as a good backbone material that can replace YSZ in the fuel electrode. Marina *et al* used porous $Ce_{0.6}Gd_{0.4}O_{1.8}$ as the fuel electrode in a cell which was tested in humidified

H_2 and CH_4 atmospheres in the temperature range of 800 °C–1015 °C [34]. No carbon deposition was observed after 1000 h in methane at 1000 °C. However, $Ce_{0.6}Gd_{0.4}O_{1.8}$ has poor electrocatalytic activity for direct methane oxidation, thus it was suggested that a catalyst for fuel oxidation should be added to the backbone [34]. The addition of Ni to the CGO to create a Ni/CGO composite electrode has therefore been widely investigated for reformate fuels [35–38]. In addition to its high sulfur tolerance, Mewafy *et al* found that Ni oxidation had a minor effect on a Ni/CGO electrode compared to Ni/YSZ when tested in the electrolysis mode with the introduction of O_2 [37]. In addition to CGO, SDC has also been investigated, and a Ni/SDC composite fuel electrode showed particularly promising performance for intermediate-temperature SOCs (IT-SOCs) [39–43].

ScSZ has been widely studied due to its strong suppression of carbon deposition [44–48]. ScSZ has also a higher ionic conductivity than that of YSZ (0.3 S cm^{-1} for ScSZ and 0.14–0.16 S cm^{-1} for 8YSZ (8 mol.% yttria-stabilized zirconia) at 1000 °C) [46, 49]. Ukai *et al* evaluated the performance of SOFC cells with Ni/ScSZ and Ni/YSZ cermet fuel electrodes and found that in humidified methane (CH_4/H_2O at a ratio of 97:3) at 1273 K, a cell with a Ni/ScSZ electrode showed a lower electrode overpotential and better stability than one with conventional Ni/YSZ [44, 45].

$SrTiO_3$-based perovskite oxides are also interesting fuel electrode materials for SOCs, given their high chemical stability in reducing atmospheres and high tolerance to sulfur [50, 51]. To enhance its relatively low conductivity, $SrTiO_3$ doped with different elements has been developed by various research groups. For example, La-doped $SrTiO_3$ (LST) was investigated and proven to have high electronic conductivity and excellent stability when exposed to high-sulfur gasified fuels [51–53]. Marina *et al* also found that $La_xSr_{(1-x)}TiO_3$ ($x = 0.1$–0.4) sintered on YSZ was dimensionally and chemically stable when exposed to oxidation–reduction cycles at 1000 °C. In addition, they found that the electronic conductivity of LST strongly depends on its sintering conditions, i.e. LST sintered at 1650 °C in H_2/Ar (2:98) (dense samples, 92.2%–74.6% of theoretical density when $x = 0.1$–0.4) showed 80–360 S cm^{-1} at 1000 °C and $pO_2 = 10^{-18}$ atm [51]. Nb-doped $SrTiO_3$ (STN) was first studied by Irvine *et al* in 1996 [54]. They found that dense pellets of STN with 25% Nb ($Sr_{0.875}Ti_{0.75}Nb_{0.25}O_3$) had the highest electrical conductivity, achieving a value of 5.6 S cm^{-1} at 930 °C ($pO_2 = 10^{-18}$ atm). Following this study, other groups further modified STN and reported its promising electrocatalytic activity and tolerance to oxygen, carbon, and sulfur-containing fuel gases for various SOC applications [55, 56].

Sr- and Mn-doped lanthanum chromite ($LaCrO_3$) perovskite, $(La,Sr)(Cr,Mn)O_3$ (LSCM), has also been considered to be a potential fuel electrode material due to its high redox stability and sulfur tolerance in hydrocarbon-containing gases [57]. $La_{0.75}Sr_{0.25}Cr_{0.5}Mn_{0.5}O_3$ was first studied by Tao and Irvine in 2003, when it was tested at 850 °C–950 °C with either H_2 or CH_4 as the fuel gas [58]. It showed comparable electrochemical performance to that of a Ni/YSZ electrode at 900 °C, with a low R_p value of 0.2 Ω cm^2 in a 97:3 mixture of H_2/H_2O. In addition, no carbon deposition was observed after it was tested in a 97:3 mixture of CH_4/H_2O at 925 °C at a current density of 0.14 A cm^{-2} for 4 h. Further modifications to

LSCM have been conducted recently, focusing on its use in the SOEC application [57, 59, 60]. For example, Lim *et al* reported that their LSCM/CGO fuel electrode had better performance and sulfur tolerance than that of a Ni/YSZ electrode for steam and co-electrolysis [57].

Metal-supported SOFCs with FeCr-based fuel electrodes have also attracted interest, driven by their low cost, excellent mechanical strength, and high tolerance to thermal cycling [4, 61]. However, this type of electrode experiences the challenges of corrosion during durability testing, especially under conditions of high fuel utilization [61].

Although various innovative fuel electrode materials have been developed to achieve high catalytic activities, the performance of these materials has still not reached the same level as that of the state-of-the-art Ni/YSZ electrode. To further improve electrode performance, novel approaches to engineering the electrode microstructure have attracted significant attention. One effective approach is the infiltration method, which introduces electrocatalysts into the existing electrode backbone. The details of the infiltration method will be elaborated in section 8.3.

8.3 Introduction to infiltration

The traditional way to integrate a fuel electrode into a planar solid oxide cell is to deposit a layer on the electrolyte (in the case of electrolyte- or oxygen-electrode-supported cells) or a fuel electrode support (in the case of fuel-electrode-supported cells) using high processing temperatures in order to form a good bond and contact [8]. For example, high-temperature sintering of a Ni(O)/YSZ fuel electrode from the cermet powder at 1300 °C–1400 °C was found to be essential in order to decrease the electrode ohmic and polarization resistances [62]. The integration methods include slurry painting, screen-printing, tape casting, etc [63].

Although many materials have been developed and modified as fuel electrodes, as outlined in the previous section, they still have limitations in terms of durability, electrochemical activity, chemical stability, and fuel flexibility. In the past few years, electrode engineering approaches for fabricating nanostructured fuel electrodes have focused on maximizing the TPB of the fuel electrode (specifically, the transport of ions/electrons/gases), decoupling the functionality of the fuel electrode, and facilitating the selection of a wide range of materials with specific functionality through low-temperature integration. Among these engineering approaches, the infiltration or impregnation of the selected functional materials is the primary approach for the fabrication of nanostructured fuel electrodes. It has the advantages of a simple integration process, low cost, and a wide selection of materials and it does not require high-temperature sintering, which is needed for traditional fuel electrode integration. Several reviews summarizing this fabrication method have been published [8, 64–67]. The process is as follows (see figure 8.2): a liquid precursor solution consisting of stoichiometric amounts of metal salts, surfactant, and solvent, is introduced into the porous backbone structure using capillary forces, followed by a heat treatment to decompose the salts and form the desired phase on the backbone. The desired phase of the infiltrates can take the form of nanoparticles or a

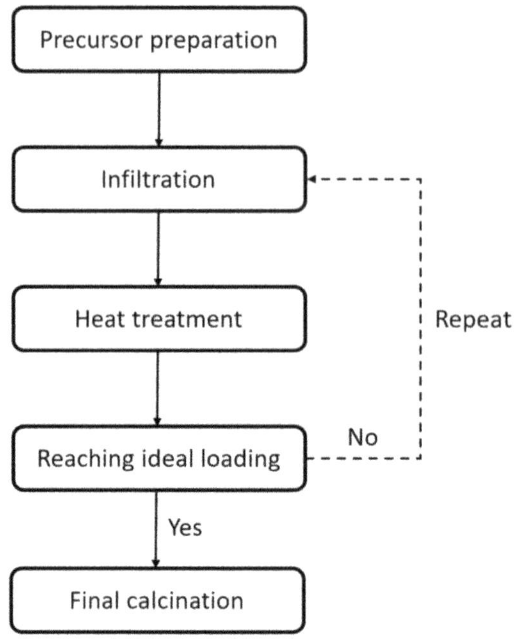

Figure 8.2. Flow diagram of the different steps in the infiltration process.

continuous thin layer on the surface of the backbone. In order to integrate the desired amount of the functional material, multiple cycles of the infiltration process may be necessary. The infiltration process can be facilitated under ambient conditions or under vacuum [68]. The combination of the infiltrate and the backbone materials usually enhances the electrocatalytic activity and/or the ionic/electronic conductivity of the fuel electrode [63].

In contrast to conventional processing techniques, the infiltration method uses relatively low heat treatment temperatures to form the electrocatalytically active nanostructured phases on the electrode backbone. This suppresses the possible reactions between the infiltrate and the backbone materials, enabling the use of a wide variety of materials that cannot usually be employed in the conventional processing method [63]. In addition, the low-temperature treatment also controls the excessive grain growth of the nanoparticles formed during multiple infiltration cycles and calcination steps and helps to minimize the TEC mismatch between the electrode materials and the electrolyte.

8.3.1 The structures of the infiltrated materials

Three types of functional infiltrate structure were presented by Kan *et al* [63], namely: distributed nanoparticles, core–shell structures, and multicomponent/multiphase structures (figure 8.3).

The morphological control of nanoparticles through backbone surface engineering, infiltrate precursor chemistry, and process parameters are the most commonly used strategies in the infiltration method. The approaches mentioned above involve

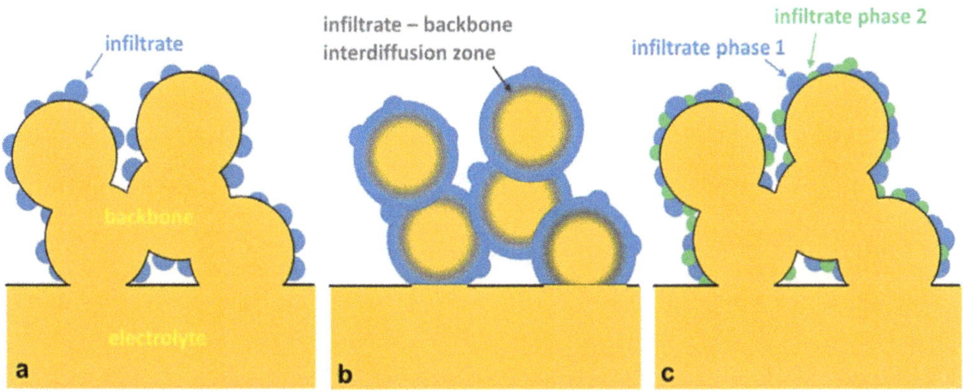

Figure 8.3. Schematic of the three functional infiltrate structures. Reproduced from [63] with permission of The Royal Society of Chemistry.

the engineering of the backbone structure, i.e. achieving high porosity, enhancing the backbone surface area, and optimizing the infiltration solutions in terms of physical properties and chemistry [63, 69]. An example of backbone engineering is that of Kungas *et al* who restructured the YSZ backbone through an etching treatment in high-concentration hydrofluoric acid (HF) to increase its surface area. They found that YSZ modified using a $La_{0.8}Sr_{0.2}FeO_3$ (LSF) infiltrate showed a much lower ASR than that of an electrode with an untreated YSZ backbone [70]. The combination of a nanostructured electrode backbone surface and nanosized catalysts obtained using a low-temperature heat treatment can significantly increase the TPB active area of the electrode.

The core–shell structure for electrodes, also known as a new component layer or an interdiffused layer at the backbone surface, is formed by the interdiffusion or the reaction between the infiltrate species and the backbone materials driven by the high temperatures at the interface. Thin films can provide a continuous catalytic network on the grain surface, thus improving the electrochemical properties and the morphological stability of the electrode [63]. However, this scheme always requires a high-temperature reactive heat treatment and highly depends on the properties of the materials used. For example, Chen *et al* infiltrated a Sr and Co solution into a $Sm_{0.2}Ce_{0.8}O_{1.9}$ (SDC) porous scaffold, followed by a high-temperature treatment at 800 °C–1100 °C. They found that when the firing temperature was higher than 1000 °C, a high electronic and ionic conductive thin film coating developed over the SDC scaffold as Sm and Ce cations from the SDC backbone diffused and incorporated into the infiltrated $SrCoO_3$ [71]. On the other hand, $LaSrCoFeO_3$ (LSCF) is widely used as an infiltrated oxygen reduction catalyst due to its high catalytic activity and MIEC property. However, if it reacts with a zirconia-based electrolyte at >700 °C, it can form $SrZrO_3$, which has poor electronic/ionic conductivity [72]. It is therefore vital to consider both the interfacial reaction and product properties when designing a core–shell electrode.

The formation of the multicomponent or multiphase compound during the infiltration process also depends on the selected precursors and infiltration process parameters (such as the heat treatment temperature). In addition to the pure phase of the infiltrated material, secondary phases with high electrochemical activity may also form during heat treatment. As an example, Samson *et al* heat treated a $Ce_{0.9}Gd_{0.1}O_{1.95}$ (CGO10) backbone with an $La_{0.6}Sr_{0.4}CoO_{3-\delta}$ (LSC40) infiltrate at a temperature range of 350 °C–900 °C and observed that multiple secondary phases occurred instead of pure LSC during the heat treatment process [73]. An optimal heat treatment temperature of 600 °C was found, at which the symmetrical cell reached the lowest polarization resistances of 0.044 Ω cm^2 and 2.3 Ω cm^2 at 600 °C and 400 °C, respectively.

8.3.2 Methods of infiltration

The integration of active materials into the backbone structure can be achieved through the infiltration process using different methods. The following section briefly describes the different methods used.

8.3.2.1 Wet infiltration

In this method, excess infiltration solution is used for the infiltration and the backbones are sometimes dipped or soaked in infiltration solution. Mass diffusion is the main mass transport mechanism in this process. The excess liquid is drained or evaporated using further heat treatments. Due to the use of excess precursor solution, it is difficult to control the volume of liquid infiltrated. The volume of the infiltration solution is often dependent on the backbone surface area. The use of excess solution and the subsequent draining/evaporation lead to a time-consuming process, and thus the infiltration process is very slow. The drying/process parameters together with the chemical stability of the solution in contact with backbone surfaces can reorganize the structure of the infiltrated material. To facilitate the good adhesion of the infiltrate material and to obtain a uniform coating/structure, it is beneficial to have chemical interactions between metal ions (from the solution) and the backbone surface.

8.3.2.2 Dry/wet incipient infiltration

This method is also known as incipient wet infiltration, as the volume of the infiltration precursor solution is empirically determined. The volume of the precursor solution corresponds to the volume of the pores in the backbone. In this process, capillary forces are the main mass transport mechanism. This process is more suitable when the metal ions from the precursor solution have weaker interaction with the surface of the backbone. The result of this is that the deposition/coating of metal ions exceeds the number of absorption sites available on the backbone surface. The mass transport is based on capillarity and weak interactions between the metal ions and the backbone surface can result in the blockage of the pores in the backbone. Furthermore, the empirically determined precursor solution volume coupled with weaker backbone interactions can result in limited or slow redistribution of the metal ion (infiltrate) solution.

8.3.2.3 Selective reaction infiltration

The reactive infiltration process involves a reaction between the dissolved metal ions and the backbone surface, which forms the desired compound after the infiltration process. Any of the infiltration methods can be used to achieve reactive infiltration. Due to the requirement for a reaction with the dissolved metal ions, the selected precursors, backbone material, temperature, and time are critical parameters which control the composition, coating quality, and structure of the infiltrated compound. As the reaction rates with the backbone surface are time and temperature controlled, the degree of infiltration is proportional to the time taken. This also makes this process slower than the other methods. Because of the reaction with the backbone surface, the wetting of the backbone surface by the precursor solution is not critical, although good wetting behavior helps to initiate a uniform reaction on the backbone surface. Furthermore, the wetting behavior changes with time, as the reaction product starts building up on the backbone surface. The reaction rate can be controlled by time, temperature, and catalytic agents.

8.3.2.4 Gas phase infiltration

This process consists of various types of vapor-phase deposition method—broadly divided into physical vapor and chemical vapor methods. This process requires that the infiltrate precursor be in gaseous phase. Diffusion processes are the critical mechanisms that control the features of the infiltrated material. By careful design of the precursor chemistry, it is possible to precisely control the stoichiometry of the intended infiltrate material. Through an appropriate selection of the gaseous precursors, temperature, and time of deposition, it is possible to react the precursors with the backbone and form a reactive coating on the backbone surface (reactive infiltration). Due to the high purity of the gaseous precursors, it is possible to achieve better control over the purity of the infiltrated compound. Furthermore, due to the nature of the deposition process, the infiltration process can be controlled much better than other infiltration processes. However, due to the requirements for gaseous precursors and a large and expensive infrastructure, this process is not commonly used for solid oxide cells.

8.2.3.5 Melt infiltration

This process is also known as melt intrusion or solvent-free infiltration (due to the melting of the precursors directly into the backbone and the nonuse of solvents, respectively). The infiltration mechanism is dependent on the capillary forces of the backbone. The appropriate melt material for the available porosity needs to be carefully optimized, as the excess melt material deposits on the surface and can thus become difficult to remove. Furthermore, to protect the melt material, it may be necessary to control the melting atmosphere. The higher temperatures involved in the melting of the precursors can result in unwanted reactions with the backbone surface, though these can be advantageous in the case of the reactive infiltration process. It is also important to consider the potential changes to the backbone surface caused by the heat applied during melting. Due to the melting requirement and possible (unwanted) reactions with the backbone, the choice of precursors is limited.

8.2.3.6 Slurry infiltration

This method is also known as the washcoat process and is widely used in the catalyst industry. The use of this process for the infiltration of electrocatalysts into SOCs is rather new. The slurry particle size and pore size/pore volume of the backbone are the critical parameters in this infiltration process. Furthermore, the viscosity of the slurry, solid loading, wetting behavior, pH, and backbone surface chemistry are other important parameters used to control the infiltration process. Following the slurry infiltration, drying and heat treatment processes are necessary to obtain the infiltration coating. Because the electrocatalyst is infiltrated in the form of particles, further decomposition processes can be avoided. In addition, unwanted reactions with the backbone can also be avoided. On the other hand, the use of reactive infiltration with this process is quite challenging; much higher temperatures are needed because the precursor consists of particles.

8.3.3 Processing parameters

As mentioned in the previous section and figure 8.2, the infiltration process usually involves the following steps: preparation of the precursor solution, integration of the precursor solution into the backbone, decomposition of the integrated solution, repetition of the infiltration cycle until the desired loading is reached, and final heat treatment. Several processing parameters need to be taken into consideration during the infiltration process.

8.3.3.1 Precursor chemistry

Precursor solutions usually consist of metal salts, a solvent, and a surfactant.

- **Salts**

  Metal nitrate salts are most commonly used for the precursor solutions. In addition, salts of chlorides, acetates, and sulfates also have been used.
- **Solvents**

  The solvent is the carrier of the precursor solution. The commonly used solvents include water and ethanol. Lou *et al* have studied how different mixtures of solvents affect the wetting angles of the precursor solutions on the electrode substrates; they found that ethanol could effectively lower the surface tension of the backbone [74].
- **Surfactants**

  The surfactant plays a critical role in the precursor; it facilitates the phase formation of the metal oxides, controls the particle sizes through suitable micelles, and improves the uniformity of the catalyst layer on the backbone [63, 74]. The type of surfactant can affect phase formation. For example, Nicholas *et al* infiltrated a $Sm_{0.5}Sr_{0.5}CoO_{3-\delta}$ (SSC) nitrate solution into a CGO backbone using different additives, namely Triton X-100 and citric acid. They found that the SSC solution with a citric acid additive produced a pure SSC phase and a more homogenous distribution during the drying process [75]. The common surfactants used for infiltration are co-polymers such as P123, citric acid, and the Triton X series. The amount of surfactant in the precursor impacts the viscosity of the solution. As found by Jiang *et al* [8], low-viscosity solutions have the advantage of forming a more uniform catalyst layer. However, their main disadvantage is that a large number of repeated infiltration cycles are needed to obtain a sufficient catalyst loading [63]. A highly viscous solution reduces the number of infiltration cycles required, but the penetration of such a solution into the porous structure could be a major problem.

- **Concentration**

  The concentration of metal ions in the precursor solution is varied to achieve specific morphological characteristics, coating thickness, and particle distribution on the backbone surface. The selection of the concentration is often decided by the solubility of the metal salt in a chosen solvent and the intended characteristics of the infiltrated particles/coating on the electrode surface. The benefit of using a highly concentrated solution is that the number of infiltration cycles required to reach the desired loading is reduced. This can also result in a morphological change of the nanoparticles on the backbone, for example, from uniformly dispersed particles to those completely covering the backbone, potentially blocking pores.

8.3.3.2 Heat treatment temperatures, cycles

The temperature of the heat treatment used after each infiltration step depends on the precursor composition and/or the backbone. The desired catalyst loading and the concentration of the precursor determine the infiltration cycles. Table 8.1 summarizes the infiltration parameters which have been applied or studied on various fuel electrodes.

8.4 Infiltrated fuel electrodes

In the following section, we review previous work on infiltrated fuel electrodes. The published reports are grouped according to the fuel electrode type, i.e. metallic or cermet-based fuel electrodes and oxide-based fuel electrodes.

8.4.1 Cermet-based electrodes

Ni-based metallic materials and cermets are the most widely used fuel electrodes. However, their performance is limited by carbon deposition, sulfur poisoning, the oxidation or grain growth of Ni particles, etc. The infiltration technique has been used to overcome these issues.

8.4.1.1 Improving cell performance

MIEC materials are effective infiltration materials that can be used to improve SOFC fuel electrode performance for the HOR. For instance, Jiang *et al* studied the performance of SOFC cells (Ni/YSZ-YSZ-LSM) with Ni/YSZ fuel electrodes infiltrated with 0.75–4.4 M $Y_{0.03}Zr_{0.97}O_2$, $Sm_{0.2}Ce_{0.8}O_2$ and CeO_2 nitrate solutions with the assistance of a vacuum [68]. For the HOR, an MIEC $(Sm,Ce)O_2$-infiltrated

Table 8.1. Summary of the processing parameters used for fuel electrode infiltration.

Fuel electrode backbone	Metal salt used in precursor	Solvent	Surfactant	Concentration	Heat treatment temperatures, °C	Cycles	Reference
NiO/YSZ	$Ce(NO_3)_3{\cdot}6H_2O$	1:1 vol% water–ethanol	–	0.4–2 M*	750	–	Ozmen [76]
Ni/YSZ;Ni/TiYSZ	$Ni(NO_3)_2$, $Gd(NO_3)_3$, $Ce(NO_3)_3$	Ethanol	–	3 M	320	1–5*	Mo [77]
Ni/YSZ	$Ce(NO_3)_3{\cdot}6H_2O + Gd(NO_3)_3{\cdot}6H_2O$	Water	PE-L62	0.3 M	300	9	Tong [78]
Ni/YSZ	Ni/CGO: $Ni(NO_3)_2{\cdot}6H_2O + Gd(NO_3)_3{\cdot}6H_2O + Ce(NO_3)_3{\cdot}6H_2O$;LSNC: $La(NO_3)_2{\cdot}6H_2O + Sr(NO_3)_2 + Ni(NO_3)_2{\cdot}6H_2O + Cr(NO_3)_2{\cdot}9H_2O$	Ethanol	Citric acid	4 M; 1 M	100 (to evaporate ethanol), 320	1; 5	Lu [79]
NiO/YSZ	Fe precursor solution	Urea	–	N/A	80 (to decompose urea), 400 (to remove organic compounds)	Several	Jeong [80]
Ni/YSZ	$Ce(NO_3)_3{\cdot}6H_2O + Gd(NO_3)_3{\cdot}6H_2O$	Water	PE-L62	0.3 M	300	Several	Ovtar [81]
NiO/YSZ	$Ba(Ac)_2$	N/A	–	0.3 M	800	–	Zheng [82]
Ni/YSZ	$Gd(NO_3)_3{\cdot}6H_2O + Ce(NO_3)_3{\cdot}6H_2O$	Water; propionic acid	Triton X-100; none	1 M; 0.5 M	500; 800 (at final stage)	–	Williams [83]
Ni/YSZ	$Ni(NO_3)_2{\cdot}4H_2O + Gd(NO_3)_3{\cdot}6H_2O + Ce(NO_3)_3{\cdot}6H_2O$	Water	Pluoronic P123	3 M	350	3	Kiebach [84]
NiO/YSZ	$SnCl_2{\cdot}2H_2O$; $Ba(NO_3)_2$	Ethanol; water	–	N/A	500; 600	–	McIntyre [85]
Ni/YSZ	$Y(NO_3)_3{\cdot}6H_2O + Sr(NO_3)_2 + ZrCl_2O{\cdot}8H_2O$	Deionized (DI) water	–	0.1–1 M*	600, 900, 1300	–	Jin [86]
Ni/YSZ	$Y(NO_3)_3$, $ZrO(NO_3)_2$, $Ce(NO_3)_3$, $Sm(NO_3)_3$	N/A	–	0.75–4.4 M*	1000	–	Jiang [68]
Ni/ScSZ	$Gd(NO_3)_3 + Ce(NO_3)_3$	N/A	–	0.5–2.5 M*	450–850*	5	Timurkutluk [87]

(Continued)

Table 8.1. *(Continued)*

Fuel electrode backbone	Metal salt used in precursor	Solvent	Surfactant	Concentration	Heat treatment temperatures, °C	Cycles	Reference
NiO	$Ce(NO_3)_3{\cdot}6H_2O + Gd(NO_3)_3{\cdot}6H_2O$	N/A	–	3 M	850	2	Wang [88]
Ni/CGO	$Sr(NO_3)_2 + Ce(NO_3)_3{\cdot}6H_2O$ + Yb $(NO_3)_3{\cdot}5H_2O$	DI water	–	0.5 M	1300	*	Yasutake [89]
Ni/CGO	$Pd(NO_3)_2{\cdot}6H_2O$	Water	–	0.1 M	700	2–8*	Babaei [90]
FeCr/3YSZ	Nitrate salts	Water	N/A	N/A	350	–	Nielsen [91]
STN94	Ce and Gd nitrates; Co/Ni/Fe nitrates	Water	N/A	3 M (CGO)	N/A	–	Drasbæk [92]
STN94	$Ce(NO_3)_3{\cdot}6H_2O + Gd(NO_3)_3{\cdot}6H_2O$; $Ni(NO_3)_2{\cdot}6H_2O$; $Pd(NO_3)_2{\cdot}6H_2O$; $H_{12}N_6O_6Pt$; $RuCl_3{\cdot}xH_2O$	Water	Pluronic P123	0.2–0.75 M	350	Several	Hussain; Ramos [93, 94]
STN94	$Ce(NO_3)_3{\cdot}6H_2O + Gd(NO_3)_3{\cdot}6H_2O$	Ethanol	Polymetric surfactant	1 M	350	3	Blennow [95]
LST/CGO	$Ce(NO_3)_3{\cdot}6H_2O + Gd(NO_3)_3{\cdot}6H_2O$; $Ni(NO_3)_2{\cdot}6H_2O$; $RhCl_3{\cdot}nH_2O$; $H_2PtCl_6{\cdot}6H_2O$; $Pd(NO_3)_2{\cdot}nH_2O$;	Water	–	N/A	1000	–	Futamura [96]
LSTF0.7 ($La_{0.3}Sr_{0.7}Ti_{0.3}Fe_{0.7}O_3$)	$Ce(NO_3)_3{\cdot}6H_2O$	Water	–	N/A	450	8	Xu [97]
LSCM/YSZ	$Ce(NO_3)_3{\cdot}6H_2O + Gd(NO_3)_3{\cdot}6H_2O + Pd(NO_3)_2$	DI water or ethanol*	–	2 M	500	Several	Spiridigliozzi [98]
LSCM/YSZ	$La(NO_3)_3{\cdot}6H_2O + Fe(NO_3)_3{\cdot}9H_2O + Co(NO_3)_2{\cdot}6H_2O$	Water	Ethylene glycol	0.3 M;0.6 M*	900	–	Khameneh [99]
LSCM/YSZ	$Ce(NO_3)_3{\cdot}6H_2O + Gd(NO_3)_3{\cdot}6H_2O + Pd(NO_3)_2$	Water	–	2 M	500	Several	Yue [100]
LSCM/YSZ	$Pd(NO_3)_2$	N/A	–	0.25 M	700	*	Jiang, Ye [101, 102]
LSCM/CGO	$Ce(NO_3)_3{\cdot}6H_2O + Mn(NO_3)_2$	DI water	Citric acid	0.5 M	600	–	Zhang [103]
LCCM/CGO	$Pd(NO_3)_2$	N/A	–	0.05 M	700	*	Babaei [104]
LSCM	$Cu(NO_3)_2{\cdot}2.5H_2O$	Water	–	1 M	175	*	Xing [105]

LSGM	$Ni(NO_3)_2$	Water	–	5 M	700 or 450*	12	Miller [106]
SFM ($Sr_2Fe_{1.5}Mo_{0.5}O_{6-\delta}$)	CGO	N/A	N/A	N/A	800	*	Lv [107]
YSZ	$La(NO_3)_3 \cdot 6H_2O + Sr(NO_3)_2 + Sc(NO_3)_3 \cdot 6H_2O + Mn(NO_3)_3 \cdot 6H_2O$*	DI water	Citric acid	N/A	450	Several	Sengodan [108]
YSZ	$Ni(NO_3)_2 \cdot 6H_2O$; $ZrO(NO_3)_2 \cdot xH_2O + (NH_4)_2Ce(NO_3)_6$	Isopropyl alcohol	–	2 M; N/A	450	Several	Cimenti [109]
YSZ	$La(NO_3)_3 \cdot 6H_2O + Sr(NO_3)_2 + Cr(NO_3)_3 \cdot 9H_2O + Mn(NO_3)_2 \cdot xH_2O$	DI water	Citric acid	N/A	477	10–20*	Kim [110]
YSZ	$La(NO_3)_3 \cdot 6H_2O + Sr(NO_3)_2 + Cr(NO_3)_3 \cdot 9H_2O + Mn(NO_3)_2 \cdot xH_2O$	DI water	Citric acid	N/A	477	Several	Bidrawn [111]
YSZ	$La(NO_3)_3 \cdot 6H_2O + Sr(NO_3)_2 + [CH_3CH(O\text{-})CO_2 \cdot NH_4]_2 \cdot Ti(OH)_2$	DI water	–	N/A	450	15	Lee [112]
YSZ	$Ni(NO_3)_2 \cdot 6H_2O$	Ethanol	–	0.3 M	500	10	Busawon [113]
YSZ	Ceria and Cu salt	Water (ceria); water or water with urea (Cu salt) *	–	N/A	450	Several	Jung [114]
ScSZ	$La(NO_3)_3 \cdot 6H_2O + Ca(NO_3)_2 \cdot 4H_2O + Fe(NO_3)_3 \cdot 9H_2O + C_{10}H_5NbO_{20} \cdot 6H_2O$	DI water	Citric acid	0.5 M	450	Several	Liu [115]
CGO	$Sr(NO_3)_2 + Fe(NO_3)_3 \cdot 9H_2O + (NH_4)_6Mo_7O_{24} \cdot 4H_2O$	Ethanol	Citric acid + ethylene glycol	N/A	400	Several	Skafte [116]
CGO	$Ni(NO_3)_2$	Ethanol	–	2 M	500	10	Duboviks [117]

* Cell/electrode performance changes with parameter changes.

cell showed lower polarization resistance (R_p) and extensive TPB areas, and the fine particles of (Sm,Ce)O_2 effectively inhibited Ni particle agglomeration during the final high-temperature sintering (to achieve good bonding between the Ni/YSZ electrode and the YSZ electrolyte) and reduction. Ozmen *et al* used the polymerized norepinephrine (pNE)-assisted dip infiltration method to introduce CeO_2 catalysts into both electrodes of a Ni/YSZ-YSZ-LSM/CGO SOFC cell [76]. Prior to being immersed in a cerium nitrate hexahydrate solution, the cell was dip coated in pNE to obtain a polymerized surface. Microscopic analysis indicated that the pNE treatment caused finer CeO_2 particles to be deposited, enhancing the morphology of the electrode surfaces. During 300 h of testing at 750 °C and at 0.325 A cm^{-2} in 3% humidified H_2, a 1 M CeO_2 precursor-infiltrated cell (with a total loading of 3 mg per 5 cm^2 fuel electrode) had the lowest R_p, exhibiting a substantial decrease from 0.335 to 0.275 Ω cm^2.

CGO is another widely studied MIEC material that has been used to modify the cell performance of Ni-based SOFCs. Timurkutluk *et al* studied the effects of various infiltration concentrations and calcination temperatures on cell performance [87]. They infiltrated 0.5–2.5 M CGO nitrate solutions into both electrodes of a Ni/ScSZ-ScSZ-LSF cell and heat treated them at various temperatures from 450 °C to 850 °C. A cell infiltrated with a 1.5 M CGO solution heat treated at 550 °C exhibited the highest power density of 1.343 W cm^{-2} at 700 °C. According to SEM images, higher-concentration solutions led to lower porosity in the backbone, restricting the gas flow. In addition, higher calcination temperatures caused the grain growth of CGO nanoparticles, reducing the TPB density and blocking the pores. Williams *et al* used an inkjet printing method to infiltrate two different CGO precursors into a Ni/YSZ backbone with either water or propionic acid (PPA) as a solvent [83]. PPA-based ink was found to have a smaller contact angle and was likely to lead to a better and more uniform distribution of CGO nanoparticles (see figure 8.4). When full cells using this electrode were tested, no significant difference was found between the performance of the infiltrated cell and that of a reference non-infiltrated cell. However, in a symmetric cell (with only the fuel electrodes) aging test at 600 °C, the infiltrated cell had a much lower ASR value than that of the reference cell and showed stable performance after 50 h of aging under 4% humidified H_2.

Mo *et al* compared Ni/YSZ-YSZ-Ni/YSZ and Ni/TiYSZ-YSZ-Ni/TiYSZ (TiYSZ:3 mol% TiO_2 doped into YSZ) symmetrical cells infiltrated with three different metal nitrate precursors in ethanol solvent, namely 3 M Ni, CGO, and Ni/CGO, and studied the effect of different loadings on the cell performance [77]. A decrease in R_p was

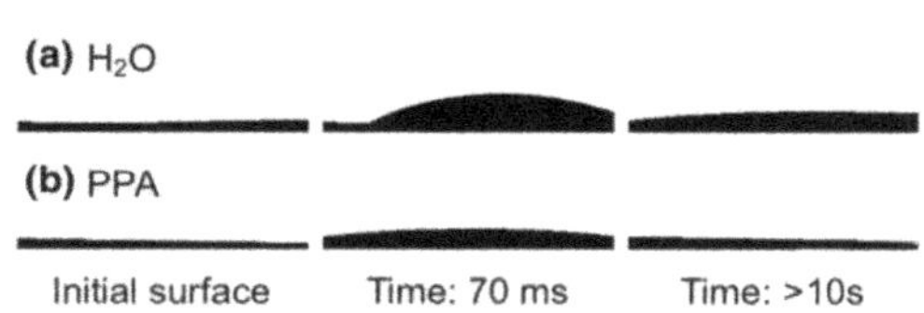

Figure 8.4. Schematic illustration of the wetting behavior of the infiltration solutions on a Ni/YSZ fuel electrode surface over time: (a) a CGO solution with water, and (b) a CGO solution with PPA [83] Copyright © 2017, The Author(s). With permission of Springer.

observed at both 700 °C and 800 °C in 50% humidified H_2 for all infiltrated cells, among which the Ni/CGO infiltrate provided the most significant performance improvement. A distribution of relaxation times (DRT) analysis showed that the CGO nanoparticles mainly promoted the charge transfer process.

Lu *et al* separately evaluated the effects of the co-infiltration of Ni with the mixed conducting oxides $La_{0.6}Sr_{0.3}Ni_{0.15}Cr_{0.85}O_{3-\delta}$ (LSNC) and $Gd_{0.1}Ce_{0.9}O_{2-\delta}$ (CGO) on SOFC cell performance at 700 °C–800 °C [79]. They infiltrated a Ni/YSZ electrode with nitrate solutions of 4 M Ni with 1 M LSNC or 4 M CGO. As shown in the SEM images (figure 8.5), both oxide phases (LSNC and CGO) effectively inhibited Ni particle coarsening at 800 °C by physically isolating the Ni. A Ni/CGO-infiltrated cell demonstrated excellent electrochemical performance with reduced R_p and higher power density at all temperatures tested. At 700 °C, the power density increased by 56% and R_p decreased by 41% in comparison to the corresponding values for a

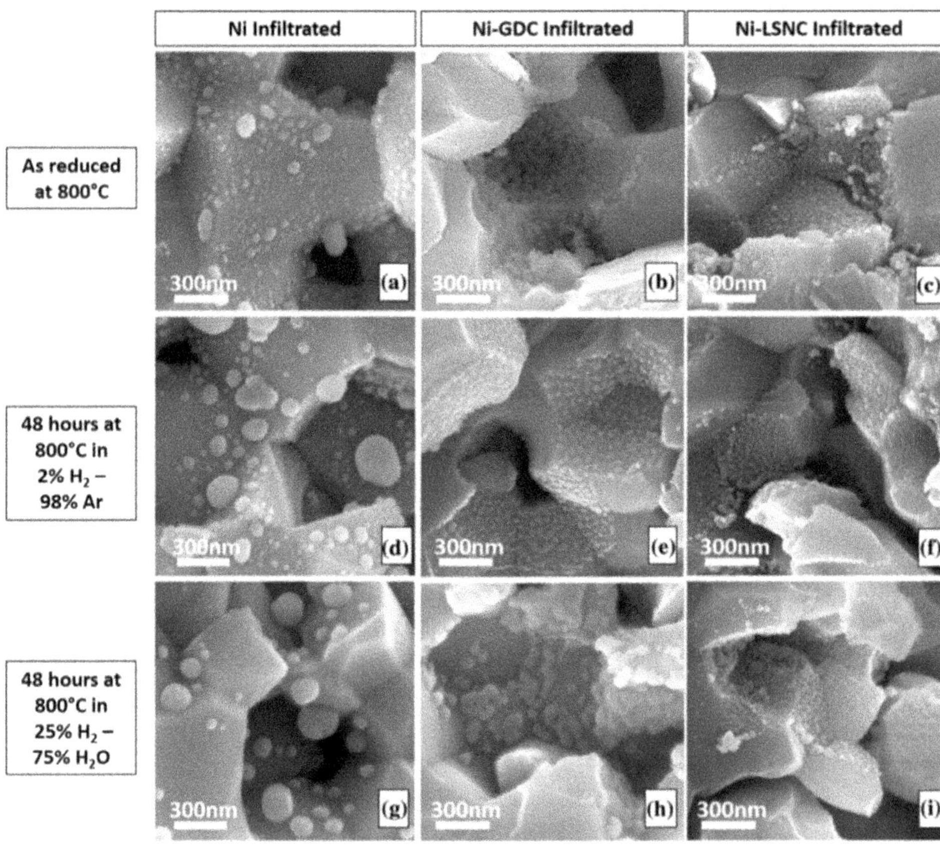

Figure 8.5. SEM images of a fractured cross section of the fuel electrode active layers infiltrated with Ni ((a), (d), and (g)), Ni/CGO ((b), (e), and (h)), and Ni/LSNC ((c), (f), and (i)). Infiltrated nanoparticles (a)–(c) in the reduced state at 800 °C, (d)–(f) after exposure to H_2/Ar (at a ratio of 2:98) for 48 h, and (g)–(i) after exposure to H_2/H_2O (at a ratio of 25:75) for 48 h [79] Copyright © 2019, The Minerals, Metals & Materials Society. With permission of Springer.

non-infiltrated cell. The fuel electrode microstructure after electrochemical testing showed that Ni/LSNC nanoparticles reduced the pore volume of the backbone by 28.6% and thus increased the mass transfer resistance by 40.1%. The infiltration of Ni/CGO decreased the pore volume by only 17.1%.

In addition to improving single-cell performance, the infiltration method has also been applied to stacks. In a study by Kiebach *et al*, a ten-cell stack was infiltrated with a 3 M Ni/CGO solution on the Ni/YSZ fuel electrode and 3 M CGO on the LSM/YSZ oxygen electrode. The stack was then tested at 680 °C–860 °C at the open-circuit voltage (OCV) and at 20 A [84]. It was found that the performance of the LSM/YSZ oxygen electrode was significantly improved by CGO infiltration, whereas infiltrating Ni/CGO had little effect on the Ni/YSZ fuel electrode.

Based on the above findings, MIEC catalysts such as CGO are widely considered for use as infiltrates that can improve the performance of Ni-based fuel electrodes.

8.4.1.2 Improving carbon tolerance

As mentioned previously, Ni-based fuel electrodes are considered unsuitable for the direct utilization of hydrocarbon fuels due to the high catalytic effect of Ni on carbon cracking and severe carbon deposition. The infiltration of other electrocatalysts into Ni-based cermets has been considered as a way of enhancing their carbon tolerance. Wang *et al* studied a CGO-infiltrated Ni fuel electrode in humidified methane with 3% H_2O at 800 °C in a three-electrode arrangement with Pt paste as the reference and counter electrodes [88]. The porous NiO layer was infiltrated with 3 M $Ce_{0.8}Gd_{0.2}O_2$ followed by heat treatment at 850 °C. The infiltration cycle was repeated twice, reaching a total loading of 1.4–1.7 mg cm^{-2} (per 0.5 cm^2 of fuel electrode area). When exposed to humidified CH_4 at the OCV, the CGO-infiltrated Ni electrode displayed the lowest R_p compared to that of a pure Ni electrode and a traditional Ni/CGO cermet electrode. The infiltrated electrode also showed good stability at a current loading of 20 mA cm^{-2} and R_p decreased during 3 h of testing. Based on SEM images, they concluded that CGO nanoparticles infiltrated into the Ni fuel electrode were more effective in oxidizing the deposited carbon than the Ni/CGO cermet and that they thereby increased electrode stability and performance. Babaei *et al* infiltrated a Ni/CGO electrode with 0.1 M palladium nitrate solution, reaching a total PdO loading of 0.04–0.15 mg cm^{-2} [90]. Under 3% humidified H_2, the infiltrated electrode showed a significant reduction in the low-frequency part of the resistance corresponding to the adsorption and diffusion processes when tested in a three-electrode arrangement. In addition, the calculated activation energy (E_a) decreased significantly with increasing PdO loadings. Their study showed that the Pd/PdO nanoparticles had a catalytic impact on the oxidation reaction of hydrocarbon fuels (ethanol and methane), but carbon deposition over the Ni/CGO cermet was still inevitable.

Yasutake *et al* evaluated the carbon tolerance of an SOFC cell (SCYb@Ni/CGO-ScSZ/SDC-LSM/ScSZ) with a proton conductor $SrCe_{0.95}Yb_{0.05}O_{3-\delta}$ (SCYb)-infiltrated Ni/CGO electrode during a degradation test in dry CH_4 at 900 °C [89].

The accelerated degradation tests comprised ten cycles with on–off current in each cycle. The similar *i–V* (current–voltage) characteristics observed after each cycle showed that the SCYb@Ni/CGO makes a negligible difference in terms of carbon deposition, compared to a bare Ni/CGO fuel electrode. This is probably due to the high resistance of Ni/CGO toward carbon deposition. However, a cell with an SCYb-infiltrated fuel electrode showed a maximum 50% increase in power density compared to the Ni/CGO electrode when tested in either humidified H_2 (1% H_2O) or dry CH_4 atmospheres. The authors proposed that two steps are involved in power generation in a dry CH_4 atmosphere, namely, a carbon deposition step and a carbon removal step that takes place via an oxidation reaction and/or a water-reforming reaction. Jin *et al* reported that an electrochemically improved SOFC cell (SZY@Ni/YSZ-YSZ-LSM) was produced by infiltrating the proton conductor $SrZr_{0.95}Y_{0.05}O_{3-\delta}$ (SZY) into a Ni/YSZ fuel electrode [86]. Examination of the *i–V* curves obtained at 900 °C shows that the single-cell power density in H_2 and dry CH_4 atmospheres increases with increasing SZY loadings. An accelerated degradation test consisting of 20 cycles was also carried out on the infiltrated cell in dry CH_4 with different SZY loadings, and the optimal amount of SZY was found to be SZY:NiO = 1:14 (mass ratio).

McIntyre *et al* separately investigated the carbon tolerance of an SOFC cell (Ni/YSZ-YSZ-LSM/YSZ) with Sn- and BaO-infiltrated Ni/YSZ fuel electrodes exposed to methane at the OCV [85]. The Ni/YSZ electrode was infiltrated by a $SnCl_2{\cdot}2H_2O$ ethanol solution or a $Ba(NO_3)_2$ aqueous solution followed by calcination, reaching a loading of 1 wt.% with respect to Ni. *In situ* Raman spectra showed that BaO@Ni/YSZ fuel electrodes had the least carbon accumulation both in the electrochemical active sites and on the top surface of the electrode, whereas Sn only inhibited carbon deposition in the active interface between the fuel electrode and the electrolyte; non-infiltrated Ni/YSZ had carbon accumulation throughout the electrode. They also found that after the electrode was exposed to CH_4 at 730 °C for 600 s, steam was the most effective reforming agent compared to CO_2 and O_2 for removing the deposited carbon from the fuel electrode for all three electrodes (see figure 8.6). However, electrochemical impedance spectroscopy (EIS) measurements in H_2 under OCV before and after exposure to CH_4 showed that the cells with either Sn- or BaO-infiltrated Ni/YSZ electrodes had higher R_p values and lower power densities than the non-infiltrated Ni/YSZ cell.

Sun *et al* tested an SOFC cell (Ni/YSZ-YSZ-LSM) with a 0.5 wt.% Ru-infiltrated Ni/YSZ fuel electrode in ethanol at 750 °C [118]. The cell's performance was enhanced by the Ru-infiltrated layer, reaching a maximum power density of 369 mW cm^{-2} compared to 312 mW cm^{-2} for a non-infiltrated cell. They further reported no significant cell voltage degradation and no carbon deposition during 1000 min of stability testing.

8.4.1.3 Improving sulfur resistance

It is well known that the sulfur present in hydrocarbon fuels poisons Ni-based electrodes and decreases their performance and lifetime significantly, especially at intermediate temperatures. For example, Matsuzaki and Yasuda reported that the

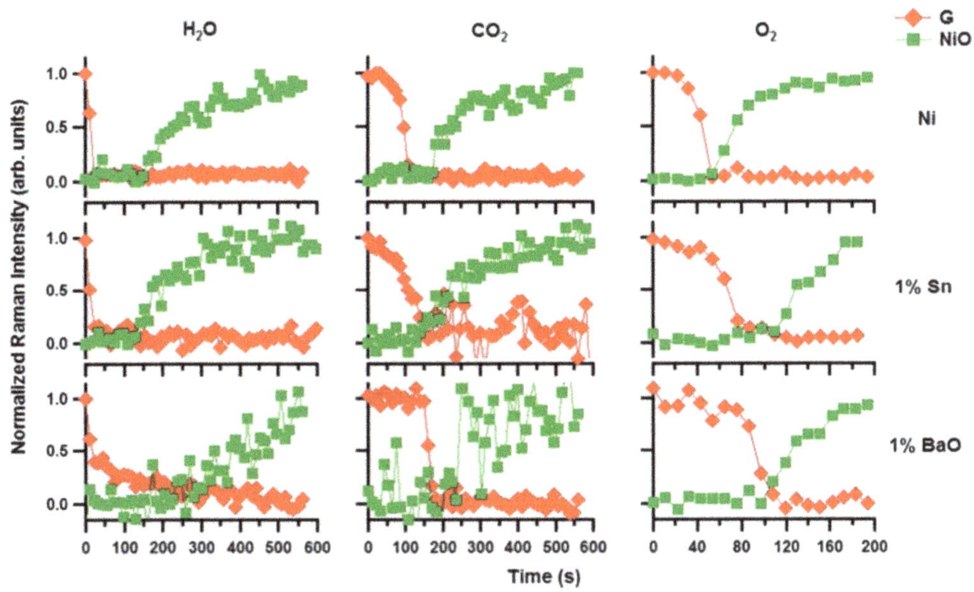

Figure 8.6. Raman kinetics data collected during the introduction of the reforming agents H_2O, CO_2, and O_2. The top, middle, and bottom rows are the spectral data collected from a non-infiltrated Ni/YSZ electrode, a 1 wt.% Sn-infiltrated electrode, and a 1 wt.% BaO-infiltrated electrode, respectively. The decrease in G peak intensity (red) at 1561 cm^{-1} and the increase in the NiO signal (green) at 1070 cm^{-1} represent the oxidation of carbon and Ni, respectively. Reprinted with permission from [85]. Copyright (2015) American Chemical Society.

R_p of a Ni/YSZ electrode increased drastically when the H_2S concentration exceeded 0.05 ppm at 750 °C [21]. The infiltration method has therefore been applied to improve the sulfur tolerance of Ni-based fuel electrodes. Sholk studied the sulfur tolerance of a full cell (SDC@Ni/YSZ-YSZ-LSM/YSZ) in humidified H_2 with 40 ppm of H_2S at 700 °C at 0.4 A cm^{-2} [119]. The infiltrated cell maintained a power density of 220–240 mW cm^{-2} for 500 h, while the voltage of the non-infiltrated cell immediately dropped to 0 V after it was exposed to H_2S. In addition, after the H_2S exposure was stopped, the cell performance of the infiltrated cell recovered in 20 min, reaching 98% of its value before exposure.

8.4.1.4 Improving electrolysis performance

In the last few years, the infiltration method has been applied to Ni/YSZ electrodes to improve cell performance and durability in the electrolysis mode. Zheng *et al* infiltrated 6.9–8.9 wt.% $BaCO_3$ into the Ni/YSZ side of an SOEC cell (Ni/YSZ-YSZ-LSM/YSZ). As compared to a cell with a bare Ni/YSZ electrode, the infiltrated cell showed a reduced R_p at 800 °C during either steam electrolysis, CO_2 electrolysis, or co-electrolysis [82]. A DRT analysis of the three-electrode installation revealed that $BaCO_3$ nanoparticles decrease the resistance at a middle frequency near 10^3 Hz under all testing conditions, indicating that $BaCO_3$ effectively accelerates the surface charge transfer reaction. In addition, $BaCO_3$ reduces the resistance at the low-frequency peak near 10 Hz, which is related to the adsorption/dissociation process for CO_2 electrolysis and co-electrolysis, indicating that $BaCO_3$ also accelerates the processes of CO_2 adsorption/dissociation. Jeong *et al* used vacuum-assisted infiltration to introduce an Fe-oxide precursor solution into a Ni/YSZ backbone, followed by the *in situ* alloying of Ni–Fe at 800 °C in humidified H_2 with 3% steam [80]. They reported that the Ni–Fe cell alloyed via infiltration showed lower E_a and R_p values than a standard cell (Ni/YSZ-YSZ-CGO-LSCF/CGO) and that Ni–Fe increased the CO selectivity of the cell under co-electrolysis. The cell was tested in co-electrolysis at −0.4 A cm^{-2} and at 800 °C for 500 h and had a degradation rate of less than 3.4%/500h.

Ovtar and Tong *et al* used CGO solutions to infiltrate Ni/YSZ electrodes to improve the cell's (Ni/YSZ-YSZ-CGO-LSCF/CGO) electrocatalytic activity and durability [78, 81]. A 0.3 M CGO nitrate solution with PE-L62 as a surfactant was introduced to a Ni/YSZ fuel electrode at a total backbone loading of 125 mg cm^{-3}. The cells were tested under steam electrolysis at 800 °C and at −1.25 A cm^{-2}. The CGO-infiltrated cell showed excellent durability and had a degradation rate as low as 66 mV kh^{-1}, while a non-infiltrated cell showed a degradation rate of 699 mV kh^{-1} under the same conditions. The authors concluded that the infiltrated CGO nanoparticles drastically extended the reaction sites from the TPB points to Ce-containing particles on either Ni or YSZ grains [81]. In another study by Tong *et al*, a similar infiltration procedure was applied with a total CGO loading of 145 mg cm^{-3} into Ni/YSZ. A cell using this material was operated for steam electrolysis at 750 °C (see figure 8.7), showing voltage degradation rates of 2%/kh at −1.25 A cm^{-2} and 0.8%/kh at −1 A cm^{-2}, more than ten times smaller than those of a non-infiltrated cell. When operated in the reversible mode at ±0.5 A cm^{-2}, no obvious degradation was observable in the infiltrated cell [78].

8.4.1.5 Metal-based electrodes

Metal-supported SOFC fuel electrodes (usually FeCr-based) have also been extensively studied for use as electrode backbones due to their excellent mechanical and thermal stability. Blennow *et al* studied a $Ce_{0.8}Gd_{0.2}O_{1.9}$ (CGO20) and Ni co-infiltrated metal-supported button SOFC with 100 vol% $Sr_{0.99}Ti_{0.9}Nb_{0.1}O_3$ (STN99) and 70 vol% STN99/30vol% FeCr as fuel electrode backbones [120]. At 650 °C in H_2/H_2O (at a ratio of 96:4), the cell with STN99/FeCr (at a ratio of 70:30) had lower R_s and R_p than the cell with 100 vol% STN99. Durability testing was carried out at 650 °C and at 0.25 A cm^{-2} for 500 h, and the cell with CGO20+Ni@ STN99/FeCr (70:30) electrode showed a degradation rate of 9%/kh. Another durability test was conducted under the same conditions on a button cell with a FeCr/YSZ (50:50 in vol%) fuel electrode infiltrated with CGO20/Ni [121]. This cell showed better stability and had a voltage degradation of 4.5%/kh during 1000 h of galvanostatic testing. They also tested a button cell with the CGO20 infiltrated STN99/FeCr electrode under conditions of low fuel and oxygen utilization (<10%), which resulted in a degradation rate of 3.5%/kh [122].

A porous electrode theory (PET) was presented by Nielsen *et al* to explain the impedance of a CGO/Ni-infiltrated FeCr/3YSZ fuel electrode [91]. From the measured impedance spectra, they found that CGO/Ni loading mainly affects the high-frequency resistance, i.e. the oxygen ion charge transfer process between the

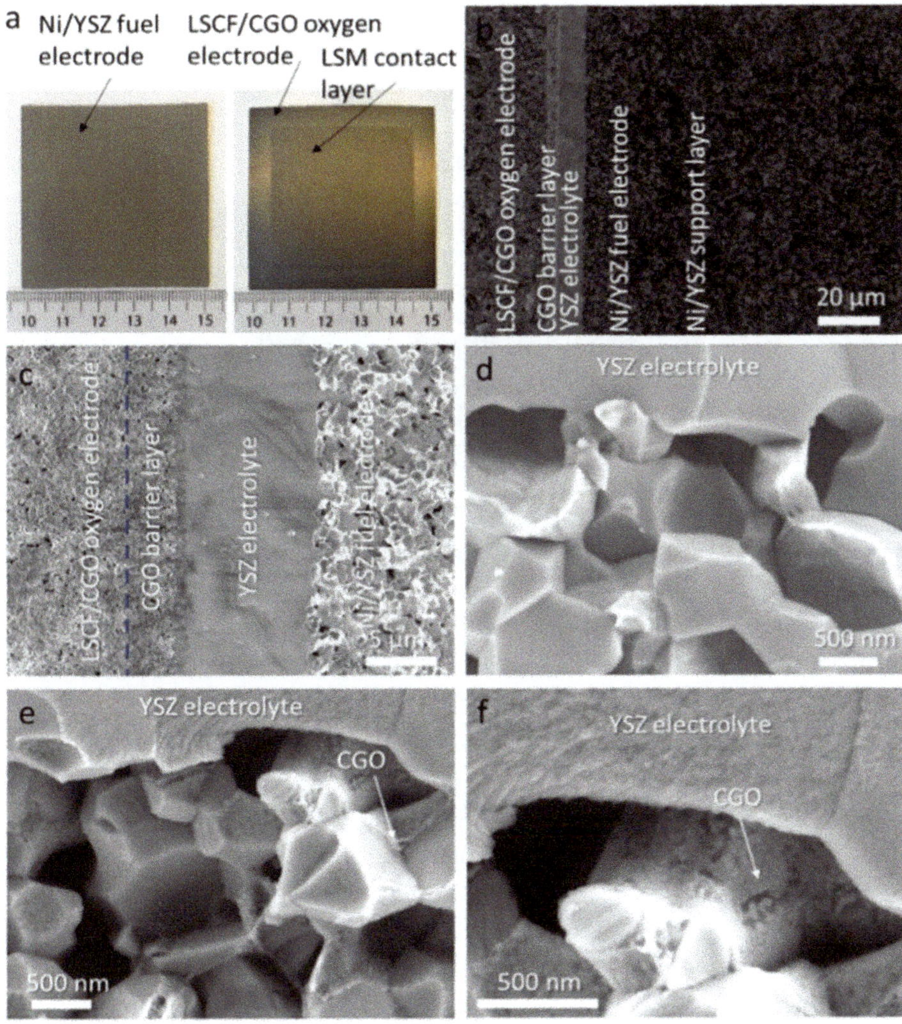

Figure 8.7. Images of the cell produced by Tong *et al*: (a) photograph of the single cell after reduction. (b) and (c) SEM images of the cell cross section after reduction. (d) Bare Ni/YSZ active layer. (e) and (f) Ni/YSZ active layer infiltrated with CGO nanoparticles. Reprinted from [78], © The Electrochemical Society. Reproduced by permission of IOP Publishing.

fuel electrode and the ScYSZ electrolyte. The TEM images in figure 8.8 show that the infiltrated nanosized Ni is distributed in the CGO matrix, thus stabilizing CGO against grain growth and giving the infiltrated nanoparticles robust performance. Sudireddy *et al* compared the electrochemical performance and durability of an MS-SOFC (FeCr/YSZ-ScYSZ-CGO-LSC) infiltrated with Ru/CGO and Ni/CGO at 650 °C [123]. The Ru/CGO-infiltrated cell showed slightly better performance than that of the Ni/CGO-infiltrated cell under 20% humidified H_2; their ASRs were 0.322 Ω cm^2 and 0.453 Ω cm^2, respectively. Both cells showed similar and poor durability at 0.25 A cm^{-2}. Agglomeration of the nanoparticles,

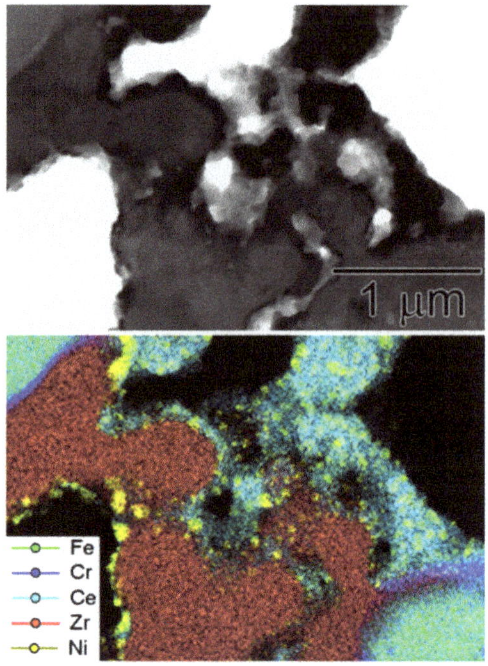

Figure 8.8. TEM images of the Ni/CGO-infiltrated FeCr/YSZ fuel electrode. Reprinted from [91], Copyright (2012), with permission from Elsevier.

cracking of the infiltrated coating, and the non-percolation of the electrochemically active network were concluded to be the cause of the degradation. Thus, the amount of electrocatalyst and the microstructure of the backbone are important areas for improvement.

8.4.2 Oxide-based electrodes

In addition to enhancing the electrode performance, the infiltration technique can also be utilized to fabricate the fuel electrode. Before infiltration can take place, an electrode backbone layer has to be prepared. This can be realized by sintering a green layer containing the pore formers (e.g. graphite powder, starch, etc) used to create the porous electrode backbone [65]. This green layer is often sintered together with the electrolyte layer, thus forming the porous electrode backbone on top of the dense electrolyte in one sintering step. The electrode backbone (scaffold), made of the same material as the electrolyte, such as YSZ, ScSZ, or CGO, is then infiltrated with electrocatalysts and heat treated at a low temperature (see figure 8.9). A significant amount of electrochemically active material loading is usually required to obtain sufficient conductivity and electrocatalytic activity.

8.4.2.1 Electrode backbone materials with a fluorite structure (YSZ, ScSZ, CGO)

The fabrication of traditional Ni/YSZ cermets requires high-temperature sintering (1300 °C–1400 °C) and reduction (1000 °C), which inevitably cause the

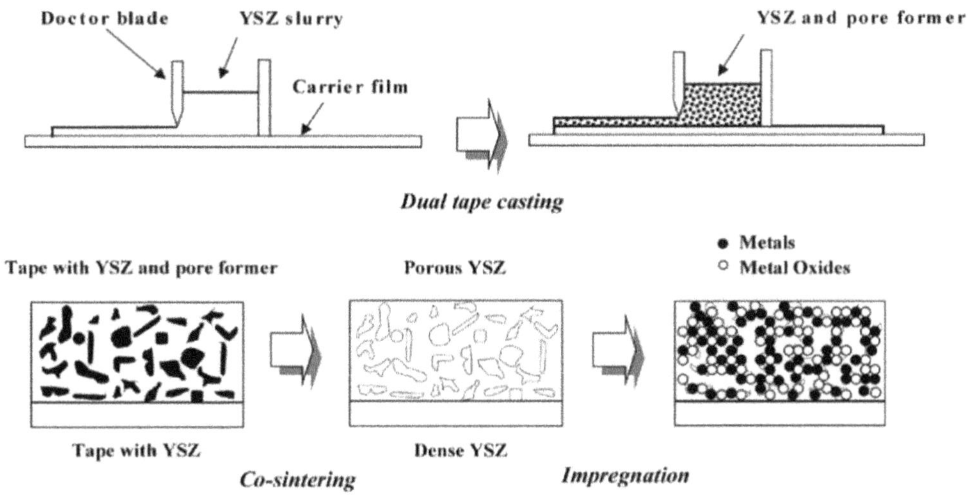

Figure 8.9. Diagram of the tape-casting procedures used to create the dense electrolyte and the porous electrode backbone in a single sintering step [124] John Wiley & Sons, copyright 2000 WILEY-VCH Verlag GmbH, Weinheim, Fed. Rep. of Germany.

agglomeration of fine Ni particles [62, 68, 125]. One way to solve this problem is to infiltrate the Ni nitrate solution into the porous YSZ backbone in order to increase the active TPB area. Busawon *et al* studied the redox stability of a Ni@YSZ cermet at 800 °C; the composite was prepared by infiltrating 12–16 wt.% Ni into a porous YSZ structure. A high conductivity of 360 S cm^{-1} was measured for 12 wt.% Ni@YSZ at room temperature. After one redox cycle at 800 °C, the conductivity dropped by 20% to 287 S cm^{-1} at room temperature, which is comparable to that of other alternative fuel electrodes [113]. Cimenti *et al* compared the SOFC cell (YSZ-YSZ-LSM/YSZ) performances of Ni and Ni/ZDC ($Zr_{0.35}Ce_{0.65}O_{2-\delta}$)-infiltrated YSZ porous fuel electrodes in humidified H_2 and pure methanol atmospheres at 800 °C [109]. The infiltrated Ni loading reached 40 wt.% of the YSZ backbone, and the ZDC loading was approximately 30 wt.%. They found that the Ni/ZDC@YSZ electrode had little carbon formation after being tested for 12 h in methanol at 0.6 V and showed better durability than the Ni@YSZ electrode in both humidified H_2 and methanol atmospheres.

Other studies have infiltrated YSZ backbones with different electrocatalysts. For example, Kim *et al* infiltrated 45 wt.% $La_{0.8}Sr_{0.2}Cr_{0.5}Mn_{0.5}O_3$ (LSCM) with 0.5–1 wt.% Pd, Rh, or Ni into a YSZ backbone to improve the electrode catalytic activity of an SOFC cell (LSCM/YSZ-YSZ-LSF/YSZ) [110]. They found that after high-temperature calcination at 1200 °C, a uniform LSCM layer was formed on the YSZ backbone surface and that after it was reduced in humidified H_2 at 800 °C, the LSCM layer broke into small particles with a high TPB length. The infiltrated SOFC cells were drastically improved by 0.5 wt.% Pd, reaching a power density of 500 mW cm^{-2} at 700 °C in humidified H_2 (3% H_2O). In another study, the same authors compared the performance of an LSCM/YSZ-based electrode further infiltrated with different catalysts, i.e. Pt, Pd, Ni, and CeO_2 [126]. After it was exposed to humidified methane at 800 °C for 24 h, a Pt@LSCM/YSZ electrode was found to have the highest carbon tolerance in hydrocarbon fuels. They also reported that co-infiltrating CeO_2 together with Ni or Pt can effectively suppress carbon deposition in humidified methane. Bidrawn *et al* examined an SOEC cell (Pd+CZY +LSCM@YSZ-YSZ-LSF@YSZ) with infiltrated YSZ as a fuel electrode for CO_2 electrolysis [111]. In their study, 45 wt.% LSCM together with 0.5 wt.% Pd and 5 wt.% $Ce_{0.48}Zr_{0.48}Y_{0.04}O_2$ (CZY) catalysts were infiltrated into a YSZ porous scaffold which was used as a fuel electrode. The nearly constant slopes of the *i–V* curves in electrolysis mode indicated that the impedance was nearly independent of current for various CO_2/CO mixtures at both 700 °C and 800 °C, with total impedances of 0.63 Ω cm^2 and 0.36 Ω cm^2, respectively.

Sc-doped LSM oxides (LSSM) have been reported to show comparable performance to that of LSCM along with good structural and chemical stability and are also of interest for use as the fuel electrode material. Sengodan *et al* investigated a $La_{0.8}Sr_{0.2}Sc_xMn_{(1-x)}O_{3-\delta}$ (x = 0.1, 0.2, and 0.3)-infiltrated YSZ fuel electrode for intermediate-temperature solid oxide fuel cells (IT-SOFCs), which were tested at 650 °C–750 °C [108]. The electrical conductivity of the LSSM@YSZ composite electrode was found to decrease with an increase in the Sc concentration in LSSM; with the highest conductivity of 0.48 S cm^{-1} was achieved at 700 °C under 10^{-23} atm for $x = 0.2$ ($La_{0.8}Sr_{0.2}Sc_{0.1}Mn_{0.9}O_{3-\delta}$).

Lee *et al* infiltrated 45 wt.% $La_{0.2}Sr_{0.7}TiO_3$ (LST) into a YSZ backbone together with 0.5 wt.% Pd and 5 wt.% ceria catalysts and tested an SOFC cell using this electrode (LST+Pd+CeO_2@YSZ-YSZ-LSF@YSZ) in humidified H_2 (3% H_2O) [112]. The calcination temperature was found to play an important role in the surface interaction between the LST catalyst and the YSZ backbone. The highest power density was reached when the calcination was performed at 1100 °C. Compared with other studies of LSCM/YSZ composites, they found that the cell with the LSCM-infiltrated YSZ electrode showed better performance than the one with LST-infiltrated YSZ, even though LST has higher conductivity than LSCM. One possible reason for this is that LSCM deposits exhibit an almost ideal morphology on the YSZ scaffold, i.e., a film layer with pores, which significantly increases the active areas. Jung *et al* fabricated the Cu+CeO_2@YSZ fuel electrodes of an SOFC cell (Cu+CeO_2@YSZ-YSZ-LSM/YSZ) by infiltrating aqueous ceria nitrate solution and Cu nitrate or nitrate/urea solutions into a YSZ backbone [114]. The *i–V* curves and EIS showed that both the infiltration solution and the annealing temperature affect cell performance (see figure 8.10). SEM images show that the Cu nitrate/urea infiltration solution resulted in a more homogenous infiltrated layer on the YSZ backbone surface than that resulting from the Cu nitrate solution. In addition, the agglomeration of Cu particles occurs when the annealing temperature is greater than 700 °C, and this has a significant impact on the conductivity of the infiltrated YSZ composite.

Liu *et al* investigated the performance of an SOEC cell (LCFNb@ScSZ-ScSZ-LSM/ScSZ) for CO_2 electrolysis at high temperatures from 700 °C to 850 °C [115]. A ScSZ fuel electrode scaffold was infiltrated with a $La_{0.9}Ca_{0.1}Fe_{0.9}Nb_{0.1}O_{3-\delta}$ (LCFNb) precursor solution followed by heat treatment at 450 °C for 5 min in

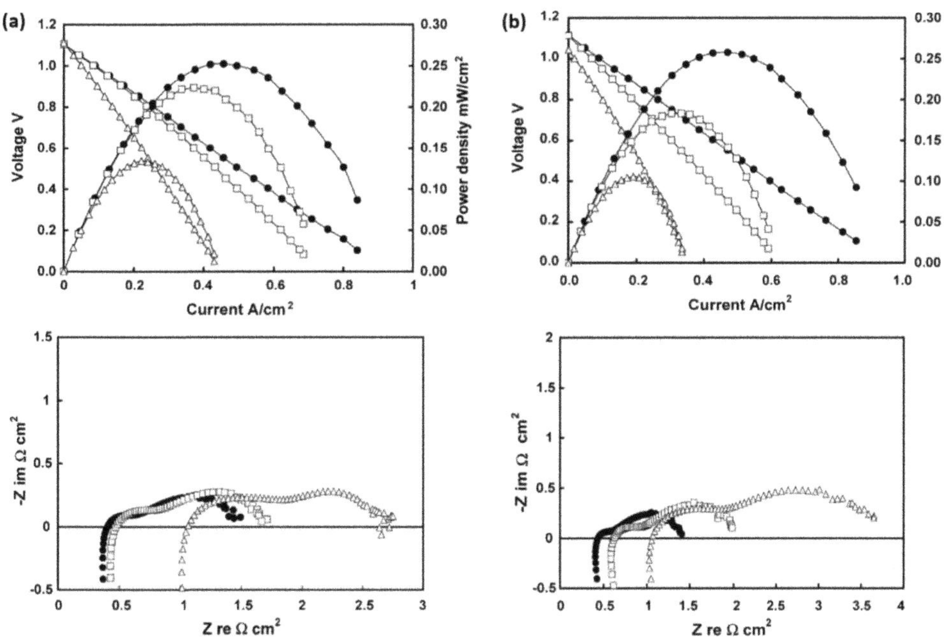

Figure 8.10. *i–V* curves and EIS measured at 0.3 A cm^{-2} and at 700 °C under humidified H_2 for a Cu and CeO_2 infiltrated cell using (a) a Cu nitrate solution and (b) a Cu nitrate/urea solution after calcination at 700 °C for 2 h (circle), 800 °C for 3 h (square) and 900 °C for 2 h (triangle). Adapted from [114], Copyright (2005), with permission from Elsevier.

each cycle, reaching a total loading of 30 wt.% of the ScSZ scaffold. The cell showed good stability in pure CO_2 at various applied voltages from the OCV to 2 V, and the lowest R_p of 0.113 Ω cm^2 was obtained at 2 V at 800 °C, which is superior to the R_p values of many perovskite and Ni-based cells. Skafte *et al* tested symmetrical cells (CGO-ScSZ-CGO) with a CGO10 backbone co-infiltrated with 10–16 vol.% $Sr_{0.99}Fe_{0.75}Mo_{0.25}O_{3-\delta}$ (single perovskite SFM) and 8–10 vol.% $C_{0.8}Gd_{0.2}O_{1.9}$ (CGO20) in both CO/CO_2 and H_2/H_2O atmospheres [116]. The SFM +CGO20@CGO10 electrode showed excellent stability in CO/CO_2 (at a ratio of 90:10) at 750 °C and through redox cycles at 650 °C and 750 °C. Graves *et al* also investigated symmetrical cells with CGO backbone electrodes co-infiltrated with two MIEC catalysts, $SrFe_{0.75}Mo_{0.25}O_{3-\delta}$ (single perovskite SFM) and $Ce_{0.8}Pr_{0.2}O_{2-\delta}$ (CPO). The electrode performance was tested at 650 °C and 700 °C in humidified H_2 [127]. A low R_p of ~0.1 Ω cm^2 at 700 °C in H_2/H_2O (at a ratio of 80:20) was reported based on the measured EIS, demonstrating the high activity of the SFM+CPO@CGO electrode.

Electrolysis testing has been conducted in recent years using modified fluorite-based electrodes. Duboviks *et al* fabricated a Cu-infiltrated CGO electrode as an alternative to Ni-based electrodes for the electrolysis of carbonaceous mixtures [117]. Symmetrical cells with Cu-infiltrated CGO electrodes were compared with Ni-infiltrated cells in a CH_4/CO_2 (70:30) gas mixture. Raman spectra showed that the Cu-infiltrated electrode was more carbon tolerant in biogas mixtures at 600 °C than the Ni-infiltrated cells, whereas in humidified H_2, they had similar performance.

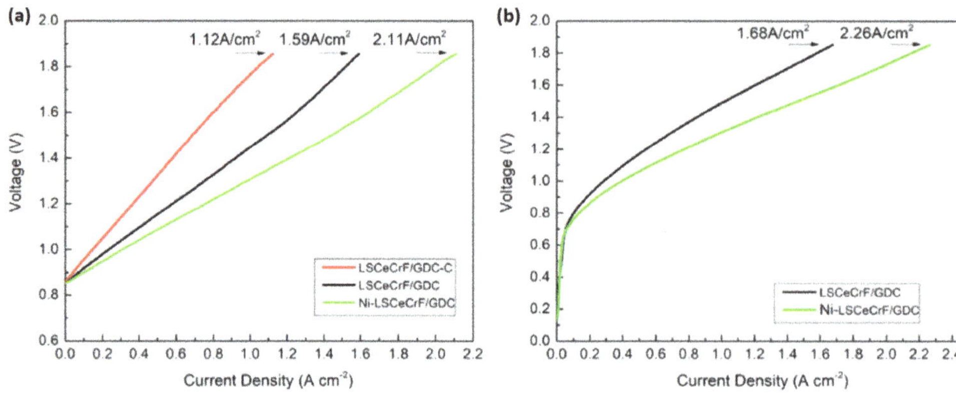

Figure 8.11. *i–V* curves of SOEC cells with different fuel electrodes at 850 °C in atmospheres of (a) CO_2/CO (70:30) and (b) pure CO_2. Adapted from [128], Copyright (2020), with permission from Elsevier.

Ding *et al* also fabricated SOEC cells (Ni+LSCeCrF@CGO-YSZ-LSCF@CGO) with perovskite-infiltrated CGO backbone electrodes for high-temperature CO_2 electrolysis [128]. $(La_{0.65}Sr_{0.3}Ce_{0.05})_{0.9}(Cr_{0.5}Fe_{0.5})_{0.85}Ni_{0.15}O_{3-\delta}$ (Ni-LSCeCrF) at a loading of 20–30 wt.% was deposited on a CGO scaffold and calcined at 900 °C to form the catalytic particles. The *i–V* curves in figure 8.11 show that a cell with the Ni-LSCeCrF@CGO electrode had high catalytic activity at 850 °C in pure CO_2 compared to those with LSCeCrF/CGO electrodes fabricated by either the conventional infiltration method or the wet infiltration method. When tested at 850 °C and at 1.6 V in pure CO_2, the cell exhibited a degradation of 149 m A cm^{-2} h^{-1}, and after 100 h the current density stabilized at 1.01 A cm^{-2}. Their previous study reported that the current density of a cell with a conventional LSCeCrF/CGO composite electrode dropped from 0.7 A cm^{-2} to 0.65 A cm^{-2} after 24 h at 0.7 V at 850 °C in a CO_2/CO (70:30) mixture [129]. It is clear that the infiltrated Ni-LSCeCrF@CGO electrode showed much better performance and durability than that of the conventional LSCeCrF/CGO electrode.

8.4.2.2 Electrode backbone materials with a perovskite structure

STN has been extensively studied as an SOFC fuel electrode material due to its high electronic conductivity. However, its use is limited by its insufficient electrocatalytic activity for H_2 oxidation. Infiltrating catalytic materials with high ionic conductivity and catalytic activity into STN backbones has thus been investigated; among these materials, CGO seems to be the most popular. Blennow *et al* studied CGO-infiltrated STN as a fuel electrode in a symmetrical cell tested in the temperature range of 600 °C–850 °C [95]. The R_p of the CGO-infiltrated STN was found to be smaller than that of a Ni/YSZ electrode, even at 850 °C. This is due to its low activation energy (0.5–0.7 eV) compared to that of Ni/YSZ cermet (0.8 eV). The infiltrated CGO@STN electrode also demonstrated great redox stability during redox cycling due to the good dimensional stability of the STN backbone in both oxidizing and reducing atmospheres. Due to their high stability when exposed to hydrocarbon fuels and their high catalytic activity, noble metal catalysts have also been infiltrated or co-infiltrated together with CGO into STN backbones. Hussain *et al* infiltrated binary and ternary electrocatalysts with various combinations of Pt, Ru, Pd, Ni, and CGO into STN backbones and tested the performance of symmetrical cells (STN-ScYSZ-STN) using these electrodes at 400 °C–600 °C [93]. Among the binary electrocatalysts, a Pt-CGO-infiltrated STN electrode showed the lowest R_p (0.16 Ω cm^2) at 600 °C in H_2/H_2O (97:3). Adding Ni to the Pt+CGO electrocatalysts further improved the electrode performance, and the lowest R_p of 0.11 Ω cm^2 and E_a of 0.85 eV were achieved by a Ni+Pt+CGO@STN electrode. The infiltrated nanoparticles were all uniformly distributed on the STN backbone, as shown in figure 8.12.

Ramos *et al* characterized STN/YSZ symmetrical cells co-infiltrated with *M*/CGO (*M* = Ni, Ru, Pd) electrocatalysts in various H_2/H_2O atmospheres at 650 °C–850 °C [94]. A Ru/CGO-infiltrated electrode had negligible degradation during 200 h of durability testing in H_2/H_2O (at a ratio of 50:50) at 850 °C with a stable R_p of 0.02 Ω cm^2. In comparison, Ni/CGO-infiltrated STN showed the highest degradation; its R_p value increased from 0.1 to 0.25 Ω cm^2 after the same testing period. In another test at 0.125 A cm^{-2} for 300 h, Ru/CGO-infiltrated STN also showed a significantly lower degradation of 0.04 mV h^{-1} than that of Ni/CGO-infiltrated STN (0.5–0.6 mV h^{-1}) [130]. The difference in electrode degradation might be due to

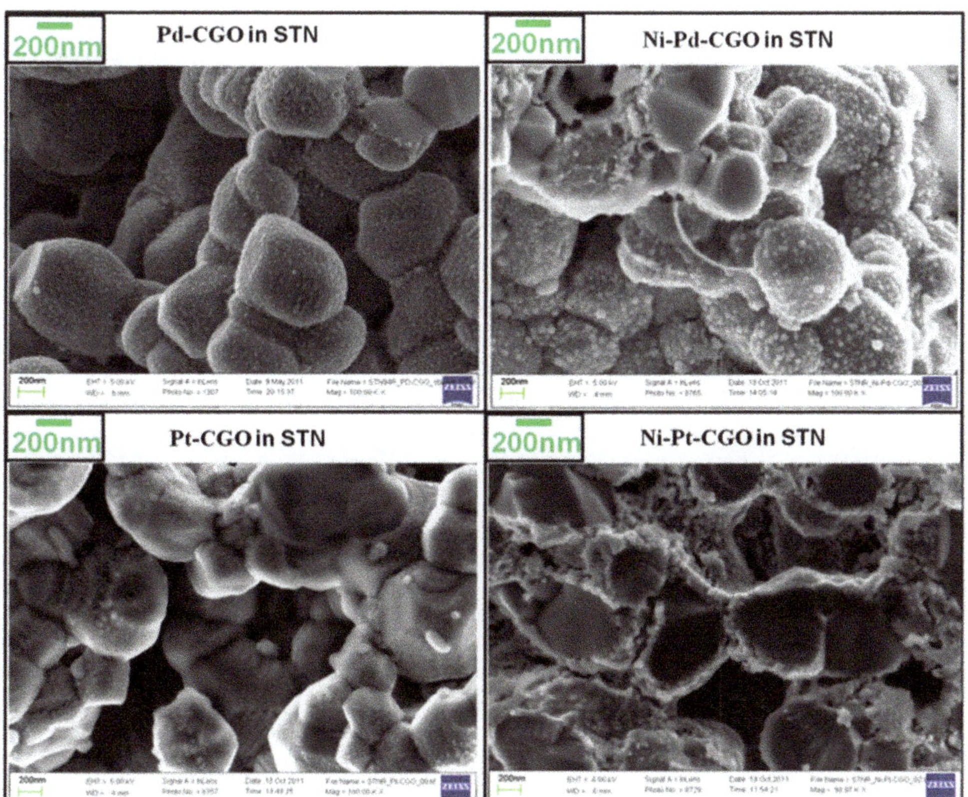

Figure 8.12. SEM images of fractured cross sections of post-tested STN fuel electrodes infiltrated with different infiltrates. Reprinted from [93], Copyright (2012), with permission from Elsevier.

the different degradation mechanisms that occur during testing, during which the Ni/CGO infiltrates may suffer from various degradation processes, such as particle growth and Ni oxidation. Recently, Drasbæk *et al* also studied the coking tolerance of symmetrical cells with infiltrated STN electrodes through *in situ* synchronous Raman spectroscopy and EIS [92]. When exposed to pure methane at 750 °C, they found that a Ni+Co+CGO@STN electrode had the highest coking resistance. This electrode showed also a stable R_p and a low degradation rate of 4.08 Ω cm^2 kh^{-1} during 300 h of durability testing in H_2/H_2O (50:50) at 650 °C and at the OCV.

LST perovskite oxides have also been explored as electrode materials due to their high electronic conductivity and good stability in SOC environments. Nielsen *et al* investigated metal-supported SOFC cells (Ni+CGO@LSFNT/ScYSZ/FeCr-ScYSZ-CGO-LSC) with Ni/CGO-infiltrated $La_{0.4}Sr_{0.4}Fe_{0.03}Ni_{0.03}Ti_{0.94}O_3$ (LSFNT)/ScYSZ/FeCr composite fuel electrodes at 650 °C–750 °C [131]. The results of performance tests showed that an infiltrated single cell with a 55% LSFNT, 40%FeCr, and 5% ScYSZ backbone was found to have the best peak power density when the temperature was higher than 700 °C (0.77 W cm^{-2} at 700 °C and 1.07 W cm^{-2} at 750 °C), while a cell with a higher FeCr content in its backbone (20%LSFNT/60%FeCr/20%ScYSZ) had the best performance at 650 °C with a peak power density of 0.52 W cm^{-2}.

Muzaffar *et al* doped Zn and Ca into LST and investigated the performance of an infiltrated $La_{0.2}Sr_{0.25}Ca_{0.45}Ti_{0.95}Zn_{0.05}O_{3-\delta}$ (LSCTZ) backbone in symmetrical cells at 650 °C–850 °C [132]. In comparison to Co- or Fe-infiltrated LSCTZ, the Ni/CGO-infiltrated backbone had the best performance with the lowest R_p and E_a under all testing conditions (see figure 8.13). In addition, the Ni/CGO-infiltrated LSCTZ was found to have lower R_p when the electrodes were tested in H_2/H_2O (50:50) than in H_2/H_2O (97:3) at 750 °C–850 °C.

The co-infiltration of noble metals (Rh, Pt, or Pd) with CGO into LST/CGO fuel electrode backbones was studied by Futamura *et al* for high-fuel-utilization SOFC operations [96]. Each electrolyte-supported SOFC single cell (*M*+CGO@LST/

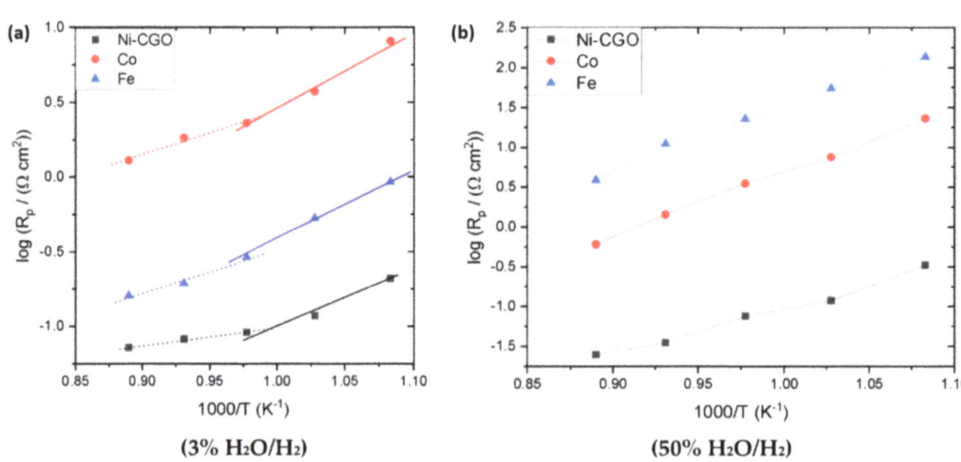

Figure 8.13. Arrhenius plots of the R_p of infiltrated cells as a function of different temperatures in atmospheres of: (a) H_2O/H_2 (3:97), and (b) H_2O/H_2 (50:50). Reproduced from [132]. CC BY 4.0.

CGO-ScSZ-LSM/ScSZ-LSM) was infiltrated with *M*/CGO (*M* = Ni, Rh, Pt, or Pd) and investigated in highly humid H_2 at 800 °C. Compared to the *i–V* performance and power density of a cell with a Ni/CGO co-infiltrated fuel electrode, cells with noble metal/CGO co-infiltrated fuel electrodes showed similar *i–V* performance and power density in 3% humidified H_2, while at higher oxygen partial pressures, the noble metal/CGO co-infiltrated cells showed slightly higher tolerance to Ni oxidation. Cost analysis suggests that infiltrating noble metals can be cost-effective as long as the loading is small.

Xu *et al* examined R-SOC (reversible SOC) cells (CeO_2@LSTF-ScSZ-LSM/ScSZ) with CeO_2-infiltrated $La_{0.3}Sr_{0.7}Ti_{0.3}Fe_{0.7}O_{3-\delta}$ (LSTF) fuel electrodes in various CO/CO_2 atmospheres at 700 °C–850 °C [97]. Cyclic stability testing was carried out at 800 °C in CO/CO_2 (50:50) in the voltage range of 0.5–1.5 V for ten cycles (see figure 8.14). A slow degradation behavior was observed in the EIS measurements before and after reversible operation at 0.5 V and 1.5 V. Based on the SEM images, the authors concluded that the cell degradation might result from a fuel electrode microstructure change caused by the agglomeration of the infiltrated particles.

As an alternative Ni-free SOC fuel electrode material, LSM-based perovskite has comparable performance to that of Ni/YSZ cermet. It benefits from the high conductivity of LSCM and good stability in reducing atmospheres. The infiltration of Pd into LSCM backbones to increase their catalytic activity has been widely investigated for applications related to hydrocarbon fuels. Jiang *et al* studied PdO@LSCM/YSZ composite fuel electrodes for methane and ethanol fuels in SOFCs at 800 °C [101]. A 0.25 M $Pd(NO_3)_2$ solution was infiltrated into a LSCM/YSZ fuel electrode which was then heated at 700 °C to form PdO, reaching a total loading of 0.36–0.46 mg cm^{-2} (for a 0.4 cm^2 fuel electrode area). A three-electrode arrangement was used with Pt paste as the counter and reference electrodes. EIS and *i–V* curves showed that the infiltrated PdO nanoparticles significantly reduced R_p and increased the cell's power output in CH_4 and C_2H_5OH atmospheres compared to the corresponding values for a non-infiltrated LSCM/YSZ fuel electrode (see figure 8.15).

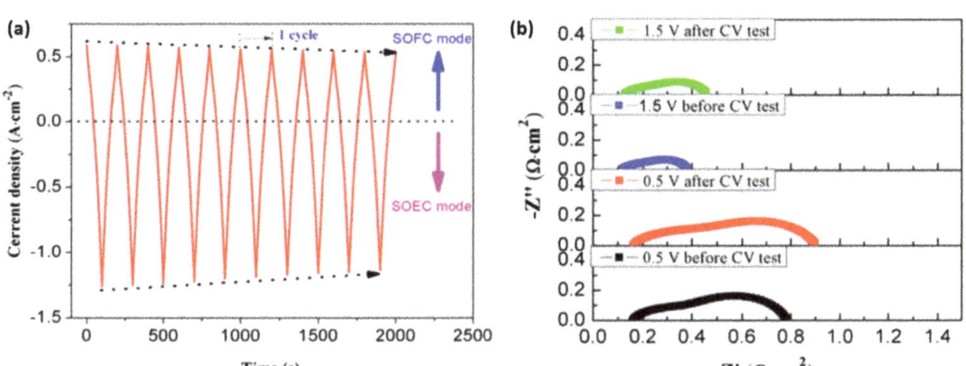

Figure 8.14. (a) Cyclic stability testing of the infiltrated cell at 800 °C in CO/CO_2 (50:50) in the voltage range of 0.5–1.5 V for ten cycles, and (b) EIS data before and after cyclic stability tests at 0.5 V and 1.5 V. Adapted from [97], Copyright (2017), with permission from Elsevier.

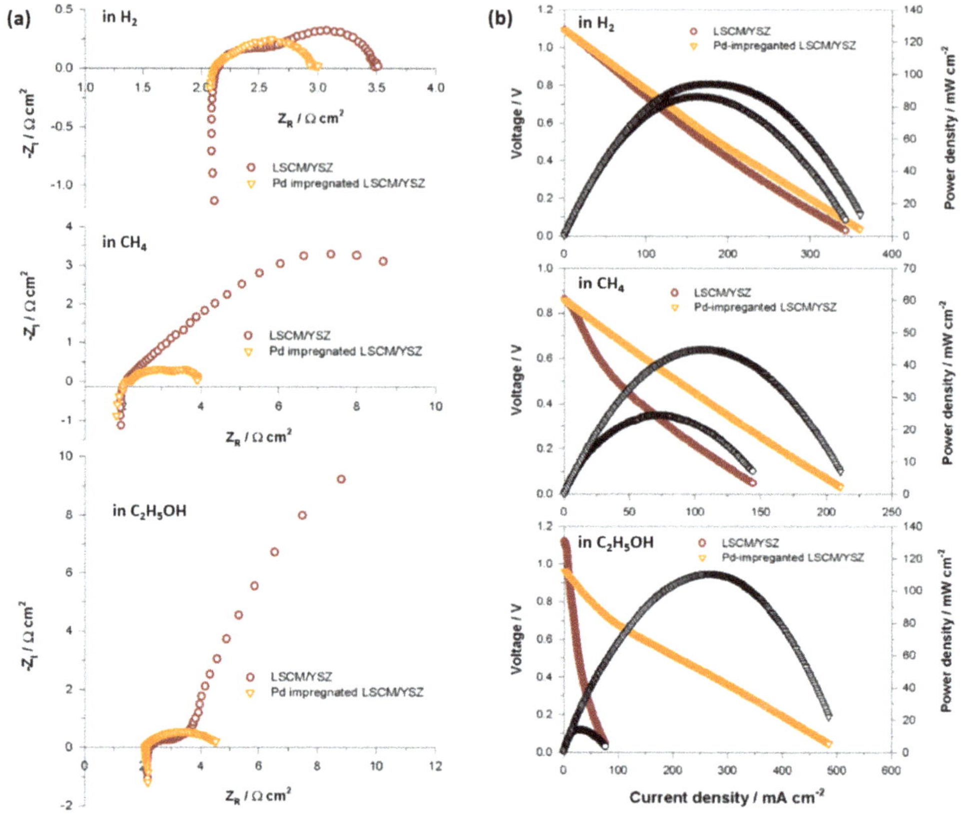

Figure 8.15. EIS data (a) and *i–V* curves with power density (b) of the LSCM/YSZ fuel electrode with and without Pd infiltration at 800 °C in different atmospheres. Adapted from [101], Copyright (2008), with permission from Elsevier.

In particular, the power output increased eight times for the infiltrated fuel electrode, reaching a power density of 111 mW cm^{-2} in C_2H_5OH. The size of the PdO particles increased slightly after the electrode was tested in methane, and no carbon deposition was observed. Ye *et al* also investigated the effects of different Pd loadings in PdO@LSCM/YSZ composite fuel electrodes in SOFCs for the direct utilization of methane at 800 °C [102]. Electrochemical results from the three-electrode arrangement showed that infiltrating a small amount of PdO (0.1 mg cm^{-2}) significantly decreased the R_p in CH_4/H_2O (97:3) by three times, from 6.5 Ω cm^2 for a non-infiltrated fuel electrode to 1.96 Ω cm^2 for a PdO@LSCM/YSZ electrode. Further increasing the Pd loading from 0.1–0.66 mg cm^{-2} barely changed the value of R_p. When the loading was more than 0.66 mg cm^{-2}, both R_p and R_s increased. The measured EIS spectra indicated that the infiltrated PdO nanoparticles mainly impact the low-frequency arc. The coexistence of Pd (under reducing environment) and PdO/PdO_x may promote the surface exchange and diffusion of oxygen species, thereby accelerating CH_4 decomposition in the LSCM/YSZ backbone.

Babaei *et al* modified a LCCM/CGO backbone by infiltrating a Pd catalyst into it and tested an electrode made from this material in humidified H_2 and methane at

750 °C–850 °C [104]. The effects of the LCCM/CGO backbone sintering temperatures and the PdO loading on the electrode performance were studied in detail in the three-electrode arrangement. It was found that the optimum sintering temperature of 1300 °C yielded a minimum R_s of 2.13 Ω cm^2 and an R_p of 0.47 Ω cm^2 at 850 °C in H_2/H_2O (97:3). With regard to the PdO loading, it was found that a higher PdO loading of 0.06 mg cm^{-2} effectively decreased R_p in both CH_4 and H_2 atmospheres. During 20 h of durability testing at 25 mA cm^{-2} and at 850 °C in humidified methane, a 0.06 mg cm^{-2} Pd@LCCM/CGO electrode showed a stable polarization potential of 0.815 V after 900 min, and no carbon deposition was observed in the post-mortem SEM analysis.

Modified LSCM-based electrodes have also been tested in the electrolysis mode. Yue and Irvine fabricated SOEC cells with CGO-infiltrated LSCM/YSZ fuel electrodes for CO_2 electrolysis [100]. The cells were prepared using a three-electrode configuration, with LSCM as the working electrode on one side of YSZ and LSM/ScSZ as both counter and reference electrodes. The CGO@LSCM electrode was further modified by infiltrating 0.5 wt.% Pd and showed excellent electrochemical performance in a CO_2/CO (70:30) mixtures at the OCV and at 900 °C with an R_p of only 0.22 Ω cm^2, compared to that of 0.42 Ω cm^2 for a CGO@LSCM composite electrode and 0.91 Ω cm^2 for a screen-printed LSCM/CGO electrode. In addition, the SOEC cell with the Pd+CGO@LSCM electrode also showed good durability at −0.15 A cm^{-2} at 900 °C. A stable cell potential of around 1.54 V was observed during 23 h of operation in a CO_2/CO (50:50) atmosphere, and R_p had barely changed by the end of the stability testing.

Spiridigliozzi *et al* studied the difference between aqueous and ethanol CGO infiltration solutions on the performance of a CGO@LSCM/YSZ-YSZ-ScSZ-LSM SOEC cell for CO_2 electrolysis [98]. It was found that a cell with a CGO-ethanol-solution-infiltrated LSCM/YSZ electrode had lower R_s and R_p compared to that of an CGO-aqueous-solution-infiltrated cell at 700 °C–800 °C at the OCV in CO_2/CO (70:30) mixtures. SEM images have revealed morphological differences between the nano-CGO particles made using different infiltration solutions. The nanoparticles made using the aqueous solution tend to agglomerate and fill up the porous surface after each calcination step. On the other hand, the non-aqueous CGO solution is more favorable for infiltration due to its well-dispersed nanoparticles and its prohibition of grain growth. Khameneh *et al* studied the performance of a $LaFe_{0.6}Co_{0.4}O_{3-\delta}$ (LFC)-infiltrated LSCM/YSZ fuel electrode for steam and co-electrolysis at 750 °C–850 °C [99]. A three-electrode configuration was used for electrochemical measurements with Pt as the counter and reference electrodes. EIS measurements showed a significant decrease in the R_p of the LFC@LSCM/YSZ electrode in both steam electrolysis and co-electrolysis mode (see figure 8.16). For example, at 850 °C and −1 V in 9% CO_2+20% steam gas mixtures, 0.6 M LFC-infiltrated LSCM/YSZ showed a significantly reduced R_p of 0.33 Ω cm^2 as compared to 1.34 Ω cm^2 for a non-infiltrated electrode. It was concluded that the infiltrated LFC nanoparticles mainly improved the middle- to low-frequency processes, i.e. the surface diffusion of absorbed or activated species as well as oxygen ion diffusion.

Zhang *et al* infiltrated $Ce_{0.9}Mn_{0.1}O_{2-\delta}$ (CMO) nanoparticles into a LSCM/CGO composite fuel electrode to improve the electrocatalytic activity of an SOEC cell

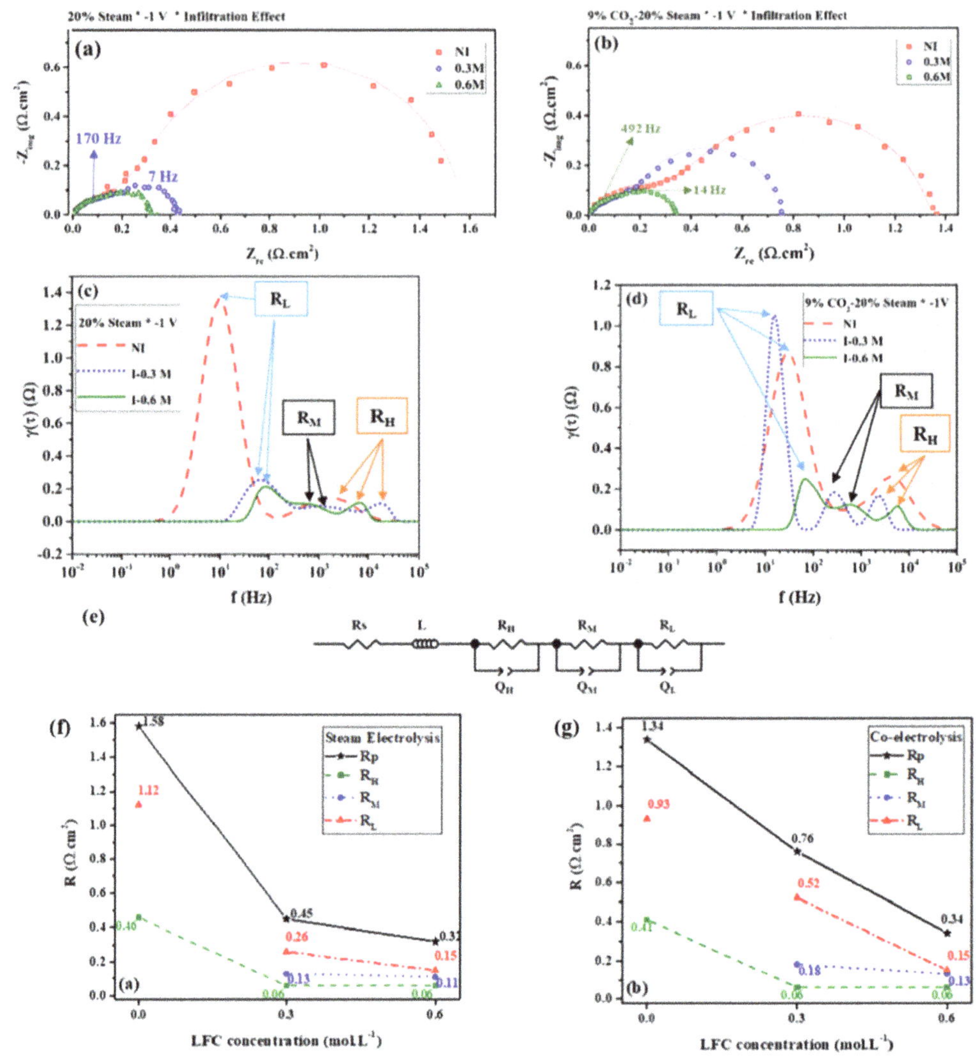

Figure 8.16. EIS and DRT plots of the non-infiltrated cell and the infiltrated cell at an applied voltage of −1 V in different test modes: (a), (c) steam electrolysis and (b), (d) co-electrolysis. (e) Equivalent circuit model for complex nonlinear least squares (CNLS) fitting, and the corresponding resistance data of different processes in (f) the steam electrolysis mode and (g) the co-electrolysis mode. Adapted from [99], Copyright (2018), with permission from Elsevier.

(LSCM/CGO-YSZ-LSM/ScSZ) for CO_2 electrolysis [103]. EIS at different applied voltages (1–1.6 V) at 800 °C demonstrated that a cell with a CMO@LSCM/CGO electrode significantly reduced the value of R_p compared to that of a cell with a conventional LSCM/CGO electrode from 4.09 Ω cm^2 (without infiltration) to 1.26 Ω cm^2 (infiltrated with CMO) both at 1 V, while R_s remained more or less unchanged. Furthermore, the CMO@LSCM/CGO-YSZ-LSM/ScSZ cell presented a stable current density of ~0.2 A cm^{-2} during 50 h of potentiostatic durability testing at 1.4 V for CO_2

reduction. DRT analysis also showed that the CMO catalysts mainly decreased the low-frequency resistance (0.1–10 Hz), indicating the promotion of CO_2 adsorption on the electrode surface and the dissociation of the carbonate intermediate processes at the TPB (see figure 8.17).

In addition to CGO, metals or metal oxides (in the form of nitrate solutions) are also considered for use as infiltrates to promote the electrical conductivity of LSCM-based electrodes. Xing *et al* infiltrated a 1 M Cu nitrate solution into a porous LSCM backbone and evaluated the performance of an SOEC cell (Cu@LSCM-LSGM-LSCF) that used this material for co-electrolysis at 750 °C–850 °C [105]. The *i–V* curves obtained showed that the current density of the infiltrated cell increased from 1.31 to 1.82 A cm^{-2} at 1.65 V in the electrolysis mode when the CO_2 content increased from 0% to 37.5% at 750 °C. In addition, R_s decreased significantly from 1.1 Ω cm^2 (for a non-infiltrated cell) to 0.25 Ω cm^2 (for the infiltrated cell) with H_2O/H_2 (50:50) as the fuel, indicating that the Cu nanoparticles effectively improved the electrical conductivity of the LSCM fuel electrode. Durability testing at −0.33 A cm^{-2} and 750 °C was also carried out with a gas mixture of $H_2O/H_2/CO_2$ at a ratio of 50:25:25, and the infiltrated cell showed a stable cell voltage of ~1.05 V during the 50 h test period. Millier *et al* studied SOFC cell (Ni@LSGM/SLT-LSGM-LSCF/CGO) performance using a Ni-infiltrated fuel electrode at intermediate temperatures with different calcination temperatures [106]. Based on

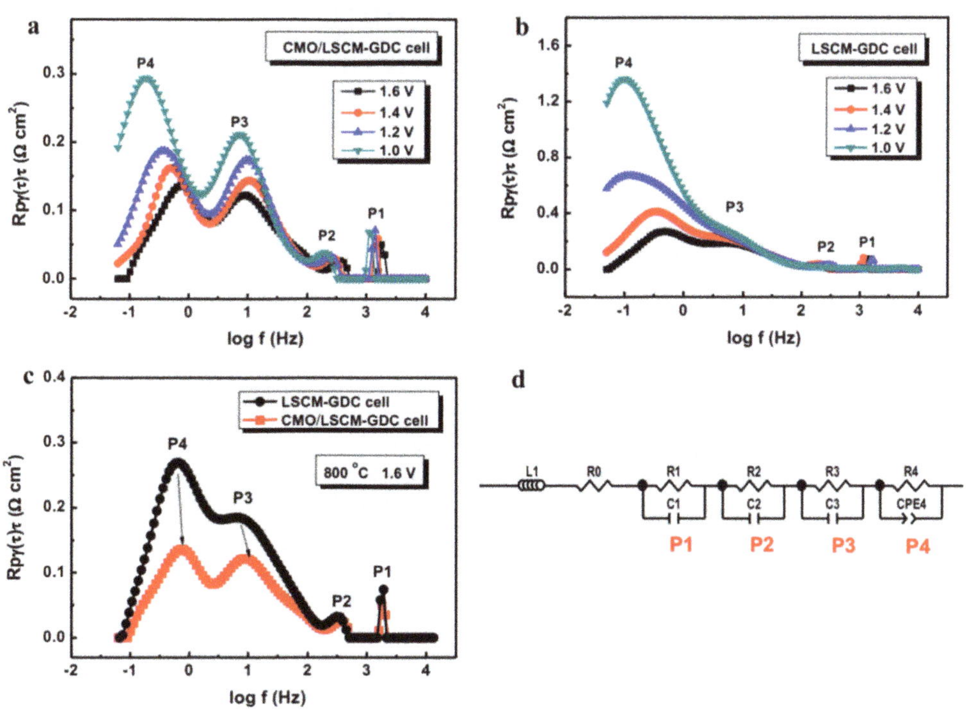

Figure 8.17. DRT plots of (a) a CMO@LSCM/CGO cell, and (b) a non-infiltrated LSCM/CGO cell at different applied voltages. (c) Comparisons of the DRT plots of two cells at 800 C and 1.6 V, and (d) an equivalent circuit model for CNLS fitting. Reprinted from [103], Copyright (2018), with permission from Elsevier.

SEM images, they found that lowering the calcination temperature from 700 °C to 450 °C can reduce the initial NiO particle size from 95 nm to 60 nm. However, during 550 h of testing at 650 °C at the OCV under H_2/H_2O (97:3), the R_p values continuously increased by a factor of two from ~0.08 Ω cm^2 to 0.16 Ω cm^2, which might be due to the inevitable Ni coarsening and a decrease in the TPB length, as also calculated from the electrode coarsening model.

$Sr_2Fe_{1.5}Mo_{0.5}O_6$ (double perovskite SFM) has been explored as a potential SOEC fuel electrode due to its high conductivity and excellent redox stability. Lv *et al* studied the performance of SOEC cells for CO_2 electrolysis at 800 °C. The cells consisted of a CGO@SFM fuel electrode with a CGO loading of 0–19.2 wt.%, a YSZ electrolyte, and an LSM/YSZ oxygen electrode [107]. Microstructural images of the backbone with different CGO loadings are presented in figure 8.18, among

Figure 8.18. SEM images of cross sections of (a), (b): the SFM-CGO-YSZ-LSM/YSZ cell, (c) the bare SFM fuel electrode, and SFM fuel electrodes infiltrated with CGO loadings of (d) 3.2 wt.%, (e) 6.4 wt.%, (f) 12.8 wt.%, (g) 16 wt.%, and (h) 19.2 wt.%. Reprinted from [107], Copyright (2018), with permission from Elsevier.

which figure 8.18(f) with 12 wt.% CGO loading has the most uniform distribution and the most continuous network of CGO on the SFM backbone. When the CGO loading was more than 12.8 wt.%, the infiltrated nanoparticles began to aggregate (figures 8.18(g) and (h)). The cell with a 12.8 wt.% CGO-infiltrated SFM reached the highest current density of 0.446 A cm^{-2} at 1.6 V, compared to 0.282 A cm^{-2} for the cell with bare SFM. Durability testing at 1.2 V and 800 °C was conducted on the cell with 12.8 wt.% CGO-infiltrated SFM, showing a degradation rate of 0.456 mA cm^{-2} h^{-1} with a faradaic efficiency of 96% during 55 h of testing.

8.5 Conclusions and outlook

In this chapter, the processing characteristics of the infiltration technique and the performance of the resulting nanostructured fuel electrodes have been reviewed. In short, the infiltration method improves the performance of conventional fuel electrodes by forming nanosized electrocatalysts on the surface of the porous electrode backbone, thus extending the reaction sites. As reported in the literature, applying the infiltration technique to both cermet-based or oxide-based fuel electrodes results in the following improvements: enhanced electrode/cell performance (particularly at intermediate temperatures), improved electrode/cell durability under current, enhanced carbon and/or sulfur tolerance in hydrocarbon fuel atmospheres (particularly for Ni-cermet fuel electrodes), greater control of the electrode microstructure than that available using conventional electrode fabrication techniques, more flexibility in the selection of electrode materials, and further cost reduction. The mechanisms responsible for the performance enhancement achieved by infiltration have also been extensively studied, and are generally ascribed to the following: the creation of extra TPB active sites for species adsorption and electro-reduction, the enhancement of oxygen ion transport, the prohibition of agglomeration and the grain growth of backbone particles (e.g. Ni), the provision of high catalytic activity or sufficient ionic and/or electronic conductivity, etc.

Though great achievements have been obtained in the fabrication of high-performance infiltrated electrodes, the development of the infiltration technique remains still at the lab scale, at a technology readiness level (TRL) of three to four. So far, most of the studies of infiltrated electrodes have been realized at the button cell level. Only a few studies have reported the performance of large planar cells or stacks. The most challenging aspects of infiltration include the optimization of the processing steps/cycles, the industrialization of this technique, and the long-term microstructural stability of the infiltrated catalyst nanoparticles. Different research groups have studied different aspects of the infiltration process. However, no common protocol for the selection of catalysts or infiltration processing parameters has been established for fabricating infiltrated electrodes. Even with the above limitations, the infiltration method is probably still the most effective way to fabricate nanoengineered electrodes with large numbers of catalytically active sites without affecting the microstructure of the electrode backbone. The future development of the infiltration technique should focus on controlling the catalyst/backbone morphology, minimizing chemical interactions during catalyst formation and operation, and further optimizing the processing steps, so that this promising technique can be industrialized on a large scale for various SOC applications.

References

[1] Singhal S C and Kendall K 2003 Introduction to SOFCs *High Temperature and Solid Oxide Fuel Cells* (Amsterdam: Elsevier), 1–22

[2] Hauch A, Kungas R, Blennow P, Hansen A B, Hansen J B, Mathiesen B V and Mogensen M B 2020 Recent advances in solid oxide cell technology for electrolysis *Science* **370** eaba6118

[3] Ishihara T, Sammes N M and Yamamoto O 2003 Electrolytes *High Temperature and Solid Oxide Fuel Cells* (Amsterdam: Elsevier) pp 83–117

[4] Mahmud L S, Muchtar A and Somalu M R 2017 Challenges in fabricating planar solid oxide fuel cells: a review *Renew. Sustain. Energy Rev.* **72** 105–16

[5] Li G, Gou Y, Qiao J, Sun W, Wang Z and Sun K 2020 Recent progress of tubular solid oxide fuel cell: from materials to applications *J. Power Sources* **477** 228693

[6] Williams M C, Vora S D and Jesionowski G 2020 Worldwide status of solid oxide fuel cell technology *ECS Trans.* **96** 1–10

[7] Jiang S P and Chan S H 2004 A review of anode materials development in solid oxide fuel cells *J. Mater. Sci.* **39** 4405–39

[8] Jiang S P 2012 Nanoscale and nano-structured electrodes of solid oxide fuel cells by infiltration: advances and challenges *Int. J. Hydrog. Energy* **37** 449–70

[9] Pihlatie M H, Kaiser A, Mogensen M and Chen M 2011 Electrical conductivity of Ni–YSZ composites: degradation due to Ni particle growth *Solid State Ion.* **189** 82–90

[10] Talebi T, Sarrafi M H, Haji M, Raissi B and Maghsoudipour A 2010 Investigation on microstructures of NiO–YSZ composite and Ni–YSZ cermet for SOFCs *Int. J. Hydrog. Energy* **35** 9440–47

[11] Yi Z, Zhai F, Tian H and Xie Z 2012 Preparation and performance of Ni-YSZ/YSZ co-ceramic using multiple gel-tape casting *KEM* **512–515** 1555–8

[12] Yao W and Croiset E 2013 Ni/YSZ pattern anodes fabrication and their microstructure and electrochemical behavior changes in H_2–H_2O environments *J. Power Sources* **226** 162–72

[13] Hauch A, Brodersen K, Chen M and Mogensen M B 2016 Ni/YSZ electrodes structures optimized for increased electrolysis performance and durability *Solid State Ion.* **293** 27–36

[14] Mogensen M B, Hauch A, Sun X, Chen M, Tao Y, Ebbesen S D, Hansen K V and Hendriksen P V 2017 Relation between Ni particle shape change and Ni migration in Ni-YSZ electrodes—a hypothesis *Fuel Cells* **17** 434–41

[15] Mogensen M B, Chen M, Frandsen H L, Graves C, Hauch A, Hendriksen P V, Jacobsen T, Jensen S H, Skafte T L and Sun X 2021 Ni migration in solid oxide cell electrodes: review and revised hypothesis *Fuel Cells* **21** 415–29

[16] Finnerty C M, Coe N J, Cunningham R H and Ormerod R M 1998 Carbon formation on and deactivation of nickel-based/zirconia anodes in solid oxide fuel cells running on methane *Catal. Today* **46** 137–45

[17] Lin Y, Zhan Z and Barnett S A 2006 Improving the stability of direct-methane solid oxide fuel cells using anode barrier layers *J. Power Sources* **158** 1313–6

[18] Timmermann H, Sawady W, Campbell D, Weber A, Reimert R and Ivers-Tiffée E 2008 Coke formation and degradation in SOFC operation with a model reformate from liquid hydrocarbons *J. Electrochem. Soc.* **155** B356

[19] Kan H and Lee H 2010 Sn-doped Ni/YSZ anode catalysts with enhanced carbon deposition resistance for an intermediate temperature SOFC *Appl. Catalysis* **B 97** 108–14
[20] Li C, Shi Y and Cai N 2013 Carbon deposition on nickel cermet anodes of solid oxide fuel cells operating on carbon monoxide fuel *J. Power Sources* **225** 1–8
[21] Matsuzaki Y 2000 The poisoning effect of sulfur-containing impurity gas on a SOFC anode: part I. Dependence on temperature, time, and impurity concentration *Solid State Ion.* **132** 261–9
[22] Gong M, Liu X, Trembly J and Johnson C 2007 Sulfur-tolerant anode materials for solid oxide fuel cell application *J. Power Sources* **168** 289–98
[23] Horita T, Kishimoto H, Yamaji K, Brito M B, Xiong Y, Yokokawa H, Hori Y and Miyachi I 2009 Effects of impurities on the degradation and long-term stability for solid oxide fuel cells *J. Power Sources* **193** 194–8
[24] Hauch A and Mogensen M 2010 Ni/YSZ electrode degradation studied by impedance spectroscopyEffects of gas cleaning and current density *Solid State Ion.* **181** 745–53
[25] Mogensen M, Jensen K V, Jørgensen M J and Primdahl S 2002 Progress in understanding SOFC electrodes *Solid State Ion.* **150** 123–9
[26] Ping X Y, Meng B, Liang W K, Zheng Q and Li C 2021 Electrical conductivity of Y_2O_3-doped CeO_2 based composite ceramics by spark plasma sintering: the effects of a second phase of $CeAlO_3$ *Solid State Ion.* **365** 115653
[27] Leonide A, Sonn V, Weber A and Ivers-Tiffée E 2008 Evaluation and modeling of the cell resistance in anode-supported solid oxide fuel cells *J. Electrochem. Soc.* **155** B36
[28] Ignatiev A, Chen X, Wu N, Lu Z and Smith L 2008 Nanostructured thin solid oxide fuel cells with high power density *Dalton Trans.* **40** 5501–6
[29] Wachsman E D and Lee K T 2011 Lowering the temperature of solid oxide fuel cells *Science* **334** 935–9
[30] Park J H *et al* 2016 Impact of nanostructured anode on low-temperature performance of thin-film-based anode-supported solid oxide fuel cells *J. Power Sources* **315** 324–30
[31] Knibbe R, Traulsen M L, Hauch A, Ebbesen S D and Mogensen M 2010 Solid Oxide electrolysis cells: degradation at high current densities *J. Electrochem. Soc.* **157** B1209
[32] Chen M *et al* 2013 Microstructural degradation of Ni/YSZ electrodes in solid oxide electrolysis cells under High current *J. Electrochem. Soc.* **160** F883–91
[33] Hjalmarsson P, Sun X, Liu Y-L and Chen M 2014 Durability of high performance Ni–yttria stabilized zirconia supported solid oxide electrolysis cells at high current density *J. Power Sources* **262** 316–22
[34] Marina O 1999 A solid oxide fuel cell with a gadolinia-doped ceria anode: preparation and performance *Solid State Ion.* **123** 199–208
[35] Kim-Lohsoontorn P and Bae J 2011 Electrochemical performance of solid oxide electrolysis cell electrodes under high-temperature co-electrolysis of steam and carbon dioxide *J. Power Sources* **196** 7161–8
[36] Riegraf M, Yurkiv V, Costa R, Schiller G and Friedrich K A 2017 Evaluation of the effect of sulfur on the performance of nickel/gadolinium-doped ceria based solid oxide fuel cell anodes *ChemSusChem* **10** 587–99
[37] Mewafy B *et al* 2019 Influence of surface state on the electrochemical performance of nickel-based cermet electrodes during steam electrolysis *ACS Appl. Energy Mater.* **2** 7045–55
[38] Sciazko A, Shimura T, Komatsu Y and Shikazono N 2021 Ni-GDC and Ni-YSZ electrodes operated in solid oxide electrolysis and fuel cell modes *J. Therm. Sci. Technol.* **16** 1–10

[39] Maric R 1998 High-Performance Ni-SDC cermet anode for solid oxide fuel cells at medium operating temperature *Electrochem. Solid-State Lett.* **1** 201
[40] Uchida H, Suzuki S and Watanabe M 1999 High performance electrode for medium-temperature solid oxide fuel cells *Electrochem. Solid-State Lett.* **6** A174
[41] Yin Y, Zhu W, Xia C and Meng G 2004 Gel-cast NiO–SDC composites as anodes for solid oxide fuel cells *J. Power Sources* **132** 36–41
[42] Misono T *et al* 2006 Ni-SDC cermet anode fabricated from NiO–SDC composite powder for intermediate temperature SOFC *J. Power Sources* **157** 754–7
[43] Chen M, Kim B H, Xu Q, Nam O J and Ko J H 2008 Synthesis and performances of Ni–SDC cermets for IT-SOFC anode *J. Eur. Ceram. Soc.* **28** 2947–53
[44] Ukai K 2001 Current status of SOFC development using scandia doped Zirconia *Electrochem. Soc. Proc.* **2001–16** 375–83
[45] Sumi H *et al* 2004 Performance of nickel-scandia-stabilized zirconia cermet anodes for SOFCs in 3% H_2O–CH_4 *Solid State Ion.* **174** 151–6
[46] Sakai N, Yamaji K, Horita T, Xiong Y P and Yokokawa H 2005 Rare-earth materials for solid oxide fuel cells (SOFC) *Handbook on the Physics and Chemistry of Rare Earths* vol 35 (Amsterdam: Elsevier) pp 1–43
[47] Ke K *et al* 2006 Effect of oxide on carbon deposition behavior of CH_4 fuel on Ni/ScSZ cermet anode in high temperature SOFCs *Solid State Ion.* **177** 541–7
[48] Matsui T, Iida T, Kikuchi R, Kawano M, Inagaki T and Eguchi K 2008 Carbon Deposition over Ni–ScSZ anodes subjected to various heat-treatments for internal reforming of solid oxide fuel cells *J. Electrochem. Soc.* **155** B1136
[49] Zhang C *et al* 2007 Ionic conductivity and its temperature dependence of atmospheric plasma-sprayed yttria stabilized zirconia electrolyte *Mater. Sci. Eng.* B **137** 24–30
[50] Shao Z and Tadé M O 2016 *Intermediate-Temperature Solid Oxide Fuel Cells.* (Berlin: Springer)
[51] Mukundan R, Brosha E L and Garzon F H 2004 Sulfur tolerant anodes for SOFCs *Electrochem. Solid-State Lett.* **7** A5
[52] Marina O A, Canfield N L and Stevenson J W 2002 Thermal, electrical, and electrocatalytical properties of lanthanum-doped strontium titanate *Solid State Ion.* **149** 21–8
[53] Zhou X, Yan N, Chuang K T and Luo J 2014 Progress in La-doped $SrTiO_3$ (LST)-based anode materials for solid oxide fuel cells *RSC Adv.* **4** 118–31
[54] Irvine J T S, Slater P R and Wright P A 1996 Synthesis and electrical characterisation of the perovskite niobate-titanates *Ionics* **2** 213–6
[55] Blennow P, Hansen K K, Wallenberg L R and Mogensen M 2009 Electrochemical characterization and redox behavior of Nb-doped $SrTiO_3$ *Solid State Ion.* **180** 63–70
[56] Sudireddy B R, Blennow P and Nielsen K A 2012 Microstructural and electrical characterization of Nb-doped $SrTiO_3$–YSZ composites for solid oxide cell electrodes *Solid State Ion.* **216** 44–9
[57] Lim C K, Liu Q, Zhou J, Sun Q and Chan S H 2017 Sulfur tolerant LSCM-based composite cathode for high temperature electrolysis/Co-electrolysis of H_2O and CO_2 *Fuel Cells* **17** 464–72
[58] Tao S and Irvine J T S 2003 A redox-stable efficient anode for solid-oxide fuel cells *Nat. Mater* **2** 320–3
[59] Jin C, Yang C, Zhao F, Cui D and Chen F 2011 $La_{0.75}Sr_{0.25}Cr_{0.5}Mn_{0.5}O_3$ as hydrogen electrode for solid oxide electrolysis cells *Int. J. Hydrog. Energy* **36** 3340–6

[60] Li Y, Gan Y, Wang Y, Xie K and Wu Y 2013 Composite cathode based on Ni-loaded $La_{0.75}Sr_{0.25}Cr_{0.5}Mn_{0.5}O_{3-\delta}$ for direct steam electrolysis in an oxide-ion-conducting solid oxide electrolyzer *Int. J. Hydrog. Energy* **38** 10196–207

[61] Blennow P *et al* 2011 Manufacturing and characterization of metal-supported solid oxide fuel cells *J. Power Sources* **196** 7117–25

[62] Jiang S P 2003 A comparative study of fabrication and performance of Ni/3 mol % $Y2O_3$–ZrO_2 and Ni/8 mol % $Y2O_3$–ZrO_2 cermet electrodes *J. Electrochem. Soc.* **150** E548–59

[63] Kan W H, Samson A J and Thangadurai V 2016 Trends in electrode development for next generation solid oxide fuel cells *J. Mater. Chem.* A **4** 17913–32

[64] Jiang S P 2006 A review of wet impregnation—an alternative method for the fabrication of high performance and nano-structured electrodes of solid oxide fuel cells *Mater. Sci. Eng.* A **418** 199–210

[65] Gorte R J and Vohs J M 2009 Nanostructured anodes for solid oxide fuel cells *Curr. Opin. Colloid Interface Sci.* **14** 236–44

[66] Liu Z, Liu B, Ding D, Liu M, Chen F and Xia C 2013 Fabrication and modification of solid oxide fuel cell anodes via wet impregnation/infiltration technique *J. Power Sources* **237** 243–59

[67] Connor P A *et al* 2018 Tailoring SOFC electrode microstructures for improved performance *Adv. Energy Mater.* **8** 1800120

[68] Jiang S P, Duan Y Y and Love J G 2002 Fabrication of high-performance Ni/Y_2O_3–ZrO_2 cermet anodes of solid oxide fuel cells by ion impregnation *J. Electrochem. Soc.* **149** A1175–83

[69] Vohs J M and Gorte R J 2009 High-performance SOFC cathodes prepared by infiltration *Adv. Mater.* **21** 943–56

[70] Küngas R, Kim J-S, Vohs J M and Gorte R J 2011 Restructuring porous YSZ by treatment in hydrofluoric acid for use in SOFC cathodes: restructuring porous YSZ by treatment in HF *J. Am. Ceram. Soc.* **94** 2220–24

[71] Chen D, Yang G, Ciucci F, Tadé M O and Shao Z 2014 3D core–shell architecture from infiltration and beneficial reactive sintering as highly efficient and thermally stable oxygen reduction electrode *J. Mater. Chem.* A **2** 1284–93

[72] Sase M, Ueno D, Yashiro K, Kaimai A, Kawada T and Mizusaki J 2005 Interfacial reaction and electrochemical properties of dense (La,Sr) $CoO_{3-\delta}$ cathode on YSZ (1 0 0) *J. Phys. Chem. Solids* **66** 343–8

[73] Samson A, Søgaard M, Knibbe R and Bonanos N 2011 High performance cathodes for solid oxide fuel cells prepared by infiltration of $La_{0.6}Sr_{0.4}CoO_{3-\delta}$ into Gd-doped ceria *J. Electrochem. Soc.* **158** B650

[74] Lou X, Liu Z, Wang S, Xiu Y, Wong C P and Liu M 2010 Controlling the morphology and uniformity of a catalyst-infiltrated cathode for solid oxide fuel cells by tuning wetting property *J. Power Sources* **195** 419–24

[75] Nicholas J D and Barnett S A 2010 Measurements and modeling of $Sm_{0.5}Sr_{0.5}CoO_{3-x}$–$Ce_{0.9}Gd_{0.1}O_{1.95}$ SOFC cathodes produced using infiltrate solution additives *J. Electrochem. Soc.* **157** B536–41

[76] Ozmen O, Lee S, Hackett G, Abernathy H, Zondlo J W and Sabolsky E M 2021 Efficient and controlled nano-catalyst solid-oxide fuel cell electrode infiltration with poly-norepinephrine surface modification *J. Power Sources* **485** 229–32

[77] Mo B, Rix J, Pal U, Basu S and Gopalan S 2020 Improving SOFC anode electrocatalytic activity using nanoparticle infiltration into MIEC compositions *J. Electrochem. Soc.* **167** 134506

[78] Tong X, Hendriksen P V, Hauch A, Sun X and Chen M 2020 An up-scalable, infiltration-based approach for improving the durability of Ni/YSZ electrodes for solid oxide cells *J. Electrochem. Soc.* **167** 024519

[79] Lu Y, Gasper P, Nikiforov A Y, Pal U B, Gopalan S and Basu S N 2019 Co-infiltration of nickel and mixed conducting $Gd_{0.1}Ce_{0.9}O_{2-\delta}$ and $La_{0.6}Sr_{0.3}Ni_{0.15}Cr_{0.85}O_{3-\delta}$ phases in Ni-YSZ anodes for improved stability and performance *JOM: J. Miner., Metals Mater. Soc.* **71** 3835–47

[80] Jeong H, Kim S, Bae Y, Yoon K J, Lee J and Hong J 2019 Effect of Fe infiltration to Ni/YSZ solid-oxide-cell fuel electrode on steam/CO_2 co-electrolysis *Int. J. Energy Res.* **43** 4949–58

[81] Ovtar S, Tong X, Bentzen J J, Thydén K T S, Simonsen S B and Chen M 2019 Boosting the performance and durability of Ni/YSZ cathode for hydrogen production at high current densities via decoration with nano-sized electrocatalysts *Nanoscale* **11** 4394–406

[82] Zheng M, Wang S, Yang Y and Xia C 2018 Barium carbonate as a synergistic catalyst for the H_2O/CO_2 reduction reaction at Ni–yttria stabilized zirconia cathodes for solid oxide electrolysis cells *J. Mater. Chem.* A **6** 2721–9

[83] Mitchell-Williams T B *et al* 2017 Infiltration of commercially available, anode supported SOFC's via inkjet printing *Mater. Renew. Sustain. Energy* **6** 12

[84] Kiebach R *et al* 2016 Infiltration of SOFC stacks: evaluation of the electrochemical performance enhancement and the underlying changes in the microstructure *Fuel Cells* **16** 80–8

[85] McIntyre M D, Kirtley J D, Singh A, Islam S, Hill J M and Walker R A 2015 Comparing *in situ* carbon tolerances of sn-infiltrated and bao-infiltrated Ni-YSZ cermet anodes in solid oxide fuel cells exposed to methane *J. Phys. Chem.* C **119** 7637–47

[86] Jin Y, Yasutake H, Yamahara K and Ihara M 2010 Improved electrochemical properties of Ni/YSZ anodes infiltrated by proton conductor SZY in solid oxide fuel cells with dry methane fuel: dependence on amount of SZY *Chem. Eng. Sci.* **65** 597–602

[87] Timurkutluk B, Timurkutluk C, Mat M D and Kaplan Y 2011 Anode-supported solid oxide fuel cells with ion conductor infiltration *Int. J. Energy Res.* **35** 1048–55

[88] Wang W, Jiang S P, Tok A I Y and Luo L 2006 GDC-impregnated Ni anodes for direct utilization of methane in solid oxide fuel cells *J. Power Sources* **159** 68–72

[89] Yasutake H, Jin Y, Yamahara K and Ihara M 2010 Effect of proton conductor $SrCe_{0.95}Yb_{0.05}O_{3-\delta}$ on electrochemical characteristics of nickel/gadolinium-doped ceria anode in dry methane *J. Electrochem. Soc.* **157** B1370–5

[90] Babaei A, Zhang L, Tan S L and Jiang S P 2010 Pd-promoted $(La,Ca)(Cr,Mn)O_3$/GDC anode for hydrogen and methane oxidation reactions of solid oxide fuel cells *Solid State Ion.* **181** 1221–8

[91] Nielsen J, Klemensø T and Blennow P 2012 Detailed impedance characterization of a well performing and durable Ni:CGO infiltrated cermet anode for metal-supported solid oxide fuel cells *J. Power Sources* **219** 305–16

[92] Drasbæk D B, Traulsen M L, Sudireddy B R and Holtappels P 2019 Combining transition metals—an approach towards high-performing coking tolerant solid oxide fuel cell anodes *ECS Trans.* **91** 1953–61

[93] Hussain A M, Høgh J V T, Zhang W and Bonanos N 2012 Efficient ceramic anodes infiltrated with binary and ternary electrocatalysts for SOFCs operating at low temperatures *J. Power Sources* **216** 308–13

[94] Ramos T, Veltze S, Sudireddy B R and Holtappels P 2014 Impedance and stability of M/CGO (M: Ni, Pd, Ru) Co-infiltrated Nb-doped $SrTiO_3$ SOFC anodes *ECS Electrochem. Lett.* **3** F5–6

[95] Blennow P, Kammer Hansen K, Wallenberg L R and Mogensen M 2008 Strontium titanate-based composite anodes for solid oxide fuel cells *ECS Trans.* **13** 181–94

[96] Futamura S, Muramoto A, Tachikawa Y, Matsuda J, Lyth S M, Shiratori Y, Taniguchi S and Sasaki K 2019 SOFC anodes impregnated with noble metal catalyst nanoparticles for high fuel utilization *Int. J. Hydrog. Energy* **44** 8502–18

[97] Xu J, Zhou X, Pan L, Wu M and Sun K 2017 Oxide composite of $La_{0.3}Sr_{0.7}Ti_{0.3}Fe_{0.7}O_{3-\delta}$ and CeO_2 as an active fuel electrode for reversible solid oxide cells *J. Power Sources* **371** 1–9

[98] Spiridigliozzi L, Di Bartolomeo E, Dell'Agli G and Zurlo F 2020 GDC-based infiltrated electrodes for solid oxide electrolyzer cells (SOECs) *Appl. Sci.* **10** 3882

[99] Khameneh M K and Babaei A 2019 Co-electrolysis of CO_2 and H_2O on $LaFe_{0.6}Co_{0.4}O_3$ promoted $La_{0.75}Sr_{0.25}Cr_{0.5}Mn_{0.5}O_3$/YSZ electrode in solid oxide electrolysis cell *Electrochim. Acta* **299** 132–42

[100] Yue X and Irvine J T S 2017 Modification of LSCM–GDC cathodes to enhance performance for high temperature CO_2 electrolysis using solid oxide electrolysis cells (SOECs) *J. Mater. Chem.* A **5** 7081–90

[101] Jiang S P, Ye Y, He T and Ho S B 2008 Nanostructured palladium–$La_{0.75}Sr_{0.25}Cr_{0.5}Mn_{0.5}O_3/Y_2O_3$–$ZrO_2$ composite anodes for direct methane and ethanol solid oxide fuel cells *J. Power Sources* **185** 179–82

[102] Ye Y, He T, Li Y, Tang E H, Reitz T L and Jiang S P 2008 Pd-promoted $La_{0.75}Sr_{0.25}Cr_{0.5}Mn_{0.5}O_3$/YSZ composite anodes for direct utilization of methane in SOFCs *J. Electrochem. Soc.* **155** B811–8

[103] Zhang X, Song Y, Guan F, Zhou Y, Lv H, Wang G and Bao X 2018 Enhancing electrocatalytic CO_2 reduction in solid oxide electrolysis cell with $Ce_{0.9}Mn_{0.1}O_{2-\delta}$ nano-particles-modified LSCM-GDC cathode *J. Catal.* **359** 8–16

[104] Babaei A, Zhang L, Tan S L and Jiang S P 2010 Pd-promoted $(La,Ca)(Cr,Mn)O_3$/GDC anode for hydrogen and methane oxidation reactions of solid oxide fuel cells *Solid State Ion.* **181** 1221–8

[105] Xing R, Wang Y, Zhu Y, Liu S and Jin C 2015 Co-electrolysis of steam and CO_2 in a solid oxide electrolysis cell with $La_{0.75}Sr_{0.25}Cr_{0.5}Mn_{0.5}O_{3-\delta}$–Cu ceramic composite electrode *J. Power Sources* **274** 260–4

[106] Miller E C, Sherman Q, Gao Z, Voorhees P W and Barnett S A 2015 Stability of nickel-infiltrated anodes in intermediate temperature SOFCs *ECS Trans.* **68** 1245–54

[107] Lv H *et al* 2019 Infiltration of $Ce_{0.8}Gd_{0.2}O_{1.9}$ nanoparticles on $Sr_2Fe_{1.5}Mo_{0.5}O_{6-\delta}$ cathode for CO_2 electroreduction in solid oxide electrolysis cell *J. Energy Chem.* **35** 71–8

[108] Sengodan S, Kim J, Shin J and Kim G 2011 Thermodynamic properties, defect analysis, and electrical conductivity of the $La_{0.8}Sr_{0.2}Sc_xMn_{1-x}O_{3-\delta}$ infiltrated into YSZ scaffolds *J. Electrochem. Soc.* **158** B1373

[109] Cimenti M, Alzate-Restrepo V and Hill J M 2010 Direct utilization of methanol on impregnated Ni/YSZ and Ni–$Zr_{0.35}Ce_{0.65}O_2$/YSZ anodes for solid oxide fuel cells *J. Power Sources* **195** 4002–12

[110] Kim G *et al* 2009 Investigation of the structural and catalytic requirements for high-performance SOFC anodes formed by infiltration of LSCM *Electrochem. Solid-State Lett.* **12** B48

[111] Bidrawn F, Kim G, Corre G, Irvine J T S, Vohs J M and Gorte R J 2008 Efficient reduction of CO_2 in a solid oxide electrolyzer *Electrochem. Solid-State Lett.* **11** B167–70

[112] Lee S, Kim G, Vohs J M and Gorte R J 2008 SOFC anodes based on infiltration of $La_{0.3}Sr_{0.7}TiO_3$ *J. Electrochem. Soc.* **155** B1179–83

[113] Busawon A N, Sarantaridis D and Atkinson A 2008 Ni infiltration as a possible solution to the redox problem of SOFC anodes *Electrochem. Solid-State Lett.* **11** B186

[114] Jung S, Lu C, He H, Ahn K, Gorte R J and Vohs J M 2006 Influence of composition and Cu impregnation method on the performance of Cu/CeO_2/YSZ SOFC anodes *J. Power Sources* **154** 42–50

[115] Liu L *et al* 2017 Composite ceramic cathode $La_{0.9}Ca_{0.1}Fe_{0.9}Nb_{0.1}O_{3-\delta}$/$Sc_{0.2}Zr_{0.8}O_{2-\delta}$ towards efficient carbon dioxide electrolysis in zirconia-based high temperature electrolyser *Int. J. Hydrog. Energy* **42** 14905–15

[116] Skafte T L, Sudireddy B R, Blennow P and Graves C 2016 Carbon and redox tolerant infiltrated oxide fuel-electrodes for solid oxide cells *ECS Trans.* **72** 201–14

[117] Duboviks V, Lomberg M, Maher R C, Cohen L F, Brandon N P and Offer G J 2015 Carbon deposition behaviour in metal-infiltrated gadolinia doped ceria electrodes for simulated biogas upgrading in solid oxide electrolysis cells *J. Power Sources* **293** 912–21

[118] Sun L, Zheng T, Hu Z, Luo L, Wu Y, Xu X, Cheng L and Shi J 2015 Nano Ru impregnated Ni-YSZ anode as carbon resistance layer for direct ethanol solid oxide fuel cells *J. Chin. Ceram. Soc.* **2** 25–9 https://kd.nsfc.gov.cn/paperDownload/1000014161741.pdf

[119] Sholklapper T Z, Kurokawa H, Jacobson C P, Visco S J and De Jonghe L C 2007 Nanostructured solid oxide fuel cell electrodes *Nano Lett.* **7** 2136–41

[120] Blennow P *et al* 2011 Metal-supported SOFC with ceramic-based anode *ECS Trans.* **35** 683–92

[121] Blennow P, Hjelm J, Klemensø T, Persson Å H, Ramousse S and Mogensen M 2011 Planar metal-supported SOFC with novel cermet anode *Fuel Cells* **11** 661–8

[122] Blennow P, Sudireddy B R, Persson Å H, Klemensø T, Nielsen J and Thydén K 2013 Infiltrated $SrTiO_3$: FeCr-based anodes for metal-supported SOFC *Fuel Cells* **13** 494–505

[123] Sudireddy B R, Nielsen J, Thydén K T S, Persson Å H and Brodersen K 2015 Investigation of novel electrocatalysts for metal supported solid oxide fuel cells—Ru:GDC *ECS Trans.* **68** 1417–26

[124] Gorte R J, Park S, Vohs J M and Wang C 2000 Anodes for direct oxidation of dry hydrocarbons in a solid-oxide fuel cell *Adv. Mater.* **12** 1465–9

[125] Jiang S P, Callus P J and Badwal S P S 2000 Fabrication and performance of Ni/3 mol% Y_2O_3–ZrO_2 cermet anodes for solid oxide fuel cells *Solid State Ion.* **132** 1–14

[126] Kim J-S, Nair V V, Vohs J M and Gorte R J 2011 A study of the methane tolerance of LSCM–YSZ composite anodes with Pt, Ni, Pd and ceria catalysts *Scr. Mater.* **65** 90–5

[127] Graves C, Martinez L and Sudireddy B R 2016 (Invited) High performance nano-ceria electrodes for solid oxide cells *ECS Trans.* **72** 183–92

[128] Ding S *et al* 2020 A-site deficient perovskite with nano-socketed Ni–Fe alloy particles as highly active and durable catalyst for high-temperature CO_2 electrolysis *Electrochim. Acta* **335** 135683

[129] Zhang Y-Q, Li J-H, Sun Y-F, Hua B and Luo J-L 2016 Highly active and redox-stable Ce-doped LaSrCrFeO-based cathode catalyst for CO_2 SOECs *ACS Appl. Mater. Interfaces* **8** 6457–63

[130] Ramos T, Veltzé S, Sudireddy B R, Jørgensen P S, Theil Kuhn L and Holtappels P 2014 Effect of Ru/CGO versus Ni/CGO co-infiltration on the performance and stability of STN-based SOFCs *Fuel Cells* **14** 1062–5

[131] Nielsen J, Persson Å H, Sudireddy B R, Irvine J T S and Thydén K 2017 Infiltrated $La_{0.4}Sr_{0.4}Fe_{0.03}Ni_{0.03}Ti_{0.94}O_3$ based anodes for all ceramic and metal-supported solid oxide fuel cells *J. Power Sources* **372** 99–106

[132] Muzaffar N, Arshad N, Drasbæk D, Sudireddy B R and Holtappels P 2019 Fabrication and electrochemical performance of Zn-doped $La_{0.2}Sr_{0.25}Ca_{0.45}TiO_3$ infiltrated with nickel-CGO, iron, and cobalt as an alternative anode material for solid oxide fuel cells *Catalysts* **9** 269

IOP Publishing

Nanoengineered Materials for Solid Oxide Cells

Katherine Develos-Bagarinao

Chapter 9

The use of anisotropic microstructures to modify solid oxide cell electrodes

Takaaki Shimura

The electrochemical performance of solid oxide cell (SOC) electrodes is critical in order to develop SOC systems that have high energy conversion efficiency. The electrochemical performance of the electrodes is governed by the transport of gas and charges (electrons and ions) and their reactions at the reaction sites. Therefore, the microstructure of the electrodes, where the transport path of the gas and the charges and the distribution of reaction sites are geometrically determined, has a crucial impact on electrochemical performance. To enhance the electrode performance, microstructural modifications of the electrolyte–electrode interface have been carried out from the mesoscale to the microscale, in an attempt to increase the density of the reaction sites near the electrolyte or the effective conductivity of the electrodes. In this chapter, such works are summarized according to the basic concept used in designing the electrode microstructure.

9.1 General guidelines for the design of solid oxide cell electrodes

9.1.1 Theoretical approach to electrode design

SOC electrodes are designed as porous structures, which easily provide sufficient transport paths for gas and charges and sufficient numbers of reaction sites. However, each transport path is tortuous due to the randomly formed electrode microstructure, which results in a larger transport resistance than the ideal one. The effective conductivities σ_i^{eff} can be calculated as follows:

$$\sigma_i^{\text{eff}} = \frac{V}{\tau}\sigma_i \ \ (i = \text{O}^{2-} \text{ or } \text{e}^{-}). \tag{9.1}$$

doi:10.1088/978-0-7503-4064-9ch9

Here, V, τ, and σ_i represent the volume fraction, tortuosity factor, and bulk conductivity, respectively. The subscripts O^{2-} and e^- represent the oxide ionic conductive and electronic conductive phases, respectively.

Generally, the overpotential arising from the concentration gradient of the gas becomes noticeable when the fuel utilization is high. Hence, under moderate operating conditions, charge transfer and electrochemical reactions dominate the overall electrode performance. The conventional materials used for SOC electrodes include nickel–ionic conductor (stabilized zirconia, doped ceria, etc.) composites for fuel electrodes and perovskite materials such as lanthanum strontium manganese, $La_xSr_{1-x}Co_yFe_{1-y}O_{3-\delta}$ (LSCF), $La_xSr_{1-x}CoO_{3-\delta}$ (LSC), or their composites with ionic conductors as air electrodes. These materials show very large electronic conductivities compared with their ionic ones. For example, the ionic conductivity of yttria-stabilized zirconia (YSZ) is five orders of magnitude lower than the electronic conductivity of nickel [1, 2]. Therefore, the resistance of the charge transport is limited by the ionic conductivity. The electrode resistance R_{rode} is modeled using the exchange current density i_0, the density of the reaction sites A (such as triple phase boundaries or double phase boundaries), and the ionic and electronic conductivities using a linear approximation of the Butler–Volmer equation, as follows [3]:

$$\frac{1}{R_{\text{rode}}} \approx \sqrt{\frac{i_0 FA}{(1/\sigma_{O^{2-}}^{\text{eff}} + 1/\sigma_{e^-}^{\text{eff}})RT}}, \tag{9.2}$$

where F, R, and T represent the Faraday constant, gas constant, and temperature, respectively. Considering that the electronic conductivity is several orders larger than the ionic conductivity, the following relationship can be derived from equation (9.2) [4]:

$$R_{\text{rode}} \propto \frac{1}{\sqrt{\sigma_{O^{2-}}^{\text{eff}} A}}, \tag{9.3}$$

which simply states that the electrode resistance can be reduced by increasing the effective ionic conductivity and the density of the reaction sites.

These approaches differ in terms of the effective reaction thickness of the electrodes. The effective thickness l^* can be described as follows, if we neglect the electronic conductive phase [5]:

$$l^* = \sqrt{\frac{RT\sigma_{O^{2-}}^{\text{eff}}}{i_0 FA}}. \tag{9.4}$$

From this relationship, it can easily be predicted that an increase in the effective conductivity results in an increase in the effective thickness, whereas an increase in the density of the reaction sites reduces the effective thickness. Thus, the electrode performance is determined by a competition between the electrochemical reaction and the ionic transport. When more electrochemical reaction sites are available, the reaction can easily occur near the electrolyte before the ion is transported far from the electrolyte. On the other hand, when ionic conductivity is enhanced, ions can be transported farther from the electrolyte, and reaction sites further from the electrolyte are effectively utilized. The use of nanoscale particles as electrode materials is a typical approach for increasing the density of reaction sites. Another approach to reducing the electrode resistance is to enlarge the value of the effective ionic conductivity.

As described in equation (9.1), the effective conductivity is defined by the bulk conductivity, volume fraction, and tortuosity factor. Among these, the bulk conductivity is determined by the material, and the volume fraction V is always less than one for porous structures. The tortuosity factor τ represents the complexity of the transport path; its minimum value of one is realized by a straight geometry. Thus, the microstructure of the electrode determines the effective conductivity. In other words, the charge transport property can be enhanced by modifying the microstructure of the electrode.

9.1.2 Modifications to the electrolyte–electrode interface or the electrode microstructure

As discussed above, the main approaches to reducing electrode resistance in terms of microstructural modification are to enhance the reaction site density or reduce the tortuosity factor. Schematic images of modifications that can realize such requirements are shown in figure 9.1. First, the conventional composite microstructure has a random distribution of ionic and electronic conductive phases, as shown in figure 9.1(a); therefore, the conductive path is tortuous and the reaction sites are randomly distributed. One approach which has been used to date is to modify the electrolyte–electrode interface. As shown in figure 9.1(b), by modifying the

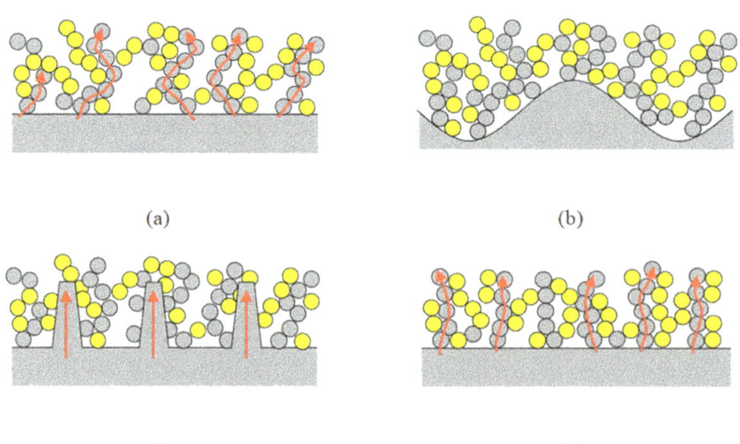

Figure 9.1. Schematic images of electrodes. The gray and yellow phases represent the ionic and electronic conductive phases, respectively. The red arrows indicate the ionic transport path from the electrolyte to the electrodes. (a) Random composite microstructure, (b) random composite structure with a modified electrolyte surface, (c) a composite microstructure with ionic conductive pillars, and (d) an aligned microstructure.

electrolyte surface structure to be 'wavy' or 'grooved,' the interfacial area between the electrode and the electrolyte can be increased, accompanied by an increased number of reaction sites near the electrolyte. Another approach is to reduce the tortuosity factor of the electrode. The tortuosity factor of the ionic conduction phase in electrodes can be decreased by constructing straight ionic conductive paths in electrodes, so that the tortuosity factor locally approaches its minimum value of one. Following the creation of pillar-like structures at the electrolyte–electrode interface, as shown in figure 9.1(c), ions can be transported within the periodic pillar structures, which reduce the tortuosity factor and supply ions to the electrode effectively. Figure 9.1(d) illustrates an approach used to enhance the effective ionic conductivity by aligning the ionic conductive phase and making its structure straight. In this chapter, both experimental and numerical investigations of such modifications are summarized.

Among these approaches, the modification of the interfacial area between the electrolyte and the electrode can barely be distinguished from the fabrication of pillar structures, depending on the interfacial geometry, because grooved electrolyte surfaces can be regarded as ionic conductive paths into the electrode. In previous papers, the purpose of such additional structures has been explained in different ways, depending on the research group. Generally, interfacial structures whose scale is larger than those of the electrodes (for example, mesoscale structures in the range of 10–100 μm) are often referred to as 'grooved' structures. On the other hand, structures whose scale is relatively close to those of the electrodes (typically on the micrometer scale) are regarded as pillar structures. Furthermore, grooved structures are often fabricated with a low aspect ratio (i.e. a relatively short height in the direction of the electrode's thickness and a relatively large scale in the plane of the electrode's surface), whereas the pillar structure is designed as a high-aspect-ratio structure. However, in this chapter, both approaches are summarized at the same time.

9.2 Numerical studies of modifications to the interfacial structure

In this section, numerical studies are reviewed and the theoretical approach used for the microstructural design is summarized. Here, the modified structure at the electrolyte–electrode interface is consistently referred to as a pillar, regardless of the scale or geometry assumed in the referenced papers. In the following, the pillar geometries are described using parameters which are determined by the typical pillar shape shown in figure 9.2.

Konno *et al* examined the effect of mesoscale pillar structures on an electrolyte surface using a two-dimensional (2D) numerical simulation [6]. They observed that the electrochemical reaction was effectively enhanced when the pillar height was larger than the original effective thickness. The pillar height was 200 μm and the pillar and bottom widths were varied from 50–300 μm. The area enhancement factor was defined as the ratio between the electrode/electrolyte interfacial area of a cell with pillars and that of one without. The electrochemical reaction was enhanced by an increase in the area enhancement factor. They further examined the origin of the

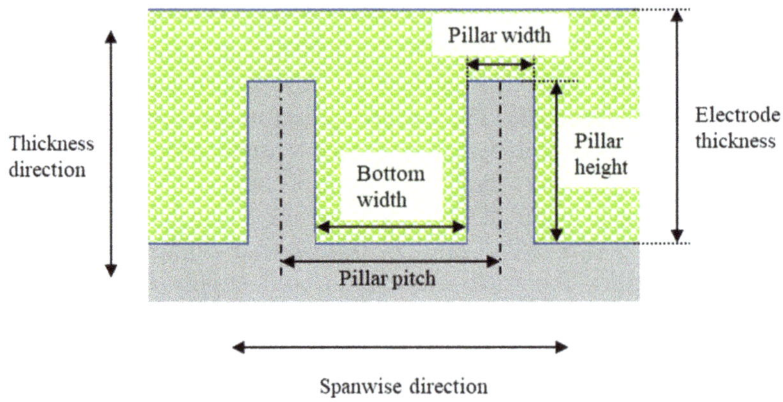

Figure 9.2. The geometry of pillar structures.

electrochemical reaction enhancement by dividing the reaction current into sections, which revealed that the enhancement effect is mainly obtained at the bottom of the pillars.

Delloro *et al* conducted a 2D simulation of pillar structures in a LSM–YSZ composite [7]. They surveyed the dependence of the enhancement of the electrochemical reaction on the effective ionic conductivity and the exchange current density using a fixed pillar geometry. Here, the effectiveness factor was defined as the ratio of the bulk ionic conductivity to the effective conductivity, which equals τ/V in equation (9.1). As shown in figure 9.3, with an increase in the effectiveness factor, i.e. lower ionic conductivity, the effect of the pillar structure on the electrochemical reaction is enhanced. When the ionic conductivity is high, i.e. on the left-hand side of the figure, the enhancement effect is negative. This implies that when the effective thickness is originally large enough, the additional pillar structure only has the negative effect of reducing the number of reaction sites. These results also revealed that the enhancement effect greatly depends on the exchange current density. With an increase in the exchange current density to 10^7–10^8 A m^{-3}, the enhancement effect increases. However, when the value increases further, the enhancement effect decreases. The authors concluded that the enhancement is obtained when the effective thickness is on the same order as that of the pillar structure, so that the electrochemical reaction is neither dominated by the exchange current density nor the ionic conduction. Thus, to design pillar structures which effectively enhance the reaction, it is important to understand the balance between the positive effect of the pillar structures in enhancing the ionic conduction and the negative effect due to the loss of reaction sites. Additionally, the authors also investigated the dependence of the enhancement effect on the pillar geometry; the results imply the existence of optimal geometric parameters.

Bertei *et al* conducted a 2D numerical simulation of pillar structures [8]. They concluded that the positive effect of the pillar structures could only be obtained when the effective ionic conductivity was less than half that of the bulk conductivity. This statement is similar to the main conclusion of Dellero *et al* as mentioned above:

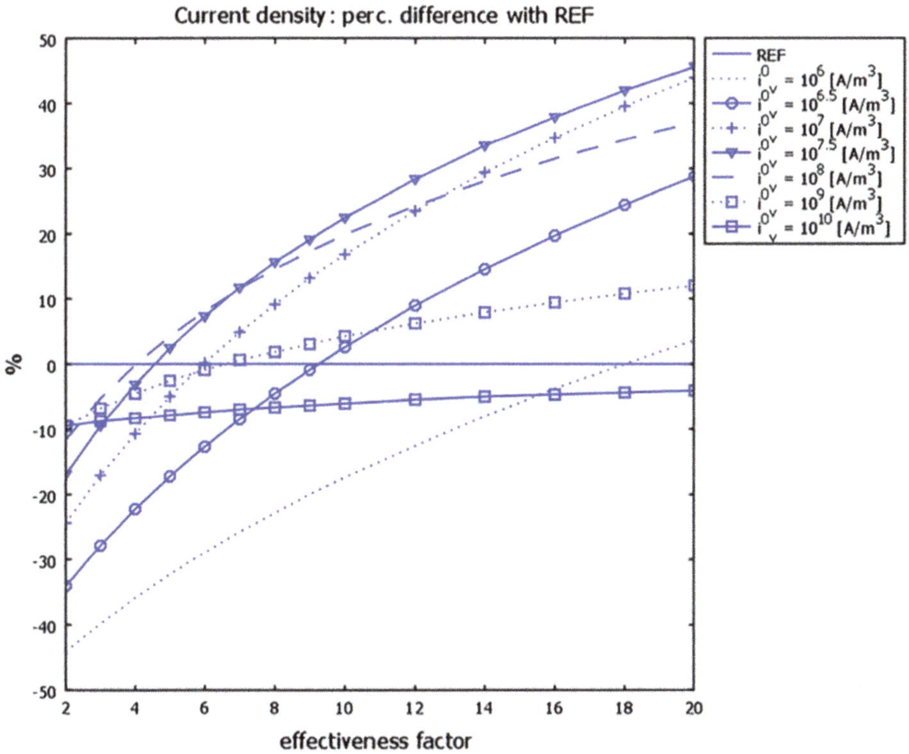

Figure 9.3. Dependence of the electrochemical enhancement effect (right axis) on the effectiveness factor of the ionic conductivity (bottom axis) and the exchange current density (see caption) [7] Copyright © 2012, Springer Science Business Media New York. With permission of Springer.

i.e. the effective thickness should not be too large in order to obtain a benefit from the pillars [7]. They also investigated the influence of pillar geometries. They reported that when the pillar width was too large compared with the bottom width, the positive effect could not be obtained. To obtain an effective enhancement effect, they concluded that the pillar width should be smaller than the effective thickness, whereas the pillar height should be larger than the effective thickness. Thus, a long, thin pillar geometry was proposed.

The above three studies mainly investigated rectangular pillar geometries at the mesoscale under the assumption of a uniform electrode microstructure. This is valid when the pillar size is large enough compared with the characteristic size of the electrode microstructure. However, when we consider downsizing the pillar structure, the influence of the electrodes' local microstructure cannot be neglected. As already mentioned, the microstructure of the electrode is conventionally random, and its conduction paths for gas and charges are complicated. As part of the recent progress in the three-dimensional microstructural characterization of SOC electrodes, its actual microstructure has been quantitatively analyzed using focused ion beam-scanning electron microscopy (FIB-SEM) reconstruction [9–16] and x-ray tomography [17–21]. The microstructure obtained then enables the electrochemical reaction in the actual electrode microstructure to be predicted using three-dimensional electrochemical simulations [22–26]. The effect of the pillar geometry has recently been investigated via 3D numerical simulations that used the actual 3D microstructure and virtually constructed pillar structures.

Shimura *et al* conducted a 3D numerical simulation using the reconstructed Ni–YSZ composite fuel electrode with virtually added pillar structures (figure 9.4) using the lattice Boltzmann method [22]. The pillar geometries were systematically varied and their influences were surveyed. Three parameters were investigated: pillar width, pillar height, and the number of pillars. Here, the total computational volume was fixed; therefore the number of pillars was varied by changing the pillar width and pillar pitch at a fixed pillar height. With the pillar structures, the electrode resistance decreases and the extent of the effective thickness can be observed. As shown in figure 9.5, as a result of the enhanced ionic conduction caused by the pillar, a high electrochemical potential field is maintained farther from the electrolyte. Generally, with an increase in each geometrical parameter, a greater reduction of the resistance is

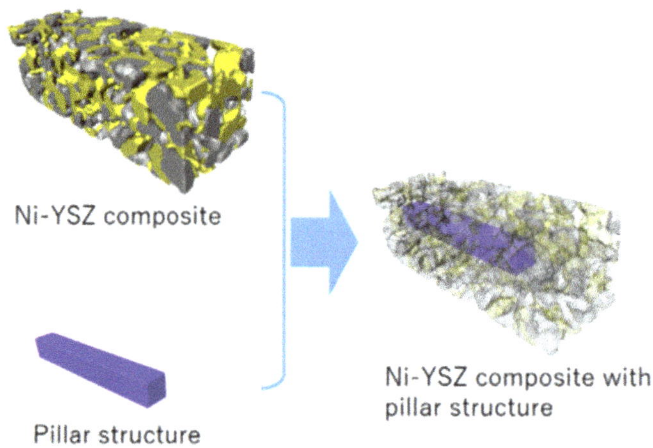

Figure 9.4. Insertion of a virtual pillar structure into a 3D Ni–YSZ microstructure obtained using FIB-SEM. Reprinted from [22], Copyright (2019), with permission from Elsevier.

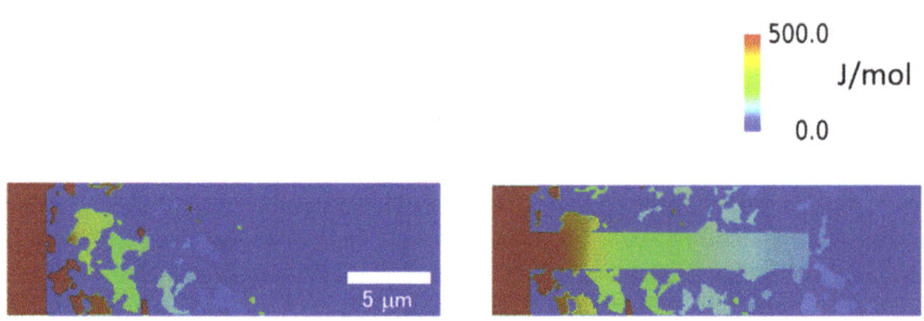

Figure 9.5. Comparison between the electrochemical potential of oxide ions in the original Ni–YSZ structure and that obtained using a pillar structure. Reprinted from [22], Copyright (2019), with permission from Elsevier.

obtained. As each parameter increases, the triple phase boundary (TPB) density decreases due to the replacement of the electrode volume by the pillar structure. On the other hand, the tortuosity factor of YSZ decreases in the thickness direction. Thus, a reduction in the resistance is obtained because the positive effect of the enhanced ionic conduction overcomes the negative effect of losing the TPB. However, when the pillar width becomes larger than a specific value, the electrode resistance increases drastically. This is due to the loss of percolation through the microstructure resulting from the replacement by the pillar structure, followed by the drastic loss of effective TPBs. Thus, by considering the local microstructure, it is suggested that the limit of the pillar size should be determined by the composite microstructure.

Using a similar approach, He *et al* conducted a 3D numerical simulation by the lattice Boltzmann method on the air electrode side with virtual pillar structures using 3D microstructures of LSCF and an LSCF–gadolinia-doped ceria (GDC) composite [25]. The enhancement effect was clearly observed in the LSCF, whereas it was small for the LSCF–GDC composite. With an increase in pillar height, a greater reduction in the electrode resistance was achieved for LSCF. On the other hand, the long pillars resulted in a negative effect for LSCF–GDC. Due to the large ionic conductivity of GDC, the LSCF–GDC composite has a higher effective thickness than the LSCF. Therefore, the effect of the pillar structure is more pronounced in LSCF. This is consistent with the discussion above in terms of the relationship between the effectiveness of the pillar structure and the effective thickness. A loss of percolation was also observed in the LSCF phase of the composite microstructure, which counteracted the enhancement effect of the pillars. Furthermore, the authors also conducted a 3D numerical simulation using a nanoscale microstructure which was constructed by resizing the original structure. As shown in figure 9.6, due to the enhancement of reaction sites, the effective thickness decreased drastically. The enhancement effect of the additional pillar structures was larger for the nanostructures than for the original structure. This can be ascribed to the reduced effective thickness due to the nanoscale structure, to which the enhanced ionic conduction contributes more significantly. In the nanoscale composite, the enhancement effect due to the pillars was observed, but it was lower

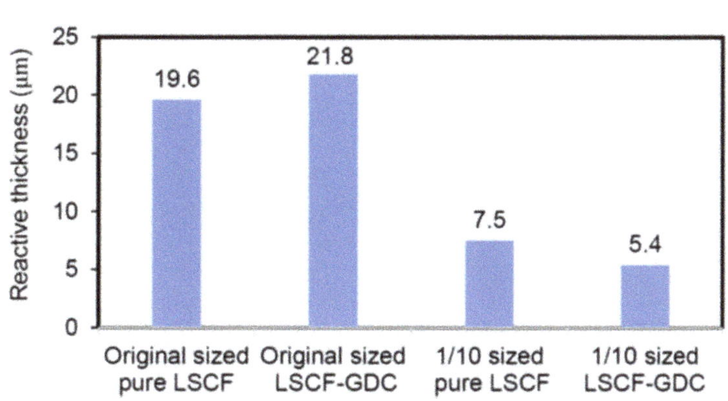

Figure 9.6. Comparison of the effective thicknesses of the original microstructures and those of the nanosized microstructures. Reprinted from [25], Copyright (2019), with permission from Elsevier.

than for the LSCF. This can be ascribed to the very small effective thickness due to the increased number of reaction sites in the nanoscale composite, where the ionic conduction enhancement effect could not compensate for the loss of active reaction sites, as discussed above by Delloro *et al* [7] and Bertei *et al* [8].

Thus, by using the actual 3D microstructure, the influence of the local microstructural effect can be taken into account in the simulation, which supports our understanding of the effects of the pillar structure. Numerical simulation is a very effective tool for evaluating the effects of the pillar structure in detail, since this cannot be carried out by simple mathematical modeling. Furthermore, the various structures of the pillars can be tested numerically without the difficulty involved in the manufacturing process.

Chueh *et al* investigated various pillar structures and the influence of their geometries on the performance enhancement effect by 2D numerical simulation [27]. The pillar geometries tested are shown in figure 9.7. They reported that a thin and long rectangular pillar structure is preferred, as previously mentioned [8]. Furthermore, they reported that the enhancement effect of the rectangular pillar shape is highly dependent on the electrode thickness. They also observed that the addition of fine structures on top of the pillar structures created a negative effect because it disturbed the gas diffusivity. They found that trapezoidal pillars could moderately enhance the performance; however, these would require high accuracy in an actual fabrication process.

Although several basic guidelines have been presented thus far, it is difficult to propose an optimized pillar structure for each electrode microstructure. Iwai *et al* conducted a 2D optimization calculation for pillar structures using the level-set method coupled with electrochemical simulation [28]. As shown in figure 9.8, the sharp pillar geometry changed into a rounded one during the optimization calculation. They found that the enhancement effect was increased by about 7.1% as a result of the optimization. Furthermore, they surveyed the dependence of optimized pillar geometries on the microstructural parameters of the electrode. When electrodes with different gas diffusivities were assumed, the pillar geometries changed to different shapes depending on the gas diffusivity. When the gas diffusivity was larger, the pillar structure became complicated, whereas flat pillar geometries were obtained for low diffusivity conditions so as not to hinder the gas diffusion. When the complexity of the solid phase of the electrode was varied,

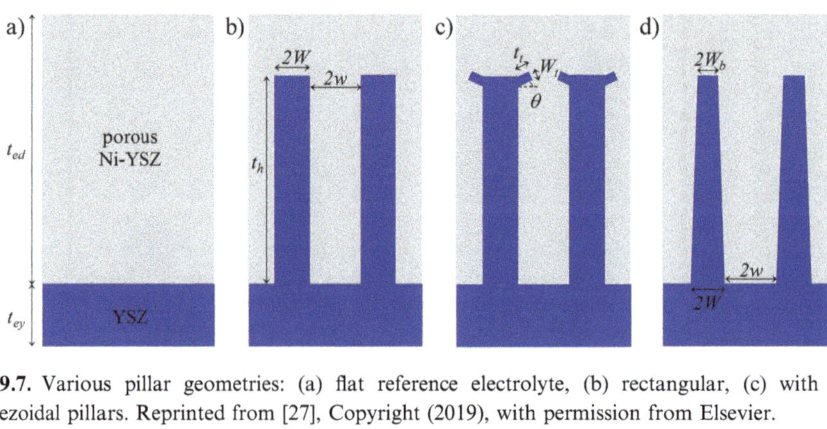

Figure 9.7. Various pillar geometries: (a) flat reference electrolyte, (b) rectangular, (c) with fins, and (d) trapezoidal pillars. Reprinted from [27], Copyright (2019), with permission from Elsevier.

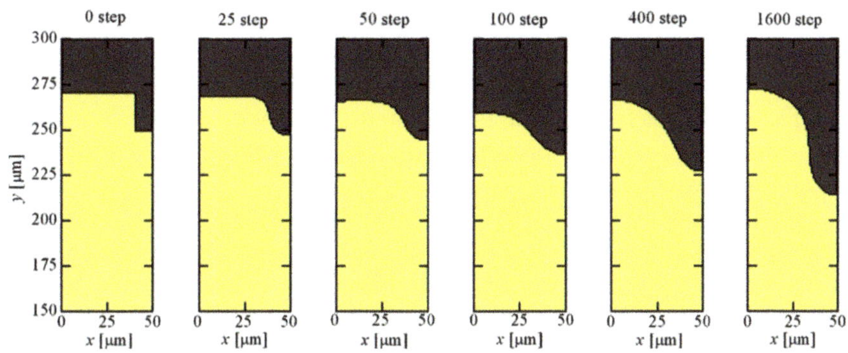

Figure 9.8. Time evolution of a 2D pillar structure during an optimization calculation. Reprinted from [28], Copyright (2010), with permission from Elsevier.

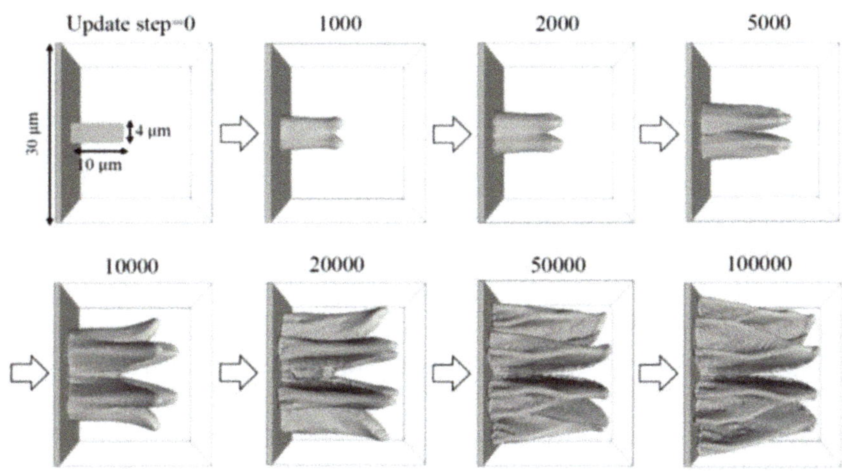

Figure 9.9. Time evolution of a GDC pillar structure on an LSCF air electrode during optimization calculations. Reprinted from [29], Copyright (2019), with permission from Elsevier.

different optimal pillar geometries were obtained. Thus, the optimal pillar geometry is highly dependent on the gas and charge transport properties.

He *et al* carried out a 3D optimization calculation for pillar structures using the adjoint method [29]. In their study, the GDC pillar structure of an LSCF air electrode was optimized by setting the reaction current as an objective function. Homogeneous double phase boundary (DPB) and TPB densities were assumed in the simulation. The interface between the GDC pillar and the LSCF was expressed by the level-set function. A simple pillar structure with a square cross section was selected as the starting structure, which was updated by repeating the optimization calculation. As shown in figure 9.9, the pillar structure showed an evolution over time for repeated optimization calculations. Finally, the reaction current increased by 70%. Furthermore, they calculated additional simulations by changing the effective ionic conductivity and the exchange current densities of the DPBs and the TPBs. Depending on the given values, the optimized pillar structures differed. The enhancement effect under optimal conditions became larger when the conductivity or the exchange current density was larger. As discussed above, it has been reported that when the conductivity or the exchange current density is very large, the pillar structure often has a fairly small or even negative effect on the electrochemical performance. However, these optimization results show the potential for further improvement that can be realized by the proper design of the pillar structure when the ionic conductivity or the reaction activity is initially high.

Thus, various research has been performed using techniques ranging from relatively simple 2D model calculations to 3D simulations including the actual microstructure and optimization processes. The pillar structure has also developed from simple geometries which can be easily imagined to more complicated ones. This knowledge is of great help in the actual design of pillar geometries and in resolving manufacturing issues, as introduced below.

9.3 The fabrication and manufacture of structures at the electrolyte–electrode interface

Along with the theoretical and mechanical approaches described above, fabrication and manufacturing advances have been made in the area of of pillar structures, not only in the SOFC research field, but also in various research fields such as micromachining and microprocessing. The fabrication methods widely vary and are difficult to be distinguish from each other. Here, the fabrication methods are categorized into four approaches: cutting, additive, molding, and others.

9.3.1 The cutting approach

Cutting is a simple approach to the fabrication of pillar structures on flat substrates; in this approach, the surface of the substrate is excavated using a trench or spot geometry, etc. Here, we first introduce the laser machining method and then describe some of the other methods.

9.3.1.1 Laser machining

One of the major ways to fabricate interfacial microstructures is to use a laser. Larrea *et al* fabricated patterned electrolytes by laser machining [30]. We emphasize here that their research motivation was to fabricate an electrolyte which achieved low ohmic resistance and mechanically sufficient strength that could also be applied in electrolyte-supported cells. They fabricated a hexagonal mesoscale pattern by laser machining a 150 μm thick YSZ electrolyte as shown in figure 9.10. The size of the structure and the thickness of the electrolyte at the bottom were 230 μm and 20 μm, respectively. They fabricated a SOC cell using the processed electrolyte and compared its electrochemical performance to that of a cell with a flat electrolyte surface. The improvement in the area-specific resistance was consistent with the reduction in the average thickness of the electrolyte.

Zhang *et al* modified the electrode–electrolyte interface using laser machining [31]. They modified the air electrode substrate with a laser, then added a coating of the YSZ electrolyte layer. They tried two kinds of structure. The first was a roughened surface created by laser machining the entire area. The second was a patterned electrolyte with

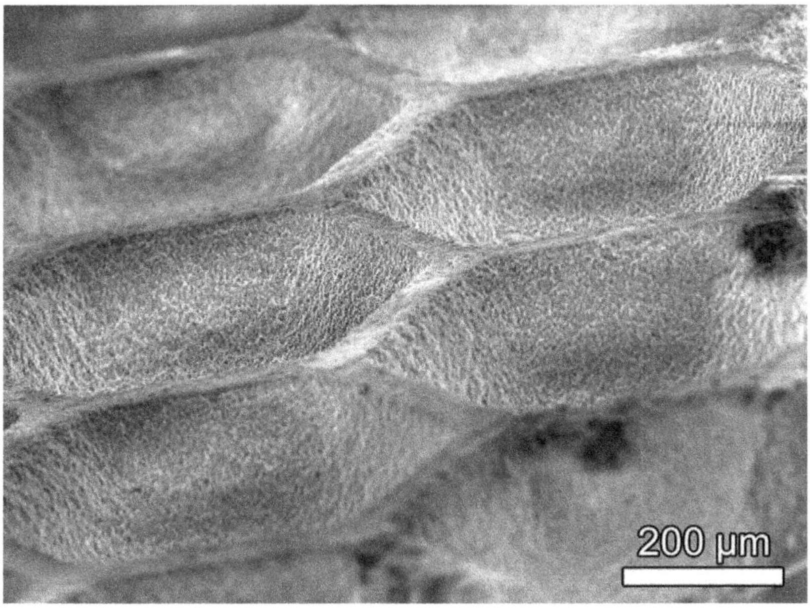

Figure 9.10. YSZ electrolyte with a hexagonal pattern fabricated by laser machining. Reprinted from [30], © The Electrochemical Society. Reproduced by permission of IOP Publishing.

a spot geometry. The roughness of the surface was on the order of one micrometer. The maximum power density increased by about 30%–50% in accordance with the roughness. The enhancement effect was due to the decrease in the electrolyte thickness and the enlarged electrolyte–electrode interfacial area. The patterned cell, which had a scale of around 100 μm, showed about a 50% increase in power density. Cai *et al* conducted laser processing of a YSZ electrolyte to fabricate a mesoscale dimpled structure with a diameter of about 50 μm on the surface [32]. After the fuel and air electrodes were applied on both sides of the electrolyte, electrochemical measurements were carried out. Reductions were achieved in both ohmic and polarization resistances, which can be ascribed to a partial reduction in the electrolyte thickness and an increase in the electrolyte–electrode interface, respectively.

In addition to the fabrication process used to obtain patterned structures, the method used to deposit the electrodes onto the surface is also important. Shimura *et al* fabricated YSZ pillar structures by laser ablation [22]. The pillar pitch was set at 5 μm, which is a similar order of magnitude that that of the conventional electrode microstructural scale. To avoid damage to the pillar structure due to the volume change caused by the reduction of NiO to Ni, the Ni–YSZ nanocomposite was deposited onto the pillar structures by sputtering, as shown in figure 9.11. Short-term degradation tests were then carried out. The sputtered Ni–YSZ resulted in a stable electrochemical performance compared with those of sputtered Ni or screen-printed Ni–YSZ. Cebollero *et al* fabricated a dimple-like structure on an electrolyte surface using a pulsed laser [33]. The size of the structure was 28 μm, and the electrodes were attached by dip coating. They observed a reduction of about 30% in

Figure 9.11. YSZ pillar structure in a sputtered Ni–YSZ composite. Reprinted from [22], Copyright (2019), with permission from Elsevier.

the polarization resistance, although some difficulty was encountered while adhering the electrode to the patterned electrolyte surface. They investigated this problem in their following research by modifying the content of the electrode dispersion used for dip coating [34]. They fabricated a patterned structure at a smaller size (a diameter of 24 μm and a depth of 22 μm) as shown in figure 9.12. The fabrication of LSM–YSZ electrodes by dip coating using a dispersion with high solid loadings resulted in reductions in the ionic transport and catalytic reaction resistances, whereas the gas diffusion was inhibited by the low porosity. When the electrode dispersion was changed to one with a lower solid loading, the gas diffusivity was maintained, whereas the reductions in the ionic transport and the catalytic activity resistance became weak.

9.3.1.2 Other methods

Konno *et al* applied a blasting method to fabricate 2D groove structures on an electrolyte at the mesoscale; they fabricated a Ni–YSZ fuel electrode by screen printing, as shown in figure 9.13 [6]. According to electrochemical measurements, a cell with the grooved structure showed an enhancement in the current density of around 20%. However, this enhancement effect was smaller than that expected from the mean electrolyte thickness. The authors referred to the possibility of larger concentration overpotential in the cell with grooves. In another study by the same group based on numerical simulations, it was reported that the concentration overpotential offset the positive effect of the pillar structures, especially when the electrode porosity was low [35]. This is consistent with several similar remarks about the influence of pillar structures on gas diffusion, which were described in section 9.2.

Chao *et al* adopted a lithographic technique to fabricate a thin electrolyte with pillar structures [36]. First, silica nanoparticles were deposited onto a silicon substrate.

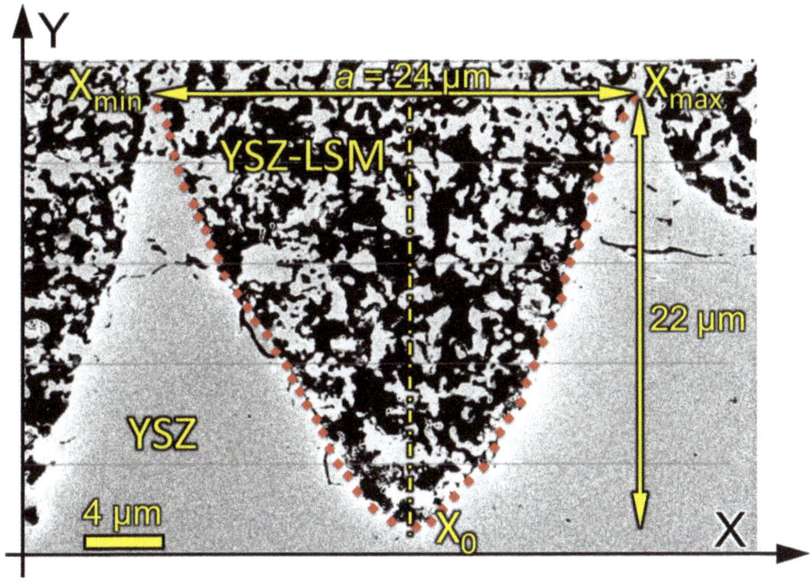

Figure 9.12. Patterned structure on a YSZ electrolyte with a dip-coated LSM–YSZ electrode. Reprinted from [34], Copyright (2019), with permission from Elsevier.

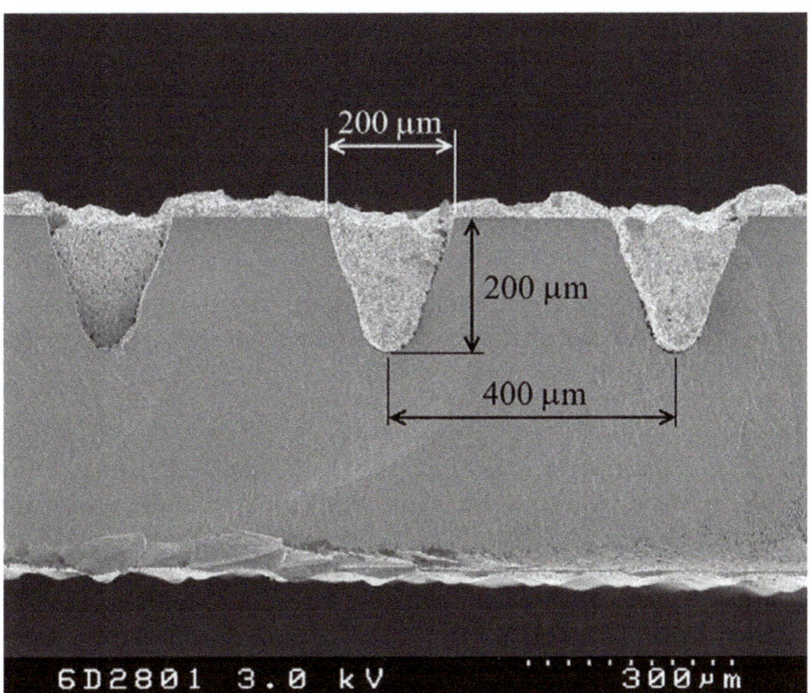

Figure 9.13. SEM image of part of the grooved area of a cell filled with a Ni–YSZ fuel electrode. Reprinted from [6], Copyright (2010), with permission from Elsevier.

Figure 9.14. SEM image of pillar structures fabricated by a lithographic technique and followed by the deposition of a Pt electrode. Reprinted with permission from [36]. Copyright (2011) American Chemical Society.

A metal pattern was then prepared by sputtering aluminum in between the nanoparticles, and the nanoparticles were then removed. Subsequently, trenches were fabricated by dry etching, and the aluminum masks were removed by wet etching. YSZ was deposited onto the fabricated structure by atomic layer deposition (ALD). After the silicon substrate was removed by wet etching, platinum electrodes were deposited on both sides as the electrodes as shown in figure 9.14. The polarization resistance decreased by 30% compared to that of a cell without pillar structures.

9.3.2 Additive approach

Additive processing is another simple idea that can be used to realize pillar structures. In this section, we introduce several papers on pillar structures fabricated by adding extra volume to flat substrates.

9.3.2.1 Spray deposition

Spray deposition techniques have been adopted to obtain rough surfaces. Wang *et al* deposited YSZ particles on a substrate by flame spraying, so that only the surfaces of the particles melted [37]. They reported that the surface morphology of the sprayed substrate could be controlled by the spraying distance. LSM was then deposited on the surface by atmospheric plasma spraying. Compared with that of a flat cell surface, the polarization resistance decreased by about 40% due to the enlarged interfacial area. Dai *et al* deposited YSZ slurries with pore formers by spray deposition to fabricate a rough surface microstructure [38]. An interfacial structure that had a similar scale to that of the pore former (20 μm) was fabricated. NiO–YSZ and $LaBaCo_2O_{5+\delta}$ (LBCO) were then screen printed to form the fuel and air electrodes, respectively, as shown in figure 9.15. Reductions in both the ohmic and polarization resistances were realized. The maximum power density doubled compared to that of a cell with a flat interface. In particular, the enhancement effect was larger on the air electrode side. This was explained as being due to the porous interface, which provided additional gas channels and more TPBs as reaction sites in the LBCO electrode, the original reaction sites of which were only DPBs.

9.3.2.2 Three-dimensional printing

Three-dimensional printing has been applied to directly control the patterning of pillar structures. Farandos *et al* fabricated pillar structures on an electrolyte using colloidal dispersions of YSZ and inkjet printing [39]. The minimum scale was reduced to 35 μm by optimizing the ink composition. They fabricated two kinds of structure: micropillar arrays and square lattices on NiO–YSZ substrates, as shown in figure 9.16. An LSM air electrode was prepared on the opposite side and the cells were tested using CO_2 electrolysis. Masciandaro *et al* fabricated a 3D geometry at the mesoscale on an electrolyte using a stereolithographic technique in which a specific geometry can be printed layer by layer [40]. Their main purpose was to fabricate a thin electrolyte while also providing sufficient mechanical strength. An electrolyte with a honeycomb supporting structure was constructed. Its ohmic resistance was lower than that of a conventional cell, which was ascribed to the

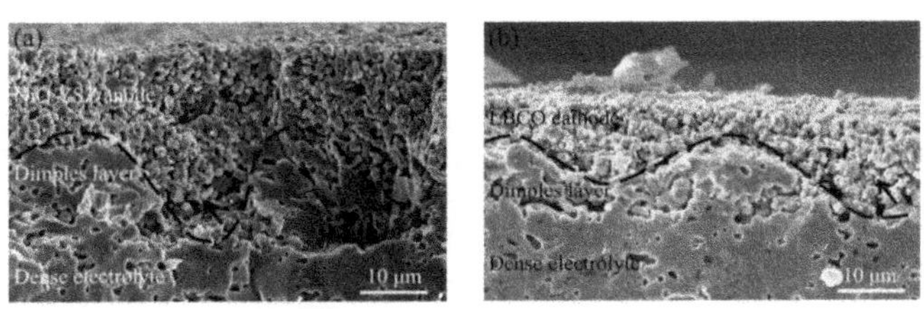

Figure 9.15. Dimpled surface with (a) NiO–YSZ and (b) LBCO. Reprinted from [38], Copyright (2015), with permission from Elsevier.

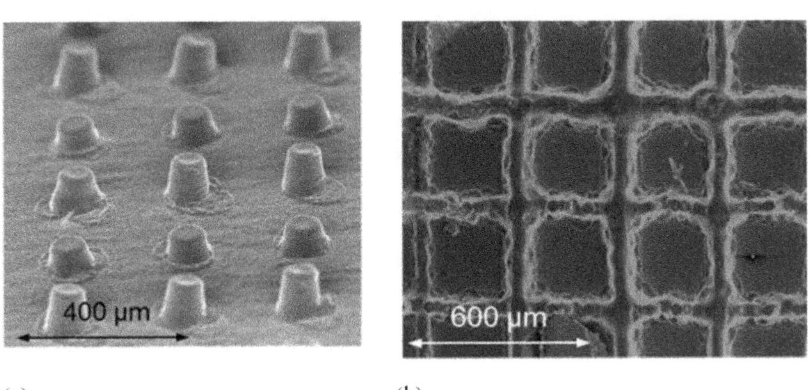

Figure 9.16. YSZ structures on electrolytes. (a) Pillar arrays and (b) square lattices. Reproduced from [39]. CC BY 4.0.

smaller thickness of the electrolyte. Salari *et al* also adopted the 3D printing technique to fabricate pillar structures on the electrolyte [41]. They combined 3D printing of the electrolyte layer with tape casting of the LSM air electrode to fabricate symmetrical cells with large pillar structures on the millimeter scale. They reported reductions in both the ohmic and polarization resistances.

Seo *et al* fabricated an interfacial structure by microextrusion printing [42]. Its structure was two-dimensional, which was a continuous groove on the fuel electrode substrate. A characteristic of their work was that the modification was applied to the fuel electrode with a very thin electrolyte layer. Therefore, the interfacial structure was modified from the fuel electrode side to the air electrode side, as shown in figure 9.17. They observed a decrease of about 40% in the total resistance, which was much larger than the value predicted for the interfacial area enlargement. This reduction effect was realized from reductions in both the ohmic and activation resistances. They also conducted a numerical simulation and stated that the large reduction in the ohmic resistance was due to not only to the interfacial area enlargement but also to the nonuniform thickness of the electrolyte, which led to a nonuniform distribution of ionic current, as shown in figure 9.18 [43]. On the other hand, they also concluded that the reduction in the activation overpotential became smaller with the inhomogeneity of the ionic current distribution in the electrolyte. The reduction effects on the ohmic and activation resistances become larger and smaller with an increase in the current density, respectively. Due to the larger effect on the reduction in the ohmic resistance, the total reduction effect became larger at higher current densities. Furthermore, they

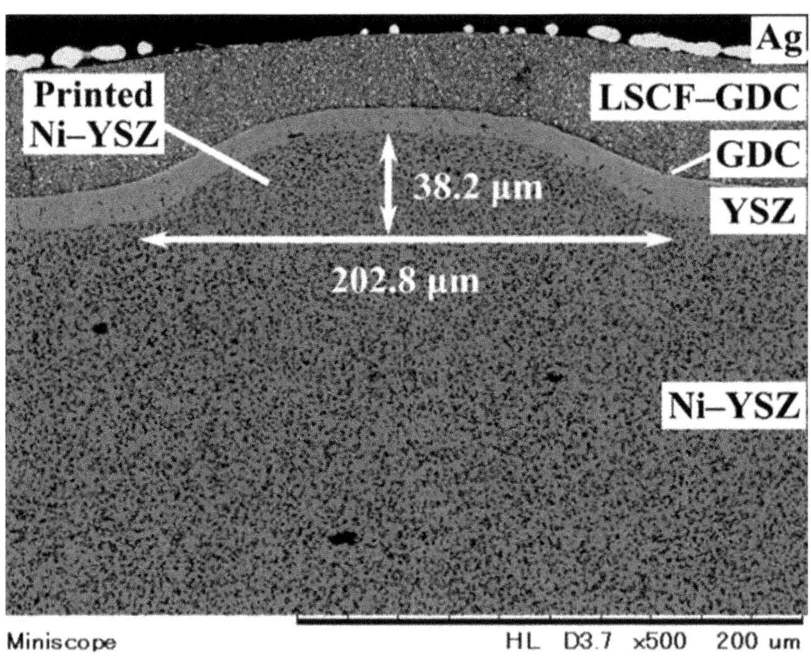

Figure 9.17. A modification of the interfacial structure achieved by microextrusion printing. Reprinted from [42], Copyright (2020), with permission from Elsevier.

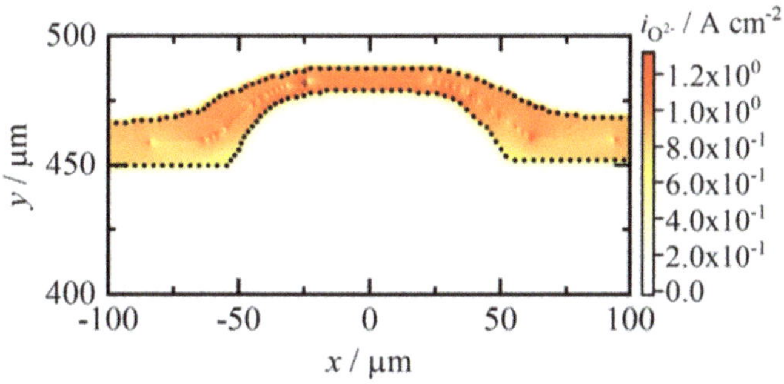

Figure 9.18. The nonuniform distribution of ionic current in the electrolyte. Reprinted from [43], Copyright (2021), with permission from Elsevier.

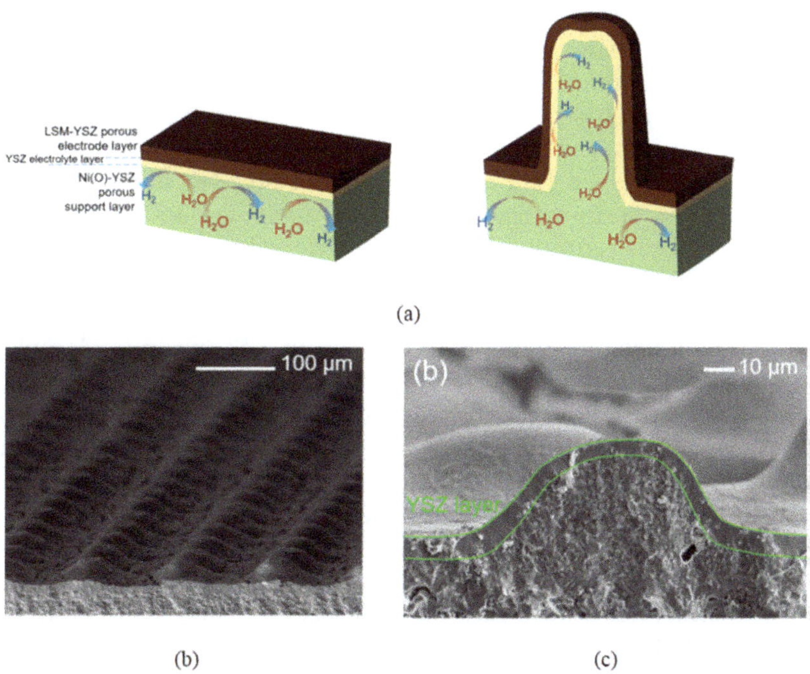

Figure 9.19. (a) A schematic illustration of reference (left) and 3D printed (right) pillar structures. (b) and (c) Three-dimensional pillar structures fabricated using YSZ electrolyte deposition. Reprinted from [45], Copyright (2022), with permission from Elsevier.

additionally infiltrated GDC nanoparticles into the LSCF, which resulted in a further reduction in the resistance, making this material especially suitable for low-temperature operations [44].

Jang *et al* fabricated a NiO–YSZ fuel electrode support with pillar structures using 3D inkjet printing [45]. Their geometry was also modified for both electrode sides, as in the case of Seo *et al* [42]. The difference between the two studies is that Seo *et al* [42] adopted a two-dimensional, mesoscale geometry, whereas Jang *et al* fabricated the structure three dimensionally and at a scale on the order of 10 μm. A schematic of the structure is shown in figure 9.19(a). Jang *et al* evaluated the electrochemical performance of the cell in the fuel cell, steam electrolysis, and CO_2 electrolysis modes, and the performance enhancement factors obtained were 1.5, 3, and 4–5, respectively.

9.3.3 Molding

Molding is an effective manufacturing method for reproducing specific geometries and fabricating numerous structures using a single process. Here, we introduce some reports in which pillar structures were fabricated from molds.

Xu *et al* fabricated an interfacial microstructure using a micropowder imprinting method [46]. First, molds were prepared by transcribing micropatterns onto polymer sheets. The molds were then impressed on the YSZ powder, which was pre-pressed as a sheet. Fuel and air electrodes were then attached on both sides and the electrochemical performance of the cell was evaluated. Xu *et al* fabricated two different patterns and compared the resulting cells with a flat-interface cell. When the pillar structure was added to the substrate, the cell showed better performance than that of the flat cell, which was ascribed to the extended interfacial area. When the interfacial structure was fabricated by making holes in the substrate, the cell showed even better performance, which was attributed to both the extended interfacial area and the reduction in the electrolyte volume.

Okabe *et al* fabricated a pillar structure at the 10 μm scale using a UV nanoimprint lithography technique [47]. They first prepared a master mold via lithography using a negative-tone photopolymer on a Si substrate. A replica mold was then prepared from the master mold using a photopolymer and polyethylene substrate. Pillar structures were fabricated from GDC powders on the YSZ substrate. A GDC slurry was first placed on the YSZ substrate and then impressed by the replica mold. The slurry was subsequently exposed to UV radiation, and then the mold was released. The imprinted structures were sintered at 1450 °C. By optimizing the slurry composition, pillars of less than 10 μm were fabricated without cracking, as shown in figure 9.20. Using this approach, Akama *et al* fabricated fuel electrode pillar structures using a $La_{0.75}Sr_{0.25}Co_{0.50}Mn_{0.50}0O_3$–GDC composite [48]. They realized a small pillar diameter of 4.14 μm.

Timurkutluk *et al* adopted a simple method to fabricate mesoscale patterns on electrolytes: pressing commercial metal mesh onto electrolytes [49]. First, a green

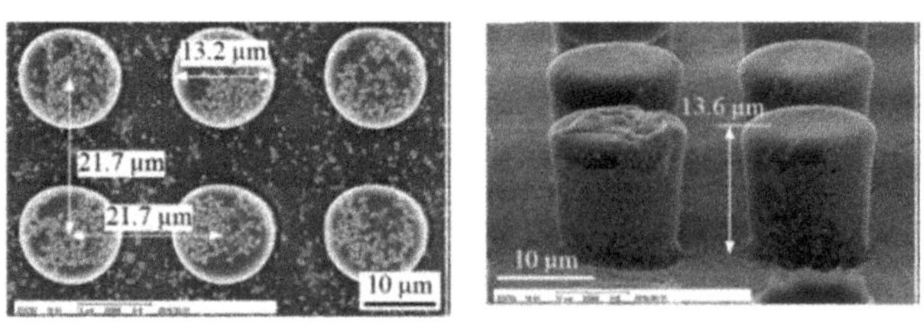

(a) (b)

Figure 9.20. (a) Top and (b) perspective views of GDC pillars fabricated by UV nanoimprint lithography. Reproduced from [47].

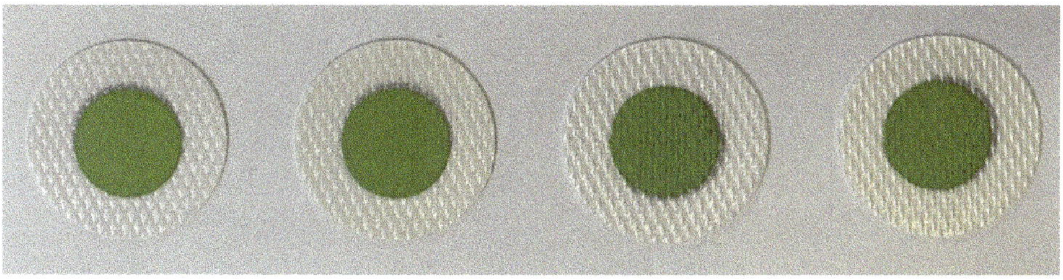

Figure 9.21. Patterned electrolytes produced by metal mesh pressing with NiO–YSZ fuel electrodes. Reprinted from [49], Copyright (2021), with permission from Elsevier.

substrate of YSZ was prepared by tape casting and lamination. The metal mesh was then impressed on it, followed by isostatic pressing. Finally, NiO–YSZ fuel electrodes and LSM–YSZ air electrodes were screen printed onto the patterned electrolytes, and their electrochemical performances were measured. The cells with patterned electrolytes are shown in figure 9.21. The average height of the structure was varied by changing the temperature during the mesh pressing. The patterned cell showed better performance than that of a cell with flat electrolyte surfaces, achieving a maximum power density enhancement of 38%. When the structure height was lower (~40 μm), the cell showed better performance than with higher structures (~100 μm). The reason for this difference was ascribed to the prevention of gas diffusion between the deeper structures, which was again similar to the findings of other research mentioned above. The same research group later attempted to fabricate patterned cells with various mesh patterns; they achieved reductions in both the ohmic and polarization resistances due to the partial reduction in the electrolyte thickness and the increase in the interfacial area [50].

9.3.4 Other approaches

Finally, we introduce a couple of studies in which modifications to the electrolyte–electrode interface were carried out using different approaches from those mentioned already.

Pan *et al* modified the interfacial structure using two kinds of powder, the diameters of which differed by orders of magnitude [51]. By mixing fine and coarse YSZ powders, they were able to control the roughness of the electrolyte film surface. The roughness varied in accordance with the mixing ratio of the two powders. Larger amounts of the coarse powder resulted in greater roughness; however, excessive mixing negatively affected the densification. The electrolyte film was formed on a NiO–YSZ substrate; LSM was deposited as an air electrode and infiltrated with samarium-doped ceria (SDC). The cell with the modified interfacial structure showed a reduction in polarization resistance and an enhancement in power density of 40%.

Instead of interfacial modification, Nagato *et al* directly controlled the electrode microstructure using a magnetic field, so that its effective conductivity was enhanced [52]. To control the microstructure of the fuel electrode using a magnetic field, Ni

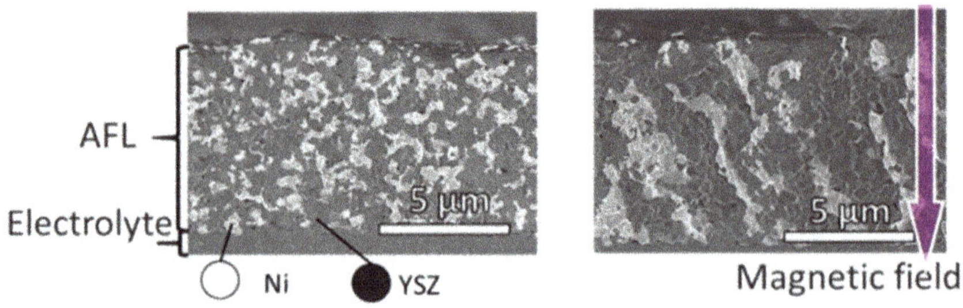

Figure 9.22. Cross-sectional images of Ni–YSZ fuel electrodes produced with and without the use of a magnetic field. Reprinted from [52], © The Electrochemical Society. Reproduced by permission of IOP Publishing.

metal was used as the starting material instead of NiO. A Ni–YSZ slurry was screen printed on a YSZ electrolyte; a magnetic field was then applied during the drying process. The fuel electrodes were deposited symmetrically with respect to the electrolyte, and electrochemical measurements were carried out. Cross-sectional images of the electrodes without and with the magnetic field are shown in figure 9.22. The Ni phase appears to have been aligned in the thickness direction by the magnetic field, which was also expected to contribute to the alignment of the YSZ phase. The two microstructures were reconstructed by FIB-SEM observations and their microstructural characteristics were evaluated. As a result of applying the magnetic field, the tortuosity factor of Ni decreased by 57%, accompanying a reduction of the tortuosity factor in the YSZ by 21%. On the other hand, the TPB density decreased by 42% as a result of the alignment. The electrochemical measurement results for the reference cells showed that the polarization resistance decreased when the Ni volume fraction was reduced, whereas the electrodes with the magnetic field showed almost equivalent polarization resistances regardless of the Ni volume fraction, which can be ascribed to the balance between the enhanced ionic conductivity and the reduction in TPB density. Thus, the electrochemical performance can be maintained by the aligned microstructure when less Ni is used, which would reduce the mismatch of the thermal expansion coefficients between the electrolyte and the electrodes.

9.4 The fabrication of microstructures to enhance gas diffusion

Thus far, we have reviewed modifications of the interfacial microstructure performed at the electrolyte–electrode interfaces or electrodes in order to enhance ionic conductivity. As already discussed in section 9.1.1, ionic conduction often limits the overall cell performance. However, in cases of high fuel utilization due to operation at high current densities or under fuel starvation conditions, gas diffusion becomes the rate-determining process. In such cases, the effective gas diffusivity, which can be defined in a similar manner to equation (9.1), becomes quite influential in the electrode design process. Here, we introduce studies focused on the modification of microstructures to enhance gas diffusion.

The major approach to enhancing gas diffusivity in the electrodes is the application of the phase-inversion method, which creates characteristic asymmetrical pores in the substrates that promote gas diffusion [53–57]. The process used in the phase-inversion method is schematically shown in figure 9.23. First, a substrate such as a NiO–YSZ solution is prepared by mixing powder with a solvent and a binder, which is then immersed in a non-solvent (e.g. water) bath. Through the process of the exchange of water with molecules of the solution, asymmetric pore structures are formed. The pore structure formed by the phase-inversion method is characterized by macroscale finger-like pores, as shown in figure 9.24, which are expected to provide gas diffusion paths deep in the effective area of the electrode near the electrolyte. The microstructure of these finger-like pores has been reported to be highly dependent on the amounts of solvent and solid phases, the species of the solvent and binder, and the species of non-solvent [58–62]. It has been reported that the overall cell performance is not necessarily related to the gas diffusivity, but is also highly dependent on the reaction site density near the electrolyte [58, 59, 61]. Thus, this trade-off relationship between the gas diffusion and electrochemical reaction is similar to the reports of pillar structure modifications at the electrolyte–electrode interfaces as already discussed in

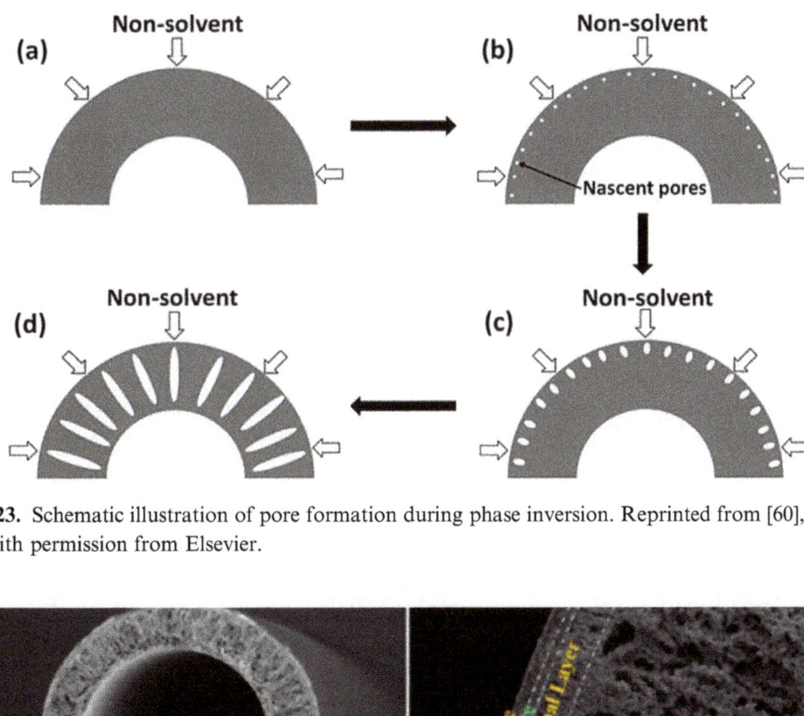

Figure 9.23. Schematic illustration of pore formation during phase inversion. Reprinted from [60], Copyright (2014), with permission from Elsevier.

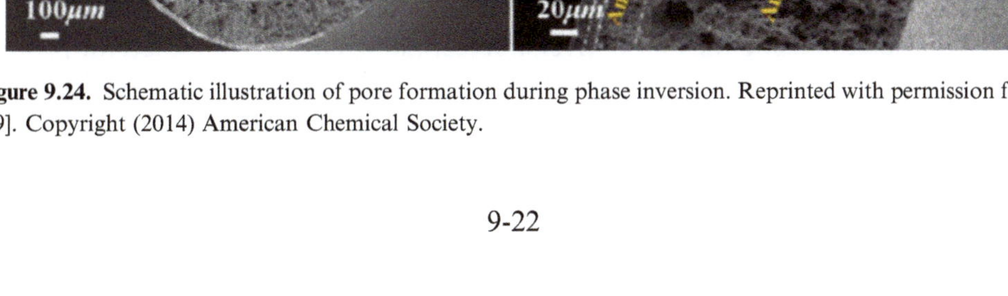

Figure 9.24. Schematic illustration of pore formation during phase inversion. Reprinted with permission from [59]. Copyright (2014) American Chemical Society.

the earlier sections. The effectiveness of the enhanced gas diffusivity on the cell performance was evaluated by Kishimoto *et al*; in their study, a cell with elongated pore structures fabricated by the phase-inversion method showed superior performance to that of a cell fabricated by the conventional method due to the reduction in concentration overpotential, even under low fuel supply conditions or in high current density regions [63].

Lu *et al* analyzed a microstructure fabricated by the phase-inversion method using 3D x-ray tomography and modeling [64]. In order to evaluate the effect of the macroscale pores formed by the phase-inversion process, the scale of which differs greatly from that of microscale pores, 3D microstructures were evaluated at different resolutions. First, characteristic microstructural parameters such as tortuosity factor and gas permeability were calculated from the fine, local microstructure. Coarse 3D structures were then used to simulate the macroscale diffusion phenomena, coupled with the effective value obtained from the fine structure. They reported that the macroscale pore structure enhances mass transport and permeability by 250% and two to three orders of magnitude, respectively. They furthermore developed a multiscale modeling scheme that included the effects of molecule–wall interactions [65]. Using the physical parameters obtained, electrochemical simulations were carried out and the results were compared with those obtained from experiments. It was reported that the simulated results showed good agreement with the experimental ones, showing a 70% reduction in the concentration overpotential for the cell produced by the phase-inversion method via the enhanced diffusion of reactants. The same group also worked on a parametric study of diffusion and pore parameters [66]. The length of the macroscale pore structure and the pore diameter were quite influential in enhancing the gas diffusion and reducing the concentration overpotential. Based on an electrochemical simulation which accounted for the trade-off relationship between enhanced gas diffusivity and TPB density, an optimal design was suggested.

Various cells have been produced using the phase-inversion method. Li *et al* produced anode-supported cells (ASCs) by co-sintering all layers including the NiO–GDC current-collecting layer [67]. They reported that the cell performance was improved by the microscale pores which connected the current-collecting layer to the fuel electrode and the reduction of ohmic resistance due to the uniform current collection. To enhance cell stability during reduction and oxidation (redox) cycles, Wang *et al* produced a cell consisting of dual fuel electrode layers of Ni–Fe alloy and Ni–YSZ by the phase-inversion method [68]. A support structure fabricated by the phase-inversion technique was impregnated with catalytic nanoparticles to further enhance its electrochemical performance. Yang *et al* fabricated a NiO–YSZ tubular support using the phase-inversion method and impregnated the electrodes with SDC [69, 70]. Meng *et al* impregnated a YSZ support fabricated by the phase-inversion method with Ni [71]. Han *et al* impregnated a NiO–YSZ support structure with Ni and Fe catalysts [72].

Cell fabrication using the phase-inversion technique has been adopted for proton-conducting SOCs. Zhao *et al* fabricated ASCs with NiO–$BaZr_{0.1}Ce_{0.7}Y_{0.1}Yb_{0.1}O_{3-d}$ (BZCYYb) composite electrodes and a BZCYYb electrolyte and their electrochemical performances were measured [73]. Shi *et al* fabricated ASCs with NiO–$BaZr_{0.3}Ce_{0.5}Y_{0.2}O_{3-d}$ (BZCY) composite electrodes and a BZCY electrolyte, and the cell performance was evaluated by comparing it to that of a conventional cell. The polarization resistance decreased as a result of applying the support structure by phase inversion, but the ohmic resistance increased due to the increase in the pore phase. The advantage of the cell fabricated by the phase-inversion method became more prominent when it was supplied with large-molecule gases such as hydrocarbons.

In addition, cells fabricated using the phase-inversion method have been applied in the electrolysis mode. Liu *et al* fabricated a cell with a Ni–YSZ support by the phase-inversion method and tested it in the steam electrolysis mode [74]. Li *et al* evaluated a cell with a NiO–YSZ substrate fabricated by the phase-inversion method in the CO_2 electrolysis mode [67]. Compared with a cell fabricated by the conventional tape-casting method, the cell produced using the phase-inversion method showed better performance due to the efficient gas transport of both reactant and product.

Another attempt to fabricate a microstructure to facilitate gas diffusion used the freeze-drying tape-casting method. Chen *et al* fabricated a NiO–GDC support using the freeze-drying tape-casting method and evaluated its microstructure using 3D x-ray tomography [75]. The microstructure of the support is shown in figure 9.25. They found a large porosity of 42% and a change in the micropore diameter from the

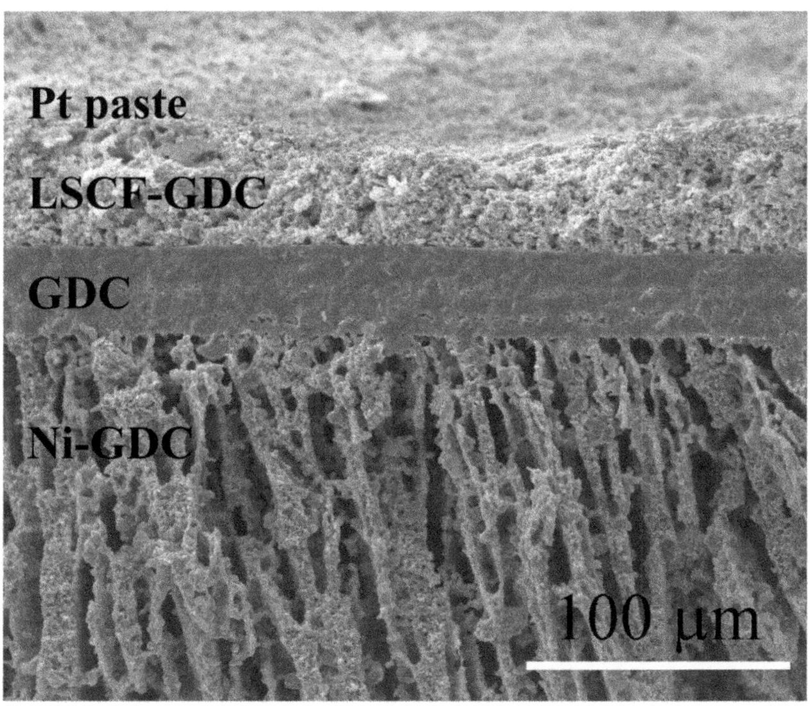

Figure 9.25. Microstructure of a cell with a Ni-GDC support fabricated by the freeze-dry tape-casting method. Reprinted with permission from [75]. Copyright (2014) American Chemical Society.

current collector side (~20 μm) to the electrolyte side (~10 μm). They also reported the anisotropic microstructure of the pore phase according to the value of the tortuosity factor. The tortuosity factor was about 1.3 in the thickness direction, whereas the values in the other two directions were more than twice as large. Thus, gas diffusion was enhanced in the thickness direction. They also reported an enhancement of cell performance in comparison to that of a cell whose support was fabricated by the dry pressing method and lacked anisotropic pore structures. Yang *et al* fabricated a cell using a YSZ support for the electrodes by the freeze-drying tape-casting method, which was impregnated with Ni-SDC and LSM-SDC to form the fuel electrode and the air electrode, respectively [76]. The cell was operated for the co-electrolysis of H_2O and CO_2.

9.5 Summary

In this chapter, we have reviewed interfacial microstructural modifications that are mainly performed in order to enhance the effective ionic conductivity of SOC electrodes. According to both theoretical and experimental approaches, the modification of interfacial microstructures using grooved patterns or the insertion of pillar structures shows great potential for enhancing ionic conduction and electrochemical performance. It has been shown that the geometrical parameters are critical to the design of the structure. Guidance for the design of the appropriate structures is mainly being revealed by theoretical and numerical approaches. The manufacturing processes, on the other hand, have some difficulty in achieving the required precision of the pattern geometries. However, as a result of recent developments in manufacturing technologies, relatively small structures have successfully been fabricated. Furthermore, along with the recent demands for high fuel utilization and electrolysis modes, new approaches have been proposed, including the modification of the pore phase to enhance gas diffusivity. Thus, microstructural modification will be greatly beneficial in enhancing SOC performance in the future, supported by the development of computation and manufacturing technologies.

References

[1] Shin E, Ahn P, Seo H, Jo J, Kim S, Woo S, Yu J H, Mizusaki J and Lee J 2013 Polarization mechanism of high temperature electrolysis in a Ni–YSZ/YSZ/LSM solid oxide cell by parametric impedance analysis *Solid State Ion.* **232** 80–96

[2] Anselmi-Tamburini U, Chiodelli G, Arimondi M, Maglia F, Spinolo G and Munir Z 1998 Electrical properties of Ni/YSZ cermets obtained through combustion synthesis *Solid State Ion.* **110** 35–43

[3] Costamagna P, Costa P and Antonucci V 1998 Micro-modelling of solid oxide fuel cell electrodes *Electrochim. Acta* **43** 375–94

[4] Shikazono N, Sakamoto Y, Yamaguchi Y and Kasagi N 2009 Microstructure and polarization characteristics of anode supported tubular solid oxide fuel cell with co-precipitated and mechanically mixed Ni-YSZ anodes *J. Power Sources* **193** 530–40

[5] Costamagna P, Costa P and Arato E 1997 Some more considerations on the optimization of cermet solid oxide fuel cell electrodes *Electrochim. Acta* **43** 967–72
[6] Konno A, Iwai H, Inuyama K, Kuroyanagi A, Saito M, Yoshida H, Kodani K and Yoshikata K 2011 Mesoscale-structure control at anode/electrolyte interface in solid oxide fuel cell *J. Power Sources* **196** 98–109
[7] Delloro F and Viviani M 2012 Simulation study about the geometry of electrode–electrolyte contact in a SOFC *J. Electroceram.* **29** 216–24
[8] Bertei A, Tariq F, Yufit V, Ruiz-Trejo E and Brandon N P 2017 Guidelines for the rational design and engineering of 3D manufactured solid oxide fuel cell composite electrodes *J. Electrochem. Soc.* **164** F89–98
[9] Wilson J R, Kobsiriphat W, Mendoza R, Chen H-Y, Hiller J M, Miller D J, Thornton K, Voorhees P W, Adler S B and Barnett S A 2006 Three-dimensional reconstruction of a solid-oxide fuel-cell anode *Nat. Mater.* **5** 541–4
[10] Allart M, Christien F and Le Gall R 2013 Ultra-fast sulphur grain boundary segregation during hot deformation of nickel *Acta Mater.* **61** 7938–46
[11] Shimura T, Jiao Z and Shikazono N 2017 Polarization characteristics and microstructural changes of solid oxide fuel cell and solid oxide electrolysis cell fuel electrodes *J. Electrochem. Soc.* **164** F1158–64
[12] Shimura T, Jiao Z, Hara S and Shikazono N 2014 Quantitative analysis of solid oxide fuel cell anode microstructure change during redox cycles *J. Power Sources* **267** 58–68
[13] Sciazko A, Komatsu Y, Yokoi R, Shimura T and Shikazono N 2022 Effects of mass fraction of $La_{0.9}Sr_{0.1}Cr_{0.5}Mn_{0.5}O_{3-\delta}$ and $Gd_{0.1}Ce_{0.9}O_{2-\delta}$ composite anodes for nickel free solid oxide fuel cells *J. Eur. Ceram. Soc.* **42** 1556–67
[14] Sciazko A, Komatsu Y, Shimura T and Shikazono N 2020 Multiscale microstructural evolutions of nickel–gadolinium doped ceria in solid oxide fuel cell anode *J. Power Sources* **478** 228710
[15] Sciazko A, Shimura T, Komatsu Y and Shikazono N 2021 Ni-GDC and Ni-YSZ electrodes operated in solid oxide electrolysis and fuel cell modes *J. Therm. Sci. Technol.* **16** 1–10
[16] Iwai H *et al* 2010 Quantification of SOFC anode microstructure based on dual beam FIB-SEM technique *J. Power Sources* **195** 955–61
[17] Shen X, Zhang Y, Shen T and Khajorntraidet C 2017 Spark advance self-optimization with knock probability threshold for lean-burn operation mode of SI engine *Energy* **122** 1–10
[18] Lu X, Tjaden B, Bertei A, Li T, Li K, Brett D and Shearing P 2017 3D characterization of diffusivities and its impact on mass flux and concentration overpotential in SOFC anode *J. Electrochem. Soc.* **164** F188–95
[19] Karen Chen-Wiegart Y C, Cronin J S, Yuan Q, Yakal-Kremski K J, Barnett S A and Wang J 2012 3D non-destructive morphological analysis of a solid oxide fuel cell anode using full-field X-ray nano-tomography *J. Power Sources* **218** 348–51
[20] Hubert M, Laurencin J, Cloetens P, da Silva J C, Lefebvre-Joud F, Bleuet P, Nakajo A and Siebert E 2016 Role of microstructure on electrode operating mechanisms for mixed ionic electronic conductors: from synchrotron-based 3D reconstruction to electrochemical modeling *Solid State Ion.* **294** 90–107
[21] Heenan T M M, Lu X, Robinson J B, Iacoviello F, Brett D J L and Shearing P R 2018 Thermally driven SOFC degradation in 4D: part II. Macroscale *J. Electrochem. Soc.* **165** F932–41

[22] Shimura T, Nagato K and Shikazono N 2019 Evaluation of electrochemical performance of solid-oxide fuel cell anode with pillar-based electrolyte structures *Int. J. Hydrog. Energy* **44** 12043–56
[23] Shimura T, He A and Shikazono N 2019 Evaluation of $La_{0.57}Sr_{0.38}Co_{0.2}Fe_{0.8}O_{3-\delta}$ electrode performance degradation based on three-dimensional microstructure reconstruction and electrochemical simulation *J. Electrochem. Soc.* **166** F821–30
[24] Shikazono N, Kanno D, Matsuzaki K, Teshima H, Sumino S and Kasagi N 2010 Numerical assessment of SOFC anode polarization based on three-dimensional model microstructure reconstructed from FIB-SEM images *J. Electrochem. Soc.* **157** B665
[25] He A, Shimura T, Gong J and Shikazono N 2019 Numerical simulation of $La_{0.6}Sr_{0.4}Co_{0.2}Fe_{0.8}O_3$–$Gd_{0.1}Ce_{0.9}O_{1.95}$ composite cathodes with micro pillars *Int. J. Hydrog. Energy* **44** 6871–85
[26] Kim Y T, Jiao Z and Shikazono N 2017 Evaluation of $La_{0.6}Sr_{0.4}Co_{0.2}Fe_{0.8}O_3$–$Gd_{0.1}Ce_{0.9}O_{1.95}$ composite cathode with three dimensional microstructure reconstruction *J. Power Sources* **342** 787–95
[27] Chueh C C, Bertei A and Nicolella C 2019 Design guidelines for the manufacturing of the electrode–electrolyte interface of solid oxide fuel cells *J. Power Sources* **437** 226888
[28] Iwai H, Kuroyanagi A, Saito M, Konno A, Yoshida H, Yamada T and Nishiwaki S 2011 Power generation enhancement of solid oxide fuel cell by cathode–electrolyte interface modification in mesoscale assisted by level set-based optimization calculation *J. Power Sources* **196** 3485–95
[29] He A, Onishi J and Shikazono N 2020 Optimization of electrode–electrolyte interface structure for solid oxide fuel cell cathode *J. Power Sources* **449** 227565
[30] Larrea A, Sola D, Laguna-Bercero M A, Peña J I, Merino R I and Orera V M 2011 Self-supporting thin yttria-stabilised zirconia electrolytes for solid oxide fuel cells prepared by laser machining *J. Electrochem. Soc.* **158** B1193
[31] Zhang Y, Cai G, Gu Y, Ge L, Zheng Y, Chen H and Guo L 2018 Modifying the electrode–electrolyte interface of anode supported solid oxide fuel cells (SOFCs) by laser-machining *Energy Convers. Manag.* **171** 1030–7
[32] Cai G, Zhang Y, Dai H, He S, Ge L, Chen H and Guo L 2017 Modification of electrode/electrolyte interface by laser micro-processing for solid oxide fuel cell *Mater. Lett.* **195** 232–5
[33] Cebollero J A, Lahoz R, Laguna-Bercero M A and Larrea A 2017 Tailoring the electrode–electrolyte interface of Solid Oxide Fuel Cells (SOFC) by laser micro-patterning to improve their electrochemical performance *J. Power Sources* **360** 336–44
[34] Cebollero J A, Laguna-Bercero M A, Lahoz R, Silva J, Moreno R and Larrea A 2019 Optimization of laser-patterned YSZ-LSM composite cathode–electrolyte interfaces for solid oxide fuel cells *J. Eur. Ceram. Soc.* **39** 3466–74
[35] Konno A, Iwai H, Saito M and Yoshida H 2011 A corrugated mesoscale structure on electrode–electrolyte interface for enhancing cell performance in anode-supported SOFC *J. Power Sources* **196** 7442–9
[36] Chao C C, Hsu C M, Cui Y and Prinz F B 2011 Improved solid oxide fuel cell performance with nanostructured electrolytes *ACS Nano* **5** 5692–6
[37] Wang X M, Li C J, Li C X and Yang G J 2010 Microstructure and electrochemical behavior of a structured electrolyte/LSM-cathode interface modified by flame spraying for solid oxide fuel cell application *J. Therm. Spray Technol.* **19** 311–6
[38] Dai H, He S, Chen H and Guo L 2016 A novel method of modifying electrolyte surface at mesoscale for intermediate-temperature solid oxide fuel cells *Ceram. Int.* **42** 2045–50

[39] Farandos N M, Kleiminger L, Li T, Hankin A and Kelsall G H 2016 Three-dimensional inkjet printed solid oxide electrochemical reactors. I. Yttria-stabilized Zirconia Electrolyte *Electrochim. Acta* **213** 324–31

[40] Masciandaro S, Torrell M, Leone P and Tarancón A 2019 Three-dimensional printed yttria-stabilized zirconia self-supported electrolytes for solid oxide fuel cell applications *J. Eur. Ceram. Soc.* **39** 9–16

[41] Salari F, Badihi Najafabadi A, Ghatee M and Golmohammad M 2020 Hybrid additive manufacturing of the modified electrolyte–electrode surface of planar solid oxide fuel cells *Int. J. Appl. Ceram. Technol.* **17** 1554–61

[42] Seo H, Iwai H, Kishimoto M, Ding C and Saito M 2020 Microextrusion printing for increasing electrode–electrolyte interface in anode-supported solid oxide fuel cells *J. Power Sources* **450** 227682

[43] Seo H, Kishimoto M, Nakagawa T, Iwai H and Yoshida H 2021 Mechanism of improved electrochemical performance of anode-supported solid oxide fuel cells by mesostructural modification of electrode–electrolyte interface *J. Power Sources* **506** 230107

[44] Seo H, Kishimoto M, Ding C, Iwai H, Saito M and Yoshida H 2020 Improvement in the electrochemical performance of anode-supported solid oxide fuel cells by meso- and nano-scale structural modifications *Fuel Cells* **20** 570–9

[45] Jang I and Kelsall G H 2022 Fabrication of 3D NiO-YSZ structures for enhanced performance of solid oxide fuel cells and electrolysers *Electrochem. Commun.* **137** 107260

[46] Xu Y, Tsumori F, Osada T and Miura H 2013 Improvement of solid oxide fuel cell by imprinted micropatterns on electrolyte *Micro Nano Lett.* **8** 571–4

[47] Okabe T, Kim Y, Jiao Z, Shikazono N and Taniguchi J 2018 Fabrication process for micropatterned ceramics via UV-nanoimprint lithography using UV-curable binder *Jpn. J. Appl. Phys.* **57** 106501

[48] Akama R, Okabe T, Sato K, Inaba Y, Shikazono N, Sciazko A and Taniguchi J 2020 Fabrication of a micropatterned composite electrode for solid oxide fuel cells via ultraviolet nanoimprint lithography *Microelectron. Eng.* **225** 111277

[49] Timurkutluk C, Altan T, Yildirim F, Onbilgin S, Yagiz M and Timurkutluk B 2021 Improving the electrochemical performance of solid oxide fuel cells by surface patterning of the electrolyte *J. Power Sources* **512** 230489

[50] Timurkutluk C, Altan T, Onbilgin S, Yildirim F, Yagiz M and Timurkutluk B 2022 Mesh patterned electrolyte supports for high-performance solid oxide fuel cells *Int. J. Energy Res.* **46** 10689–703

[51] Pan W P, Lü Z, Zhang Y H, Huang X Q, Wei B, Wang Z H and Su W H 2012 Improved electrodes/electrolyte interfaces for solid oxide fuel cell by using dual-sized powders in electrolyte slurry *Fuel Cells* **12** 732–8

[52] Nagato K, Shintani K, Shimura T, Shikazono N and Nakao M 2019 Magnetic alignment of anode microstructure in solid oxide fuel cell *J. Electrochem. Soc.* **166** F144–8

[53] Droushiotis N, Doraswami U, Kanawka K, Kelsall G H and Li K 2009 Characterization of NiO-yttria stabilised zirconia (YSZ) hollow fibres for use as SOFC anodes *Solid State Ion.* **180** 1091–9

[54] Droushiotis N, Othman M H D, Doraswami U, Wu Z, Kelsall G and Li K 2009 Novel co-extruded electrolyte-anode hollow fibres for solid oxide fuel cells *Electrochem. Commun.* **11** 1799–802

[55] Jin C, Liu J, Li L and Bai Y 2009 Electrochemical properties analysis of tubular NiO-YSZ anode-supported SOFCs fabricated by the phase-inversion method *J. Memb. Sci.* **341** 233–7

[56] Wei C C and Li K 2008 Yttria-stabilized zirconia (YSZ)-based hollow fiber solid oxide fuel cells *Ind. Eng. Chem. Res.* **47** 1506–12
[57] Yang C, Li W, Zhang S, Bi L, Peng R, Chen C and Liu W 2009 Fabrication and characterization of an anode-supported hollow fiber SOFC *J. Power Sources* **187** 90–2
[58] Ren C, Liu T, Mao Y, Maturavongsadit P, Luckanagul J A, Wang Q and Chen F 2014 Effect of casting slurry composition on anode support microstructure and cell performance of MT-SOFCs by phase inversion method *Electrochim. Acta* **149** 159–66
[59] Liu T, Ren C, Fang S, Wang Y and Chen F 2014 Microstructure tailoring of the nickel oxide-yttria-stabilized zirconia hollow fibers toward high-performance microtubular solid oxide fuel cells *ACS Appl. Mater. Interfaces* **6** 18853–60
[60] Ren C, Liu T, Maturavongsadit P, Luckanagul J A and Chen F 2015 Effect of PEG additive on anode microstructure and cell performance of anode-supported MT-SOFCs fabricated by phase inversion method *J. Power Sources* **279** 774–80
[61] Liu T, Ren C, Zhang Y, Wang Y, Lei L and Chen F 2017 Solvent effects on the morphology and performance of the anode substrates for solid oxide fuel cells *J. Power Sources* **363** 304–10
[62] Ren C, Zhang Y, Xu Q, Tian T and Chen F 2020 Effect of non-solvent from the phase inversion method on the morphology and performance of the anode supported microtubular solid oxide fuel cells *Int. J. Hydrog. Energy* **45** 6926–33
[63] Kishimoto M, Masuyama A, Iwai H and Yoshida H 2019 Dual-resolution microstructural analysis of anisotropic pore structure in sofc anode fabricated by phase-inversion tape casting *ECS Trans.* **91** 1861–9
[64] Lu X, Taiwo O O, Bertei A, Li T, Li K, Brett D J L and Shearing P R 2017 Multi-length scale tomography for the determination and optimization of the effective microstructural properties in novel hierarchical solid oxide fuel cell anodes *J. Power Sources* **367** 177–86
[65] Lu X, Li T, Bertei A, Cho J I S, Heenan T M M, Rabuni M F, Li K, Brett D J L and Shearing P R 2018 The application of hierarchical structures in energy devices: new insights into the design of solid oxide fuel cells with enhanced mass transport *Energy Environ. Sci.* **11** 2390–403
[66] Lu X, Bertei A, Heenan T M M, Wu Y, Jl D and Shearing P R 2019 Multi-length scale microstructural design of micro-tubular solid oxide fuel cells for optimised power density and mechanical robustness *J. Power Sources* **434** 226744
[67] Li T, Wu Z and Li K 2015 High-efficiency, nickel-ceramic composite anode current collector for micro-tubular solid oxide fuel cells *J. Power Sources* **280** 446–52
[68] Wang M, Li N, Wang Z, Chen C and Zhan Z 2022 Electrochemical performance and redox stability of solid oxide fuel cells supported on dual-layered anodes of Ni-YSZ cermet and Ni–Fe alloy *Int. J. Hydrog. Energy* **47** 5453–61
[69] Yang C, Jin C and Chen F 2010 Performances of micro-tubular solid oxide cell with novel asymmetric porous hydrogen electrode *Electrochim. Acta* **56** 80–4
[70] Yang C, Jin C and Chen F 2010 Micro-tubular solid oxide fuel cells fabricated by phase-inversion method *Electrochem. Commun.* **12** 657–60
[71] Meng X, Yan W, Yang N, Tan X and Liu S 2015 Highly stable microtubular solid oxide fuel cells based on integrated electrolyte/anode hollow fibers *J. Power Sources* **275** 362–9
[72] Han Z, Yang Z and Han M 2018 Optimization of Ni-YSZ anodes for tubular SOFC by a novel and efficient phase inversion-impregnation approach *J. Alloys Compd.* **750** 130–8

[73] Zhao F, Jin C, Yang C, Wang S and Chen F 2011 Fabrication and characterization of anode-supported micro-tubular solid oxide fuel cell based on $BaZr_{0.1}Ce_{0.7}Y_{0.1}Yb_{0.1}O_{3-\delta}$ electrolyte *J. Power Sources* **196** 688–91

[74] Liu T, Wang Y, Zhang Y, Fang S, Lei L, Ren C and Chen F 2015 Steam electrolysis in a solid oxide electrolysis cell fabricated by the phase-inversion tape casting method *Electrochem. Commun.* **61** 106–9

[75] Chen Y, Zhang Y, Baker J, Majumdar P, Yang Z, Han M and Chen F 2014 Hierarchically oriented macroporous anode-supported solid oxide fuel cell with thin ceria electrolyte film *ACS Appl. Mater. Interfaces* **6** 5130–6

[76] Yang C, Li J, Newkirk J, Baish V, Hu R, Chen Y and Chen F 2015 Co-electrolysis of H_2O and CO_2 in a solid oxide electrolysis cell with hierarchically structured porous electrodes *J. Mater. Chem.* A **3** 15913–9

IOP Publishing

Nanoengineered Materials for Solid Oxide Cells

Katherine Develos-Bagarinao

Chapter 10

Synchrotron x-ray radiation for solid oxide cell material characterization

Jérôme Laurencin, Maxime Hubert, Federico Monaco, Aline Léon and Dario Ferreira Sanchez

Solid oxide cells (SOCs) constitute an emerging technology for energy conversion; however, they still need to be improved to enhance their efficiency and lifetime. To support the realization of this objective, the development of SOCs must rely on advanced characterization techniques to probe the properties of SOC materials from the micrometer to the nanometer scale. In this context, x-ray radiation offers ways to characterize these materials down to the nanoscale using a large variety of methods. In this chapter, we briefly introduce and discuss the fundamentals of the different synchrotron x-ray computed tomography (XCT) methods. We also present an overview of the main x-ray spectroscopic and scattering techniques including x-ray fluorescence and diffraction (XRF and XRD), x-ray photoelectron spectroscopy (XPS) and x-ray absorption spectroscopy (XAS). For each of these techniques, the main advantages of using synchrotron x-ray radiation are discussed with respect to laboratory equipment. The application of these techniques to SOC technology is illustrated using a selection of relevant and representative studies.

10.1 Introduction

Solid oxide cells (SOCs) are electrochemical devices that work at high temperatures. As a result of their flexibility, they can be operated either in the fuel cell mode (i.e. as solid oxide fuel cells (SOFCs)) or in the electrolysis mode (i.e. as solid oxide electrolysis cells (SOECs)). SOCs are classically composed of a dense electrolyte made of yttria-stabilized zirconia (YSZ) sandwiched between two porous electrodes. The hydrogen electrode is typically a cermet made of nickel and YSZ (Ni–YSZ), while the oxygen electrode is based on lanthanum strontium cobalt ferrite (LSCF) [1]. A barrier layer of gadolinia-doped ceria (GDC) is usually added between the electrolyte and the oxygen electrode to limit the reaction between YSZ and LSCF.

doi:10.1088/978-0-7503-4064-9ch10

The thickness of one of these layers is increased so that it can be used as the mechanical support for the whole cell. The two classical configurations are hydrogen-electrode-supported cells and electrolyte-supported cells. Electrolyte-supported cells have higher mechanical stability and durability, while hydrogen-electrode-supported cells have been reported to be a good compromise between manufacturing cost, performance, and mechanical robustness.

Microstructural optimization, including the nanoscale optimization of electrode material architecture, are performed to improve cell efficiency. For instance, advanced manufacturing techniques such as infiltration or exsolution allowed obtaining electrode nanostructures with enhanced electrochemical activities [2–4]. Moreover, new materials are currently being developed to increase the cell lifetime by retaining high performance, especially for cells operated in the electrolysis mode [5]. Indeed, SOC durability is still insufficient and remains an issue for the industrialization of this technology. At present, the rate of cell performance degradation is around $\approx$1%/kh in the fuel cell mode, while it can reach $\approx$5%/kh when operated with an electrolysis current [6].

SOC degradation has several origins, such as delamination, mechanical damage within the cell layers, microstructural evolution of the electrodes, and reactivity and interdiffusion between cell components [7]. For instance, Ni particles are not stable in operation and are likely to undergo substantial agglomeration [8]. Moreover, a significant Ni migration away from the electrolyte interface is triggered when the cells are operated in the electrolysis mode [9]. In addition, the oxygen electrode materials tend to decompose in operation at high temperatures [10]. For example, strontium tends to segregate at the LSCF surface, leading to the formation of a passivating strontium oxide film [10]. The release of Sr from the LSCF phase is generally associated with the precipitation of a secondary phase (i.e. $SrZrO_3$) in the pores of the barrier layer [11]. Gadolinium is also liable to diffuse in the electrode, leading to the formation of a so-called interdiffusion layer (IDL) [12]. Mechanical damage can also be generated in the electrodes during the startup and shutdown of the system. As an example, microcracks have been observed in the YSZ backbone of the cermet when Ni is re-oxidized from its reduced state [13].

The underlying driving forces controlling all these degradation phenomena are still not precisely understood. They involve physicochemical processes that take place at the nanoscale in the bulk of materials, at the interfaces between cell components, and on the electrode surfaces. They are also strongly related to the local electrochemical reactions that extend from the electrolyte interface into the electrode thickness over a distance of few micrometers. Moreover, the distribution of current density, partial pressures, and overpotentials along the cell length can induce macroscopic heterogeneities in the cell degradation.

A combination of advanced characterizations techniques at different length scales is thus required to better understand the cell degradation mechanisms. It is also needed to assess the properties of new materials or nanostructured electrodes. In this respect, techniques based on x-ray radiation complement other methods such as atomic probe tomography, transmission electron microscopy, and secondary-ion mass spectrometry. On the one hand, laboratory x-ray equipment is well adapted to macroscopic characterizations that can be used to investigate heterogeneities in the cell layers. On the other hand, synchrotron x-ray radiation offers the possibility of probing material evolution down to the nanoscale using a large variety of techniques that can be combined to address the main SOC degradation phenomena.

Several synchrotron XCT methods have been successfully applied for 3D electrode reconstruction. They allow investigating microstructural evolution, such as Ni coarsening or mechanical damage induced in the electrodes. XRD and XRF have been coupled to study interdiffusion, reactions between cell layers, and phase decomposition. The evolution of physicochemical material properties can be studied by XAS, while XPS is well adapted to surface passivation investigations.

The first section of this chapter is dedicated to advanced synchrotron XCT methods for electrode reconstruction, while the second section is devoted to advanced x-ray spectroscopic and scattering techniques for structural and chemical analyses of SOCs using synchrotron x-ray radiation.

10.2 Reconstructing electrode microstructures using x-ray computed tomography

10.2.1 A brief description of XCT techniques

X-rays were discovered in 1895 by Wilhelm Roentgen [14], who radiographed his wife's hand (figure 10.1). The interest generated by these new rays quickly attracted the attention of many researchers. One of the first uses of x-rays was carried out by Marie Curie for medical diagnosis during the First World War. If several radiographs or 2D projections are acquired at different angles of rotation, it becomes theoretically possible to reconstruct the studied object in 3D. However, computational resources are required to perform the raw data postprocessing using the well-known Fourier slice theorem [15]. It was only in 1979 that Allan M Cormack and Godfrey N Hounsfield received the Nobel Prize in Physiology or Medicine for the development of computer-assisted tomography [16]. The use of XCT techniques is now widespread for industrial and medical applications. In this way, reconstructions of SOC porous electrodes have successfully been obtained using laboratory XCT systems [17–19]. For instance, 3D images of typical Ni–YSZ cermets have been recently obtained with a large field of view of 16 μm and a voxel size of 33 nm [18]. Nevertheless, it is worth mentioning that the voxel size does not represent the spatial resolution that must be assessed on 3D images using different approaches such as the line profile method or Fourier shell correlation [20]. Since the ceramics are highly absorbent toward x-rays, synchrotron radiation at high energies presents several advantages for the material characterization of SOCs. Indeed, they allow investigating large volumes at high resolution in a short acquisition time. In addition, the temporal and spatial coherence of synchrotron radiation sources enable the use of the different tomography methods described hereafter.

These imaging techniques are based on the interaction between x-rays and matter [21]. This interaction is described by the complex refractive index n defined by the following relation:

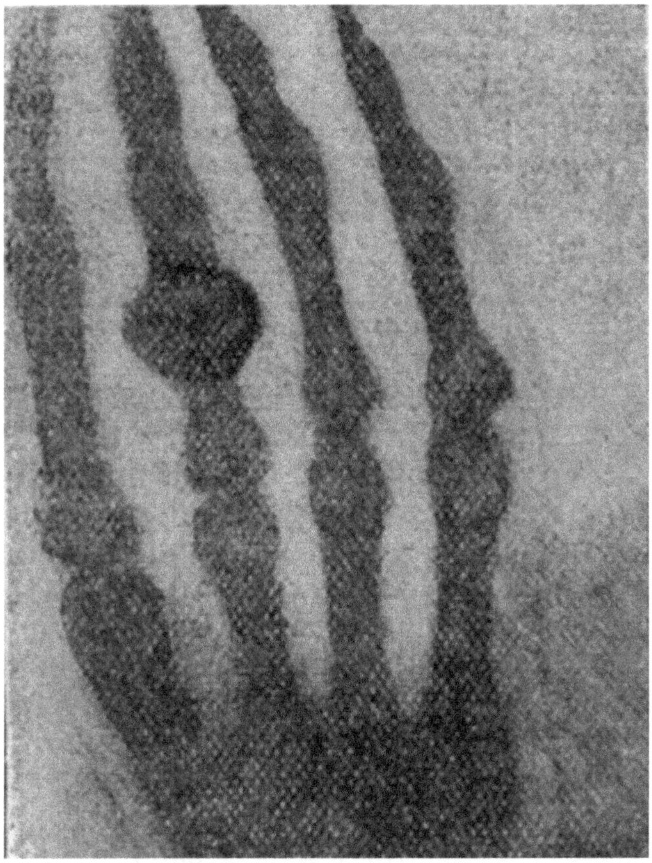

Figure 10.1. First x-ray radiograph of a human hand with a ring [14] Copyright © 1896, Nature Publishing Group. With permission of Springer.

$$n = 1 - \delta + i\beta, \tag{10.1}$$

where δ is the refractive index decrement. The coefficient β is the absorption index, which is linked to the linear attenuation coefficient μ and the wavelength λ_0 of the incident x-ray beam:

$$\mu = \frac{4\pi}{\lambda_0}\beta \tag{10.2}$$

The coefficient μ is a material property characterizing the x-ray absorption of each material. These coefficients are used to express the transmittance T, which is defined as the ratio between the incident and the transmitted waves, E_o and E_{trans}:

$$T = \frac{E_{trans}}{E_o} = \exp\{-B(x, y)\} \times \exp\{i\Delta\varphi(x, y)\}, \tag{10.3}$$

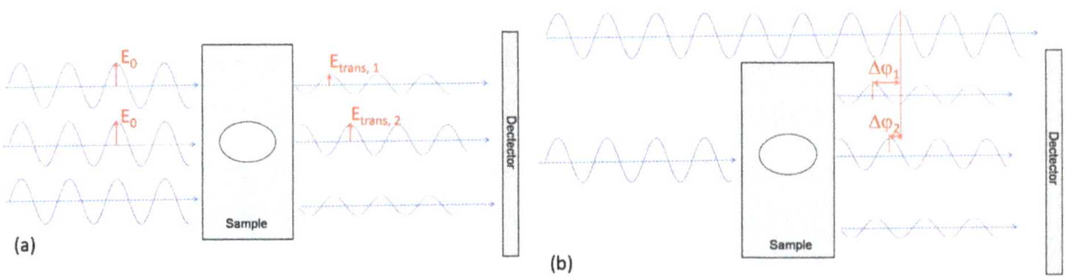

Figure 10.2. Schematic representation of the x-ray amplitude attenuation that takes place in absorption tomography (a) and the phase shift that takes place in propagation-based imaging tomography (b) for a sample with a single pore.

where $\Delta\varphi(x,y)$ represents the phase shift of the x-ray wave in its direction of propagation x. The well-known Beer–Lambert law is given by the square modulus of T and expresses the transmitted intensity across a heterogeneous sample:

$$|T|^2 = \frac{I_{\text{trans}}}{I_0} = \exp\{-2B\} \quad \text{with} \quad B = \frac{2\pi}{\lambda_0}\int_{x=0}^{x=\ell} \beta(x, y)dx, \tag{10.4}$$

where I_0 denotes the incident intensity and l is the sample thickness.

In classical absorption tomography, the transmitted intensity is recorded on the detector at different angles of rotation. According to equation (10.4), these radiographs correspond to a projection that integrates information about the x-ray attenuation through the sample thickness caused by the different phases (figure 10.2(a)). It is worth mentioning that the absorption coefficients of air and SOC materials are very different. In this case, x-ray with small wavelength (i.e. high energy) can be used to characterize porous electrodes made of a single material. Therefore, according to equation (10.4), the transmitted intensity is enhanced in such a way that thick samples can be analyzed in order to obtain large reconstructions. However, for composite electrodes such as Ni–YSZ cermet, the absorption coefficients of the two solids are close to each other making difficult the distinction between the solid phases in the projections. To overcome this issue, the sample needs to be scanned twice, i.e. below and above the absorption edge of one of the electrode materials. It can be observed that the first absorption edge of classical SOC materials arises at an intermediate energy. For instance, the Ni K edge appears at 8.3 keV, limiting the x-ray energy used for the experiment and hence the volume of the reconstruction. Despite this drawback, some authors have reconstructed typical Ni–YSZ electrodes reaching volumes of around 10^3 μm^3 with voxel sizes in the range of 20–50 nm [13, 22–30].

Other tomographic techniques take advantage of a change in the wavelength in the material λ_m, which is a function of the refractive index decrement as follows:

$$\lambda_m = \frac{\lambda_0}{(1-\delta)} \quad \text{with} \quad \delta \approx 1.3 \times 10^{-6}\rho_m\lambda_0, \tag{10.5}$$

where ρ_m is the material's mass density [31]. As illustrated in figure 10.2(b), this change in the wavelength induces a phase shift or retardation at the sample outlet:

$$\Delta\varphi(x, y) = \int_{x=0}^{x=\ell} \left(\frac{2\pi}{\lambda_0} - \frac{2\pi}{\lambda_m} \right) dx = \frac{2\pi}{\lambda_0} \int_{x=0}^{x=\ell} \delta(x, y) dx. \tag{10.6}$$

This phase retardation also contains information about the sample's composition that can be used in the reconstruction. However, as shown in equation (10.4), information about the phase shift is lost in the transmitted intensity. To circumvent this difficulty, several solutions have been proposed. The so-called crystal interferometry and analyzer-based imaging are two common techniques that require the development of complex experimental setups [32, 33]. Moreover, the spatial resolution of the final reconstruction is relatively low and unsuitable for the characterization of SOC materials. Propagation-based imaging is an alternative approach that does not require any specific instrumentation and allows high resolutions to be achieved [34, 35]. This method takes advantage of the diffraction phenomena induced by all the interfaces in the sample. They result in interference fringes that can be measured at the detector under near-field conditions. The detector is thus positioned at a 'short' distance from the sample so that the Fresnel approximation is valid. Holotomography is one of the phase-contrast imaging methods based on this approximation [36, 37]. Under these conditions, it can be shown that the intensity measured by the detector is proportional to the second derivative of the phase shift, according to the so-called transport of intensity equation [38]:

$$I_{\text{detector}}(x, y) \propto \nabla \cdot [I_{\text{trans}}(x, y) \nabla \varphi(x, y)]. \tag{10.7}$$

Using this equation, it is possible to retrieve the phase shift from the measured intensity and hence to reconstruct the sample with different projections [39]. In this case, four measurements are combined by changing the distance between the sample and the detector in order to improve the image quality (figure 10.3(a)) [37]. In addition, random motion can be added to the sample during acquisition to avoid artifacts [40].

Near-field ptychography is another method for which the detector is also positioned at a 'short' distance from the sample, where the diffraction pattern can be approximated by the Fresnel diffraction equation. In this particular case, 16 radiographs are recorded at different positions in the plane orthogonal to the x-ray beam (figure 10.3(b)) [41]. As in the case of absorption tomography, the data for holotomography and near-field ptychography are acquired in full field, meaning the sample is entirely contained in the x-ray beam (figures 10.3(a) and (b)). In other words, the projections are obtained at each angle without scanning. Finally, far-field ptychography consists of moving the detector far away from the sample, where the diffraction pattern can be described by the Fraunhofer equation. In this approach, the data are collected by scanning the sample, which is positioned at the x-ray's focal plane (figure 10.3(c)) [42]. Compared to holotomography, ptychographic methods do not require prior knowledge of a material's properties (δ and β) for the phase retrieval treatment.

The phase shift becomes significant compared to the amplitude attenuation when the ratio δ/β between the refractive index decrement and the absorption coefficient is high. As shown in figure 10.4 for typical SOC materials, this ratio increases steeply

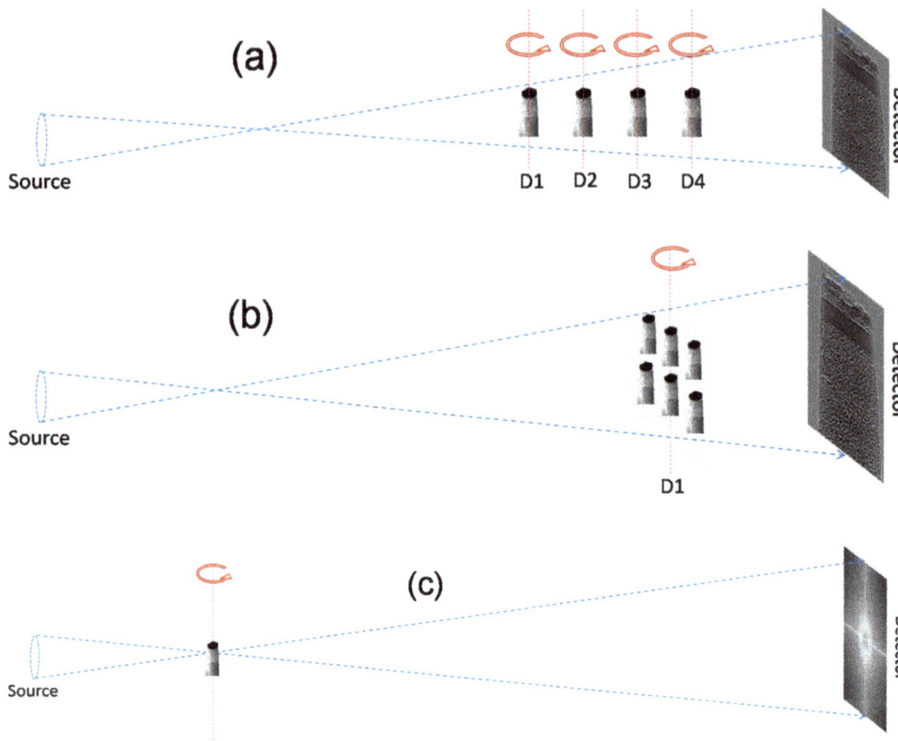

Figure 10.3. A schematic representation of holotomography with four measurements at four distances in the x-ray beam (a), near-field ptychography with 16 measurements in the orthogonal plane at one distance (b), and far-field ptychography, for which the sample is scanned (c).

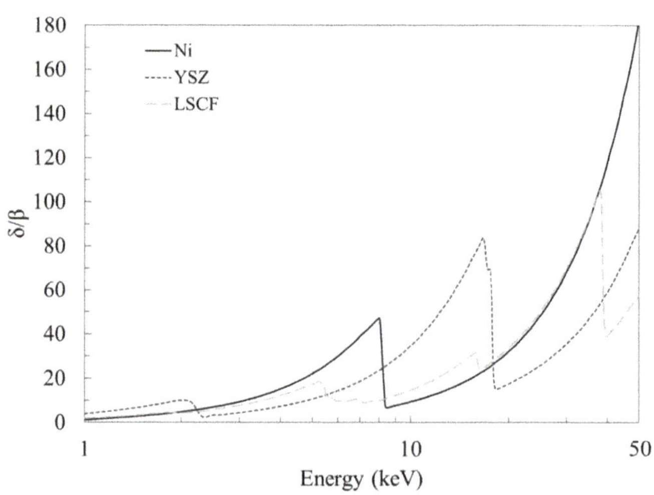

Figure 10.4. The δ/β ratio plotted as a function of the beam energy for three typical SOC materials [54].

with increasing x-ray energies. Therefore, the XCT methods based on x-ray phase imaging become particularly relevant at high energies compared to those used in absorption tomography. These methods thus offer the possibility of obtaining large reconstructions while retaining high resolution. Using holotomography, Ni–YSZ cermet and LSCF electrodes have been reconstructed with a large volume of $50 \times 25^2 \times \pi$ μm^3 and a voxel size of 25 nm [8, 9, 40, 43–50]. A spatial resolution of 50 nm was obtained for 3D images using the line profile method [49]. Near-field ptychography and holotomography were compared in reference [50] using analyses of the same Ni–YSZ electrode. It has been shown that near-field ptychography allows improving the global image quality and offers a better signal-to-noise ratio. Far-field ptychography has been also used to characterize SOC electrodes with a typical voxel size of $\approx$18 nm and reconstructed volumes of $\approx 10^3$ μm^3 [51–53].

Tomographic experiments require the preparation of a thin sample extracted from the complete cell in order to avoid the total absorption of x-rays by the electrode materials. A first approach consists of cutting off small pieces of the cell using a diamond or wire saw. Then, the sample size needs to be reduced by micropolishing equipment to obtain thin lamella or pyramidal samples [29, 43]. However, the geometry of these samples is not perfectly suitable for tomographic experiments. Focused ion beam (FIB) using Ga or plasma FIB (pFIB) working with xenon can be used to mill samples to a suitable cylindrical shape (figure 10.5). The method consists of several steps with a final milling stage performed at low current to obtain a clean and undamaged surface [49]. Xe pFIB allows preparing large samples with typical diameters of 60–80 μm in a reasonable time, while the diameter is usually limited to 10 μm using Ga FIB. A method that combines laser micro-machining and FIB milling has recently been proposed for cost-effective sample preparation [55].

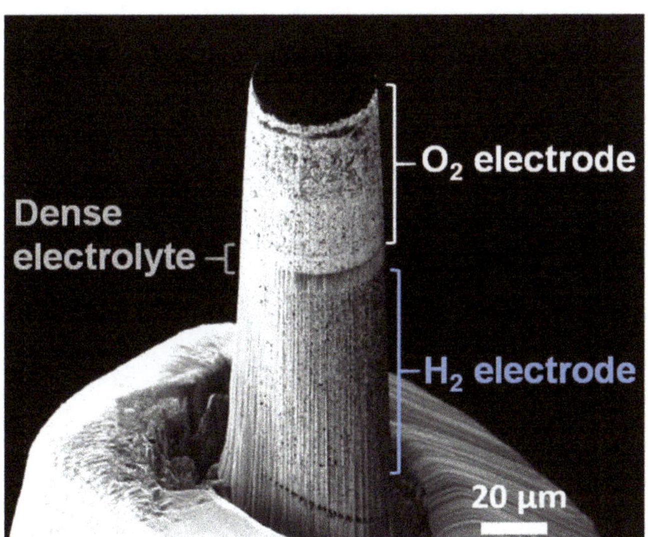

Figure 10.5. Cylindrical pillars prepared by Xe pFIB for a typical SOC.

The XCT techniques are non-destructive methods that can be used to analyze the same before and after *ex situ* experiments. For instance, reconstructed volumes of the same microstructure have been obtained by synchrotron absorption tomography [29] and far-field ptychography [52] to characterize Ni agglomeration at high temperatures. From this point of view, XCT can be considered as a complementary method to the widespread focused ion beam scanning electron microscopy (FIB-SEM) tomography [56–72]. Indeed, this destructive technique consists in milling and imaging successive slices of material in order to obtain a stack of SEM images. This technique has been recently extended using a Xe pFIB to reach large reconstructions [73, 74].

Figure 10.6 summarizes the main tomographic studies devoted to the SOC electrodes. The voxel size is plotted as a function of the volume of the reconstructed electrode for the different tomography techniques. It is worth recalling that the voxel size is not the image spatial resolution. This information is usually not specified in the published articles. Nevertheless, it can be roughly considered that at least two voxels are needed to describe the interfaces in 3D images. As shown in figure 10.6, the FIB-SEM allows reaching high resolutions with typical reconstructed volumes of less than than 15^3 μm^3. On the other hand, the phase-contrast imaging techniques allow obtaining large volume of 40^3 μm^3 with intermediate resolutions. The characteristics of absorption tomography reconstructions lie in between the two.

The spatial resolution required to characterize SOC electrodes can be roughly estimated by considering that the particle diameters must be described by at least ten voxels [75]. According to this criterion, a spatial resolution of around 50 nm is needed to describe accurately all the geometrical features of typical electrodes [49]. From this point of view, all these methods are rather well adapted to the characterization of SOC material. Nevertheless, a higher resolution than 50 nm would be necessary to precisely image the very fine morphology of nanostructured electrodes

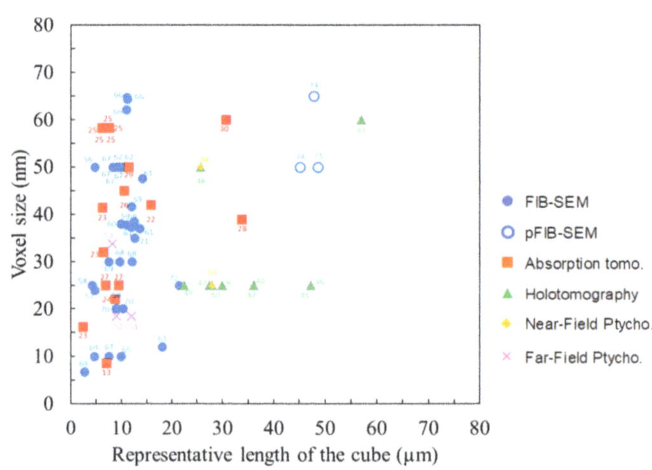

Figure 10.6. Voxel size plotted as a function of the volume of the reconstructed electrode for the different tomographic techniques. The numbers near each point correspond to the publications given in the references.

or local damage such as nanoporosities. FIB-SEM tomography has a clear advantage in the characterization of such fine geometries even though the synchrotron XCT techniques are continuously progressing towards higher resolutions.

The active or functional layer of an electrode usually exhibits a rather fine microstructure. In this case, it has been shown that the minimum volume that can represent this heterogeneous medium is around 10^3–15^3 μm^3 [75–78]. This so-called representative volume element (RVE) is necessary to obtain good statistics on the microstructural properties measured on the reconstructions (phase volume fraction, tortuosity factor, etc). This constraint on the RVE size can easily be fulfilled by all the techniques. However, the electrode substrates used to support thin active layers and electrolyte layers generally have a coarser microstructure with large pores to promote gas diffusion. Therefore, the typical RVE of these microstructures has been estimated to be around 35^3–40^3 μm^3 [43, 79, 80]. In this case, the phase-imaging techniques and Xe pFIB-SEM tomography are especially suitable to obtain relevant reconstructions.

10.2.2 Illustrations using representative studies

As introduced in the previous section, x-ray computed tomography (XCT) has been extensively used to characterize the fine microstructures of SOCs with a particular focus on cermet-based hydrogen electrodes. Using a laboratory x-ray source coupled with condenser lenses and Fresnel zone plate optics, transmission x-ray microscopy (TXM) has been applied for the first time to investigate the porosity distribution inside a Ni–YSZ hydrogen electrode [22]. Nevertheless, it has been shown that laboratory sources are inadequate to study the different solid phases inside the electrodes. Therefore, researchers have moved to synchrotron-based radiation that provides more brilliant x-rays to fully characterize the microstructure of SOC electrodes. With this objective, several studies have used synchrotron-based TXM above and below the Ni absorption edge (8333 eV) to distinguish between Ni and YSZ based on the different absorption contrasts at the two energies. This type of characterization has been extensively used at beamline 32-ID of the Advanced Photon Source (APS, USA) and beamline X8C of the National Synchrotron Light Source (NSLS, USA). As a result of the better contrast available using these sources, it has been possible to fully characterize the functional layer of the hydrogen electrode (i.e. the region close to the electrolyte) and to obtain the microstructural parameters (such as specific surface areas, volume fractions, and triple phase boundary densities) [23, 24, 28]. The same tomographic technique has been used to investigate Ni coalescence in the fuel cell mode at different temperatures up to 700 h [81, 82]. These microstructural characterizations have suggested that Ni particle coarsening is based on a thermally activated sintering process. In addition, it has been possible to estimate the parameters of a power-law model for the coarsening of Ni and the corresponding decrease in triple phase boundary lengths during relatively short-term ageing. These results have been coupled with electrochemical characterization to estimate the effects of Ni agglomeration on cell performance during long-term operation.

More recent studies have taken advantage of the non-destructivity of x-ray analysis to study microstructural evolution of the Ni–YSZ by TXM tomography before and after *ex situ* thermal ageing [29, 83]. Finally, the same technique has been used to characterize the microstructural evolution in the cermet during redox cycling [13]. Despite the relatively poor contrast between YSZ and NiO, the high resolution offered by this technique allowed detecting microcracks created in the YSZ backbone during reoxidation. This damage has been ascribed to the mechanical stresses arising in the ceramic structure due to the volume expansion associated with the Ni and NiO transformation [13]. These microcracks are suspected to decrease the ionic conductivity of the ceramic network, degrading the overall cell performance.

In addition to these studies based on absorption contrast, synchrotron-based phase-contrast x-ray nanotomography has also been extensively used to characterize SOC materials. In particular, nanoholotomography available at the ID16A-NI beamline of the European Synchrotron Radiation Facility (ESRF, France) [36] has been adapted to characterize the microstructures of typical LSCF and Ni–YSZ electrodes [43–50]. The microstructural parameters extracted from the reconstructed volumes have been used to develop detailed microkinetic models that describe the reaction mechanism in the electrodes [49]. Finally, holotomography has been also used to study the degradation of the hydrogen electrodes in SOCs after long-term operation in the fuel cell mode and the electrolysis mode [8, 9]. Comparing the Ni microstructural evolutions after ageing for several thousands of hours in both modes has confirmed that the coarsening of nickel is a purely thermally activated phenomenon which is independent of the polarization. The impact of Ni agglomeration on the loss of triple phase boundaries and the overall cell performance has also been highlighted, suggesting that this phenomenon tends to stabilize during very long-term operation. This result has been related to the effect of the YSZ backbone, which remains unaltered during operation at high temperature and inhibits further coarsening of Ni [8]. In addition, the evolution of the Ni distribution in the functional layer has been investigated by comparing different types of electrode microstructure aged in the fuel cell mode and the electrolysis mode up to 15 000 h [9]. These characterizations have shown that in contrast to Ni agglomeration, Ni migration is strongly dependent on the operating mode. Indeed, it has been observed that a loss of Ni only occurs in electrolysis mode within the first few micrometers and specifically in the region that is electrochemically active (i.e. only in the polarized area of the electrode) (figure 10.7). As a result of this type of experimental result, some mechanisms have been proposed to explain the Ni depletion that occurs in the electrolysis mode [7, 84]. Moreover, 3D reconstructions of different types of electrodes allowed suggesting that finer microstructures should be less prone to suffering from the loss of Ni at the electrode–electrolyte interface.

Finally, it has to be mentioned that other phase-contrast nanotomography methods using a ptychographic approach for phase retrieval have recently been used to characterize SOC materials. In particular, near-field ptychography carried out at ID16A-NI of the ESRF has been used to demonstrate the possibility of this type of characterization for highly absorbent materials [48]. The results of this technique have been compared with those of the holotomography for SOC

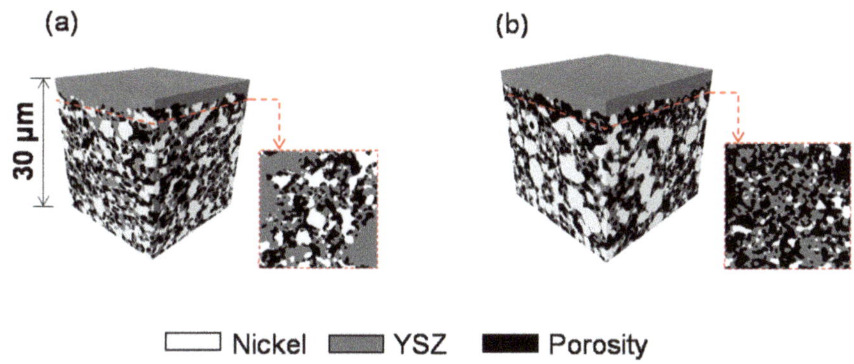

Figure 10.7. 3D segmented rendering and 2D cross sections extracted for a pristine Ni–YSZ electrode (a) and an aged electrode in SOEC mode during 1000 h at 800°C and at -0.5 A cm^{-2} (b).

applications [50]. In parallel, far-field ptychography carried out at the cSAXS beamline of the Paul Scherrer Institute (PSI, Switzerland) has been used to study the impact of coarsening and redox cycling [51–53]. Taking advantage of the non-destructive nature of this x-ray characterization technique, the effect of thermal annealing for several hours at 850 °C was studied using the same sample, highlighting the losses of Ni connectivity and triple phase boundary density [52]. Moreover, the effects of the oxidation and reduction of Ni on the microstructure have also been investigated using far-field ptychography in *ex situ* experiments. This investigation confirmed the formation of microcracks in the YSZ skeleton after reoxidation [51].

10.3 Advanced x-ray spectroscopy and scattering techniques for structural and chemical analyses

An understanding of the operation and degradation phenomena in SOCs at the relevant scales is essential in order to design innovative solutions such as new alternative materials and/or original microstructural optimizations. Physical and chemical heterogeneities are usually observed at various length scales in typical SOC component materials. These heterogeneities can affect the macroscopic physical properties of the cell materials and their stability in operation. In this context, the need for 'chemical microscopes' is growing rapidly, not only to visualize the chemistry in space (three dimensions) but also in time (four dimensions). In particular, it is critical to develop standard methodologies that can be used to investigate the local structure and chemical/elemental composition at the micrometer and sub-micrometer scales.

10.3.1 X-ray fluorescence and diffraction

10.3.1.1 A brief overview of the two techniques

Both techniques, i.e. XRF spectroscopy and XRD, are of great interest in an extremely broad number of scientific disciplines including geology, catalysis, energy, cultural heritage, medicine, pharmacy, biology, agriculture, materials science

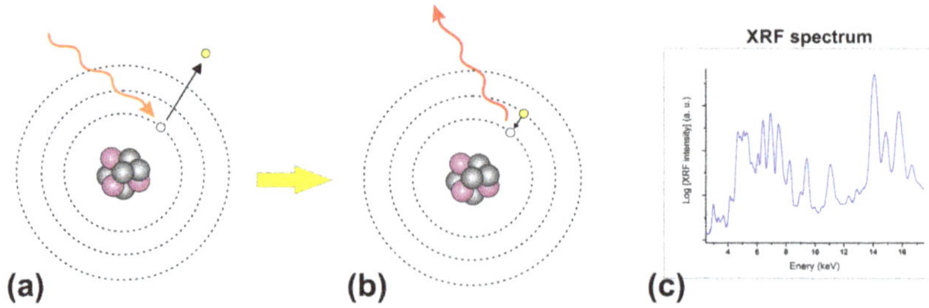

Figure 10.8. Schematic of the XRF principle: a photon with energy higher than the absorption edge of the material is absorbed by the atom, causing a core electron to be ejected from the electron cloud (a). An outer-shell electron relaxes to the hole left by the core electron and emits a fluorescence photon that has a specific energy that depends on the material and the energy levels involved in the transition (b). The fluorescence spectrum records the emitted photons as a function of the energies of the chemical components of the sample (c).

technology, and metallurgy. While XRF allows the chemical/elemental composition of materials to be assessed, XRD provides an understanding of how atoms are arranged in solid matter.

Any material can produce a fluorescence signal by interacting with an x-ray photon that has an energy equal to or higher than the energy of a core electron of the elements present in the analyzed sample. As shown in figure 10.8(a), the photon energy is transferred to core electrons, which are ejected from the atom. The resulting core hole can later be filled by an electron of an external atomic shell, leading to the emission of a photon with a characteristic energy equal to the difference between the core and the shell energies (figure 10.8(b)). Different elements have different characteristic atomic energies that can be seen as an atomic fingerprint. This allows the exact identification of the elements present in the material by analyzing the XRF spectrum, (i.e. the XRF intensity as a function of the emitted characteristic energies of elements in a material excited by an x-ray source: figure 10.8(c)). The XRF spectra are usually analyzed and modeled using standard programs such as PyMCA [85].

When the elements in a material are organized in crystalline form (i.e. the presence of a large range of atomically coherent spatial arrangements), XRD can be applied to characterize the high degree of atomic arrangement. XRD patterns are acquired in a very similar way to that of XRF, i.e. by illuminating the sample with an x-ray beam. Here, the photons are described as waves in contrast to their treatment as particles in XRF spectroscopy. In XRD, the scattering intensity is acquired and analyzed as a function of the scattered angle. As shown in figure 10.9(a), coherent x-ray waves are scattered by the crystalline atomic planes. The supplementary path for an x-ray from one plane to another is equal to $2d\sin\theta$, where d is the interplanar distance and θ the scattering angle. Constructive interference can happen when the additional length is equal to an integer multiple of the wavelength λ according to Bragg's law:

$$\lambda = \frac{2d\ \sin\theta}{n}. \tag{10.8}$$

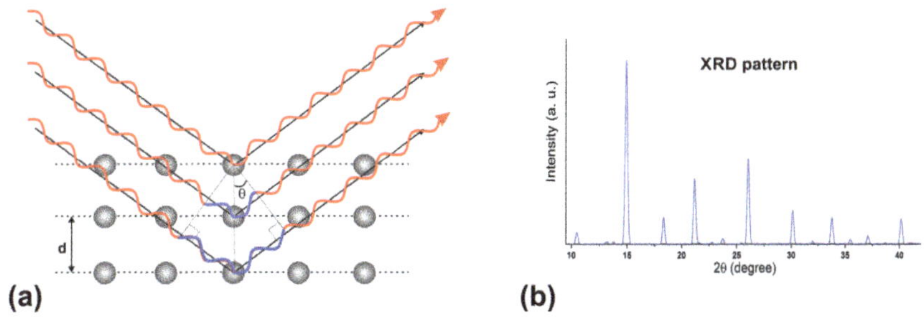

Figure 10.9. Schematic of the XRD principle: a coherent x-ray diffracted by a crystalline sample creates constructive interference at a specific angle depending on the photon wavelength and the interplanar distance in the lattice (a). Constructive peaks as a function of the diffraction angle associated with the different crystalline phases inside a powder-like sample (b).

The coherent constructive interference corresponds to the so-called diffraction angle. A pattern of diffraction peaks as a function of the scattering angle (typically called a diffractogram) can be analyzed to characterize the crystalline structure (figure 10.8(b)). This relatively simplistic approach can be applied in powder-like XRD (pXRD), which is the case when the x-ray probe size is significantly larger than the size of the crystallites in the sample. Using a typical diffractogram, a given crystalline structure can be identified as well as other crystalline characteristics such as stress, volume of the unit cell, and atomic site occupancy.

Using a synchrotron source facility, XRF and XRD can be acquired simultaneously from a material to image its chemistry by detecting its elements (through XRF) and by obtaining knowledge about the organization of these elements in the material (through XRD). Moreover, due to the development of efficient x-ray focusing optics that has taken place over the last 20 years, this simultaneous and combined analysis can be carried out with good spatial resolution. The typical range of the resolution currently varies from tens of micrometers for large samples down to tens of nanometers for micron-sized samples. A typical setup for this combined XRF and XRD analysis using a focused x-ray beam is illustrated in figure 10.10. It can be seen that the XRD detector is placed behind the sample in such a way that the diffraction signal is recorded in transmission mode. In this case, ring diffraction patterns are obtained if enough crystal grains with different orientations are probed by the x-ray beam (the azimuthal integration allows the diffractogram to be obtained as represented in figure 10.9(b)). By scanning a sample in the focal plane of the focused x-ray beam, XRF spectra and XRD patterns can be simultaneously collected at each sample position to obtain 2D maps. The integration time per location is typically in the order of 100 ms at synchrotron facilities and the availability of a high photon flux allows relatively large areas to be investigated in a relatively short time.

This approach allows for non-destructive *in situ*/operando experiments [86, 87]. Moreover, due to the considerable penetration power of hard x-rays, investigations are feasible for 'chemical tomography.' The XRF+pXRD computed tomography approach is typically applied by analyzing both the fluorescence and diffraction

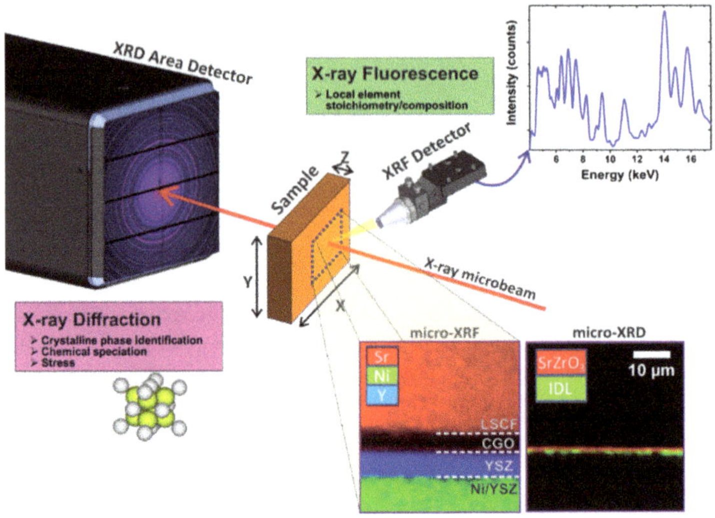

Figure 10.10. Typical setup for a combined microdiffraction and microfluorescence experiment (corresponding to the microXAS beamline at the PSI).

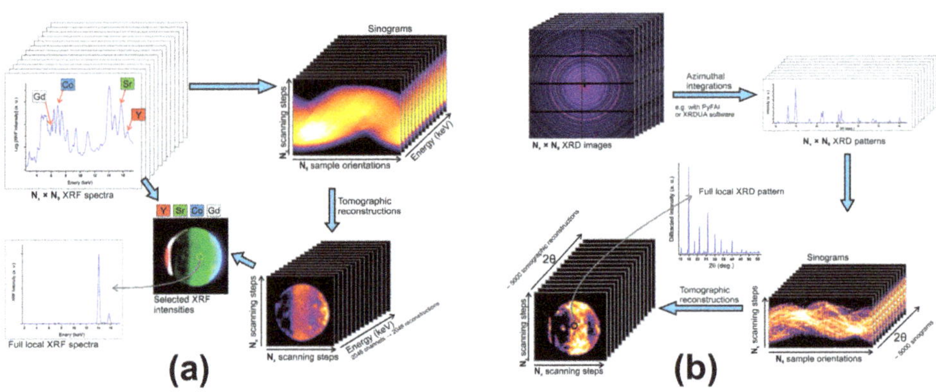

Figure 10.11. Flowchart for fluorescence (a) and powder-like diffraction (b) tomographic analyses.

signals at different rotational angles of the sample. For each angle, XRF spectra and XRD patterns are acquired along a line scan or in 2D (for 2D or 3D 'chemical tomography,' respectively). This method (illustrated in figure 10.11) is used to obtain a 2D reconstructed slice of the sample. At each position of the line scan, the XRF signal can be analyzed for each energy of the recorded spectra (figure 10.11(a)). For this purpose, the raw data are processed in order to plot the XRF signal as a function of the lateral position (i.e. the scanning step) and the sample's orientation (i.e. the rotational angle). These 2D images are commonly called sinograms. The intensity distribution of each sinogram can be retrieved in real space through a number of possible reconstruction algorithms. As a result, the local XRF spectra can be spatially deconvoluted and resolved in 3D. The postprocessing of pXRD signals can be quite similar, as shown in figure 10.10(b), but requires the additional step of reducing the 2D XRD pattern (i.e. the ring diffraction pattern) to 1D XRD through the procedure of azimuthal integration. The resulting sinograms of the scattered intensities as a function of the scattering diffraction angle can also be retrieved in real space through the use of reconstruction algorithms.

It is worth recalling that x-rays are strongly absorbed by SOC materials even at relatively high energies. This characteristic limits the maximum size of the specimens. Therefore, the sample geometry should be optimized for this type of characterization, especially when the data for the XRF and XRD techniques are acquired together using a micro/nano beam. Indeed, on the one hand, for a reliable pXRD signal, the sample should be thick enough in such a way that the diffracted x-rays traverse a sufficient number of crystal grains with different orientations. On the other hand, for the XRF signal, it is important to minimize the phenomenon of self-absorption of the fluorescence x-rays. Indeed, the x-rays photoemitted by this type of material can often be reabsorbed by the material itself, resulting in artifacts such as local inhomogeneities in the chemical composition that must be minimized/corrected. For tomographic experiments, cylindrical specimens remain the most appropriate geometry, as for the absorption/phase-contrast imaging techniques (see figure 10.5). For 2D acquisition, thin lamella can be prepared with a limited thickness but with relatively large horizontal and vertical dimensions (figure 10.12). This geometry allows a representative zone along the interface to be investigated, while the thickness of the sample is a trade-off between x-ray absorption and the need to probe enough crystallographic grains for diffraction. In both cases, the most convenient sample preparation technique for synchrotron-based characterization at the micro- or nanoscale is based on FIB or pFIB.

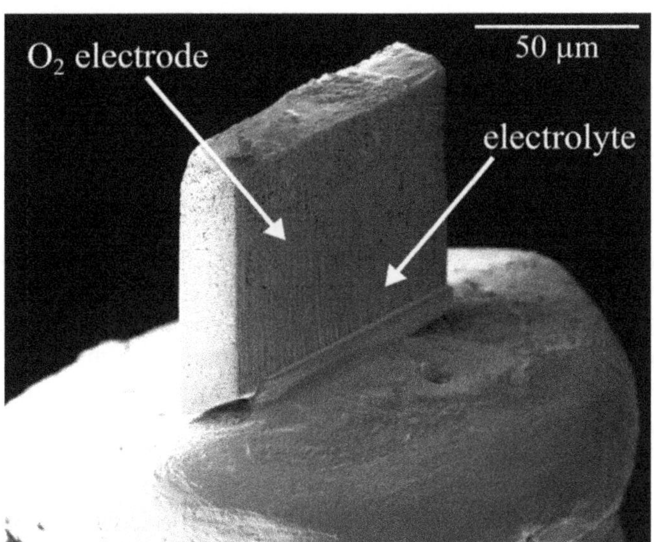

Figure 10.12. Typical thin lamella prepared using a Xe^+ plasma focused ion beam for micro XRD and micro XRF analysis of a classical SOC oxygen electrode.

10.3.1.2 Examples of representative studies

Many articles have been already published using classical synchrotron XRD characterizations. Some of these studies were dedicated to *in situ* experiments. For example, Heenan *et al* [88] measured the Ni lattice parameters and thermal strain for Ni–YSZ and Ni–GDC cermets during thermal cycles at the X04SA-MS beamline of the PSI. The stresses induced by thermal gradients were also measured using Ni–YSZ half-cells [89] at beamline I12 (the Joint Engineering, Environmental, and Processing (JEEP) beamline) at the Diamond Light Source. Regarding the oxygen electrode, Sr and Co segregation at the surface of thin LSCF films has been characterized in terms of temperature and applied potential by total external reflection x-ray fluorescence at the Advanced Photon Source (APS, beamline 12-BM) [90].

In contrast, only a very limited number of studies have been devoted to imaging the chemistry and the crystallographic structure of SOC materials in 2D or 3D. Li *et al* [86] used XRD computed tomography to study the phase distribution through a full cross section of a 'micro-monolithic' during thermal cycling. The experiment was carried out at beamline station ID15 of ESRF at high energy (90 keV) with a spot size of 40 μm × 20 μm. The stability of classical LSCF-CGO oxygen electrodes has been investigated at the microXAS beamline (PSI) in *ex situ* experiments [11, 12, 91]. A combination of 2D cartography and diffraction tomography using an x-ray microbeam (1×0.59 μm^2) at high energy (>17 keV) has been used to study the reaction between an LSCF-based oxygen electrode and an electrolyte material [11, 91]. These characterization techniques were first used on a pristine LSCF-based electrode to identify the formation of $SrZrO_3$ at the interface between the GDC barrier layer and the YSZ electrolyte after sintering [91]. These measurements were subsequently applied to oxygen electrodes extracted from cells aged in electrolysis mode and fuel cell mode in order to investigate the effect of the ageing conditions. In addition, the decomposition of LSCF-based oxygen electrodes has been studied using the same techniques, focusing on the coupled effects of operating temperature and polarization. In this case, thin lamellae extracted from pristine and aged samples were characterized and 2D maps containing the chemical and crystalline distribution at the submicron scale were obtained. This study has confirmed the larger degree of electrode decomposition that occurs in the electrolysis mode with respect to fuel cell operation, highlighting the strong impact of the operating temperature. Moreover, characterization at the barrier layer/electrolyte interface after sintering and after operation in the electrolysis mode has highlighted the formation of the so-called interdiffusional layer [12]. Finally, the same characterization techniques have been used at the microscale to study the material decomposition of innovative nickelate electrodes based on lanthanum and praseodymium ($La_{2-x}Pr_xNiO_{4+\delta}$ (LPNO)) as a function of polarization. It has been shown that these materials decompose in different crystalline phases after ageing, even though their electrochemical performances remain satisfactory [92].

These studies have demonstrated the importance of the high-resolution characterization, which gives important insights into the physicochemical degradation of oxygen electrodes in SOCs. Nevertheless, improving the resolution could provide even more precise information, which could be beneficial in studies of the local evolution of the chemical composition and crystalline structure after ageing. In this context, a 2D nanocharacterization of SOC material was recently carried out at the ID16B-NA beamline of the ESRF. In this study, 2D XRF maps with nanometric resolution were acquired to visualize the interdiffusion of the chemical elements in the cell. This technique has been applied to a cathode and an electrolyte-supported cell operated over long term (>5000 h) at different operating temperatures and at different current densities. Such a measurement allowed identifying an accumulation of gadolinium at the electrolyte/diffusion barrier interface; this layer was detected after sintering and seemed to increase with the operating time. In contrast, the secondary phase based on Sr at the electrolyte/diffusion barrier present after sintering does not evolve during long-term operation [93].

10.3.2 X-ray photoelectron spectroscopy and x-ray absorption spectroscopy

10.3.2.1 A brief overview of the two techniques

Knowledge of the surface, interface, and bulk chemistries of energy materials is essential in order to improve their properties and in turn increase their lifetimes under realistic operating conditions. To meet this objective, two complementary techniques are preferred, namely XPS spectroscopy and XAS spectroscopy. XPS is a surface-sensitive technique that gives information about the composition and chemical state of the outermost atomic layers of a sample, while XAS spectroscopy is a local structural technique used to investigate bulk properties in terms of the chemical state and local atomic environment of an element in the material. In the following, the two spectroscopy techniques are briefly described.

10.3.2.1.1 X-ray photoelectron spectroscopy

XPS, also called electron spectroscopy for chemical analysis (ESCA) or photoelectron spectroscopy (PES), is an analysis method used to investigate the electronic and chemical structure of a material's surface. To meet the need to perform measurements under realistic working conditions, the XPS technique has evolved over the years from measurements made under ultrahigh vacuum (in order to obtain a clean sample surface and to detect electrons emitted from the investigated material at high resolution) to near-ambient-pressure (NAP) XPS using soft x-rays (x-rays with energies of less than 1500 eV) that is surface sensitive (the sample is exposed to several mbar of pressure) and ambient-pressure XPS using hard x-rays (ambient-pressure hard x-ray photoelectron spectroscopy (APHAXPES)) that is bulk sensitive. Compared to laboratory x-ray sources, synchrotron radiation photon sources offer high intensity and brightness, energy tunability, and high spectral resolution in the acquisition of XPS spectra. Several reviews of this technique are available [94–97], and the following summary describes the basics of the technique.

The principle of photoelectron spectroscopy is based on the photoelectric effect, in which a sample is irradiated with monochromatic photons and photoelectrons are emitted from the atomic core levels with the creation of a core hole. A photoionized atom can relax by filling such a hole with an electron from a higher level, which is accompanied by the emission of characteristic x-rays. The emitted photoelectrons are analyzed with respect to their kinetic energies. The basic equation that describes this effect is as follows:

$$E_K = h\upsilon - E_B - \phi_S, \tag{10.9}$$

where E_K is the measured energy of the photoelectron, $h\upsilon$ is the x-ray energy, E_B is the binding energy of the core–shell electron in relation to the Fermi level, and ϕ_S is the spectrometer work function. The Fermi level is the appropriate reference for the binding energies of elemental lines and corresponds to the zero binding energy by definition.

Photoelectron spectroscopy can probe the core-level electrons as well as the Auger electrons resulting from secondary de-excitation processes and the valence-band electrons. The photoemission process dominates for heavy elements and the kinetic energies of the ejected photoelectrons are measured by the analyzer. With light elements, the Auger process occurs when a core hole is filled by an electron from a higher level without the emission of radiation. In this case, an Auger electron is ejected from an outer shell or from the valence band.

The emitted electrons are analyzed with a hemispherical analyzer composed of an electron lens, an electron analyzer, and an electron detector. Figure 10.13(a) displays the spectrometer setup used at the HIPPIE beamline of the MAX IV laboratory synchrotron facility; it also shows the different analyzers and spectrometer as well as the sample preparation chamber. The elemental lines of pure metals (Mg Kα: Cu $2p_{3/2}$ at 932.62 eV, Ag $3d_{5/2}$ at 368.22 eV, and Au $4f_{7/2}$ at 83.95 eV) are used to calibrate the binding energy scale of the spectrometer. First, the binding energies of Cu $2p_{3/2}$ and Au $4f_{7/2}$ are measured and the difference is adjusted by the electronics to be equal to the reference value of 848.67 ± 0.1 eV. Then the spectrometer work function is set to the reference binding energy of the Ag $3d_{5/2}$ line and the Fermi edge at the conduction band. The standard deviation for binding energies is less than ±0.1 eV for conductors and less than ±0.2 eV for non-conductive samples.

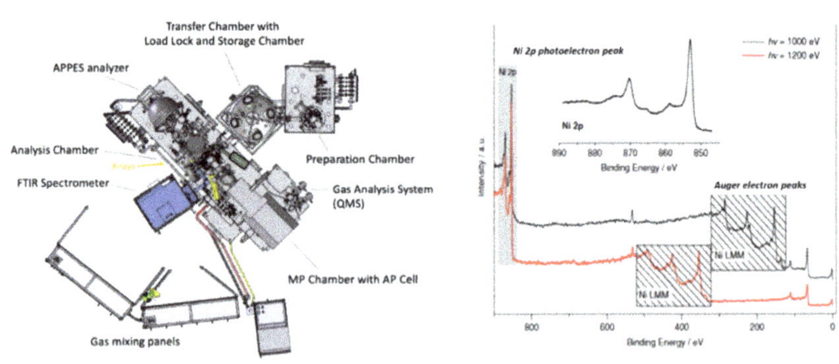

Figure 10.13. (a) AP-XPS end station at the HIPPIE beamline with the catalysis ambient-pressure (AP) setup at the MAX IV laboratory (the HIPPIE AP-XPS beamline end station with the electron hemispherical analyzer). Reproduced from [98]. CC BY 4.0. (b) Typical synchrotron x-ray photoelectron survey scan of a 50 nm Ni particle excited by photon energies of 1000 and 1200 eV [97] Springer-Verlag Berlin Heidelberg © 2015. With permission of Springer.

XPS spectra are plotted in terms of electron intensity as a function of their kinetic energy and more frequently in terms of their binding energy E_B as calculated by equation (10.9); the elements that compose a material can be identified by measuring the binding energies of the photoelectron lines. The photoelectron and Auger electron lines are recognized by recording spectra at two different x-ray energies. The Auger electron lines are the ones that shift with the x-ray energies used. Using XPS, distinct elements can be identified and atomic concentrations and chemical states can be calculated based on the chemical shift of the photoelectron lines. Figure 10.13(b) displays typical survey spectra of 50 nm Ni particles taken at two different energies to differentiate the peaks related to the photoelectrons and the Auger electrons.

10.3.2.1.2 X-ray absorption spectroscopy

XAS is of interest because it complements the results obtained from other characterization techniques by detailing the local atomic and electronic structures of a specific chemical element in the probed material. As in the case of XPS, x-rays are absorbed by matter and hence, through the photoelectric effect, electrons from inner core states are excited into empty states at energies above the ionization threshold.

During the absorption of x-rays, specific features known as absorption edges are observed when the energy of the x-rays is close to the binding energy of electrons. Several reviews [99–101] of XAS can be found in the literature, and the following summary describes the fundamentals of the technique.

An absorption spectrum is obtained by measuring the photon intensity in front of the sample (I_0) and behind the sample (I_t) as a function of the incident photon energy E. The absorption for a given sample length is given by the Lambert–Beer law:

$$\mu(E) \times d = \ln [I_0(E)/I_t(E)], \tag{10.10}$$

where μ is the absorption coefficient and d the thickness of the sample. The absorption of a reference sample with known E_0 is used for energy calibration.

Figure 10.14(a) displays a typical XAS beamline end station used for XAS. The measurement is generally performed in transmission mode with the sample placed perpendicular to the x-ray beam. In particular cases in which the concentration of the element under analysis is diluted, the fluorescence mode is used, in which the sample is placed at 45° to the beam and the fluorescence detector. Figure 10.14(b) shows a typical Ti K-edge XAS spectrum of a Ti foil. The spectrum represents the normalized absorption coefficient as a function of energy. The absorption coefficient μ has an oscillatory structure; based on the scattering processes, the x-ray absorption spectrum is divided into two parts, as follows:

(i) the x-ray absorption near-edge structure (XANES) from the energy region that goes from 50 to 100 eV beyond the absorption edge
(ii) the extended x-ray absorption fine structure (EXAFS) at energies higher than 100 eV

Features in the XANES region called pre-edge structures result from the multiple scattering of the outgoing photoelectrons by the outlying neighboring atoms.

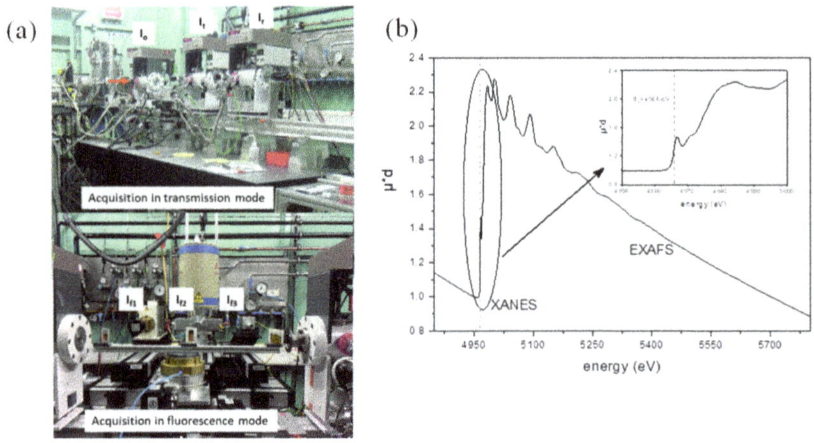

Figure 10.14. (a) The XAS experimental station of the KIT synchrotron, which is used for acquisition in either the transmission mode or the fluorescence mode. Adapted from [102]. CC BY 4.0. (b) Typical Ti *K*-edge XAS spectrum of a Ti foil and the corresponding XANES and Fourier transform EXAFS (FT-EXAFS) portions of the spectrum showing the radial distribution function.

The XANES region gives information about the transitions of the photoelectrons to the bound excited states of the absorbing atoms and thereby contains information about the coordination geometry of the absorbing atom, including interatomic distances and angles. However, the multiple scattering character of XANES does not allow all these parameters to be extracted easily. Instead, the 'fingerprinting' method or theoretical calculations are used to explain the features observed in the experimental data. The EXAFS region is defined by energies that differ from the threshold energy E_0 by more than 100 eV. The oscillatory pattern corresponds to an interferogram of the atomic arrangement surrounding the absorbing atom in which the constructive interference between the outgoing photoelectron wave function and the backscattered wave leads to maximum EXAFS oscillations and destructive interference leads to minima. Fourier transformation of EXAFS (FT-EXAFS) oscillations gives a pattern that is close to the function representing the radial distribution of atomic density. Consequently, the analysis of this part of the spectrum allows the determination of the metrical parameters characterizing the arrangement, such as the numbers and types of neighboring atoms and their distance to the absorbing atom.

10.3.2.2 Illustrations using the most representative studies

Generally, for laboratory or synchrotron-based XPS analysis, model electrodes are used that consist of dense films prepared by sputtering or pulsed laser deposition on the substrate, in contrast to the porous electrodes used in real applications. For solid oxide materials, XPS has been either used separately to probe the surface or to complement XAS and/or other characterization techniques in order to understand the degradation processes. Such a multimodal approach allows researchers to gain an understanding of the fundamental processes of the electrochemical reaction or the surface/volume chemistry of a solid oxide-based material used in the fuel cell mode or the electrolysis mode (it should be noted that data for the latter operational mode are scarce). Further, due to gaseous reactant environments, high operating temperatures (above 700 °C), and non-equilibrium conditions, *in situ* and operando techniques have increasingly been developed in recent years to provide insights into the SOC mechanisms that are present under operating conditions.

Laboratory XPS has largely been used to gain insights into the electrochemical reaction in terms of surface oxidation measurements [103, 104], the influence of the surface oxidation state on the performances of Ni–YSZ and Ni–GDC [105, 106], and elemental analysis of new electrodes or electrolyte compositions [107, 108] and also to determine the oxidation characteristics of interconnects based on ferritic stainless steels [109, 110] and glass seals [111]. Synchrotron-radiation-based XPS has opened up new potential thanks to its high intensity and brilliance, its energy tunability, and its high spectral resolution. Recent advances have been achieved due to the specific properties of synchrotron radiation. Nowadays, techniques such as NAP XPS/ambient-pressure XPS (AP-XPS) [112], APHAXPES [113], and scanning photoelectron microscopy [114] are widely used to non-destructively study particle-size dependency, surface chemistry, and the depth profile within the volume of a sample. The combination of XPS and XAS has been applied to probe the local atomistic and electronic structure of specific elements, either in an LSCF thin-film electrode or in a Ni cermet electrode exposed to O_2, H_2O, H_2, or H_2/H_2O under current. In the following, selected examples are described to highlight the complementarity of these techniques as well as the knowledge gained about the surface and bulk reactivity of specific SOC materials.

Oxygen electrodes in SOCs operated either in the fuel cell mode or the electrolysis mode are generally based on perovskites with mixed ionic and electronic conductivities. In order to understand the surface structure and the chemistry of these materials at temperatures up to 520 °C and oxygen partial pressures from 10^{-3} to 1 atm, *in situ* AP-XPS and depth-resolved XAS have been used with $La_{0.8}Sr_{0.2}CoO_{3-\delta}$ (LSC) thin films (three thicknesses: 20 nm, 45 nm, and 130 nm) prepared by pulsed laser deposition (PLD) with a gadolinia-doped ceria (GDC) buffer layer on single-crystal (001) oriented yttria-stabilized zirconia (YSZ). For comparison, polycrystalline LSC pellets were used as reference material [115]. Near-ambient and high-temperature XPS data for the LSC films were collected at beamlines 9.3.2 and 11.0.2 at the Advanced Light Source (ALS), while XAS measurements were performed at beamlines BL37XU and BL01B1 at Spring8, a facility operated by the Japan Synchrotron Radiation Research Institute (JASRI). AP-XPS performed at a partial oxygen pressure of 10^{-3} atm showed that significant changes took place in the spectra of Sr 3*d* and O 1*s* for different energies, temperatures, and structures compared to the spectra of La 4*d* and Co 3*p*. By varying the energy, one could differentiate between the lattice component of the perovskite structure and the surface component assigned to the surface secondary phases. Thereafter, by increasing the temperature from 220 °C to 520 °C, the analysis of the Sr 3*d* and O 1*s* spectra indicated changes from the lattice component toward the surface of the LSC film, while no change was observed for the LSC pellet. To complement these data, *in situ* depth-resolved XAS was performed at an oxygen partial pressure of 1 atm at the Co *K* edge for exit angles from 0.158° (surface) to 0.552° (bulk). The broadened absorption edge features and the shift of the 'white line' toward higher energies indicates that the surface structure of LSC is preferentially reduced upon increasing the temperature and more oxygen vacancies are generated at the perovskite surface than in the bulk [115]. Along the same lines, a study of $La_{0.75}Sr_{0.25}Cr_{0.9}Fe_{0.1}O_3$ (LSCrF) under the relevant gas phase conditions for the fuel cell mode of operation and the electrolysis mode of operation was carried out using AP-XPS and near-edge x-ray absorption fine structure (NEXAFS) at the TEMPO beamline at the SOLEIL synchrotron radiation facility [116]. Analysis of the La 3*d*, Sr 3*d*, Cr 2*p*, and Fe $2p_{3/2}$ spectra of LSCrF samples exposed to an O_2 atmosphere indicates that the surface of LSCrF is dominated by lanthanum, while in H_2 and H_2O atmospheres, the lanthanum is gradually replaced by strontium. In contrast, the gas atmosphere does not induce significant changes in the concentrations of Cr and Fe. Analysis of the O 1*s* spectra shows two oxygen contributions, one of which is related to the surface oxygen layer of the perovskite lattice and the other of which is related to oxygen vacancies. It was concluded that the oxide lattice structure was not significantly modified by the strontium and lanthanum cation interdiffusion induced by the gas environment. In addition, adsorbed oxygen species were not detected on the LSCrF electrodes, even including the electrode exposed to an H_2O atmosphere [116].

Only a few studies of porous oxygen electrodes have been published. The LSCF electrode of a commercial SOC cell operated in the electrolysis mode for 1000 h at -0.8 A cm^{-2} was studied using XAS at the Co/Fe *K* edge recorded at the BM30B beamline (ESRF) [117]. Post-mortem analysis indicated that the electronic configuration of Fe did not change after the operation, while the Co oxidation state was shifted towards a higher energy and exhibited changes in its EXAFS, indicating a change in the local structure around Co. Using a porous LSCF symmetrical electrode, the effects of CO_2 and H_2O were studied in operando in the fuel cell mode to analyze the Co 3*p*, Fe 3*p*, and O 1*s* evolution. XPS spectra were acquired at the U12A beamline (see figures 10.15(a) and (b)) and the Co and Fe *K*-edge XAS spectra were taken at the X18B beamline at the National Synchrotron Light source at Brookhaven National Laboratory (see figures 10.15(c) and (d)) [118]. It has been shown that under the experimental conditions (700 °C, a bias voltage of -1.0 V, under air, H_2O, and CO_2) the FT-EXAFS indicates that the local structure around Co was significantly deformed and had an elongated bond length, while the Fe local structure remained stable. In addition, the XPS data suggested that carbonate species disappear at high temperatures, while the O 1*s* spectra indicate an evolution from surface oxygen to lattice oxygen upon heating and the formation of hydroxide upon cooling.

Regarding Ni cermet electrodes, NAP XPS and NEXAFS measurements were conducted on Ni–YSZ and Ni–GDC electrodes at the ISISS beamline at the BESSY synchrotron radiation facility. The spectra of Ni 2*p* and Ce 3*d* were recorded for Ni–YSZ/YSZ/Pt and Ni–GDC/YSZ/Pt cells at 640 °C at a constant current of -100 mA cm^{-2} and at different oxygen concentrations, respectively. The results

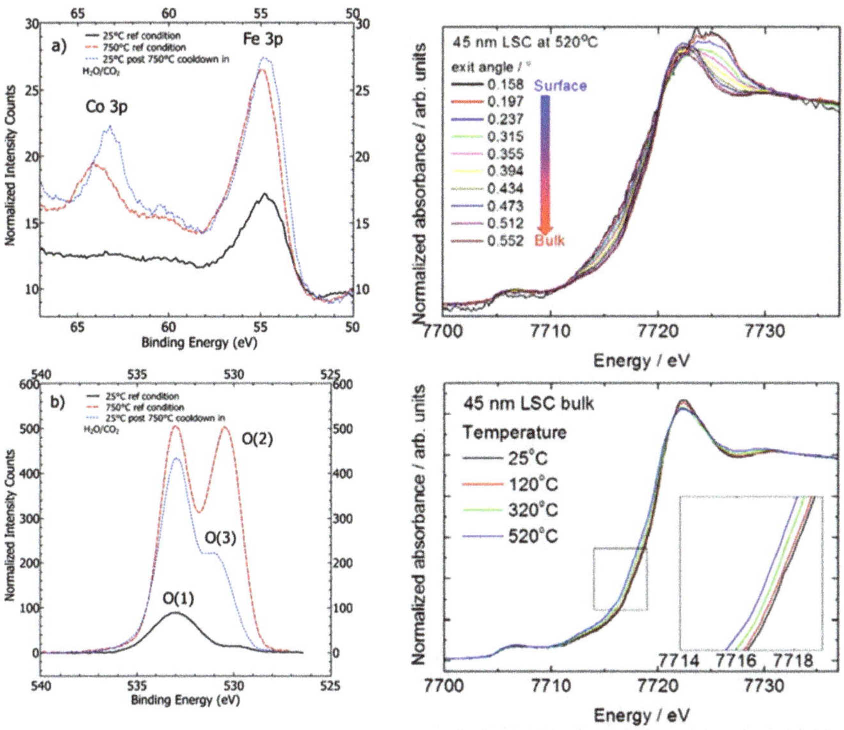

Figure 10.15. (a) In operando Co and Fe 3*p* XPS spectra of an LSCF thin-film porous electrode deposited on a single-crystal YSZ substrate [118] John Wiley & Sons. [© 2014 WILEY-VCH Verlag GmbH & Co. KGaA, Weinheim]. (b) The corresponding O 1*s* spectrum showing the features attributed to surface oxygen O(1), lattice oxygen O(2), and hydroxide oxygen species O(3) [118] John Wiley & Sons, copyright 2014 WILEY-VCH Verlag GmbH & Co. KGaA, Weinheim]. (c) and (d) Depth-resolved Co *K*-edge XAS of a 45 nm LSC film measured at different angles and temperatures under a partial oxygen pressure of 1 atm. Reprinted from [115], © The Electrochemical Society. Reproduced by permission of IOP Publishing.

showed that a NiO layer of a few nanometers thick was present on the surface of the Ni–YSZ electrode that significantly increased the overpotential for the electrolysis reaction, while the effect was minor for the Ni–GDC electrode under similar conditions [105]. To analyze the oxidation stages of nickel, a Ni/YSZ/Pt cell with a dense Ni electrode free of pinholes or pores was studied in operando using synchrotron-based AP-XPS to measure the changes in its chemical composition while the electrochemical reaction was taking place [104]. This experiment was performed at beamline 11.0.2 of the ALS to study the oxidation stages of Ni electrode in a H_2/H_2O atmosphere that mimicked fuel cell operating conditions. Figure 10.16(a) displays a schematic of the experimental cell in operation and (b) shows the corresponding applied voltage profile together with the XPS spectra acquired for Ni 3*p*. Analysis of the Ni 2*p*, Ni 3*p*, and O 1*s* XPS spectra at different stages of the bias voltage (0.4 V < E_{bias} < 1.0 V) indicates that at 700 °C and at an E_{bias} of <0.4 V, Ni remains metallic, whereas when E_{bias} > 0.4 V, a phase transition to NiO occurs at the Ni/YSZ interface [104]. More recently, in operando XPS

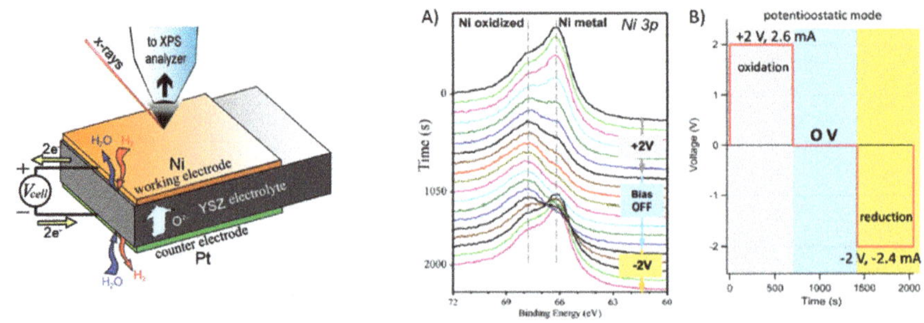

Figure 10.16. Schematic view of an cell in operation (left). (A) Ni 3*p* series showing the oxidation and the reduction of the Ni surface when the applied potential follows the sequence shown in (B). Reproduced from [104] with permission of The Royal Society of Chemistry.

studies were performed at beamline 9.3.2 of the ALS to evaluate the surface states of GDC thin-film electrodes during H_2 oxidation and H_2O electrolysis on YSZ-supported solid oxide electrochemical cells [119]. Two-dimensional mapping of the O 1*s* spectra was acquired at three different cell potentials for Ni/GDC and Au/GDC electrodes at $pH_2 = pH_2O = 0.26$ mbar and $T = 620$ °C. Analysis of binding energy shifts revealed the active electrochemical sites and allowed the fundamental charge transfer parameters such as the exchange current density to be derived.

10.4 Conclusions

This chapter has shown that several techniques based on synchrotron x-ray radiation can be used for the precise characterization of SOC materials. Different XCT methods have recently been adapted to electrode materials and microstructures. Reconstructions combining a large field of view and a spatial resolution of a few tens of nanometers have been obtained using propagation-based imaging methods (i.e. ptychography and holotomography). In contrast to the more conventional destructive FIB-SEM techniques, XCT offers the possibility of scanning the same sample before and after *ex situ* experiments. XRD and XRF are usually combined for structural and chemical analyses and provide high resolutions in the range of 50–500 nm. These two complementary techniques are especially relevant to study the phase decomposition in the thin electrode active layer, the reactivity and formation of secondary phases at the interfaces, and interdiffusion between the cell components. To date, the XPS and XAS methods based on synchrotron radiation have not been extensively applied to the characterization of SOC materials. The published studies have revealed the nature of the species adsorbed on the electrode surface and its potential passivation after operation as well as the evolution of the chemical states of the active-layer elements.

The current upgrade of synchrotron sources to the fourth generation allows improving the spatial resolution of XCT imaging and decreasing the time required for data acquisition. A multimodal approach with complementary techniques in the same experimental setup should create new opportunities for the characterization of

SOC materials. For instance, far-field ptychography could be associated with XRF tomography to obtain 3D reconstructions combining morphological and chemical information in the same dataset. Thus, the distribution of secondary phases in the electrode microstructure could be more easily revealed. The development of specific setups for *in situ* and operando experiments should constitute a breakthrough in the characterization of SOC materials. However, the management of cell environments operating at high temperatures and the reducing conditions on the hydrogen electrode side is very challenging for operando experiments.

References

[1] Mogensen M B 2020 Materials for reversible solid oxide cells *Curr. Opin. Electrochem.* **21** 265–73

[2] Yoon K J, Biswas M, Kim H-J, Park M, Hong J, Kim H, Son J-W, Lee J-H, Kim B-K and Lee H-W 2017 Nano-tailoring of infiltrated catalysts for high-temperature solid oxide regenerative fuel cells *Nano Energy* **36** 9–20

[3] Jiang Z, Xia C and Che F 2010 Nano-structured composite cathodes for intermediate-temperature solid oxide fuel cells via an infiltration/impregnation technique *Electrochim. Acta* **55** 3595–605

[4] Cao T, Kwon O, Gorte R J and Vohs J M 2020 Metal exsolution to enhance the catalytic activity of electrodes in solid oxide fuel cells *Nanomaterials* **10** 2445

[5] Nechache A and Hody S 2021 Alternative and innovative solid oxide electrolysis cell materials: a short review *Renew. Sustain. Energy Rev.* **149** 111322

[6] Skafte T L, Hjelm J, Blennow P and Graves C R 2016 Quantitative review of degradation and lifetime of solid oxide cells and stacks *Proc. 12th European SOFC & SOE Forum B05* (Lucerne, Switzerland: European Fuel Cell Forum) 8–27 https://orbit.dtu.dk/en/publications/quantitative-review-of-degradation-and-lifetime-of-solid-oxide-ce

[7] McPhail S J *et al* 2022 Addressing planar solid oxide cell degradation mechanisms: a critical review of selected components *Electrochem. Sci. Adv.* **2** e2100024

[8] Hubert M, Laurencin J, Cloetens P, Morel B, Montinaro D and Lefebvre-Joud F 2018 Impact of nickel agglomeration on solid oxide cell operated in fuel cell and electrolysis modes *J. Power Sources* **397** 240–51

[9] Monaco F, Hubert M, Vulliet J, Ouweltjes J P, Montinaro D, Cloetens P, Piccardo P, Lefebvre-Joud F and Laurencin J 2019 Degradation of Ni-YSZ electrodes in solid oxide cells: impact of polarization and initial microstructure on the Ni evolution *J. Electrochem. Soc.* **166** F1229–42

[10] Koo B, Kim K, Kim J K, Kwon H, Han J W and Jung W 2018 Sr Segregation in perovskite oxides: why it happens and how it exists *Joule* **2** 1476–99

[11] Laurencin J, Hubert M, Ferreira Sanchez D, Pylypko S, Morales M, Morata A, Morel B, Montinaro D, Lefebvre-Joud F and Siebert E 2017 Degradation mechanism of $La_{0.6}Sr_{0.4}Co_{0.2}Fe_{0.8}O_{3-\delta}/Gd_{0.1}Ce_{0.9}O_{2-\delta}$ composite electrode operated under solid oxide electrolysis and fuel cell conditions *Electrochim. Acta* **241** 459–76

[12] Monaco F, Ferreira-Sanchez D, Hubert M, Morel B, Montinaro D, Grolimund D and Laurencin J 2021 Oxygen electrode degradation in solid oxide cells operating in electrolysis and fuel cell modes: LSCF destabilization and inter-diffusion at the electrode/electrolyte interface *Inter. J. Hydrog. Energy* **46** 31533–49

[13] Nakajo A *et al* 2015 Characterization of cracks and their effects on the effective transport pathways in Ni-YSZ anodes after reoxidation using x-ray nanotomography *ECS Trans.* **68** 1069–81

[14] Röntgen W C 1896 On a new kind of rays *Science* **3** 227–31

[15] Hsieh J 2003 *Computed Tomography: Principles, Design, Artifacts, and Recent Advances* vol 114 (Bellingham, WA: SPIE Press) pp 55–117

[16] Di Chiro G and Rodney A B 1979 The 1979 Nobel prize in physiology or medicine *Science* **206** 1060–2

[17] Shearing P R, Gelb J and Brandon N P 2010 X-ray nano computerised tomography of SOFC electrodes using a focused ion beam sample-preparation technique *J. Eur. Ceram. Soc.* **30** 1809–14

[18] Heenan T M M, Bailey J J, Lu X, Robinson J B, Iacoviello F, Finegan D P, Brett D J L and Shearing P R 2017 Three-phase segmentation of solid oxide fuel cell anode materials using lab based X-ray nano-computed tomography *Fuel Cells* **17** 75–82

[19] Heenan T M M, Finegan D P, Tjaden B, Lu X, Iacoviello F, Millichamp J, Brett D J L and Shearing P R 2018 4D nano-tomography of electrochemical energy devices using lab-based X-ray imaging *Nano Energy* **47** 556–65

[20] Holler M, Diaz A, Guizar-Sicairos M, Karvinen P, Färm E, Härkönen E, Ritala M, Menzel A, Raabe J and Bunk O 2014 X-ray ptychographic computed tomography at 16 nm isotropic 3D resolution *Sci. Rep.* **4** 3857

[21] Paganin D 2006 *Coherent X-ray Optics* (Oxford: Oxford University Press) https://global.oup.com/academic/product/coherent-x-ray-optics-9780198567288

[22] Izzo J R, Joshi A S, Grew K N, Chiu W K S, Tkachuk A, Wang S H and Yun W 2008 Nondestructive reconstruction and analysis of SOFC anodes using X-ray computed tomography at sub-50 nm resolution *J. Electrochem. Soc.* **155** B504–8

[23] Grew K N, Chu Y S, Yi J, Peracchio A A, Izzo J R, Hwu Y, De Carlo F and Chiu W K S 2010 Nondestructive nanoscale 3D elemental mapping and analysis of a solid oxide fuel cell anode *J. Electrochem. Soc.* **157** B783–92

[24] Shearing P R, Gelb J, Yi J, Lee W K, Drakopoulos M and Brandon N P 2010 Analysis of triple phase contact in Ni-YSZ microstructures using non-destructive X-ray tomography with synchrotron radiation *Electrochem. Commun.* **12** 1021–24

[25] Guan Y *et al* 2011 Quantitative analysis of micro structural and conductivity evolution of Ni-YSZ anodes during thermal cycling based on nano-computed tomography *J. Power Sources* **196** 10601–5

[26] Nelson G J *et al* 2011 Comparison of SOFC cathode microstructure quantified using X-ray nanotomography and focused ion beam-scanning electron microscopy *Electrochem. Commun.* **13** 586–9

[27] Shearing P R, Bradley R S, Gelb J, Tariq F, Withers P J and Brandon N P 2012 Exploring microstructural changes associated with oxidation in Ni–YSZ SOFC electrodes using high resolution x-ray computed tomography *Solid State Ion.* **216** 69–72

[28] Cronin J S, Chen-Wiegart Y K, Wang J and Barnett S A 2013 Three-dimensional reconstruction and analysis of an entire solid oxide fuel cell by full-field transmission x-ray microscopy *J. Power Sources* **233** 174–9

[29] Chen-Wiegart Y K, Kennouche D, Cronin J S, Barnett S A and Wang J 2016 Effect of Ni content on the morphological evolution of Ni-YSZ solid oxide fuel cell electrodes *Appl. Phys. Lett.* **108** 083903

[30] Guan Y *et al* 2016 Effects of nickel oxide impurities on the microstructure and electrical properties of a nickel–yttria-stabilized zirconia anode *Int. J. Hydrog. Energy* **41** 1–11
[31] Guinier A 1994 *X-ray Diffraction in Crystals, Imperfect Crystals and Amorphous Bodies* (Mineola, NY: Courier Dover Publications)
[32] Bonse U and Hart M 1965 An x-ray interferometer *Appl. Phys. Lett.* **6** 155–6
[33] Förster E, Goetz K and Zaumseil P 1980 Double crystal diffractometry for the characterization of targets for laser fusion experiments *Krist. Tech.* **15** 937–45
[34] Snigirev A, Snigireva I, Kohn V, Kuznetsov S and Schelokov I 1995 On the possibilities of x-ray phase contrast microimaging by coherent high-energy synchrotron radiation *Rev. Sci. Instrum.* **66** 5486–92
[35] Cloetens P, Barrett R, Baruchel J, Guigay J-P and Schlenker M 1996 Phase objects in synchrotron radiation hard x-ray imaging *J. Phys. D: Appl. Phys.* **29** 133–46
[36] Cloetens P, Ludwig W, Baruchel J, Van Dyck D, Van Landuyt J, Guigay J P and Schlenker M 1999 Holotomography: quantitative phase tomography with micrometer resolution using hard synchrotron radiation x rays *Appl. Phys. Lett.* **75** 2912–14
[37] Mokso R, Cloetens P, Maire E, Ludwig W and Buffière J-Y 2007 Nanoscale zoom tomography with hard x rays using Kirkpatrick–Baez optics *Appl. Phys. Lett.* **90** 144104
[38] Turner L D, Dhal B B, Hayes J P, Mancuso A P, Nugent K A, Paterson D, Scholten R E, Tran C Q and Peele A G 2004 X-ray phase imaging: demonstration of extended conditions with homogeneous objects *Opt. Express* **12** 2960–65
[39] Paganin D, Mayo S C, Gureyev T E, Miller P R and Wilkins S W 2002 Simultaneous phase and amplitude extraction from a single defocused image of a homogeneous object *J. Microsc.* **206** 33–40
[40] Hubert M, Pacureanu A, Guilloud C, Yang Y, da Silva J C, Laurencin J, Lefebvre-Joud F and Cloetens P 2018 Efficient correction of wavefront inhomogeneities in X-ray holographic nanotomography by random sample displacement *Appl. Phys. Lett.* **112** 203704
[41] Stockmar M, Cloetens P, Zanette I, Enders B, Dierolf M, Pfeiffer F and Thibault P 2013 Near-field ptychography: phase retrieval for inline holography using a structured illumination *Sci. Rep.* **3** 1–6
[42] Pfeiffer F 2018 X-ray ptychography *Nat. Photon.* **12** 9–17
[43] Laurencin J, Quey R, Delette G, Suhonen H, Cloetens P and Bleuet P 2012 Characterisation of solid oxide fuel cell Ni–8YSZ substrate by synchrotron X-ray nanotomography: from 3D reconstruction to microstructure quantification *J. Power Sources* **198** 182–89
[44] Quey R, Suhonen H, Laurencin J, Cloetens P and Bleuet P 2013 Direct comparison between X-ray nanotomography and scanning electron microscopy for the microstructure characterization of a solid oxide fuel cell anode *Mater. Charact.* **78** 87–95
[45] Villanova J, Laurencin J, Cloetens P, Bleuet P, Delette G, Suhonen H and Usseglio-Viretta F 2013 3D phase mapping of solid oxide fuel cell YSZ/Ni cermet at the nanoscale by holographic X-ray nanotomography *J. Power Sources* **243** 841–49
[46] Lay-Grindler E, Laurencin J, Villanova J, Cloetens P, Bleuet P, Mansuy A, Mougin J and Delette G 2014 Degradation study by 3D reconstruction of a nickel–yttria stabilized zirconia cathode after high temperature steam electrolysis operation *J. Power Sources* **269** 927–36
[47] Usseglio-Viretta F, Laurencin J, Delette G, Villanova J, Cloetens P and Leguillon D 2014 Quantitative microstructure characterization of a Ni–YSZ bi-layer coupled with simulated electrode polarisation *J. Power Sources* **256** 394–403

[48] Stockmar M *et al* 2015 X-ray nanotomography using near-field ptychography *Opt. Express* **23** 12720–31
[49] Hubert M, Laurencin J, Cloetens P, Da Silva J C, Lefebvre-Joud F, Bleuet P, Nakajo A and Siebert E 2016 Role of microstructure on electrode operating mechanisms for mixed ionic electronic conductors: from synchrotron-based 3D reconstruction to electrochemical modeling *Solid State Ion.* **294** 90–107
[50] Monaco F, Hubert M, Da Silva J C, Favre-Nicolin V, Montinaro D, Cloetens P and Laurencin J 2022 A comparison between holographic and near-field ptychographic X-ray tomography for solid oxide cell materials *Mater. Charact.* **187** 111834
[51] De Angelis S, Jørgensen P S, Esposito V, Tsai E H R, Holler M, Kreka K, Abdellahi E and Bowen J R 2017 Ex-situ tracking solid oxide cell electrode microstructural evolution in a redox cycle by high resolution ptychographic nanotomography *J. Power Sources* **360** 520–27
[52] De Angelis S, Jørgensen P S, Tsai E H R, Holler M, Kreka K and Bowen J R 2018 Three dimensional characterization of nickel coarsening in solid oxide cells via ex situ ptychographic nano-tomography *J. Power Sources* **383** 72–9
[53] De Angelis S, Jørgensen P S, Tsai E H R, Holler M, Fevola G and Bowen J R 2020 Tracking nickel oxide reduction in solid oxide cells via ex situ ptychographic nanotomography *Mater. Charact.* **162** 110183
[54] Hubert M 2017 Durability of solid oxide cells: an experimental and modelling investigation based on synchrotron x-ray nano-tomography characterization *PhD Thesis* University of Grenoble Alpes, Grenoble, France https://theses.hal.science/tel-01689775
[55] Bailey J J *et al* 2017 Laser-preparation of geometrically optimised samples for X-ray nano-CT *J. Microsc.* **267** 384–96
[56] Wislon J R, Kobsiriphat W, Mendoza R, Chen H Y, Hiller J M, Miller D J, Thorton K, Voorhees P W, Adler S B and Barnett S A 2006 Three-dimensional reconstruction of a solid-oxide fuel-cell anode *Nat. Mater.* **5** 541–44
[57] Wilson J R, Gameiro M, Mischaikow K, Kalies W, Voorhees P W and Barnett S A 2009 Three-dimensional analysis of solid oxide fuel cell Ni-YSZ anode interconnectivity *Microsc. Microanal.* **15** 71–7
[58] Shearing P R, Golbert J, Chater R J and Brandon N P 2009 3D reconstruction of SOFC anodes using a focused ion beam lift-out technique *Chem. Eng. Sci.* **64** 3928–33
[59] Iwai H *et al* 2010 Quantification of SOFC anode microstructure based on dual beam FIB-SEM technique *J. Power Sources* **195** 955–61
[60] Shikazono N, Kanno D, Matsuzaki K, Teshima H, Sumino S and Kasagi N 2010 Numerical assessment of SOFC anode polarization based on three-dimensional model microstructure reconstructed from FIB-SEM images *J. Electrochem. Soc.* **157** B665–72
[61] Sumi H, Kishida R, Kim J Y, Muroyama H, Matsui T and Eguchi K 2010 Correlation between microstructural and electrochemical characteristics during redox cycles for Ni–YSZ anode of SOFCs *J. Electrochem. Soc.* **157** B1747–52
[62] Cronin J S, Wilson J R and Barnett S A 2011 Impact of pore microstructure evolution on polarization resistance of Ni-yttria-stabilized zirconia fuel cell anodes *J. Power Sources* **196** 2640–43
[63] Holzer L, Münch B, Iwanschitz B, Cantoni M and Graule T 2011 Quantitative relationships between composition, particle size, triple phase boundary length and surface area in nickel-cermet anodes for solid oxide fuel cells *J. Power Sources* **196** 7076–89

[64] Nelson G J *et al* 2011 Comparison of SOFC cathode microstructure quantified using X-ray nanotomography and focused ion beam-scanning electron microscopy *Electrochem. Commun.* **13** 586–89
[65] Vivet N, Chupin S, Estrade E, Piquero T, Pommier P L, Rochais D and Bruneton E 2011 3D microstructural characterization of a solid oxide fuel cell anode reconstructed by focused ion beam tomography *J. Power Sources* **196** 7541–49
[66] Matsui T, Kim J-Y, Muroyama H, Shimazu M, Abe T, Miyao M and Eguchi K 2012 Anode microstructural change upon long-term operation for the cathode-supported tubular-type SOFC *Solid State Ion.* **225** 50–4
[67] Duong A T and Mumm D R 2012 Microstructural optimization by tailoring particle sizes for LSM-YSZ solid oxide fuel cell composite cathodes *J. Electrochem. Soc.* **159** B39–52
[68] Joos J, Ender M, Rotscholl I, Menzler N H and Ivers-Tiffée E 2014 Quantification of double-layer Ni/YSZ fuel cell anodes from focused ion beam tomography data *J. Power Sources* **246** 819–30
[69] Kishimoto M, Lomberg M, Ruiz-Trejo E and Brandon N P 2014 Enhanced triple-phase boundary density in infiltrated electrodes for solid oxide fuel cells demonstrated by high-resolution tomography *J. Power Sources* **266** 291–95
[70] Lichtner A Z, Jauffrès D, Roussel D, Charlot F, Martin C L and Bordia R K 2015 Dispersion, connectivity and tortuosity of hierarchical porosity composite SOFC cathodes prepared by freeze-casting *J. Eur. Ceram. Soc.* **35** 585–95
[71] Trini M, Jørgensen P S, Hauch A, Bentzen J J, Hendriksen P V and Chen M 2019 3D microstructural characterization of Ni/YSZ electrodes exposed to 1 year of electrolysis testing *J. Electrochem. Soc.* **166** F158
[72] Komatsu Y, Sciazko A and Shikazono N 2021 Isostatic pressing of screen printed nickel–gadolinium doped ceria anodes on electrolyte-supported solid oxide fuel cells *J. Power Sources* **485** 229317
[73] Hsu T *et al* 2018 Mesoscale characterization of local property distributions in heterogeneous electrodes *J. Power Sources* **386** 1–9
[74] Mahbub R *et al* 2021 Quantifying morphological variability and operating evolution in SOFC anode microstructures *J. Power Sources* **498** 229846
[75] Joos J, Ender M, Carraro T, Weber A and Ivers-Tiffée E 2012 Representative volume element size for accurate solid oxide fuel cell cathode reconstructions from focused ion beam tomography data *Electrochim. Acta* **82** 268–76
[76] Metcalfe C, Kesler O, Rivard T, Gitzhofer F and Abatzoglou N 2010 Connected three-phase boundary length evaluation in modeled sintered composite solid oxide fuel cell electrodes *J. Electrochem. Soc.* **157** B1326–35
[77] Parikh H, Hilli N, De Guire M R, Heuer A H, Liu Z and Goettler R 2017 Long-term microstructural changes in solid oxide fuel cell anodes: 3D reconstruction *J. Am. Ceram. Soc.* **100** 1653–60
[78] Moussaoui H, Laurencin J, Gavet Y, Delette G, Hubert M, Cloetens P, Le Bihan T and Debayle J 2018 Stochastic geometrical modeling of solid oxide cells electrodes validated on 3D reconstructions *Comput. Mater. Sci.* **143** 262–76
[79] Jiao Z and Shikazono N 2016 3D reconstruction size effect on the quantification of solid oxide fuel cell nickel–yttria-stabilized-zirconia anode microstructural information using scanning electron microscopy-focused ion beam technique *Sci. Bull.* **61** 1317–23

[80] Moussaoui H, Debayle J, Gavet Y, Cloetens P and Laurencin J 2020 Particle-based model for functional and diffusion layers of solid oxide cells electrodes *Powder Technol.* **367** 67–81
[81] Kennouche D, Chen-Wiegart Y C K, Riscoe C, Wang J and Barnett S A 2016 Combined electrochemical and X-ray tomography study of the high temperature evolution of nickel–yttria stabilized zirconia solid oxide fuel cell anodes *J. Power Sources* **307** 604–12
[82] Kennouche D, Chen-Wiegart Y C K, Yakal-Kremski K J, Wang J, Gibbs J W, Voorhees P W and Barnett S A 2016 Observing the microstructural evolution of Ni–yttria-stabilized zirconia solid oxide fuel cell anodes *Acta Mater.* **103** 204–10
[83] Kennouche D, Chen-Wiegart Y K, Cronin J S, Wang J and Barnett S A 2013 Three-dimensional microstructural evolution of Ni–yttria-stabilized zirconia solid oxide fuel cell anodes at elevated temperatures *J. Electrochem. Soc.* **160** F1293–304
[84] Mogensen M B, Chen M, Frandsen H L, Graves C, Hauch A, Hendriksen P V, Jacobsen T, Jensen S H, Skafte T L and Sun X 2021 Ni migration in solid oxide cell electrodes: review and revised hypothesis *Fuel Cells* **21** 415–29
[85] Solé V A, Papillon E, Cotte M, Walter P and Susini J A 2007 A multiplatform code for the analysis of energy-dispersive X-ray fluorescence spectra *Spectrochim. Acta* B **62** 63–8
[86] Li T *et al* 2019 Design of next-generation ceramic fuel cells and real-time characterization with synchrotron x-ray diffraction computed tomography *Nat. Commun.* **10** 1–11
[87] Nowack L, Grolimund D, Samson V, Marone F and Wood V 2016 Rapid mapping of lithiation dynamics in transition metal oxide particles with operando X-ray absorption spectroscopy *Sci. Rep.* **6** 1–9
[88] Heenan T M M, Robinson J B, Lu X, Tjaden B, Cervellino A, Bailey J J, Brett D J L and Shearing P R 2018 Understanding the thermo-mechanical behaviour of solid oxide fuel cell anodes using synchrotron X-ray diffraction *Solid State Ion.* **314** 156–64
[89] Robinson J B *et al* 2015 Investigating the effect of thermal gradients on stress in solid oxide fuel cell anodes using combined synchrotron radiation and thermal imaging *J. Power Sources* **288** 473–81
[90] Chang K C, Ingram B, Ilavsky J, Lee S, Fuoss P and You H 2017 Synchrotron X-ray studies of model SOFC cathodes, part I: thin film cathodes *Solid State Ion.* **311** 118–26
[91] Sanchez D F, Grolimund D, Hubert M, Bleuet P and Laurencin J 2017 A 2D and 3D X-ray μ-diffraction and μ-fluorescence study of a mixed ionic electronic conductor *Int. J. Hydrog. Energy* **42** 1203–11
[92] Khamidy N I, Laurencin J, Sanchez D F, Monaco F, Charlot F and Djurado E 2020 Durability of nanostructured $LaPrNiO_{4+\delta}$electrode for solid oxide cells: electrochemical, microstructural, and structural investigation *J. Power Sources* **450** 227724
[93] Villanova J, Schlabach S, Brisse A and Léon A 2019 X-ray fluorescence nano-imaging of long-term operated solid oxide electrolysis cells *J. Power Sources* **421** 100–8
[94] Watts J F and Wolstenholme J 2019 *An Introduction to Surface Analysis by XPS and AES* 2nd edn (New York: Wiley)
[95] Briggs D and Seah M P 1990 *Practical Surface Analysis: Auger and X-ray Photoelectron Spectroscopy* (Chichester: Wiley)
[96] Schild D 2008 X-ray photoelectron spectroscopy *Hydrogen Technology* (Berlin: Springer) pp 575–601
[97] Doh W H, Papaefthimiou V and Zafeiratos S 2015 Applications of synchrotron-based X-ray photoelectron spectroscopy in the characterization of nanomaterials *Surface Science Tools for Nanomaterials Characterization* (Berlin: Springer) pp 317–66

[98] Zhu S *et al* 2021 HIPPIE: a new platform for ambient-pressure x-ray photoelectron spectroscopy at the MAX IV laboratory *J. Synchrotron Radiat.* **28** 624–36

[99] Rothe J and Léon A 2008 X-ray absorption fine structure (XAFS) spectroscopy *Hydrogen Technology* (Berlin: Springer) pp 603–22

[100] Garino C, Borfecchia E, Gobetto R, van Bokhoven J A and Lamberti C 2014 Determination of the electronic and structural configuration of coordination compounds by synchrotron-radiation techniques *Coord. Chem. Rev.* **277** 130–86

[101] Aksenov V L, Koval'chuk M V, Kuz'min A Y, Purans Y and Tyutyunnikov S I 2006 Development of methods of EXAFS spectroscopy on synchrotron radiation beams *Crystallogr. Rep.* **51** 908–35

[102] Mangold S 2018 Fully automated beamline control system for XAS beamlines *J. Synchrotron Radiat.* **25** 960–66

[103] DeCaluwe S C *et al* 2010 In situ characterization of ceria oxidation states in high-temperature electrochemical cells with ambient pressure XPS *J. Phys. Chem.* C **114** 19853–61

[104] El Gabaly F, McCarty K F, Bluhm H and McDaniel A H 2013 Oxidation stages of Ni electrodes in solid oxide fuel cell environments *Phys. Chem. Chem. Phys.* **15** 8334–41

[105] Mewafy B, Paloukis F, Papazisi K M, Balomenou S P, Luo W, Teschner D, Joubert O, Le Gal La Salle A, Niakolas D K and Zafeiratos S 2019 Influence of surface state on the electrochemical performance of nickel-based cermet electrodes during steam electrolysis *ACS Appl. Energy Mater.* **2** 7045–55

[106] Kishimoto H, Shimonosono T, Yamaji K, Brito M E, Horita T and Yokokawa H 2013 Oxide component in the nickel base cermet anode: its effect on the performance of SOFCs *Electrochem.* **81** 255–58

[107] Joo J H, Merkle R and Maier J 2011 Effects of water on oxygen surface exchange and degradation of mixed conducting perovskites *J. Power Sources* **196** 7495–99

[108] Orliukas A F, Šalkus T, Kežionis A, Venckutė V, Kazlauskienė V, Miškinis J, Laukaitis G and Dudonis J 2011 XPS and impedance spectroscopy of some oxygen vacancy conducting solid electrolyte ceramics *Solid State Ion.* **188** 36–40

[109] Swaminathan S, Lee Y S and Kim D I 2016 Long term high temperature oxidation characteristics of La and Cu alloyed ferritic stainless steels for solid oxide fuel cell interconnects *J. Power Sources* **327** 104–18

[110] Kaderi A B, Hartmann H and Besmehn A 2017 Observation of oxide development from room temperature (RT) to 700°C demonstrated by *in situ* XPS of Crofer 22 APU alloy *Oxid. Met.* **88** 459–68

[111] Sasmal N, Garai M and Karmakar B 2016 Influence of Ce, Nd, Sm and Gd oxides on the properties of alkaline-earth borosilicate glass sealant *J. Asian Ceram. Soc.* **4** 29–38

[112] Crumlin E J, Mutoro E, Liu Z, Grass M E, Biegalski M D, Lee Y L, Morgan D, Christen H M, Bluhm H and Shao-Horn Y 2012 Surface strontium enrichment on highly active perovskites for oxygen electrocatalysis in solid oxide fuel cells *Energy Environ. Sci.* **5** 6081–88

[113] Takagi Y, Uruga T, Tada M, Iwasawa Y and Yokoyama T 2018 Ambient pressure hard X-ray photoelectron spectroscopy for functional material systems as fuel cells under working conditions *Acc. Chem. Res.* **51** 719–27

[114] Bozzini B, Abyaneh M K, Amati M, Gianoncelli A, Gregoratti L, Kaulich B and Kiskinova M 2012 Soft X-ray imaging and spectromicroscopy: new insights in chemical state and morphology of the key components in operating fuel-cells *Chem.–A Eur. J.* **18** 10196–210

[115] Orikasa Y, Crumlin E J, Sako S, Amezawa K, Uruga T, Biegalski M D, Christen H M, Uchimoto Y and Shao-Horn Y 2014 Surface strontium segregation of solid oxide fuel cell cathodes proved by *in situ* depth-resolved X-ray absorption spectroscopy *ECS Electrochem. Lett.* **3** F23

[116] Paloukis F, Papazisi K M, Dintzer T, Papaefthimiou V, Saveleva V A, Balomenou S P, Tsiplakides D, Bournel F, Gallet J-J and Zafeiratos S 2017 Insights into the surface reactivity of cermet and perovskite electrodes in oxidizing, reducing, and humid environments *ACS Appl. Mater. Interfaces* **9** 25265–77

[117] Lay-Grindler E, Laurencin J, Villanova J, Kieffer I, Usseglio-Viretta F, Le Bihan T, Bleuet P, Mansuy A and Delette G 2013 Degradation study of the $La_{0.6}Sr_{0.4}Co_{0.2}Fe_{0.8}O_3$ solid oxide electrolysis cell (SOEC) anode after high temperature electrolysis operation *ECS Trans.* **57** 3177

[118] Lai S Y, Ding D, Liu M, Liu M and Alamgir F M 2014 Operando and in situ x-ray spectroscopies of degradation in $La_{0.6}Sr_{0.4}Co_{0.2}Fe_{0.8}O_{3-\delta}$ thin film cathodes in fuel cells *ChemSusChem* **7** 3078–87

[119] Wang L, Yu Y, Gaskell K J, Crumlin E J, Liu Z, Eichhorn B W and Jackson G S 2020 In operando x-ray photoelectron spectroscopy studies of H_2 oxidation and H_2O electrolysis on gadolinia-doped ceria electrodes *J. Phys. Energy* **3** 014004

www.ingramcontent.com/pod-product-compliance
Ingram Content Group UK Ltd.
Pitfield, Milton Keynes, MK11 3LW, UK
UKHW051905100926
8556IPUK00005B/770

9 780750 340625